SpringerWienNewYork

Friedrich G. Barth
Patrizia Giampieri-Deutsch
Hans-Dieter Klein (eds.)

Sensory Perception

Mind and Matter

Friedrich G. Barth
Department of Neurobiology, Faculty of Life Sciences, University of Vienna,
Vienna, Austria

Patrizia Giampieri-Deutsch
Institute of Philosophy, University of Vienna, Austria

Hans-Dieter Klein
Institute of Philosophy, University of Vienna, Austria

Printed in Austria

SpringerWienNewYork is part of
Springer Science+Business Media
springer.at

Cover Illustrations: Johannes Deutsch, Vision Mahler – Draft 10b, 2004, Detail
© Johannes Deutsch, VBK Wien
Typesetting: Jung Crossmedia Publishing GmbH, 35633 Lahnau, Germany
Printing: Holzhausen Druck GmbH, 1140 Wien, Austria

Printed on acid-free and chlorine-free bleached paper
SPIN: 86147749

With 87 (mostly coloured) Figures

Library of Congress Control Number: 2011940536

ISBN 978-3-211-99750-5 SpringerWienNewYork

Prologue

This book is about *sensory perception* in a very broad sense. It aims towards the strengthening of existing bridges and the building of new ones between a wide range of disciplines traditionally separate from each other and considered to be hunting in their own different woods. The gaps between the search for the molecular mechanisms at work in sensory cells or the brain of a fruit fly and the examination of cognitive functions and capabilities in man are indeed obvious, as are the differences between the psychophysics of color vision and neurophilosophy. We are confronted with differences in regard to both the level of complexity of the systems examined and the procedures applied to analyse them.

However, despite of all these differences there are commonalities as well. The most eminent ones are the sensory systems and the brains themselves. Without them there would neither be neurophilosophy nor molecular neurobiology, which has included the so-called lower animals like flies and bees and snails into its research most successfully. Such animals may vastly differ from us humans. At the same time they share with us a surprising number of the basic properties of living organisms. These include brain functions like learning and memory and even quite advanced cognitive abilities which only a few years ago no one could have expected to find in bees, birds, or dogs.

Our book is meant to provide at least a glimpse at some of the many exciting modern developments in the study of sensory perception, both technical and conceptual. At the same time we are aware of the fact that in terms of evolution the story we would like to illuminate from various points of view is a very old one. *All living organisms* rely on information on their outside and inside worlds. Without such information proper behavior ensuring survival and procreation is impossible. It should therefore not come as a surprise to find sensory mechanisms of remarkable refinement even in bacteria and single celled protozoans.

In higher animals evolutionary selective pressures have led to a fascinating diversity of sensory and nervous systems. Eyes, ears and noses as well as many types of sense organs providing information alien to our human experience represent the windows of the brain to the outside world. They are the interfaces between an organism's environment and its behavior. As a rule all these windows are highly specialized filters. They provide the individual organism with the information on a small fraction of the physical world only. This small fraction, however, is the biologically relevant one, both in space and time.

We humans are no exception to these general rules. To explain and understand the mechanisms of information processing and the generation of meaning by our brain has been given attention for more than 2 000 years. A particular incentive for this long-lasting search may have been the fact that we do have subjective conscious experiences associated with brain activity and that we are able to describe them verbally. Unfortunately one has to admit, that despite all the brilliant advances of the modern molecular and cellular neurosciences higher functions emerging from the complex activity of systems of many neurons cannot be explained yet at the level of brain function.

Sensory perception, the topic of the present book, provides access to both the sciences of the brain and the sciences of the mind. We are convinced that classical questions of philosophy and psychology referring to perception and mind will benefit from an understanding of the insights of modern neurobiology into the mechanisms of sensory processing in animals and humans. Likewise neurobiology can and indeed should profit from a central task of philosophy, which is to keep or make terminology and concepts clear, thereby increasing their unifying power. Also, in the age of biology and in view of the overwhelming importance often attributed to it, it may be wise trying to see biology in a still larger context. This seems to be particularly relevant for functional brain research which despite its fascinating advances clearly demonstrates the present limits of biological reasoning.

Unfortunately, *perception and consciousness* have often been conflated in philosophy (as in every day language). Both neuropsychological studies of brain-damaged patients and *subliminality* studies have now led to a revision of the common assumption that perception and consciousness of perception are always inseparable. Studies of subliminality have in addition demonstrated that controlled investigations into unconscious conflicts, unconscious affect, and unconscious anxiety are possible.

Sensory perception is functionally related to an organism's reactions. The range of issues addressed by the present book therefore includes *free will,* which, defined as libertarian free will, implies free actions, unconstrained and uncaused by any physical process, and has been a topic of hot debates recently. Libertarian free will is incompatible with the concept of the physical world as a closed causal system. The more sophisticated approach of the compatibilists rejects libertarian free will and is supported by a long philosophical tradition inaugurated by Aristotle more than 2000 years ago.

There is currently no consent regarding the answer to the question whether the sciences involved in the study of sensory perception are autonomous enterprises which can interface with each other and exchange their results, or whether the findings of, say, psychology, should be reducible and indeed reduced to findings elaborated by the neurosciences. The current philosophy of mind again raises the traditional metaphysical questions but certainly requires empirical answers that only the experimental sciences can provide.

Our book strives to bring together the neurosciences with psychology, which speaks the language of cognitive experiences and with philosophy, which has been thinking and arguing about the meaning and origin of consciousness since its beginning. The debate about the possibility to explain mental phenomena neurobiologically will still have to continue for a while. The editors of this book are convinced, however, that in any case we need a patient interdisciplinary discourse between neurobiology, psychology and philosophy if we aim at an in depth understanding of the many fascinating facets of sensory perception and their relation to brain functions and cognition. Leading experts have written chapters for our book as food for thought and reflection which hopefully will contribute to promote such a discourse and also point to the big gaps still to be bridged. Some of the contributions may also help to overcome the deep rooted bias of humans to overemphasize the uniqueness of their brains and increase the awareness of the impossibility to decouple the "mind" from evolutionary biology and the question of biological fitness.

We are very grateful to the University of Vienna (in particular the faculty of Life Sciences) and the Austrian Academy of Sciences which were the main sponsors of a very successful and stimulating international symposium held in Vienna in 2008 on the same topic. This symposium initiated the idea of publishing the present book. We also thank Springer Verlag Wien New York for the help and guidance received during the preparation of this book.

Wien and Lofer, January 2011
Friedrich G. Barth
Patrizia Giampieri-Deutsch
Hans-Dieter Klein

Contents

III Perception, art, and illusion

IV Perception and memory: conscious and unconscious processes

V The history of sensory worlds

VI Neurophilosophy and free will

Basic mechanisms in sensory systems I

Introductory remarks

Senses and sensing are a characteristic property of living beings. From bacteria and unicellular protozoans all the way to vertebrates, primates and man there is a need of information on what is going on inside and outside the body in order to maintain homeostasis and to properly behave in the organism's own and specific environment. Although plants do have sensing as well, it is the heterotrophic animals which have a particularly intimate relation with their habitat. The reason is their need to spend a considerable amount of time and effort to move around in search of energy pre-packaged in the bodies of other organisms, be they plants or animals. Their highly developed sensory systems (like their muscular and nervous systems) reflect the particular demands resulting from their dependence on information about the static and in particular the dynamic properties of environmental conditions. Most sensory systems are indeed particularly well adapted to respond to dynamic rather than static stimulation.

The study of sensory systems has come a long way and made innumerable fascinating discoveries. The scope of what we want to understand is enormous and ranges from the molecular mechanisms at work when a sensory cell takes up a stimulus and turns it into a nervous signal to the old problem of what the activity of sensory systems tells us and other animals about the physical reality of the world. In this section of our book four experts address basic mechanisms of sensing, starting with processes at the cellular and molecular level and ending with the intriguing problem how the distributed organization of our brain and the lack of single loci for integrated percepts goes together with the coherent wholes we experience.

1. Sensory transduction

Stephan FRINGS of the University of Heidelberg outlines the first steps of perception, that is the ways how sensory cells detect stimuli of different forms of energy and transform their properties to electrochemical signals which are sent to the central nervous system and its brain. As will be seen there are a number of commonalities among the primary processes regarding key cell structures and transduction processes in cells serving mechanoreception, vision, and chemoreception. Obviously, questions of the sensitivity and selectivity of sensory cells and sensory organs, as treated in this chapter, are highly relevant for an understanding of what we and other animals perceive.

2. Hearing

In his chapter, Geoffrey MANLEY of the Technical University Munich deals with vertebrate hearing. He traces the evolutionary history of hearing back to the fishes and draws our attention to parallel developments in reptiles and birds. Not only are the major improvements of the hearing organs per se highlighted but also the changes and specializations of the brain pathways devoted to hearing pointed out. As so often seen when taking a comparative approach the human sensory system is not the culmination of evolution; instead, even more spectacular specializations are found in many other species reflecting the importance of hearing in their species-specific normal behavior.

3. Vision

Kristine KRUG of Oxford University introduces the reader to the principles of function found in primate vision. The emphasis is on the brain and the ways in which it transforms patterns of activity received from the eye and representing patterns of local contrast in the visual field to signals directly related to perception in the higher areas of the visual cortex. As has been known for a while various parameters of the visual stimulus like color, shape and motion are processed in parallel in different cortical areas. Experimental evidence will be provided that links the activity of individual neurons in the brain to visual motion and depth perception.

4. The binding problem

Finally, Wolf SINGER of the Max Planck Institute for Brain Research in Frankfurt /Main takes us a step further into the functional principles of our brain, stressing that even the most complex cognitive features need to be explained on the basis of neurobiological data. Evolution, the most important determinant of perception, is itself considered as an adaptive cognitive process and given particular attention as the source of our a priori knowledge of the world which to a large extent determines what and how we perceive. Despite all the exciting advances made by brain research during the last decades there are still substantial gaps in our understanding of the brain's operations. Thus a major issue still is the neuronal basis explaining the qualia of our subjective experiences, including awareness and self-consciousness. Different from previous concepts, parallelity, reciprocity, and distributedness turned out to be the principles dominating the brain's cortical connectivity. There is much less linearity and hierarchy than previously assumed and no single site onto which all information is converging to allow coherent interpretations of the world and our obviously coherent perception integrating the many different aspects of an object. An exciting hypothesis proposes a solution of an enormous problem: How are the activities of the many neurons distributed widely in the cortex and subcortical areas bound together to a particular representation? As is explained in *Wolf Singer's* chapter the temporal synchronization of oscillatory activity of widely distributed neurons, that is convergence in time rather than space, may be a key factor solving the binding problem and giving perceptions access to consciousness.

FGB

Sensory cells and sensory organs

1

Stephan Frings

Contents

Abstract

Animals possess sensory organs that collect information and convey this information to the central nervous system. This introductory chapter outlines how sensory cells perform their task of detecting adequate stimuli and how they produce electrical signals that encode information for the brain. In the course of animal evolution, the specific transduction mechanisms that operate in the various sensory cells have been optimized under intense selective pressure. The results of this process often include extreme sensitivity for the adequate stimuli and efficient signal amplification. To illustrate different solutions to the problem of detecting and encoding complex information, the sensory modalities of touch, vibration detection, hearing, vision, and olfaction are briefly introduced on the levels of key cell structures and transduction molecules. Pain perception is described as a sensory modality with very special features that differ fundamentally from those of other modalities. Thus, polymodality of sensory cells, modulation by the immune system, and suppression by endorphins represent characteristic properties of the pain system, linked to its unique protective function. This chapter is designed to direct the reader's attention to some central points of the topic. It does not deal with the subject of sensory detection in a comprehensive way. It rather highlights a set of particularly important aspects of sensory transduction that may be of interest for the interdisciplinary approach followed in this book. Questions of sensitivity, selectivity, and adaptation in sensory cells are directly related to the mode of perception that defines our view of the world. Inasmuch as our sensory organs have been shaped by evolution, our perception of reality is the consequence of evolutionary forces and constraints as well.

Stephan Frings
University of Heidelberg
Molecular Physiology
Im Neuenheimer Feld 230
69120 Heidelberg, Germany
e-mail: s.frings@zoo.uni-heidelberg.de

1. Introduction

The process of sensory perception begins when sensory cells detect stimuli in the environment. Light acts on photoreceptors, sound on mechanoreceptors, and odorants on olfactory receptors – all highly specialized cells of eyes, ears, noses that are exposed to the outside world. The sensory organs we have today are the result of millions of years of adaptation to the needs of animals in their struggle to survive. This is an important point, as all sensory systems fulfil a clear and vital purpose: the survival of an individual and the continuance of a species. There is no luxury in sensory organs; we do not perceive unnecessary things. Humans do not perceive the Earth's magnetic field because no selective pressure has ever favoured the development of a magnetic sense in the evolutionary line leading to the hominids, while such pressure did bring about an acute sense for magnetic fields in migratory birds. Thus, our specific set of sensory organs serves the vital functions of finding food, finding mates, and avoiding predators. Consequently, we are not equipped to perceive the world as it is, but we are able to survive in it (see also Chapter II, 6 by FG Barth).

This selective development of sensory organs began early in evolution, in the precambrian age, more than 550 millions ago. The oldest fossils of larger animals with robust shells date from the cambrian age, 500–550 millions years ago, and some of these animals clearly had eyes (Conway Morris 1998). In fact, eyes have probably already been used for about 300 million years, when cambrian predators like *Anomalocaris* were searching the bright and shallow seas for animals to eat, using eyes which resembled those of today's insects. And prey organisms in turn had to develop some means of perceiving an approaching *Anomalocaris* as early as possible, and to get away before it was too late. Trilobites, for example, started into the cambrian age already with well developed eyes. Although probably one of the favourite foods of *Anomalocaris*, they outlived the cambrian predator and became one of the most successful group in animal evolution (Fortey 2000). Thus, predatory animals challenged animals of prey to acquire a decent set of sensory organs, or else to perish. This enormous selective pressure has continued over millions of years, and it continues today in the intricate scenarios of co-evolution between the hunter and the hunted. It has produced eyes, ears and noses of almost unimaginable sensitivity, in fact, a sensitivity that has reached physical limits: A single photon, the movement of one atom, or a single odorant molecule can be detected by sensory cells of some animal species. This beautifully illustrates the power of evolution. Evolutionary processes can optimize living systems right to the state of absolute perfection: It would not help to develop a photoreceptor that detects less than one single photon. In fact, such a cell would be useless, as it would respond to absolute darkness. The single-photon response of vertebrate photoreceptors indeed represents a sensory system driven to perfection by the relentless forces of evolution. One of the fascinations of sensory physiology is to witness how perfection has been achieved in the various sensory organs.

2. How sensory cells work

It requires a multitude of different sensory cells to carry out all our vital functions. The brain has to be informed on every relevant detail so as to be able to coordinate these functions in a sensible way. The spectrum of information acquired by the brain is quite dazzling. The posture of the body, its supply with nutrients and oxygen, the state of the cardiovascular and digestive systems, as well as the body temperature and ion concentrations are constantly monitored by various types of sensory cells. Information about objects in the environment, their shape, colour, chemical composition, their distance and movement are collected and conveyed to the brain. This steady and complex flow of information is then integrated and used to generate expedient behaviour. However, the brain itself can only process information which is encoded in a language that consists of electrical discharges, termed action potentials. The task of a sensory cell is to convert the relevant stimuli into this language – a process that is called transduction. Transduction differs greatly between light-sensitive cells and cells that detect mechanical stimuli. But a few common principles can be outlined that illustrate the working of all sensory cells (Fig. 1).

The pivotal components of each sensory cell are specific sensory molecules that are contained in specialized cellular structures like cilia, microvilli or other membrane structures. Sensory molecules are highly specialized for their particular stimulus. If it is the right stimulus – the adequate stimulus – then the cell will respond even to very weak stimulation. The sensor may also respond to other stimuli, but not with high sensitivity. Thus, we may see stars with our photoreceptors at night in our bedroom in absolute darkness when we bump our eye against the edge of our wardrobe, thus receiving a strong mechanical stimulus. But the identity of a sensory cell and, indeed, of the entire sensory modality is defined by the *adequate* stimulus which elicits the sensory response; the ability of dim light to stimulate a sensory response defines a photoreceptor.

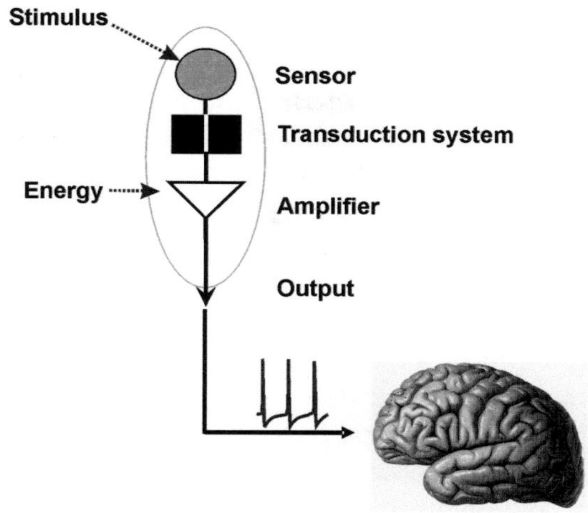

Fig. 1 Functional components of a sensory cell. An adequate stimulus acts on the cell's sensor, a structure specialized for the detection of this stimulus. The sensor triggers a transduction system which generates a chemical or electrical signal inside the cell. An amplification process increases the signal strength, using metabolic energy to boost the cellular response that is elicited by the stimulus. Finally, an output signal is generated in form of a series of electrical action potentials (spikes) that inform the brain about the detection of the stimulus

Following the uptake of the stimulus the next functional step is the transduction of the sensory signal – meaning the conversion of the extracellular stimulus into an intracellular signal. All cells operate with a certain repertoire of intracellular signals. These may be chemical or electrical signals which trigger the cell's internal responses to stimulation. There is only a limited number of such signals. Roughly ten different chemicals and basically four types of electrical signals carry such signals within all cell types of the body. In sensory cells, the sensor molecule must actuate at least one of them – for example a calcium signal or an electrical depolarization. In most cases, this task is fulfilled by ion channels residing in the plasma membrane of a sensory cell. Ion channels are proteins that can trigger both chemical and electrical signals, because ions – like the calcium ion

Ca^{2+} – can enter the cell through ion channels These ion channels are termed transduction channels as their job is to start the transduction process. Once they are activated by the stimulus, the cell can start to process the sensory signals.

Sensory transduction virtually always includes a step of signal amplification. If the adequate stimulus consists of only a few photons or a few molecules of odorant, not much energy is fed into the sensory cell. However, this little energy must be converted into a robust output signal, usually a series of electrical potential changes that can be conveyed to the brain for analysis. The difference in energy between the input and the output of a sensory cell is added to the sensory signal, thereby amplifying it. Different sensory cells have developed different amplification strategies. Not surprisingly, the most effective amplification strategies known are operating in sensory cells with high detection sensitivity.

Transduction of the stimulus energy to a first cellular response is followed by a process called encoding. Encoding leads to the generation of an electrical signal that contains the sensory information. Ideally, *all* aspects of the stimulus would be translated into the electrical code. Stimulus intensity, stimulus duration and other relevant parameters should be encoded in such a way that the brain is able to extract all this information by analysing the action-potential activity received from a sensory cell. The duration and shape of action potentials are uniform and therefore not useful for coding. The sensory information must, therefore, be encoded in the number of action potentials and in the time between them This coding principle is called frequency modulation (FM). It is a very reliable method of information coding, as we know from the excellent quality of FM-coded music and speech in radio transmission. Thus, the final task of a sensory cell is to convert the amplified sensory signal into a message encoded in a frequency modulation that is then read out and deciphered in the brain, a process that leads in most cases to perception.

3. Touch, medium flow and mechanosensitive hairs

Touching things or being touched is arguably the most basic sensory experience. Even *Paramecium* is able to register touch when it bumps into an obstacle. And, what is more important, it can properly respond to this experience. It stops, then swims backwards for a short distance, readjusts its heading, and continues to swim in the new direction – apparently to bypass the obstacle and to continue on its way. This is quite a remarkable accomplishment for a single-cell organism, and it illustrates that the processing of touch information is almost as old as life itself. Today's complex animals use all kinds of specialized structures to feel even the slightest touch. One of the most successful developments for this purpose was the combination of a hair-like structure and a sensory cell. Imagine a hair shaft, delicately suspended in a soft, elastic membrane, able to move into any direction upon the slightest touch, and connected at its base to the dendrite of a sensory cell and its sensor structure. If anything touches this hair, a force will act on the sensor and start the transduction process (Fig. 2A). We find touch-sensitive hair-like structures on the surface of insects which are able to detect with these highly-sensitive mechanoreceptors air currents that may indicate an approaching mate or predator. The combination of hair-like structures and sensory cells has been employed throughout evolution and works perfectly well in higher animals. The motile whiskers of rats are a good example. The animals can feel their way in the dark by probing their environment with their whiskers. If anything touches these hairs, sensory cells at their base are activated. Physiologists think that the input from all whiskers is integrated by the animal's brain to form an image of objects that surround the rat's head – that the rat "sees" with its whiskers. While we do not sport whiskers ourselves, we have a less sophisticated form of the hair – sensory cell combination: the hair-follicle receptors. Each hair on our forearm is equipped with a sensory cell that picks up each movement of the hair and hence contributes essential information to the touch sensation of our skin.

Although touch is such a basic, omnipresent sensory modality, we do not know much about the transduction mechanisms in mechanoreceptor cells. We know that our skin contains at least seven different types of mechanoreceptors apart from the hair-follicle receptors, and we have a good idea what they are there for. Some detect vibrations, others the touch intensity or the speed of an object moving along our skin. However, we do not know how the mechanical stimulus is converted into an output signal. The main reason for this ignorance is that our touch sensors are hidden in the skin and very difficult to study. Fortunately, one of biology's most popular model organisms, the nematode worm *Caenorhabditis elegans*, lends itself also to studies of touch reception. These animals respond to a gentle touch with an evasive movement that can be triggered by each of its six touch-sensitive cells. The underlying transduction mechanism was examined with immense effort by Martin Chalfie over a period of almost thirty years (Bounoutas and Chalfie 2007). It turned out that the transduction channel of the

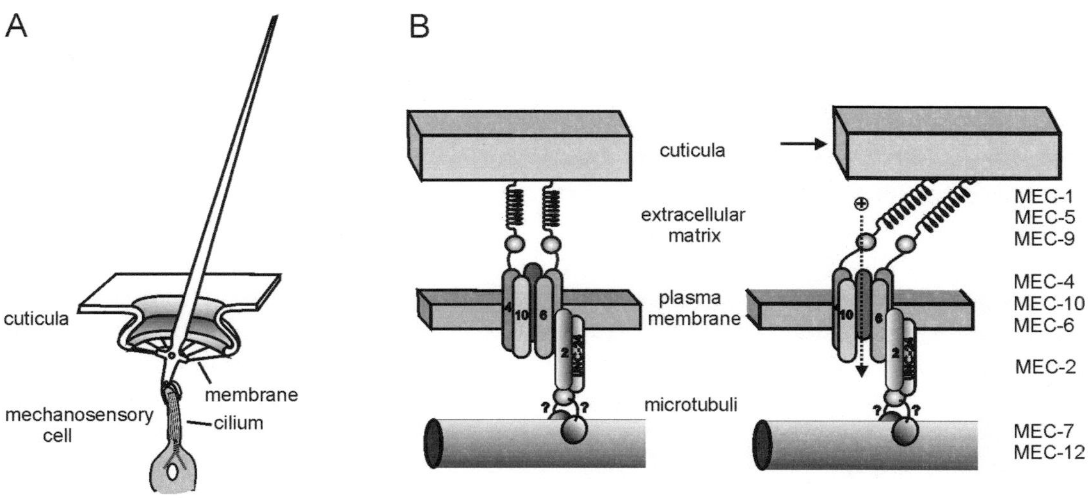

Fig. 2 Mechanosensory cells. **A** The combination of a hair and a sensory cell allows insects to detect medium flow signals like wind with extreme sensitivity. The deflection of a hair that is supported by a membrane is converted into a mechanical stimulus detected by a mechanosensory cell (modified from Müller and Frings 2009). **B** Model of the mechansosensory transduction channel that operates in the touch receptor of the nematode worm *Caenorhabditis elegans*. The channel consists of several proteins that are coassembled in the plasma membrane of the mechanosensory cell. It is tethered to the microtubule system inside the cell and to the cuticula outside of the cell. When the cuticula moves upon being touched, the transduction channel is pulled open and allows cation current to flow into the cell. This channel is an example of a multi-protein complex – in this case made of multiple MEC proteins (modified from Frings 2009)

touch-sensitive cells is a protein complex in the sensor membrane, connected to the cuticula, the skin of the worm (Fig. 2B). When something touches the cuticula, the transduction channel is pulled open and causes an electrical signal. Thus, the worm operates its touch receptors by linking the transduction channel to the cuticula, just like the insect links its mechanoreceptor cell to a hair-like structure. Conceivably our touch receptors also work with such tethered transduction channels – but this is not yet known. Despite the manifold tasks that touch receptors perform in our lives, from explorative, tactile activity to social signalling, we know little about how they work. In fact, our sense of touch is the least understood of all our senses despite its fundamental importance. Maybe the worm can help us here.

4. From vibration detection to hearing

For many animals the perception of vibrations is even more important than touch sensation. Being touched by a predator marks the moment when it is often too late to escape. Vibrations, on the other hand, travel over some distance and can alert the animal well before the predator can strike. Vibrations in water or soil are caused by animals moving around. They spread into the surroundings and warn every animal that is able to detect them. It is therefore not surprising that both fish and land animals have developed sensory organs for the detection of vibratory signals. Some of the vibration sensors found in animals are almost incredibly sensitive. Cockroaches possess inside their legs sensory organs that consist of a tiny horizontal membrane supported in air by a ring of sensory cells. Whenever the slightest vibration travels along the ground where the cockroach sits, the membrane will itself vibrate and stimulate the sensory cells. These will then transmit the information to the animal's central nervous system and brain, which will initiate an appropriate flight response. The astonishing finding about this vibration detector (the subgenual organ) is that it responds to vibration amplitudes in the range of 0.2 nm – the diameter of two hydrogen atoms. This is a movement on the scale of thermal vibrations, the trembling and shivering of all small particles, including atoms and molecules, that is caused by heat. All particles of the substrate the cockroach sits on wobble about in that range – even without any vibrations caused by another animal. Does the cockroach detect thermal vibrations? Probably not, because the animals leg averages the thermal movements of a large number of particles. Moreover, its vibration detector responds with that extreme sensitivity only to a particular frequency range of substrate vibrations. The vibration detector of the cockroach is most sensitive for frequencies around 1.4 kHz. We can envisage a cockroach sitting on the kitchen floor. If anybody enters the house, the vibrations spreading from each footstep contain a 1.4 kHz component. The cockroach picks them up and runs for cover.

A particularly successful strategy to detect mechanical stimuli was realized in fish that need an early-warning system just as urgently as insects do. It led to the development of a multi-purpose mechanoreceptor, a cell type that is used in vibration detection, in hearing and in the sense of balance, one of the pivotal achievements of animal evolution: the hair cell (Fig. 3A). Hair cells possess a tuft of hair-like extensions or villi on one side – termed stereovilli – and a device for activating neurons, a synapse, on the opposite side. If anything deflects the "hairs", the synapse will activate neurons which will then conduct the information to the brain. Fish use hair cells in their lateral line organ, a canal along each side of the fish's body with many openings to let water enter the canal lumen. Driven by pressure differences between individul openings, minute volumes of water enter the canal and deflect the stereovilli of hair cells. Fluctuations of the water flow, caused by other fish or by any currents and eddies, will thus produce sensory signals in the lateral line organ. With its lateral line organ the fish is able to collect information about what is going on around him – even in murky water or at night when vision is limited or impossible. But fish also use hair cells for keeping balance as they swim. They have an intricate fluid-filled bony structure, the vestibular organ in their heads, which uses hair cells to measure any movement of this fluid relative to the canal's wall that occurs upon acceleration of the body. Moreover, some hair cells in the vestibular organ are specialized to measure the position of the body relative to the Earth's gravitational field – a sensory information that allows the fish to swim upright. Finally, sound waves travelling through water cause vibrations in various parts of the fish, including the swim bladder and the skull. These vibrations are translated by a system of small bones into medium flow inside the inner ear which in turn stimulates hair cells. Thus, the fish depends on the signals obtained from hair cells at various parts of its body in many ways. As evolution proceeded from fish through amphibians and reptiles to mammals, hair cells remained important. Today, our vestibular organ allows us to walk upright, and the hair cells detect medium flow in our organ of Corti in the cochlea of the inner ear (Fig. 3B).

As we analyse airborne sound (see also Chapter I, 2 by GA Manley), we are still interested in the three basic parameters: the amplitude (perceived by us as the loudness), the frequency (perceived by us as the pitch), and the direction to the sound source. Of course, our ear is adapted to detect and to analyse airborne sound, vibrations in form of pressure waves travelling through the air. We pick those signals up with our outer ear and channel them towards the eardrum which is set into motion by the sound waves. This motion is transmitted by the three tiny bones of the middle ear (malleus, incus, stapes) to the cochlea, a bony tube that houses the organ of Corti (Fig. 3 C; see also Fig. 6 in Chapter II,6). The human cochlea has the form of a snail-shell with three turns. In this way, its lumen has sufficient space for the organ of Corti which is about 3 cm long. Along its entire length, hair cells are positioned in four rows, numbering approximately 3000 in each row. The cells sit on a tissue, the basilar membrane, which has a very particular property: it is stiff at the bottom of the cochlea and floppy at the top of the snail-shell. Mainly as a consequence of this gradient in stiffness, the basilar membrane responds to sound in a peculiar way: If a high-pitched tone enters the ear, it is the stiff bottom part of the organ of Corti that vibrates most. Conversely, if we hear a low-pitched tone, the upper, floppy part of the basilar membrane vibrates most. When we play a triad on the Piano, three distinct sections of the organ of Corti are set into vibration, one vibrating with the lowest tone, one with the middle tone, and one with the highest tone. All other areas of the basilar membrane remain almost motionless. The important point here is: the frequency information contained in a sound is converted into spatial information, a phenomenon termed tonotopy. In this way, the cochlear covers the entire frequency range that we can hear: 20 Hz at the top of the snail-shell and 18 kHz at its base. Wherever the basilar membrane vibrates, it stimulates the local hair cells. To obtain information about the frequency of the sound, the brain simply has to look which of the hair cells along the organ of Corti were

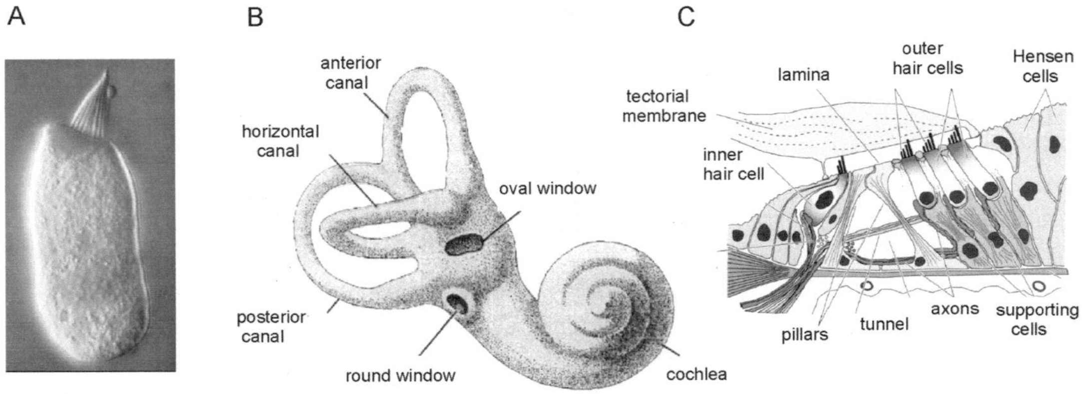

Fig. 3 Vibration detectors. **A** A hair cell isolated from the inner ear of a frog. The hair bundle on the apical pole consists of about 50 mechanosensitive stereovilli. The length of the individual villi increases from left to right; the structure with the spherical tip is the so called kinocilium, which is not mechanosensitive. **B** Schematic drawing of the inner ear showing the three canals of the vestibular organ that aid spatial orientation and balance, and the cochlea that mediates hearing. Sound enters the inner ear from the middle ear through the oval window and leaves via the round window. The inner ear houses hair cells for the detection of acceleration (vestibular organ) and for the analysis of sound (cochlea). **C** The organ of Corti is a sensory epithelium inside the cochlear coils. Hair cells are aligned in four rows, with 3000–4000 cells per row. The organ of Corti vibrates with the incoming sound; high frequencies (18 kHz) cause the largest vibrations near the oval window, low frequencies (20 Hz) near the top of the cochlea. The three rows of outer hair cells sense local vibrations and amplify them by shaking the tectorial membrane. The inner hair cells pick up these amplified vibrations and activate neurons which convey the auditory information to the brain (modified from Fain 2003)

stimulated. Each position corresponds to a frequency. The tonotopic organization of the cochlea and its precise tuning to the individual frequencies of our auditory world is endlessly fascinating to sensory physiologists, and it is by no means completely understood.

Thus, loudness correlates with the intensity of hair-cell stimulation, and frequency with the position of the hair cell along the organ of Corti. The third parameter, the direction to the source of the sound, can only be derived from the comparison of the two sensory signals generated in left and right ear. When the sound comes from the right, it hits the right ear with full intensity, but it is muffled when it reaches the left ear after having travelled through the head. Thus, there is directional information in the difference of the sound pressure levels (perceived as loudness) at the two ears. Moreover, the sound will arrive at the left ear a small fraction of a second later than on the right. This time-difference can be analyzed by the brain with amazing accuracy. Finally, the highly asymmetrical shape of our outer ears can help us to decide whether a sound comes straight from the front or straight from the back, in which case there are no differences in loudness or timing.

The sensitivity of our hearing is amazing, but the accuracy and sensitivity of hearing in many animals far surpasses ours. Especially nocturnal predators like owls, bats or leopards have driven the detection and evaluation of acoustic information to the extreme. While these animals have optimized their ears with respect to position, size, structure, and motility, the common basis for the extraordinary performance of all hearing systems is the exquisite sensititvity of hair cells. The stereovilli of hair cells respond to the slightest mechanical stimulus, even to movements that are much smaller than the diameter of a single stereovillus. How is that possible? If looked at from the side, the tuft of about fifty stereovilli tapers from one side of the cell to the other (Fig. 3A). The top of each stereovillus is connected to the side of the next longer neighbour by a protein filament termed tip link. But no tip links connect sideward to the stereovilli of equal length. In this way, the entire tuft of stereovilli is organized to move unisono, as one unit, whenever a stimulus pushes it along the tapering axis. Any movement along this axis will elicit a response with extreme sensitivity. Physiologists have employed ingenious methods to study the hair-cell response and found that the tip links are likely to be directly connected to ion channels. The slightest dislocation of the stereovilli stretches the tip links and pulls the ion channels open which, thus, act as transduction channels. They conduct potassium and calcium ions into the hair cell and trigger a cellular response that eventually leads to action potential generation in the neurons that are attached to the hair cell. When researchers examined the relation between the movement of stereovilli and the cellular response they saw their phenomenal sensitivity and found an intriguing image to illustrate it: Deflection of the stereovilli out of their resting position by 0.003° was sufficient to cause a cellular response (Hudspeth 1989). To appreciate what this means, we can envisage a stereovillus as large as the Eiffel tower, 300 m high; deflecting such a giant structure by 0.003° would cause its tip to move by no more than the width of a thumb (20 mm). In the real world, the tip of the stereovilli must move only 0.3 nm to cause a sensory response, the diameter of 3 hydrogen atoms. Thus, hair cells show a similar sensitivity to mechanical stimulation as the vibration detectors of the cockroach discussed in the previous chapter.

5. Vision

There is an amazing diversity of eyes in the animal kingdom (Land and Nilsson 2002). Eyes may be small inconspicuous spots in the skin, they may be huge, complex structures that make up more than a quarter of an animal's body, or they may look like huge globes of transparent glass waved about on thin stalks like some magic lantern. Animals may use the physical principles of diffraction or reflection, absorption dichroism, circular dichroism, frequency dispersion and light polarization to extract information on the visual world, and the depth of analysis by visual systems is really astounding. In fact, a recent theory on brain evolution suggests that the strongest impetus of early brain evolution was the wealth of information provided by eyes and the fundamental advantage for any species that was able to process that information – first came the eye, then the brain (Gehring 2002).

In all animals, the sense of vision (see also Chapter I,3 by K Krug) is mediated by a single family of proteins – the opsins. Animal opsins are not light-sensitive themselves. But they can house light-sensitive molecules, chromophores, which change their structure when hit by light of adequate wavelength (perceived by us as colour). In our eye, retinal is the chromophore. The combination of opsin and retinal is called rhodopsin, and some rhodopsins respond best to light seen as blue, others to light seen as green or red. When a chromophore is illuminated, it absorbs light, changes its structure and, hence, kicks the opsin molecule into an active state – the chromophore switches the opsin "ON". This is the start of vision. The activated rhodopsin is able to generate cellular signals that, in turn, can cause a cellular response. Sensory cells that possess rhodopsin and respond to illumination are called photoreceptors – light-sensitive sensory cells. They operate in all animal eyes, irrespective of their principle of construction. A photoreceptor provides the information that it is hit by light, a rather basic quality of information that simply distinguishes "bright" versus "dark". More useful information can be gleaned from the combination of a photoreceptor with a pigment cell that absorbs light but does not produce neuronal information itself (Fig. 4A). Such a combination tells the animal from which side the light comes, a message that may help to escape into the right direction. The interpretation of shadows and the position of light-obscuring objects is much improved when many photoreceptors are placed into a pit or even into a chamber with a small opening (Fig. 4B). In such an eye, the shadow of the predator will black out some of the photoreceptors while others stay illuminated. With the appropriate neuronal equipment, an animal can derive the form of an object from this shadow. The final addition to the eye is a lens, a translucent piece of tissue that focuses the image of an object onto an array of photoreceptors and at the same time increases the light flux by widening the hole for its access.

In our eye, the back wall is covered with about 150 million photoreceptors which together with many other cells form the light-sensitive retina. The lens is composed of translucent cells which have lost practically all large organelles which would be in the way of light entering the eye. The cells in the lens have developed a sort of low-key mode of life without nucleus and mitochondria, doing not much more than staying clear. They can live for many years in that quiet state, and they must prevent at all costs the aggregation of proteins, which would cause turbidity and a reduced optical quality of the eye. A special protein, crystallin, prevents aggregation and keeps our lenses clear, usually up to old age. What we see is projected by the lens onto our retina, where a sharp image is formed, just like on the film (or chip) of a camera. The retina converts this image into a pattern of electrical activity. This involves considerable processing inside the retina. Eventually, the information about our field of view is delivered to the brain by approximately two million neurons, the ganglion cells of the two retinae. Through a relay station, the thalamus, the information reaches the primary visual cortex in the back of the brain where the process of visual perception begins (see also Chapter I,3 by K Krug).

The eye is the only human sensory organ where a considerable amount of information processing already takes place before entering the brain circuits that eventually generate perception. The retina is a neuronal network made up of different cell types (Fig. 4C). The network performs operations which are characteristic of brain function. This is not surprising considering the fact that the retina is a developmental derivative of the brain. *Convergence:* In most parts of the retina, highly sensitive photoreceptors (so called rods) respond to minute light intensities, corresponding to a flux of 10–500 photons per second. More than 100 individual rod photoreceptors may converge onto a single target neuron. This convergence increases the chance that the target cell is activated even at very low light intensities when only a few photoreceptors detect a light signal. The incoming signals are added up to generate a stronger activation in the target cell. However, convergence reduces spatial resolution, as the output of the target cells represents the averaged activity of the entire area of the visual field covered by the converging photoreceptors. Convergence does, therefore, not occur where high spatial resolution is required. In the small central spot of the retina specialized for high spatial resolution, the fovea, there is no convergence; instead, each photoreceptor is connected to its own target cell.

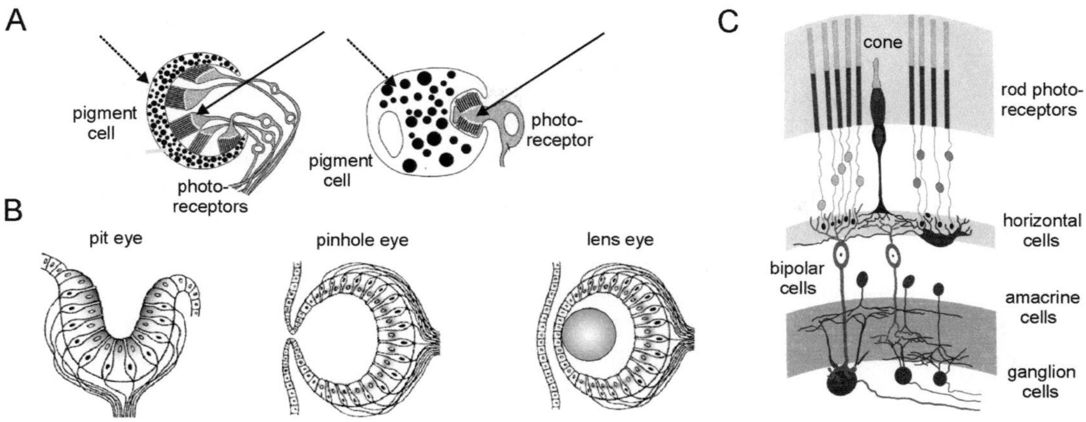

Fig. 4 Photoreceptors and eyes. **A** The combination of photoreceptor cells with pigment cells yields information about the position of the light source. The pigment cells shield the photoreceptors from the left side, so that only light coming from the right side can be seen. **B** Directional information is improved when a set of photoreceptors sits inside a pit eye. A pinhole eye produces an image on the retina, a light-sensitive tissue with many photoreceptors. Each photoreceptor can generate one point (one pixel) of the image. A translucent lens projects a sharp image onto the retina, thus strongly improving the optical quality of the eye (modified from Land and Nilsson 2002). **C** Photoreceptors in the retina pass their sensory signal through bipolar cells on to ganglion cells, which in turn send action potentials to the brain. Horizontal cells and amacrine cells connect these three cell types horizontally and enable the retina to perform complex signal processing

Lateral inhibition: One of the most important properties of a clear image is a pronounced contrast. On the level of the retina, contrast means that the activity of photoreceptors clearly marks the edge of a dark image; those on the bright side are active, those on the dark side are not. In fact, the retinal network exaggerates this difference in activity: The activated photoreceptors send an inhibitory signal to the less active ones in their vicinity, suppressing their residual activity and, hence, increase the perception of contrast.

Direction-selective activity: Perceiving motion is a complicated task for the visual system. In the retina, certain target cells of photoreceptors only respond when the image moves from right to left, but do not respond to other motion directions. Others are specialized to movements into other directions. These cells ("starburst cells") collect inputs from many other retinal cells and perform some kind of computation that results in direction selectivity – the neuronal basis of motion vision.

Colour contrast: In addition to the highly-sensitive rod photoreceptors, our eye possesses three types of cone photoreceptors with distinct wavelength specificities: "blue" cones, "green" cones, and "red" cones. These photoreceptors are concentrated in the fovea and mediate daylight colour vision. To extract and perceive colour information, the brain evaluates the relative activation of the three cone types; each wavelength in our spectrum of visible light corresponds to a discrete relation of activity in the "blue", "green", and "red" cones. In the same way as contrast is important for black-and-white images, colour images are only clear if they have good colour contrast. The retina employs the principle of lateral inhibition to increase the contrast between wavelengths corresponding to green and red, as well as to blue and yellow. Thus, colour vision is aided already by network computation in the retina.

Classification: Whenever possible, the brain operates by parallel information processing. The retina splits the visual information into two main channels: One contains the information on colour and form of an object, the other has the information about its localization and its movement. These two classes of information go separate ways through the brain; only after the analysis has been completed, they rejoin to allow the perception of an object as a whole. Thus, when visual signals leave the eye, they are already processed and prepared for further analysis in the brain.

6. Olfaction

Smelling something usually triggers an emotion (see also Chapter V,18 by R. Juette). We may find an olfactory stimulus agreeable or unpleasant. We may even experience craving or revulsion; but we seldom are indifferent towards an olfactory stimulus. Smelling seems to inevitably come with hedonic judgement. Unlike the skin, the ear, and the eye, the nose forces us to make a decision about an object that we examine by its odour. We have to decide whether the object is good for us or bad. Some of these hedonic responses are even genetically fixed. For example, the smell of a carnivorous predator makes us uneasy. In the zoological garden, even on the safe outer side of the fence, most of us feel the hairs on our necks stand on end when we smell lions or wolves. This kind of response it not learned, it is inherited from our ancestors. Even one-day old babies smile when they are offered a sniff of banana or vanilla scent, but they display disgust or start crying when challenged with the smell of rotten eggs. Thus, without having learned anything about food yet, babies know how to distinguish a "good" smell from a "bad" one. The smell of burning wood and grass has an alarming effect on almost every land-living animal, irrespective of prior experience. A rare exception seem to be humans, as tens of thousands of years of barbecuing our meat seems to have abolished the negative connotation of the smell of fire. In addition to congenital decisions on odorous objects, we learn to associate odours with the quality of objects all through our lives. We memorize olfactory experiences together with what they mean to us, and sometimes we can remember even after decades, how something or someone smelled – and whether we liked the smell or not.

Chemically speaking, the odorants which cause the perception of odours form a complex mixtures of volatile compounds, each at a different concentration and with a different smell. The aroma of coffee consists of about 800 individual odorants. We are unable to identify – let alone name – individual odorants within this mixture, nor are we interested in doing so. The relevant sensory signal for us is the impression that this unique combination makes on our olfactory system, and the pleasant feeling it generates. The total number of odorants present in our olfactory world cannot be established. We are constantly exposed to the smell of people, animals, plants, earth, cars, factories, perfumes, indeed an endless variety of olfactory objects, each of them exuding their individual combination of odorants. And whenever we try to put a number to the diversity of odorants, a chemical factory or a new strain of bacteria creates new odorous compounds – and our number will be obsolete. For the purpose of the present discussion, I take the variety in our olfactory world as unlimited.

How can a sensory system cope with such an endless number of different stimuli? Twenty years ago, Linda Buck and Richard Axel, then at Columbia University, discovered that mice possess a large gene family that encodes 1000–1300 different odorant receptor proteins (Buck and Axel 1991). This is by far the largest gene family in the genome. Dogs have roughly 900, humans 400 odorant receptor proteins. The receptor proteins are exposed to the inhaled air by olfactory receptor cells, tens of millions of which make up the olfactory epithelium at the ceiling of our nasal cavity (Fig. 5A). These cells are neurons which carry a tuft of sensory cilia, and each individual cell expresses only one odorant receptor gene. When the right odorant hits the sensory cilia, it binds to the receptor protein and induces a sensory signal that is sent to the brain. Since we have only 400 odorant receptors, each of these receptors must be able to bind not only one type of odorant but thousands of different odorants (otherwise our olfactory experience would be limited to 400 different odour molecules only). Physiological experimentation has shown this very clearly. Each olfactory receptor cell can be stimulated by many different odorants. But what kind of message is delivered to the brain by a sensory cell that does not clearly distinguish between individual odorants? When the brain receives action potential activity from an olfactory receptor neuron, this could mean: "I have detected something that may be amyl acetate *(banana)* or possibly limonene *(peppermint)*, but – depending on the concentration, which I do not know – it could also be isovaleric acid *(sweat)* or acetophenone *(almond)*." The brain cannot derive any reliable information from that kind of fuzzyness. Obviously, the read-out of a single olfactory re-

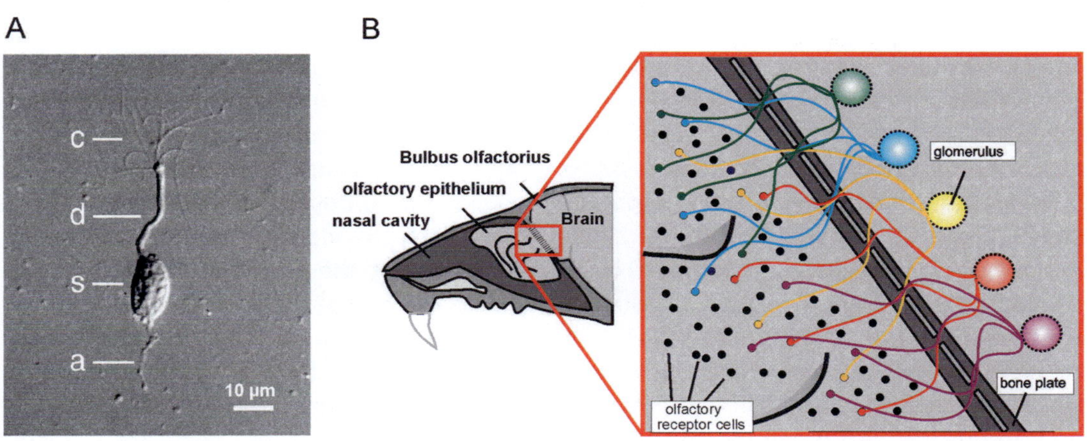

Fig. 5 Detection and sorting of olfactory information. **A** An olfactory receptor cell isolated from the olfactory epithelium of a frog. A dendrite (d) leads from the cell body (soma, (s) to the tissue surface where it carries a tuft of sensory cilia (c). The cilia present odorant receptors to the inhaled air and transduce the chemical information into a cellular signal. Output signals travel as action potentials along the axon (a) to the brain. (modified from Kleene and Gesteland 1981) **B** The mouse has approximately 1300 different odorant receptors, but only one type is expressed in each individual olfactory receptor cell. Cells with the same odorant receptor send their axons to the same spot in the olfactory bulb of the brain; cells with another receptor target a different spot. These spots are called glomeruli, and receive several thousand axons from cells expressing the same odorant receptor type. During a sniff, the activity of all glomeruli generates a spatial pattern that encodes the olfactory information for further processing by the brain (modified from Waldeck and Frings 2005)

ceptor cell does not identify an odorant for the brain. However, the olfactory system is very good at identifying odorants, although it operates with fuzzy receptors. The principle of information coding must be different.

Odour discrimination is a combinatorial process; the sensory information is contained in the combined activity of many sensory cells. This can be understood if one considers that each of our 400 odorant receptors has a slightly different structure and hence a preference for a slightly different group of odorants. Consequently, the set of receptor neurons activated by one odorant differs gradually from the set activated by another odorant. Instead of looking at the response of a single receptor neuron, the brain analyzes the *pattern of activity* that results from the fact that some receptor neurons respond and others do not. In mice, such odour-induced activity patterns can be visualized by modern brain imaging methods. This is possible because all olfactory receptor neurons with the same type of odorant receptor send their axons to the same spot in the brain, a glomerulus in the olfactory bulb (Fig. 5B), while neurons with a different receptor target a different glomerulus (Mombaerts 2006). Several thousand axons of cells with the same receptor converge onto a single glomerulus. If these cells are activated by an odorant, the combined cellular response at the level of the glomerulus is strong enough to be recorded in a brain-imaging experiment. So, a response of the olfactory bulb to an odorous stimulus is seen as the "lighting up" of several glomeruli as they are activated, while other glomeruli remain quiescent and dark. The current notion of olfactory encoding holds that olfactory information is transformed into a map of glomerular activity, a spatial code for the quality and intensity of an odour. How this map is read out by higher levels of the brain, how it is translated into perception, hedonic assessment, and memory, are exciting research topics still at an early state of understanding.

7. Pain

Pain is usually perceived as an experience that inflicts diseased or wounded people as an abnormal occurrence that causes agony and makes us suffer. Such a negative view of pain mostly results from the experience of chronic pain, long-lasting distress that may accompany inflammation or almost any kind of lesion. The physiological concept of pain is quite different. For the physiologist, the pain system provides its own sensory modality, just like hearing or vision, but specialized on stimuli that are strong enough to cause damage to our body. Such noxious stimuli are an every-day occurrence, they alarm us when the coffee is too hot to drink, when we step on a drawing pin on the floor, or when the way we lie in bed obstructs the blood flow into our hand. By warning us, the pain system constantly protects our body from damage. A striking illustration of the beneficial role that the pain system plays in our lives is the fate of people who are unable to feel pain. Patients with congenital insensitivity to pain have normal touch perception, but do not feel pain. Because they lack the alarm function, they are at constant risk of suffering damage, as they bite their tongues, break their fingers and toes, or touch hot objects, all without being alerted to the danger they are in. In severe cases, these patients may also have psychological problems, being in a continuous state of withdrawal, reminiscent of autism. The observation that these patients tend to scratch off their skin absentmindedly and are indifferent towards the wounds they inflict on themselves, promoted the hypothesis that in some of them the lack of pain perception prevents the maturation of the sense of self, the awareness of personal identity. Apparently, the pain system enables us to perceive the boundary of our body, an experience that may be necessary to develop an appropriate sense of what belongs to our person and what does not.

Noxious stimuli are detected by a set of sensory cells called nociceptors. These are neurons that have their sensory endings in the skin, in joints, muscle, and in inner organs, and they convey nociceptive signals to the central nervous system. The working of nociceptors differs in several ways from that of hair cells, photoreceptors or olfactory receptor cells. Firstly, they have a relatively low sensitivity – they respond only to stimuli strong enough to destroy living tissue. Secondly, a single nociceptor is often able to respond to several different stimuli, for example to heat exceeding 50 °C, to mechanical stimulation, and to acids. This "polymodal" behaviour points to the fact that the pain system generates a signal that primarily has to alert us to a critical situation independent of its cause. Informing us precisely about what happens to our body, whether the affected part suffers a burn, a cut, or contact with a corrosive chemical, seems to be of secondary importance. The pain system does not have to be specific but rapid. It has to provide an intense danger signal that can elicit protective behaviour as fast as possible – before tissue damage becomes too severe. A third fundamental difference between the pain system and the other senses is its intimate relation to the immune system. Whenever we suffer a cut in our skin, the immune system springs into action. Cells in the wounded area release signalling molecules that recruit immune cells to come to the aid of the local immune response. While the skin tissue is slowly repaired, the aroused immune system detects and removes bacteria and other harmful material, thus preventing a spread of the infection throughout the body. This process is termed inflammation, and it is accompanied by four standard observations: *rubor,* a reddening of the tissue caused by increased blood supply to the wounded area; *calor*, a local increase of tissue temperature; *tumor*, a local tissue swelling that occurs when the wall of blood vessels allow serum to leave the blood stream and enter the tissue during the recruiting process; and *dolor*, increased pain sensitivity. This sensitization to noxious stimuli is important as it prevents us from touching the inflamed tissue and makes us shield it from contact with anything. Sensitization is based on the ability of the immune system to change the response characteristics of nociceptors: Under the influence of inflammatory mediators, the nociceptors respond more intensely to a noxious stimulus than they do in intact tissue. We experience an amicable slap on the shoulder, that would usually cause us no pain, as quite painful when our skin is inflamed by sunburn. Thus, the immune system tunes the sensitivity of the pain system in the interest of the repair of damaged tissue.

A fourth and very striking difference that distinguishes pain perception from all other sensory modalities is the brain's ability to shut down the entire sense of pain in situations where pain perception or pain-driven behaviour may be counterproductive to survival. Pain suppression helps animals to escape a predator even after they have suffered horrible wounds like the loss of a limb. Humans also report this loss of pain perception. People who were severely wounded in car accidents or in battle report that they did not feel any pain during the first state of shock. The perception of severe pain started only later. This reflects a temporal cut of the information line. The brain presses a *veto* button and prohibits the entry of pain information. The most important site of this *veto* function is the spinal cord. Here, nociceptors pass the nociceptive information to the central nervous system through synaptic connections with spinal cord neurons. These synapses are targets for the brain's pain suppression system: The brain analyzes the situation during an assault on the body and may decide that pain perception is momentarily less important than efficient flight behaviour or, alternatively, the stabilization of vital functions like blood pressure. Under these conditions, a signal descends from the brain into the spinal cord, reaches the nociceptor synapses, and shuts them down by releasing endorphins onto them – self-made pain killers that derive their name from a fusion of "endogenous" and "morphins". Endorphins prevent the synapses from handing over the pain signal to the central nervous system and, hence, uncouple the nociceptor system from the system of pain perception. Thus, although the nociceptors scream out their message about a serious wound, the message is not heard by the brain because the critical synaptic junctions are blocked by endorphins. Pain perception can thus be functionally and anatomically separated from the nociceptive system; pain perception is an activity of the brain, not an activity of nociceptors. In medicine, the functional separation of the two systems is achieved by analgetic drugs like morphin, a substance produced by the opium poppy *Papaver somniferum*. Morphin mimics the brain's *veto* function as it binds to the nociceptor synapses in the spinal cord and prevents the passage of the nociceptive signal to the brain (see also Chapter III,11 by R. Weissensteiner et al.).

8. Synopsis

This brief introduction to sensory cells and sensory organs was designed to illustrate a few fundamental aspects about the first steps of perception: the detection of stimuli and the initial information processing that produces useful signals for the brain. The first point to consider is that sensory organs exist only for stimuli relevant for survival. Consequently, our senses enable us to perceive only a particular segment of reality. The close link between perception and survival has shaped the sensory organs and has sometimes driven their effectiveness to the physical limits. In these cases evolution has come to an end. Sensory cells have developed various sensors to detect the adequate stimulus at low intensity; specialized structures present the sensor to the outside world. Stimulation leads to transduction: The sensory cell has to translate the stimulus first into an intracellular signal, and eventually into a neuronal output signal in the shape of action potentials. On their way to the higher levels of processing in the brain, the sensory signals are sorted and spatially segregated. This allows rapid, parallel processing of information as it is channelled towards the final step of sensory function, perception.

References

Bounoutas A, Chalfie M (2007) Touch sensitivity in *Caenorhabditis elegans*. Pflügers Arch – Eur J Physiol 454: 691–702

Buck L, Axel R (1991) A novel multigene family may encode odorant receptors: a molecular basis for odor recognition. Cell 65: 175–187

Conway Morris S (1998) The crucible of creation. The Burgess shale and the rise of animals. Oxford University Press, Oxford

Fain GL (2003) Sensory transduction. Sinauer Ass., Sunderland, MA

Fortey R (2000) Trilobite. Eyewitness to evolution. Vintage Books, New York

Frings S (2009) Primary processes in sensory cells – current advances. J Comp Physiol A 195: 1–19

Gehring WJ (2002) The genetic control of eye development and its implications for the evolution of the various eye-types. Int J Dev Biol 46: 65–73

Hudspeth AJ (1989) How the ear's works work. Nature 341: 397–404

Land MF, Nilsson D-E (2002) Animal eyes. Oxford University Press, Oxford

Kleene SJ, Gesteland RC (1981) Dissociation of frog olfactory epithelium with *N*-ethylmaleimide. Brain Res 229: 536–540

Mombaerts P (2006) Axonal wiring in the mouse olfactory system. Ann Rev Cell Dev Biol 22: 713–737

Mueller W, Frings S (2009) Tier- und Humanphysiologie. Springer, Heidelberg

Waldeck C, Frings F (2005) Wie wir riechen, was wir riechen. Biol in uns Zeit 35: 302–310

Vertebrate hearing: origin, evolution and functions 2

Geoffrey A. Manley

Contents

"The manifestations of life . . . . . are so diverse that they must contain a large element of the accidental. And yet the nature of life is so uniform that it must be constrained by many necessities."

J. Bronowski, 1976. *The Ascent of Man*, BBC, p. 291.

Abstract

Using the studies of fossils and modern animals, the evolution of the sensory systems involved in hearing has been elucidated in great detail. The results trace the evolution of hearing back to our fish ancestors, through the establishment of a dedicated auditory epithelium and of independently- evolved, impedance-matching middle ears in amniotes. The resulting improvement in auditory input resulted in the expansion and specialization of auditory sensory epithelia. On the one hand these changes permitted new means of communication and other interactions with the physical environment. On the other hand the interpretation of complex sensory inputs from enlarged hearing organs necessitated the expansion and specialization of brain pathways

Geoffrey A. Manley
Technical University Munich
Institute for Zoology
Hochfeldweg 2
85003 Freising, Germany
e-mail: geoffrey.manley@wzw.tum.de

devoted to hearing. These pathways were subdivided according to the many different aspects of stimulus input (e.g. frequency, level, timing). Humans are relatively untypical mammals, in that their hearing range is restricted to relatively low frequencies but human development and upbringing enables the brain to cope with the complex processing required for speech and music appreciation. Far from being the "culmination" of some anthropomorphic evolutionary development, many other species show equal or even more spectacular specialization according to the importance of hearing in their daily lives. In modern society, the sensitivity of auditory sensory cells to damage is proving to be an inconvenient legacy of the early evolutionary trend to maximal sensitivity of sensory detection systems.

1. Introduction

The ability to detect and respond appropriately to stimuli was one of the first achievements of living organisms. The tendency to avoid noxious stimuli and move towards food sources must have been critical to survival and this ability is observed even in bacteria and single-celled organisms. In such organisms, the detection of stimulus energy is mediated by specialized protein complexes known as receptors that interact with or are ion channels in the cellular membrane and their activation leads directly to a response in the same cell. With the origin of more complex organisms and, later, nervous systems, a far-reaching division of labour between a large number of different kinds of cells that make up the body led to some cells specializing as detectors of the stimuli (receptor cells), others as integrators and transmitters of the information (nerve cells) and still others as response cells (e.g. muscle or glandular cells). The more complex the organisms, the more complex – in general – are the sense organs and the neural pathways that developed to deal with the sensory input. This included not only the development of different sense organs for stimuli of the various modalities (e.g., light, sound, odour, taste, touch), it also led to a specialization within each sensory organ to produce cells only capable of responding to specific qualities of particular stimuli (e.g. different colours of light, different frequencies of sound). This specialization enabled each cell to achieve maximal sensitivity to a narrow range of stimulus energy and this principle is abundantly realised in the inner ear.

This chapter concerns only one of the stimulus modalities, that of the reception of mechanical stimuli and, indeed, only the one kind of mechanical stimulus that we call sound. Although sound is actually quite difficult to define (since at the very-low frequency end of the sound spectrum it forms a continuum with vibration and continuous-pressure stimuli), we often think of sound in a too-limited way as those spectral components to which we humans are sensitive (between about 20 Hz and 20,000 Hz or 20 kHz; Hertz is the unit of sound pitch or frequency). Let us assume that sound consists of oscillating pressure waves in a medium, generally either air or water, that radiate out from their source, gradually losing strength as they dissipate. In general, the gas or water molecules simply move forwards and backwards, thus transmitting the pressure wave, but in the end, they finish up where they began. In other words, there is no net movement of the molecules. The average speed of their movements over the oscillations defines the sound pressure that we subjectively interpret as the loudness. Since higher sound frequencies mean faster oscillations, it is easier to achieve a high pressure at higher frequencies than at lower.

In all sensory cells that respond to sound stimuli that have been studied in detail, the receptor mechanism at the cellular level is amazingly simple and direct. Unlike some sensory stimuli, such as light, that need to activate a light-sensitive pigment in the cell before the cell membrane can respond indirectly, sound stimuli activate the receptor machinery (they *transduce* the stimulus) directly and are thus very fast. Sound stimuli can work within less than 10 microseconds, because the ion channels that are activated are themselves the receptor molecules. The transduction channels are connected by an elastic molecular chain across the space outside the cell to something that moves relative to it during a stimulus. The relative movement stretches the connector molecules and these in turn tug on the channel "gate" and, if strong enough, open that gate, letting positive ions stream through the channel and activate the cell (Fig. 1e). In vertebrate animals and their immediate ancestors, such stimuli are detected by so-called "hair" cells, which have over their apical surface a coppice-like bunch of "hairs". These hairs are thin, finger-like extensions of the cell membrane that contain a stiff framework of actin molecules (Fig. 1d). The hairs are actually very large villi and are often referred to as stereovilli (or, inaccurately, stereocilia). These stereovilli are interconnected and it is the specialized connections near their tips (so-called tip links) that are bound to ion channels. Sound stimuli reach the hair cells (that are surrounded by a fluid) via other tissues (Fig. 1 a-c) and the bundle of stereovilli is pushed back-and-forth, resulting in channel opening and cell activation (Gillespie and Walker 2001; Fig. 1).

Due to fluid viscosity, it is actually difficult to move structures through a fluid, and, for very small structures, molecular attraction forces increase this problem. Good sensitivity of hair cells is only possible because they have a built-in amplification system in the transduction channels of the stereovilli. Put briefly, this system works since the ion channels also admit small amounts of calcium ions when open. These calcium ions lead to a rapid increase in the closing probability of the channel, exerting a force that makes it easier for the channel to close again. When this force operates with the right timing in relation to the channel opening-and-closing cycle as caused by the sound stimulus, it leads to amplification of the effect of the stimulus at low levels of sound. During evolution, this mechanism was inherited by the hair cells of auditory organs and thus also contributes to their sensitivity (Hudspeth 2008).

Sounds are usually complex, in that they consist not only of one single sinusoidally-oscillating component (that would be a pure tone, which is very rare in nature and in music). To understand and dissect such complex sounds, scientists use spectrum analysers that nowadays are able to display the frequencies and amplitudes of the components present at different frequencies very quickly. In many respects, these instruments carry out the same processes that occur in sense organs for audition (hearing), as we will see below. The ability to generate precisely both simple and complex sounds has played a very important role in the scientific investigation of how ears work. Ears – sense organs specialized for the detection and analysis of sound – are found in molluscs such as *Octopus*, in insects (many kinds of ears of independent origin) and in all classes of vertebrates (Review in Manley and Ladher 2007). Since this chapter is limited in length, I will confine my description to structures relevant to the ears of vertebrate animals such as fish, amphibians, reptiles, birds and mammals. (I will note here that the term "reptile" refers to a diverse and not very closely-related group of organisms and no longer has a scientific foundation. I use it here for brevity).

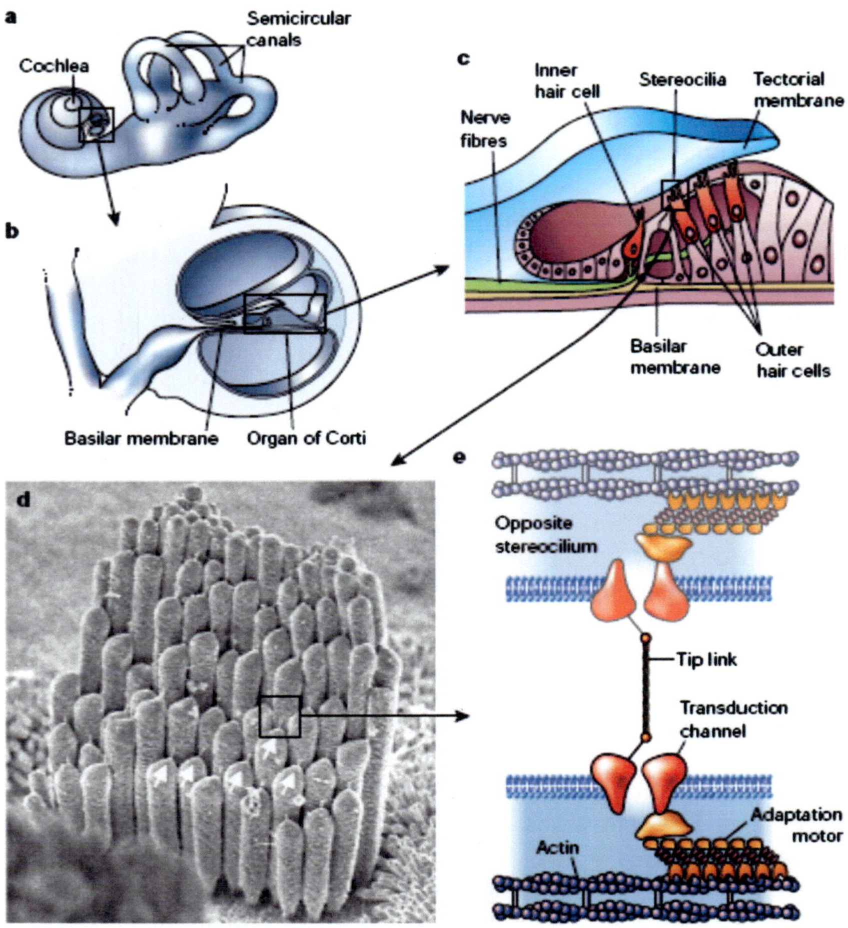

Fig. 1 The structure of the mammalian cochlea. In **a** the cochlea is shown as a whole, with the associated parts of the vestibular system. In **b** one half of a single cochlear spiral is shown in cross-section. In the centre of the three fluid-filled canals lies the basilar membrane, which supports the organ of Corti. **c** The organ of Corti shown in more detail. The hair-cell bundles are surmounted by the gelatinous mass of the tectorial membrane. **d** Scanning electron micrograph of the stereovillar bundle, showing stereovilli of graduated height. In this case, the bundle is from a chicken cochlea. The box highlights a tip link connecting two stereovilli. In **e** the tip link structure is shown schematically. It is connected to both stereovilli. We now know that the transduction channel is at the lower end of the tip link. Modified after figure 4 in Gillespie and Walker (2001). Used by permission of Nature Publishing Group

2. Origins of hearing organs

Unfortunately, my task is not made easier by the fact that vertebrate animals use several different kinds of sense organ to detect and analyse sounds. However, all of these organs contain hair cells and it is always the hair cells that are the detector elements. Very early in their evolution, vertebrate animals developed a standard set of complex sense organs. These paired sensory organs were localized in the head since, if you are moving with the head first, it is better to detect stimuli in that region, knowing, as it were, where you are going, rather than where you have been! The nasal organs developed at the front, the eyes behind those and the "ears" at the rear of the head. I have used quotation marks for ears, since these organs were initially most likely only organs of balance and not of hearing. They consisted of several fluid-filled spaces containing patches of hair cells that were covered either by a sail-like gel mass or a gel mass that contained tiny stone-like otoliths. The former hair cells were placed at the ends of semicircular canals and responded to movement of the fluid caused by head rotation, the other patches responded to linear acceleration (gravity). Compared to sound stimuli, both of these modalities involve stimuli that usually change slowly, in contrast to the rapid pressure changes of sound. In spite of this, in the evolution of various groups of fishes, one or more of these hair-cell patches developed the ability to respond to such faster changes. These were all hair cells covered by a gel containing otoliths, since otoliths have a different density to that of the surrounding fluid and when stimulated by sound can have a delay in their motion relative to the fluid. This delay creates relative movement between the hair cell stereovillar bundles and the otoliths and this forms the stimulus to the hair cells. In this way, many groups of fishes have developed sensitivity to relatively low-frequency sounds, up to about 1 kHz. In addition, a few fish groups developed ways of connecting their gas-filled buoyancy organ, the swim bladder, to their ears. The density of gas is of course very much less than that of water and sound passing through the fish induces a large movement at the interface of the fish body and the gas bladder. These movements, transmitted (e. g. by way of modified bones) to the inner ear make the ears of fish with these specializations more sensitive to sound and also sensitive to a greater frequency range, up to several kHz (review in Ladich and Popper 2004). These sound responses are, however, mediated by sense organs that also have other, vestibular functions.

In the early Palaeozoic, about 375 million years ago, the earliest land vertebrates appeared (Carroll 1988). Of the many changes in the body's physical characteristics necessary to enable these animals to live on land were modifications to the sensory organs. Sound transmitted through air essentially bounces off the body, since the densities of air and body tissue differ hugely. The physicist terms this an impedance mismatch. Although the earlier literature indicated that the first land vertebrates solved this problem for hearing in air very quickly, this has turned out not to be generally true. In fact, is was not until the early Mesozoic, 150 Million years later, that impedance matching structures – I will refer to them as tympanic middle ears – appeared in the major groups collectively known as amniotes (an amniotic membrane surrounds their embryos). Amazingly, tympanic middle ears, that is, ossicle systems attached to a tympanic membrane or eardrum, appeared within a brief geological period and independently of one another in the reptilian ancestors of (a) lizards and snakes, (b) crocodiles and birds and (c) mammals (Clack 2002). Before this, it has to be assumed that land vertebrates (like most fish) heard only very low frequencies that were of high sound pressure. Perhaps they also heard sounds through the substrate quite well. In spite of this, at least one dedicated hair-cell epithelium had developed that was devoted only to hearing. It is perhaps not surprising, due to the inferior quality and the poor amplitudes of the sound stimuli for these early land vertebrates, that this dedicated sensory epithelium, the hearing organ, was small and not very significant in the inner ear. As such soft tissues do not fossilize, assumptions about the size of organs can only be made via comparative studies of modern species from all groups. Such studies suggest that the early hearing organ was at the most about 1 mm in length and contained a few hundred hair cells (see below). By comparison, the largest of the hair-cell areas devoted to balance contains many thousands of hair cells.

Modern amphibians such as frogs differ very greatly from the earliest amphibians (the first land vertebrates) and it is impossible to know how their hearing organs compare. Many modern amphibians have two small auditory sensory organs or papillae, whereas the other land vertebrates, the amniotes, have only one. There have been suggestions that the hearing epithelia of amniotes and amphibians developed independently and at present it is not possible to decide whether this is correct. For this reason and because of space limitations I will ignore amphibian ears and simply discuss the hearing epithelia of amniotes, which are considered by all authorities to be homologous, that is, to have one single and common origin. As noted above, in contrast to fish audition, amniote hearing organs are devoted entirely to audition.

3. The origin of middle ears and hearing organs of amniotes

The hearing epithelium of amniotes in its earliest form was small and nestled in an out-pocketing of the largest space of the vestibular (balance) system, the sacculus. Its hair cells are derived in ontogeny, when the animal develops this organ, from a cell patch that also gives rise to the saccular hair cells and this is almost certainly their evolutionary origin. During the early Mesozoic, the form of the skull of amniotes and the mechanics of the jaws changed such that one bone, that had been a strut supporting the jaw area of the skull during biting, became redundant. This bone was near the jaw joint, had always made a connection to the bony capsule of the inner ear and was already presumably responsible for some relative motion between itself and the ear capsule and thus low-level auditory stimulation of the inner ear. Its redundancy permitted it to become much less robust and the tissues surrounding its outer projections to thin down to what we now refer to as the eardrum. Thus the stapes (stirrup) in one of its modern forms arose as the ossicle of single-ossicle middle ears of non-mammals such as lizards and birds (Clack 2002). Where the stapes, or columella, as it is referred to in non-mammals, connected to the inner ear, a footplate formed supported by a ring of connective tissue that enabled free movement, at least for small amplitudes. The much larger size of the eardrum compared with this footplate created an area-amplifier. Together with a lever mechanism in the ossicle itself, these factors contributed to an amplification of the pressure of the sound passing from the eardrum to the inner ear that more or less matched the loss of amplitude incurred at the interface from air to body tissue. Thus middle ears are referred to as impedance-matching devices (Fig. 2). Unfortunately, the columella forms a second-order lever system that depends on bending of its projection to the eardrum for its function and this bending makes it difficult for higher frequencies (above 6–8 kHz) to actually reach the inner ear in any strength.

In the evolution of mammals, a remarkable thing happened. Mammals derive from a separate lineage of reptilian ancestors (the synapsids) and, although their skulls were changing in a similar way to those of the other reptilian groups of the time, mammal-like reptiles went one step further. *Simulta-*

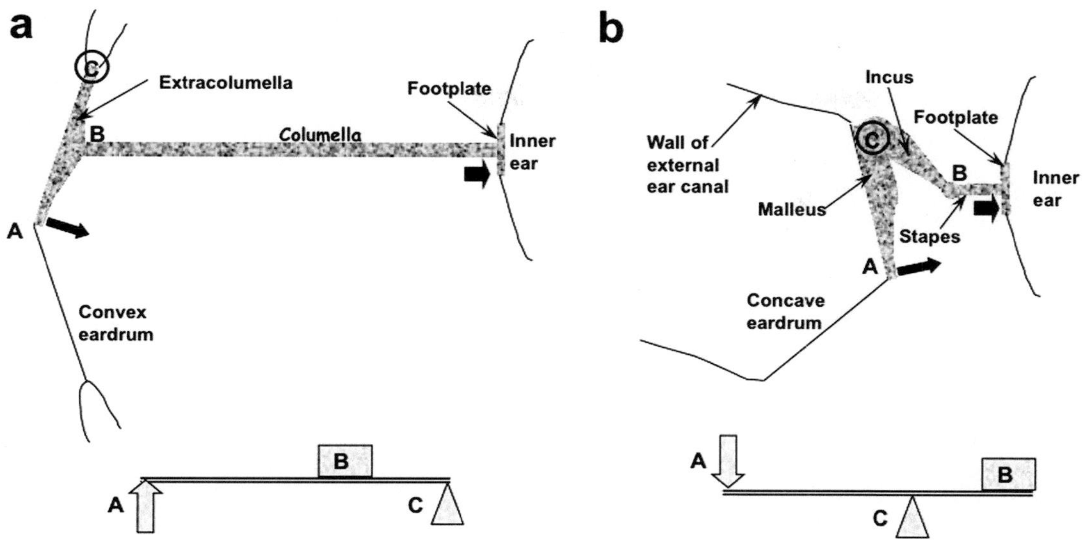

Fig. 2 Highly schematic presentation of middle ear function, comparing non-mammalian amniote **a** and mammalian **b** middle ears. In both cases, a model of the lever system involved is shown, with the capital letters corresponding to the positions of force application (A, idealised to the middle of the eardrum), load B and fulcrum C. The axis of rotation is shown as a circle around the fulcrum. The necessity for transforming a rotation of the extracolumella in **a** into a piston-like movement of the columella is enabled by a flexible area of the extracolumella where it joins the columella. The amplitude and force at the eardrum (longer black arrow) is changed by the lever into a smaller amplitude and greater force at the footplate of the columella/stapes (shorter but wider black arrow). In **b**, the two arms of the lever make a large angle (~300°) to each other, so that A and B move roughly in the same direction

neously with changing the function of the stapes, they also developed a new jaw joint. The jaw joint of all early vertebrates and all modern amniotes with the exception of mammals is made up of two bones of the upper and lower jaw that are derived from ossification (bone formation) of the cartilaginous jaws of the earliest craniates (vertebrates having skulls). This bony articulation was assisted in mammal-like reptiles by two bones derived from ossification of so-called dermal bone, i. e., superficial bones that form in deeper layers of the skin. These new articulating bones were on the same rotational axis as the primary jaw joint but outside of it. The new, or secondary, jaw joint was so efficient that the old jaw articulation became redundant and the two bones that formed it moved back and became simultaneously incorporated into the newly-emerging middle ear. As a result of this, the ossicular chain in mammals developed not as one bone, but as three – the malleus (hammer) and incus (anvil) in addition to the stapes (=columella). Co-incidentally, these bones formed a primary lever system that was much more efficient at transmitting higher frequencies than that of non-mammals. Later in mammalian evolution, after high-frequency hearing had evolved, this facility was heavily exploited for ultrasonic hearing in some groups of mammals. The presence of three ossicles in the middle ear is one of only a few features that easily define the transition in fossil remains from mammal-like reptile to true mammal.

4. The evolution of hearing organs of reptiles, birds and mammals

The evolution of both kinds of middle ear, with one or three ossicles, resulted in a great increase in both the amplitude and the frequency range of auditory input to the inner ear. Using the structure of hearing epithelia in all modern groups of amniotes as a guide, this generally led to important changes in the size and configuration of the organs of hearing themselves. They became more important sensory structures and their further development led to a proportionate increase in the size and complexity of those parts of the brain occupied with auditory processing – the auditory pathway.

Apart from two groups of reptiles (turtles and tuataras), all amniote lineages increased the size and complexity of their hearing organs (Fig. 3; Manley 2000). They became highly sensitive, in some cases reaching very close to the theoretical limit of hearing ability that is restricted, for example, by thermal molecular noise. With respect to the frequency range, reptiles generally expanded their upper hearing limits to near 6 kHz and birds up to above 10 kHz. The very obvious differences in hearing-organ structure between lizard-like reptiles on the one hand and crocodiles and their relatives the birds on the other show quite clearly that changes of inner-ear structure and improvements in inner-ear performance arose independently in each group. The same is true of mammals, since the mammalian inner ear is also unique. The ability of the mammalian middle ear to transmit higher frequencies was exploited and at least small mammals usually have upper hearing limits in excess of 50 kHz. In some families of bats, the upper limit exceeds 100 kHz. Larger mammals have larger structures throughout, including the middle ear, and the laws of physics determine a poorer performance of large structures at higher frequencies. Thus humans have an upper limit near 20 kHz, the mouse about 70 kHz. Interestingly, mammals that have returned to live in an aqueous environment can exploit a different option to overcome this limitation deriving from body size. In water, the eardrum becomes a handicap, since there is no significant impedance mismatch between the water and the animal's body. In cetaceans, the eardrum has been eliminated and the middle-ear ossicles are specialized to receive their sound input through the tissues of the head. By by-passing the eardrum, the upper limit of hearing can be far higher than in equivalently-large land mammals such as humans. Some dolphins and porpoises are known to hear even higher than most bats and – like bats – to utilize echolocation to catch their prey. Since the wavelength of sound in water is much larger than in air (almost 5 times), the detectable prey organisms that reflect the same frequencies in water are larger than those in air (i. e. fish as compared to insects).

Interestingly, in both birds (and the related crocodiles) and in mammals, a specialization of hair cells of the hearing epithelium developed. In complex animal bodies, cell specialization through a division of labour is very common indeed. The hearing organ of mammals is characterized by the presence of supporting cells that form an arch or tunnel along the middle of the organ, dividing two sets of hair cells. On its inner side is, generally, one row of so-called inner hair cells. Outside of the tunnel are found at least three rows of outer hair cells that, very curiously, are connected to far fewer nerve fibres than are the inner hair cells (Fig. 1). This pattern is fairly strictly observed by all advanced mammals (placental and marsupial mammals). In the less-derived, egg-laying mammals such as platypuses, the numbers of hair-cell rows is larger and the inner-ear hearing organ is not coiled into a spiral cochlea. In the advanced hearing epithelium, the structural specialization was obvious to the earliest anatomists, but the function(s) of the specialized hair-cell groups was, for decades, obscure and the subject of serious but also wild speculation. When it was realized that amplification at the hair-cell level (see above) was important in enabling and indeed generating the high sensitivity of hearing, things began to fall into place (Hudspeth 2008). It was shown that the outer hair cells are in fact essentially motors using specialized – and uniquely mammalian – amplification mechanisms that result in greater driving forces for the entire organ – and thus for the inner hair cells also – to low-level simulation, thus enabling greater sensitivity. This explained the nominal nervous connections of the outer hair cells, which act as the motors, but also their greater number (~80% of the hair-cell complement of the ear). There is thus a division of labour: outer hair cells are specialized as amplifiers, inner hair cells as detectors. In mammals, there is much evidence that the ancient amplifying mechanism (at the hair-cell transduction channels) was supplemented by modified membrane channels made up of the protein *Prestin*. Uniquely in the lateral membranes of outer hair cells, these proteins are present in very high numbers and when stimulated are able together to rapidly change the length of the cell. There is evidence that these length changes modify the vibration of the local organ of Corti area and, through this, amplify input sounds (review in Hudspeth 2008). How this amplifying mechanism interacts with the channel amplifiers has not yet been elucidated.

In the crocodilian-bird type of cochlea, the evidence is less complete but suggestive of a similar division of labour between the less numerous, so-called tall hair cells on the inner side of the papilla (there is no tunnel) and short hair cells on the outer side. There is a similar difference in the nerve supply; indeed short hair cells completely lack that component of the nerve supply that sends sensory information to the brain (afferent nerve fibres). In mammalian outer hair cells, this component is still there, but only just. "Still" implies here that this component in mammals may well be in the process of being lost during evolution. There is little evidence for amplification by bird short hair cells, but the comparison of ears – and sensitivities – makes it seem likely (Köppl et al. 2000).

The hearing epithelium was elongated in all of these amniote groups following the development of tympanic middle ears (Manley 2000; Manley and Köppl 1998). Birds attained intermediate lengths of the inner ear (over 10 mm in some owls) and some mammals the longest known auditory papillae (71 mm in the blue whale, more typically 10 to 35 mm). Such very long organs in therian mammals were accommodated by coiling the papilla (Fig. 3). All of these hearing epithelia function like a spectrum analyser, the various regions of the papilla respond to different sound frequencies and these response properties are graduated along the papilla's length. Thus – with one interesting exception – the hair cells of the most apical region respond to the lowest sound frequencies and those of the most basal region to the highest frequencies. The interesting exception proves that this arrangement does not depend on the orientation of the hearing organ with respect to the middle ear. In the gecko family of lizards, the frequencies are arranged the other way around (Manley et al. 1999), a result of the fascinating variety of evolutionary patterns in lizard ears (Manley 2002).

Such a distribution of responses is known as a tonotopic (place-tone) organisation and the frequencies generally change in a logarithmic fashion to linear length. In humans, for example, the frequency doubles (i. e. changes by one octave) about every 3 mm

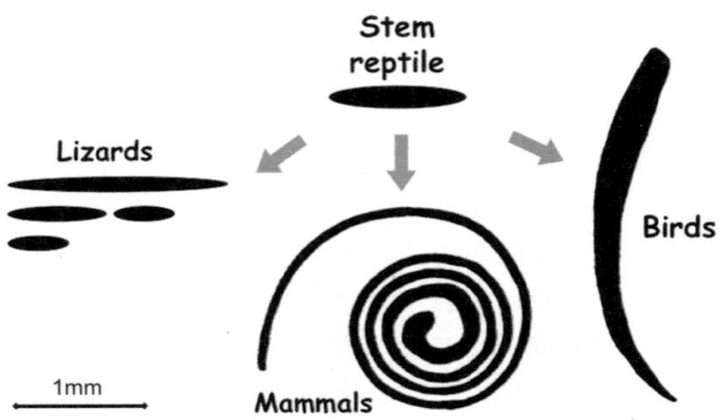

Fig. 3 A diagrammatic comparison of the surface dimensions of basilar papillae in three groups of amniotes. It is assumed that in stem reptiles, the papilla was about 1 mm long and showed no division of labour among its hair cells. Following the evolution of tympanic middle ears, the basilar papillae in these three groups tended to elongate, but independently in all three. In lizards, there resulted a great variety of forms and dimensions and a separation of low- and high-frequency hair cells. In birds and in mammals, the elongation was much greater, maximally to 11 mm in birds and 104 mm in mammals. In both groups, unique kinds of hair-cell specialisations with a division of labour developed

along the papilla. Thus in fact many more hair cells respond to a given frequency range (say, 1 kHz) within low-frequency octaves than within high-frequency octaves. Papillar elongation permitted the spreading of a given number of frequency octaves over a greater length of sensory surface, leading to an increase in the number of hair cells – and therefore nerve fibres – devoted to any given frequency range. In this way, the exactness and detail of the analysis of sound was greatly improved. Thus the most fundamental specialisation of hair cells in the inner ear relates to their highly selective responses to narrow frequency ranges, ranges that widen as the sound pressure increases (forming so-called tuning curves). The selectivity shown is the result both of the systematically varying properties of the hair cells themselves, but also of the accessory structures in which they are embedded.

Not all hair cells and not all the nerve fibres connected even to a single hair cell respond to sound with the same sensitivity. In addition, each hair cell and nerve fibre responds over a specific range of sound pressures and a certain level of electrical activity in a specific nerve fibre can be directly related to the sound pressures at the eardrum in a particular frequency range. The sensitivity of hair cells can also be adapted by the activity of so-called efferent nerve fibres emanating from the brain; these efferent fibers are able to shift the response range of hair cells to higher sound-pressure levels. Although the structural determinants underlying the coding of sound pressure differ in different vertebrate groups, the result of this differentiation is that not only is the neural input to the brain coded according to frequency, it is also coded, for example, with regard to the sound pressure level (Pickles 2008).

5. Evolutionary changes in skull structure and the processing of cues to sound localisation

In addition to the above, the neural input also delivers important information concerning the timing of external sound stimuli, e. g., their on- and offset and when changes in sound pressure occur. Although this information is surely very important in its own right, e. g., when a bird listens to the song of a territorial rival, it is also important for another reason. In order to understand this, I need to make a brief excursion back into evolutionary history and briefly discuss how the ears are coupled – or not – in the different amniote vertebrates.

We often forget that the skull structure in mammals is unique and shows several features not seen in all vertebrates. For example, mammalian (and, in spite of derogatory sayings, bird) brains are very large compared to those of reptiles and amphibians. One interesting consequence of the massive brain enlargement that followed the evolution of true mammals was that the skull enclosing the brain grew not only outwards, but downwards towards the mouth. During this time, mammals also developed a secondary palate, creating a nasal cavity above the palate and below the brain, isolating the brain case from the mouth cavity. In birds, reptiles and amphibians, the bones enclosing the brain abut directly on the buccal or mouth cavity. In these non-mammals, there was no cavity that deserved the name middle-ear cavity; mostly, the space is so large and open that one can look through the superficial eardrum and see right through the head (Fig. 4)!

Now back to my point: With this open configuration, sound that activates one eardrum will pass from that eardrum across the head and activate the opposite eardrum as well. Depending on the sound direction and frequency, the sound passing across the head can nullify the effect of sound that arrives directly at the opposite eardrum, giving a new meaning to the old adage "in one ear and out the other". Thus, if a sound comes, say, from the left, the head of a lizard might create a sound shadow that produces a small loss of 2 to 3 dB in the sound pressure that arrives at the right eardrum from outside. The sound pressure arriving through the inside, having crossed the head in the open buccal spaces, can – if the phase is right – almost completely cancel the effect on the right of the sound from outside. This means that, while the left eardrum vibrates strongly to the sound, the right eardrum would be virtually still. In my work with Christensen-Dalsgaard (Denmark), we showed that instead of the 2 dB difference between the amplitudes of the left and right eardrums expected from sound shadowing alone, the difference can be larger than 20 dB (Christensen-Dalsgaard and Manley 2005)! Thus the neural input to the brain would reflect a much larger difference between the two

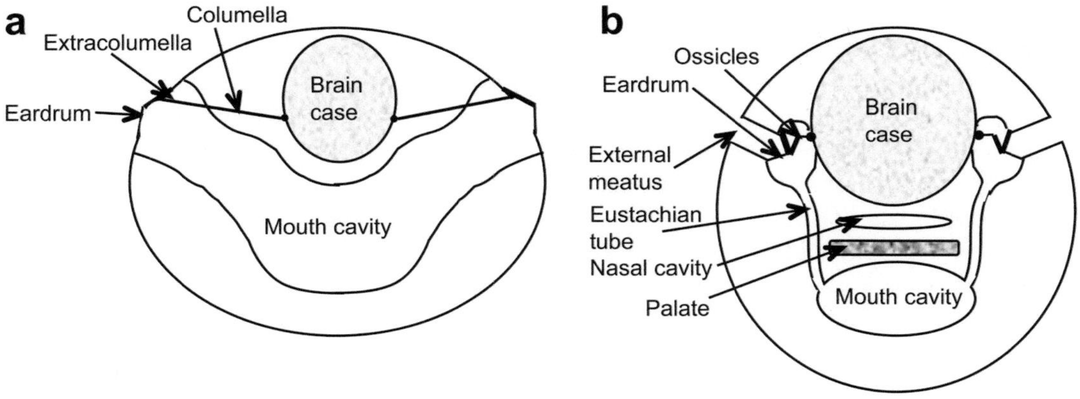

Fig. 4 Highly schematic diagram to illustrate the difference between non-mammals and mammals with regard to separation of the middle ears. In non-mammals **a**, the space behind the eardrums is widely continuous with the mouth or buccal cavity. In mammals **b**, the increase in brain size and the introduction of a secondary palate permitted only a narrow connection between the middle-ear spaces, the closable Eustachian tubes. The space behind the eardrums is generally small and often surrounded by a bony bulla

ears and this information can be used to determine the place of origin of a sound source, in other words better localise it. This phenomenon has been called a pressure-difference receiver and provides non-mammals with localisation information without the necessity of complex neural processing.

Mammalian ancestors show no evidence of the development of a wide connection between the ears. Each set of ossicles is surrounded by a small air-filled space, the middle-ear cavity and, in order to equalise air pressure with the outside world, this space is connected to the nasal/mouth cavity only by a narrow, closable canal, the Eustachian tube. The middle ears are thus totally isolated from each other, producing pressure receivers, rather than a pressure-difference receiver pair (Fig. 4). Sound directionality information provided freely and without processing by the latter is absent and the direction of a sound source had to be deduced and calculated from other sources of information. This meant an equivalent specialisation of the auditory neural pathway to deal with this new requirement. In birds, the brain also increased in size and the open connection between the ears was reduced. Birds did not evolve a secondary palate and the bony skull of birds is, like a sponge, full of spaces, creating a light structure but also providing a number of secondary pathways for sound to cross the head. How effective these connections are has yet to be determined, but in the barn owl it is known that above 6–7 kHz, the middle ears are essentially isolated from each other (Review in Manley 1990). Thus the neural specialisations described below for mammals apply to some extent – but developed independently – also to birds.

Dramatic changes in the anatomy and size of the hearing ear were accompanied by appropriate changes in the brain. In all sensory systems, the brain carries out its analysis of the information content by splitting the incoming information according to characteristic features of the stimulus (see below, section 6). The brain compares sound signals arriving at the two ears, which enables it to derive information concerning the locality of the source. The processing of information on the direction of a sound source is not trivial and this can be seen alone in the fact that many brain nuclei of the auditory pathway show very clear specialisations to deal with one or the other clue to sound direction. What are those clues? I will ignore so-called monaural clues, which arise from reflections within one external ear or pinna, although these can be significant. There are two essential clues to the direction of a sound source (1) the sound pressure difference between left and right ears caused by head shadowing and (2) the small time delay between the arrivals of the sound at the two ears. Of course, both of these cues will be at zero difference if the sound comes exactly from the front or from the back. But that is actually the point – any other location will present the ear with generally unambiguous information as to the source's location. The size of the head's sound shadow will depend on the animal's size (bigger heads produce bigger shadows) and the sound frequency (bigger heads produce useful shadows at lower frequencies). Since most mammals are actually quite small, the head shadowing is most effective at high frequencies. It is thus fortunate that by the "accidental" evolution of a three-ossicle middle ear, mammals are uniquely equipped to hear such frequencies.

In addition to sound shadowing, sound travels rather slowly (at least compared to electromagnetic radiation). In spite of that, the time sound takes to travel across a small head is very small indeed. In dry air at 20 °C, sound travels at about 343 ms^{-1}, it thus traverses a 1 cm head in about 30 microseconds (30µs). To put this in perspective, a single action potential that forms the basis for information transfer along nerve axons lasts roughly 1ms or 1000μs, i.e., thirty times longer. One begins to wonder how the brain can cope with such small time differences and to realise that, as I noted above, the neural processing of sound localisation cues is not trivial! In view of this, it is not surprising that part of the expansion of the avian and mammalian brain was devoted to the neural processing of time and sound-level differences. In well-studied cases in both birds and mammals, it has become clear that over major portions of the auditory pathway, these two cues are processed separately, as the required cell specialisations are different. At some point high in the brain, these two pathways then converge and neural maps of auditory space in the outside world are computed (see below, section 6). Compared to the visual system, where a map of the outside world is inherent in the visual signal and already present at the eye's retina, this is indeed an enormous investment in neural computing power.

Sound-level and time differences from the two ears, the binaural cues to sound locality, are only two aspects of the neural processing of sound signals, but they do in fact determine and dominate the structure of the auditory pathway in the brain (Köppl 2009). Other aspects, such as the spectral content of a sound, the relative levels of different components, the duration of signals, their amplitude oscillations, etc. all also need to be processed to produce a complete set of information of the acoustic world outside the body. Not surprisingly, the decades of scientific study of the hearing brain, while producing a plethora of extremely interesting data and theories, are still far from producing a complete picture.

6. Sound and behaviour

The importance of sound to behaviour can be summed up under two headings: (1) communication with conspecifics and other species and (2) alerting to the presence and location of enemies, food sources and other organisms. It might be expected that the extent of use of these kinds of information by different organisms will depend on their sensory and their neural equipment and there are indeed general indications that this is so.

Communication using sound can be cheap and effective. Cheap in the sense that, since ears are so sensitive, the amount of energy necessary to produce detectable sounds is very small. A bird's song, for example, consumes vastly less energy than any kind of courtship dance involving major bodily movements. In addition, a bird that dances must be seen by the object of its attention, whereas a singing bird can be heard even if not seen. It is thus less exposed to potential predators. Sound communication can also be carried out in darkness (important for many amphibians and mammals!), whereas a dance in darkness would be wasted energy. Communication using sound thus has a number of important advantages and it is hardly surprising that many vertebrates, including fishes, use this pathway in different contexts. There are, however, large differences between the various groups of land vertebrates and these are not only explicable on the basis of brain size and thus neural computational abilities.

If we compare reptiles, birds and mammals, i.e., the groups of amniotes, there are clear patterns in the use of sound communication. Birds and mammals make extensive use of sound signals, whereas most reptiles are not known to communicate with sound or, indeed, make any sounds at all. Few lizards are vocal and the exceptions prove the rule. Geckos and their relatives, the pygopods, are quite vocal and many have a small repertoire of sounds used in aggressive and mating encounters. These are mostly nocturnal animals, however, and visual signals are not useful, but they also have the most complex inner ear of all lizard families (Review in Manley 1990). It can also be concluded that the relatively high use of vocalisation by anuran amphibians (frogs and toads) is explained by their nocturnal mating habits and is not accompanied by a large brain. Crocodiles and alligators are more related to birds than to any other reptiles and they do have a small repertoire of vocal signals used in various behavioural contexts. However, neither gecko nor crocodile vocal signals compare favourably to the enormous usage made of complex vocal communication by birds and mammals. At the moment, we are not in a position to judge the extent to which the various parts of the hearing brain are devoted to the various sound components that are important in different contexts and for different reasons. It must therefore suffice to say that a large number of hair cells and nerve fibres, the necessity to generate maps of acoustic space through neural computation and the processing of species-specific and other acoustic signals adequately explains the large differences in the auditory processing capacity of the brains of birds and mammals on the one hand and amphibians and reptiles on the other hand.

The way we perceive the acoustic world is of course the result of both the selectivity of the hearing organ and the neural processing of auditory nuclei in the brain. These sensory pathways extract information in different ways and the nuclei are to some extent organized in groups, each group being dedicated to one main aspect – or feature – of auditory perception. Thus it can be seen that often within large nuclei, groups of neurons exist that deal with one particular aspect of a stimulus (Fig. 5). Two of the most obvious, general questions asked of stimuli are "what is it?" and "where is it?"

Through the previous decades, an enormous amount of information has been gathered on the processing of sound, using behavioural and electrophysiological techniques. These are, in essence, approaching the problem from diametrically opposite ends of the complexity spectrum. The behavioural experiment can often quickly tell us what the animal is able to perceive, but less on how the animal achieves this. Electrophysiological experiments on single or multiple neurones tell us only how each cell or small group of cells responds to particular stimuli, but these responses are often far from anything that can be regarded as an integrated response that builds the basis for behaviour. Nonetheless, a remarkable degree of complexity has been revealed in the responses of single neurones and this complexity not unexpectedly rises, the higher we climb in the auditory pathway. Very subtle and complex interactions between the – possibly hundreds – of inputs from other cells have been revealed, both excitatory and inhibitory. These different inputs have their own frequency- and time-dependent properties. Through studying excitatory and inhibitory inputs and, for example, reversibly inactivating other neural centres that provide input to the cell, the neuronal circuitry underlying particular response patterns can be dissected. In all cases, of course, it must be remembered that a single neuron may be a poor reflector of the global responses of a cell population and/or of a nucleus. Nonetheless, using such approaches it has been possible to learn about the rules and processes underlying auditory brain function.

A sound may be a communication signal from conspecifics, containing in its frequency structure, its loudness and its time course information such as the identity of the indi-

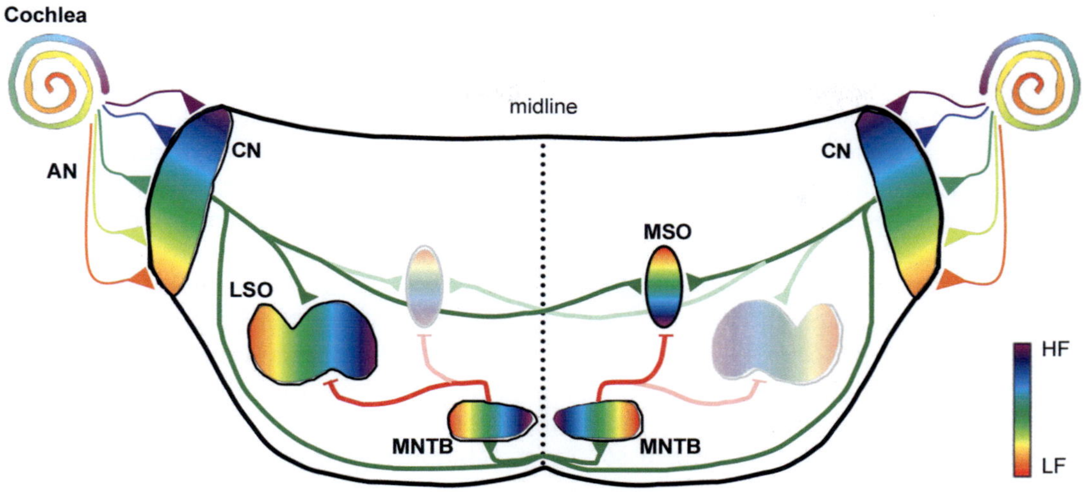

Fig. 5 Schematic drawing of primary auditory sound localization circuits in a transverse section of the mammalian brainstem. Simplified presentation of tonotopic organizations of the nuclei. For clarity, not all nuclei are shown. Except for the auditory nerve, excitatory connections are shown in green and inhibitory connections are shown in red. AN, auditory nerve; CN, cochlear nucleus; HF, high frequency; LF, low frequency. Figure 1 from Kandler et al. (2009). Used with permission of Nature Publishing Group

vidual and his or her emotional state. It may warn of the presence of a predator, of the other's intentions, status and strength, of the urgency of the signal, etc. Individuals can use the full range of sound frequencies, their combination in different time structures (e. g. frequency-swept signals), durations and time patterns to convey a huge range of information. Few vertebrate neural systems that control well-defined acoustic behaviour are known. In many respects, birdsong is an exception, because it is so widespread, because birds are easy to hold under controlled conditions and they have been studied for a long time. Like humans, many birds learn their song in an early phase of life and are able to distinguish it from very similar sounds. Birdsong offers one of the best examples of an animal behaviour that offers insights into important aspects of human behaviour, in this case language acquisition and use (Köppl et al. 2000).

This information on the character of the sound ("What is it?") can be passed to the brain using one ear alone. Information on the location of a sound source, however ("Where is it?"), is in mammals and to a large extent in birds mostly dealt with by comparing the signals at the two ears and this requires a great deal of neural processing. We can illustrate this by considering data from species such as bats and owls that, because of their way of life, have hypertrophied auditory pathways. Interestingly and, I think, remarkably, the two vertebrate groups birds and mammals have independently evolved auditory pathways that show very similar, but not identical, patterns in the way the different acoustic aspects of stimuli are processed. There are of course a limited number of possible evolutionary changes of neural processing mechanisms that can be brought into play in response to the selective pressures for good sound localisation (Köppl 2009). Studies of the barn owl have been particularly useful, since not only is this species capable of capturing a moving target on the ground in complete darkness, but the avian auditory pathway contains several nuclei that are morphologically rather flat and/

or have their organisational axis arranged such that the electrode angles used in electrophysiological studies easily fall close to these axes. Thus electrode passages through the nuclei clearly show neurones whose properties gradually and systematically change across the nucleus. In mammals, many nuclei are rather convoluted, which of course makes no problem for the animal but does make the interpretation of the organisation of the nuclei and thus the processing patterns much more difficult.

In the barn owl, it has been shown that with regard to the localisation of a sound stimulus (e. g. a mouse rustling through dead leaves), the auditory pathway through the entire brainstem consists of two sets of nuclei. One set mainly processes differences in time-of-arrival between the ears, the other mainly differences in sound level between the ears. These two aspects of interaural differences can deliver information to the owl on the azimuth (time pathway, horizontal plane) and elevation (level pathway, vertical plane) of the stimulus. The first is only possible because the owl's time pathway can handle time differences of a few tens of microseconds. Analysis in the second pathway is enhanced by the fact that the owl's skull and head is asymmetrical and sound enters one ear best from a location lower than the preferred elevation of the other ear. The result of analyses in these two partial pathways, in parallel processing, each using a wide range of sound frequencies, is ultimately combined at the level of the midbrain (inferior collicular nucleus), where a global "map" of the outside world is created (Konishi 2000). Each neuron thus responds only to sounds emanating from a very small area of outside space. There seems little doubt that the barn owl perceives this map in a very similar way to a visual map. Indeed, in early development, the auditory pathway is "calibrated" using visual input and gradually becomes independent of this input as the brain learns to correctly interpret the auditory signals. In addition, in higher nuclei these stimulus modalities are analysed through the same sets of neurones. There is little doubt that, similarly, bats "see" the outside world through the interpretation of the echo patterns resulting from their – mostly ultrasonic – vocalisations. In both of these specialised animal groups, the "map" is created before the information reaches the very highest brain levels, presumably because it is often useful to be able to react extremely quickly to the stimuli and thus catch fast-moving prey. Auditory processing in other species and less pressing aspects of analysis (e. g. conspecifics' vocalisations) presumably involves a greater proportion of processing in areas of the thalamus and – in mammals – forebrain cortex.

Thus the auditory brain can be regarded as a hierarchically organised, very large set of neuron populations that are separated into sub-pathways, that interact in an excitatory and inhibitory fashion using several neurotransmitter substances, that vary in their preferred frequency and time-response patterns and that culminate in the "highest" nuclei in global neural patterns that can specifically direct behaviour.

The human ear is a typical mammalian ear, except that the middle ear does not, as in most mammalian families, have a bulla (an enlarged bony cavity surrounding the middle-ear space into which the cochlea extends). Humans are large mammals and this correlates both with a relatively long hearing organ (about 35 mm) and a low upper-frequency hearing limit. Humans are unique in their use of music, although there are many, mainly anecdotal, accounts of a degree of musicality in animals (mainly animals that move in synchrony with a musical rhythm). It is disputed whether there is something uniquely complex about the human auditory brain, but of course being rather difficult to examine acutely regarding its neurophysiology, this discussion may continue for a very long time. It is perhaps most likely that, rather than there being anything terribly special about the human auditory pathway, it is rather those parts of the brain that receive the output of the neural analyses and computations of the auditory brain, interpret and integrate it, that are likely to be uniquely human. In addition, the brain has a good degree of plasticity, especially in younger years, and its response properties may also be moulded by experience – including musical experience. This has been dramatically shown in experiments that systematically altered the visual "calibration" of the developing auditory system in the barn owl, producing owls that later made equally systematic mistakes in targeting prey in darkness (Knudsen 2002). That early auditory experiences can influence brain function has recently been reported from human studies and there are now exciting possibilities for new therapies in areas of hearing and language deficits (review in Kandler et al. 2009).

Outlook

We have traced the evolution of mammalian and thus human hearing and noted parallel developments in the sister groups of amniotes, the reptiles and birds. The human inner ear is not unusually developed but typical of large land mammals. Extraordinarily complex brain areas contain the clues to our uniquely complex and rich usage of language and music. In order to continue to enjoy these for many years to come, we need to protect our ears. The enormous auditory sensitivity presented to mammals by the evolution of amplifying mechanisms can easily be destroyed by loud noises and loud music. The sensory hair cells themselves are easily irreversibly damaged and – at least in mammals such as humans – do not spontaneously regenerate and are thus lost for ever. The addiction of many young people to very loud music that needs to be "felt" rather than just heard is creating a generation that, in later years, will help fill the pockets of many a medical practitioner and lead to an enormous loss in the quality of life. Anything much louder than human speech is, in the long run, damaging and an insult to the evolutionary forces that have brought us these marvels of natural tissue engineering.

References

Carroll RL (1988) Vertebrate palaeontology and evolution. Freeman, New York, Oxford

Christensen-Dalsgaard J, Manley GA (2005) Directionality of the lizard ear. J Exp Biol 208: 1209–1217

Clack JA (2002) Patterns and processes in the early evolution of the tetrapod ear. J Neurobiol 53: 251–264

Gillespie PG, Walker RG (2001) Molecular basis of mechanosensory transduction. Nature 413: 194–202

Hudspeth AJ (2008) Making an effort to listen: mechanical amplification in the ear. Neuron 59: 530–545

Kandler K, Clause A, Noh J (2009) Tonotopic reorganization of developing auditory brainstem circuits. Nature Neurosci 2009, 12: 711–717

Knudsen EI (2002) Instructed learning in the auditory localization pathway of the barn owl. Nature 417: 322–328

Köppl C (2009) Evolution of sound localisation in land vertebrates. Curr Biol 19: R636-R639

Köppl C, Manley GA, Konishi M (2000) Auditory processing in birds. Curr Op Neurobiol 10: 474–481

Konishi M (2000) Study of sound localization by owls and its relevance to humans. Comp Biochem Physiol A 126: 459–469

Ladich F, Popper AN (2004) Parallel evolution in fish hearing organs. In: Manley GA, Popper AN, Fay RR (eds) Evolution of the vertebrate auditory system. pp 95–127

Manley GA (1990) Peripheral hearing mechanisms in reptiles and birds. Springer, Heidelberg, New York

Manley GA (2000) Cochlear mechanisms from a phylogenetic viewpoint. Proc Nat Acad Sci (USA) 97: 11736–11743

Manley GA (2002) Evolution of structure and function of the hearing organ of lizards. J Neurobiol 53: 202–211

Manley GA, Köppl C (1998) Phylogenetic development of the cochlea and its innervation. Current Opin Neurobiol 8: 468–474

Manley GA, Köppl C, Sneary M (1999) Reversed tonotopic map of the basilar papilla in *Gekko gecko*. Hear Res 131: 107–116

Manley GA, Ladher R (2007) Phylogeny and evolution of ciliated mechano-receptor cells. In: Hoy RR, Shepherd GM, Basbaum AI, Kaneko A, Westheimer G (eds) The senses: a comprehensive reference. Elsevier, Amsterdam, pp 1–34

Pickles, J. O. (2008) An introduction to the physiology of hearing, Third Edition. Emerald Group Publishing, Bingley, UK

Principles of function in the visual system

3

Kristine Krug

Contents

Abstract

A large fraction of the primate brain, especially of the cortex, is dedicated to the processing of visual information. A complex network of brain structures transforms electrical signals in the eye about local differences in brightness into signals which directly relate to visual perception in higher areas of visual cortex. Using electrical recordings from individual brain cells in animals, we can trace the hierarchical visual pathway from representations of the visual world by spots of lights in the eye to representations of lines and edges in primary visual cortex and more complex representations in higher cortical areas. The primate visual system also processes different visual attributes, like motion direction or colour, in parallel. This is achieved through specific rules that govern connectivity and organisation of neurons in the visual system. The nature of these image transformations directly shapes the perception of the visual world. The close relationship between visual cortical processing and perception has been an important experimental key for neuroscientists to identify brain cells that directly contribute to visual perception. Thus, our current understanding of visual processes is no longer limited merely to representations of the physical stimuli in the outside world.

Kristine Krug
Oxford University
Department of Physiology, Anatomy and Genetics
Sherrington Building
Parks Road
Oxford Ox1 3PT, UK
e-mail: kristine.krug@dpag.ox.ac.uk

1. The visual pathway: an introduction

Visual perception is a complex process, which includes tasks as diverse as object recognition and navigation in the visual world. In primates, large parts of the brain, in particular of cortex, are devoted to visual processes (Fig. 1). Over half a century of neuroscience research has identified more than 30 distinct visual areas, in which brain cells (neurons) form visual pathways that represent and process visual information. In each cortical area, neurons respond to different visual parameters, like motion, shape or colour or to complex combinations of them.

It is impossible to cover the structure and function of the whole visual system in one chapter. But there are several principles that seem to govern the processing of visual information in our brains. First there is the

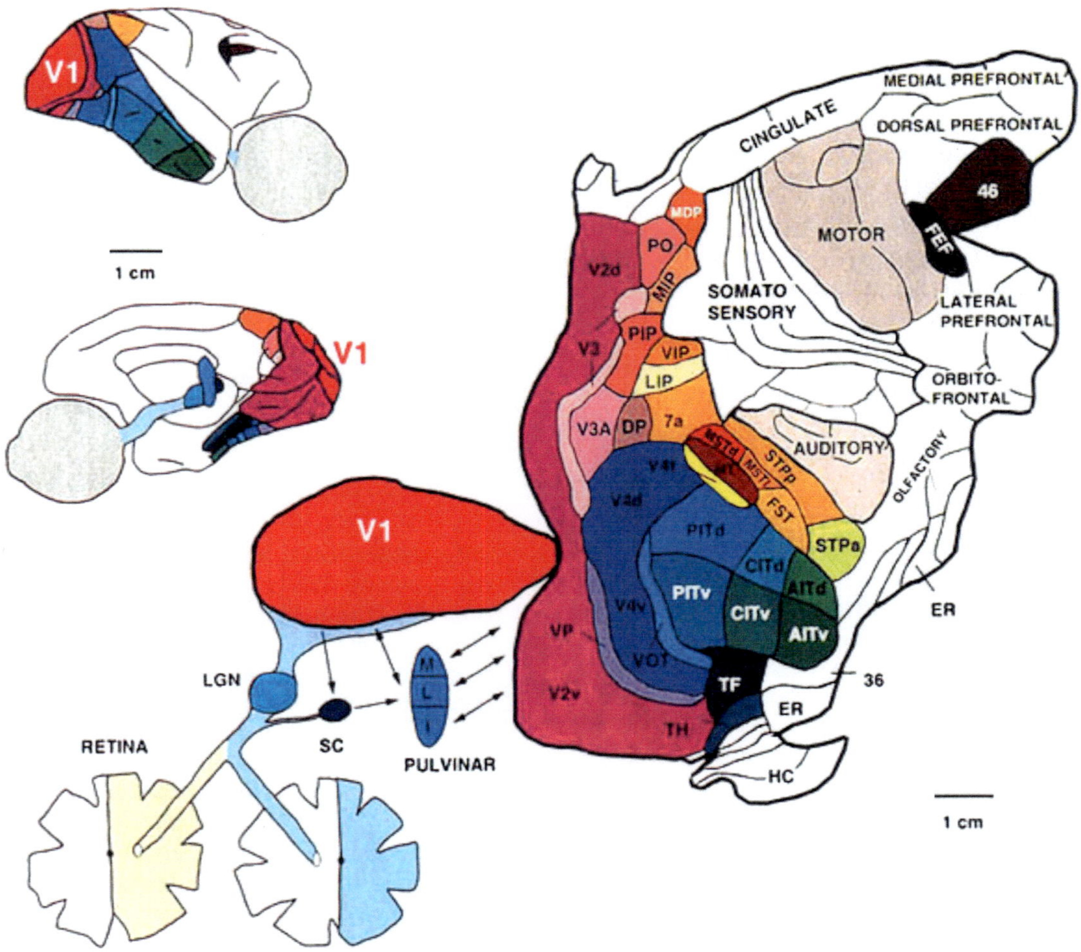

Fig. 1 An overview of the visual system in the rhesus monkey from retina to lateral geniculate nucleus (LGN) and onward to cortex. Depicted is one half of the brain. Top left shows a side view of the monkey brain and underneath it is a midline view. The large image is a representation in which brain areas have been unfolded into a flat sheet. The different visual cortical areas are coloured in shades of purple for early stages of cortical processing, blue for higher cortical areas thought to be involved in object recognition or orange for cortical areas thought to be involved in processing of visual motion for action. This chapter will investigate more closely neuronal function in the retina, in primary visual cortex (V1) and the visual motion area V5/MT (MT). Reproduced with permission from Van Essen et al. (1992) Science

hierarchical nature of the visual pathway. Anatomical connections between the retina and the lateral geniculate nucleus (LGN) and in particular between the cortical visual areas provide guidance about the flow of information in the visual system (Felleman and van Essen 1991). When we study the representations of visual information from the eyes to primary visual cortex and on to higher cortical visual areas, we encounter increasing complexity. The functional hierarchy of information processing becomes evident, particularly by considering the features of the visual world that are processed at each stage. I will describe how cells in the retina represent the local contrast between light and dark at specific points of our field of vision and how from such inputs representations of lines can be generated in primary visual cortex. Further along the visual pathway, in extrastriate visual areas, the responses of neurons can be linked directly to the perception of visual motion and depth (Parker and Newsome 1998).

Visual processing also involves parallel processes, with different visual attributes being handled in different cortical areas. There is evidence for two or more visual pathways at the level of cortex. Different sets of strongly interconnected visual areas are thought to process colour and shape information for the recognition of visual objects and visual motion information for action (Ungerleider and Mishkin 1982; Milner and Goodale 1993). I will not cover these extensive pathways in great detail. But I will provide evidence how the activity of brain cells in one of the higher cortical areas (V5/MT) can be linked to the perception of motion and depth. Visual motion and depth perception are among the most intensively studied and best understood visual processes.

Importantly, our current understanding of visual processes in the brain is not limited merely to the representation of the physical stimuli in the outside world. Electrical recordings from single cells in the visual system of the cat, and in particular from the awake behaving monkey, have revealed how the nature of information processing in the visual system directly shapes visual perception. Electrical stimulation of brain cells in behaving monkeys has demonstrated the causal link between neurons in higher areas of visual cortex and perception of the visual world (Cohen and Newsome 2004). But the influence of neural processing on perception already becomes evident in the way the retina transforms the incoming information about light levels in our environment to a signal about local contrast, so that we can effectively navigate in our visual environment at midday and at moonlight, at very different levels of illumination

2. The retina: a representation of local contrast

The retina is the site of two major transformations of visual information. First is the transduction of the pattern of light that reaches our eyes into electrical signals. This occurs in specialized cells called photoreceptors. In a densely packed sheet at the back of the eye, each photoreceptor translates the incoming light for one point on the retina into a graded electrical signal. But our retinae do more than this. Let us consider the simple task of identifying a zebra (or pedestrian) crossing on a street, so that we can safely cross a road. For most people, the visual system allows them to do this regardless of whether it is noon on a cloudless summer day or winter evening twilight. However, the amount of light that enters our eyes under these environmental conditions differs by orders of magnitude. Yet, we can see the black and white pattern of the zebra crossing in either circumstance. We will take a closer look at the output neurons of the retina called retinal ganglion cells that project to visual cortex and see that these cells signal visual information not in terms of actual brightness but as a local difference or contrast signal.

2.1 The receptive field of retinal ganglion cells: what retinal ganglion cells 'see'

It was in 1953 that the neuroscientist Stephen Kuffler first demonstrated which patterns of light activate retinal ganglion cells in mammals. He put the tip of a fine metal fibre, a microelectrode, as close as he could to retinal ganglion cells in an anaesthetized cat's eye. The output of the electrode was connected to a signal amplifier not unlike that of an ordinary 'Hi-Fi'. When the electrode was closer to one retinal ganglion cell than any others, it picked up individual, fast electrical events called action potentials or 'spikes' (Fig. 2a). Spikes from one cell can be distinguished by their size and shape from the activity of the surrounding cells. These spikes are brief changes in electrical potential across the cell's outer membrane. They travel very fast over long distances along the main processes extending from the cell, called the axons. These are the signals that convey information from the eye to higher structures in the brain and between different brain areas. Information is mainly represented through changes in spike frequency: when a neuron is activated, its spike frequency increases.

In order to test what part of the visual world a retinal ganglion cell 'looks at', Kuffler used small bright spots of light on a darker background. With such spots he systematically probed the cat's field of vision. Kuffler noted the positions of the spot that elicited an increase in spike frequency from the reti-

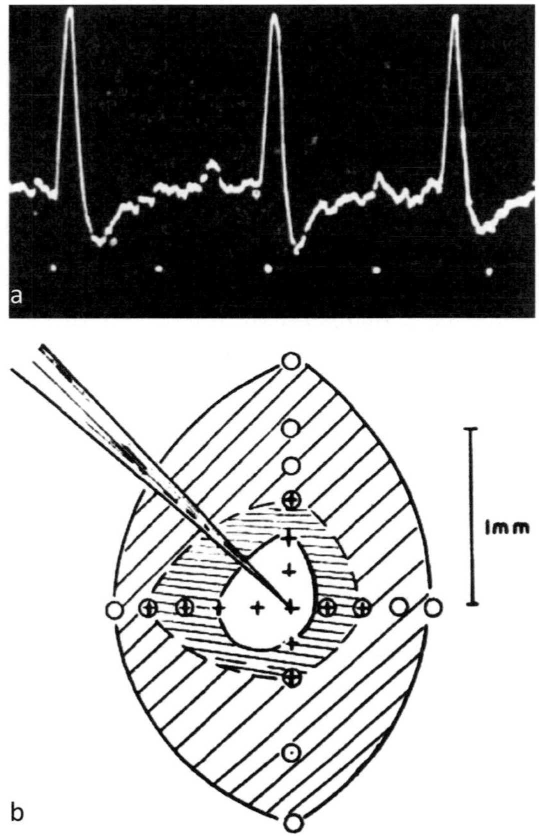

Fig. 2 **a** A recording of electrical potential changes from a retinal ganglion cell. The trace shows three rapid changes in recorded electrical voltage. This neuronal activity is termed action potentials or 'spikes'. The frequency of spikes is one of the major means for transmission of information in the nervous system. Dots underneath the trace are spaced at 1 ms. **b** The receptive field of a retinal ganglion cell. Crosses chart the area of the visual field for which the cell increased the firing of spikes when probed with a small spot of light: the 'ON-centre'. The diagonally hatched part depicts the area of the receptive field that ceased to fire spikes when the bright spot was directed there (the cell became active again when the bright spot was turned 'off'): this is the antagonistic 'OFF-surround'. In the intermediary zone (horizontally hatched) responses were mixed on-off. Reproduced with permission from Kuffler (1953) J Neurophysiol

nal ganglion cells from which he was recording. He found that many retinal ganglion cells could be activated by a small spot of light at a certain position of the visual field and that the same cells were also suppressed by light directed at the area surrounding the central spot (Fig. 2 b). Thus, centre and surround are antagonistic. The area of the visual field and the pattern of visual input that modulate a neuron's activity are termed its 'receptive field'. This type of receptive field is called a concentric 'ON-centre' receptive field. Another type of retinal ganglion cell that was found had the inverse receptive field, with an OFF-centre and an ON-surround response. Thus, retinal ganglion cells preferentially respond to spots of light on a dark background or dark spots on a light background. But can these cells respond to such local patterns of light at dusk as well as in the midday sun?

2.2 The retina signals contrast

How well retinal ganglion cells respond in different lighting conditions was investigated by Bert Sakmann and Otto Creutzfeldt (1969), again for retinal ganglion cells in the cat. They also inserted fine metal microelectrodes into the eye, so that the tip of the electrode was very close to the outside of a retinal ganglion cell. They tested the spike rate of retinal ganglion cells to spots of light of different brightness (or luminance) shone on the centre of their receptive field. They found that the brighter the centre stimulus the greater the number of spikes that were fired by ON-centre retinal ganglion cells – up to the point where the firing rate saturated, i. e. the cell could not fire any more.

Then they repeated the same experiment under different background lighting conditions. The spot they used was always lighter than the surrounding background but the background itself could be in some experiments lighter or darker. The changes in the background luminance were so great that, had the centre stimulus not been adjusted, it would be lighter than the background on some occasions but darker than the background on others.

When Sakmann and Creutzfeldt looked at the firing pattern of a retinal ganglion cell under the different background conditions, they found that the cell did not always fire the same number of spikes to the same absolute luminance of the centre stimulus. But rather the cell signalled that the centre stimulus was lighter than the surround background and by how much it was lighter (Fig. 3). Thus, retinal ganglion cells signal the luminance contrast between the centre and the surround of the receptive field. The contrast of the spot of light is its light intensity relative to the intensity of the background on which the spot lies. A spot of light has a constant contrast if the intensity of the spot is always a fixed percentage brighter or darker than the background, as the intensity of the background is increased or decreased. Different retinal ganglion cells signal the local contrast for different points in the visual field, such that the whole field of vision is covered. The underlying mechanism is thought to be like a divisive mathematical operation, such that the signals related to the centre of the retinal ganglion cell receptive field are normalized by the luminance level of the background. Much of this is probably already achieved by the photoreceptors and the horizontal cells that connect individual photoreceptors in the eye to each other.

Thus, in order for us to be able to navigate our visual environment under changing lighting conditions, the retina transforms the visual input to signal information about local differences in luminance to the brain. That is why a zebra crossing is highly visible to most of us, regardless of the lighting conditions.

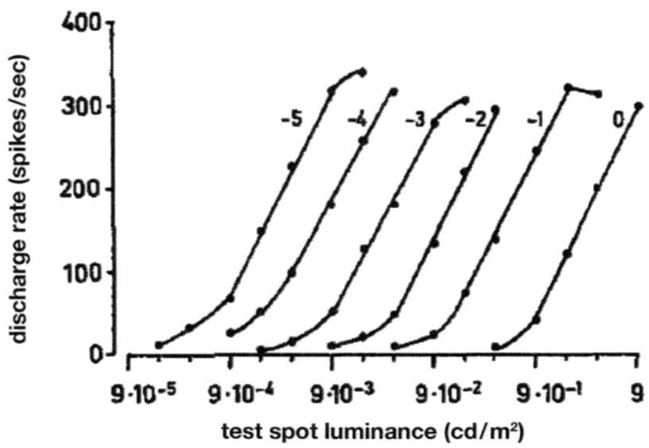

Fig. 3 The neuronal response of a retinal ganglion cell (ordinate) is plotted against the luminance of a spot of light (abscissa). The spot was always shone on the centre of the cell's receptive field. Each of the six curves was obtained with different background lighting. The background light levels are given on the top of each curve (in log cd/m^2). For each background light level, this retinal ganglion cell responded when the brightness of the test spot exceeded that of the background by a certain fraction. The number of spikes per second fired by the recorded cell increased with increasing brightness of the test spot until the response saturated. Reproduced with permission from Sakmann and Creutzfeldt (1969) Pflügers Arch

On top of this point-by-point filtering of local luminance contrast, the central part of each retina in primates also signals local differences in the wavelength in incoming light. Some retinal ganglion cells are particularly activated by small spots of red, of green or of blue light. The mechanism by which this is achieved is similar to that for luminance contrast. Some retinal ganglion cells compare locally the relative amounts of red and green light or blue and yellow light. Receptive fields are also organised in a centre-surround fashion. Therefore, the representation of the visual world that is transmitted from our eyes to visual cortex can be imagined similarly to the image on a TV screen: under close inspection the neural image is composed of individual red, blue and green points of different local contrast.

3. Primary visual cortex: from spots to lines to ordered maps

The majority of retinal ganglion cells project axons via a relay in the lateral geniculate nucleus (LGN) to the primary visual cortex at the back of the brain. Brain cells in visual cortex generally respond to more complex visual features than cells in the retina. For instance the preferred stimulus of a cortical neuron might be a line or an edge rather than a spot of light. This specialised selectivity was first discovered by David Hubel and Torsten Wiesel in their seminal experiments in the visual cortex of the cat and of the monkey (Hubel and Wiesel 1959; 1968). Before these experiments, neurophysiologists had failed to extract robust responses from visual cortical cells by using spots of light like those used with retinal ganglion cells.

3.1 The receptive field of simple cells in primary visual cortex

Hubel and Wiesel (1959) inserted an electrode into the visual cortex of the anaesthetized cat and recorded electrical activity from single cells, similarly to the experiments described in the retina. During the recording they presented to the opened eyes of the cat bar stimuli of varying length, thickness and orientation, either light on dark or dark on light. First they established the receptive field position, where a visual stimulus would elicit an electrical response, for each neuron. When they systematically changed the orientation of a bar stimulus within this receptive field, they found they could count more electrical spikes for certain orientations of the bar than for others (Fig. 4a). Often, the bar orientation orthogonal to the one that elicited the highest response triggered very few spikes. When Hubel and Wiesel plotted the number of electrical events against stimulus orientation, they found a smooth tuning function. The orientation that elicited the most spikes was termed the preferred orientation of the neuron.

In some experiments, a light bar of the correct orientation would only elicit a response when it was in a specific position within the receptive field. Such a cell is termed a simple cell. When Hubel and Wiesel mapped the receptive field of such a cell with spots of light, they could discriminate distinct zones, where the cell preferred either the spot being on or the spot being turned off. The shape of these 'ON' and 'OFF' zones matched the preferred orientation stimulus (Fig. 4b). Like retinal ganglion cells, the preferred stimulus can be predicted from the shape of the underlying receptive field.

a

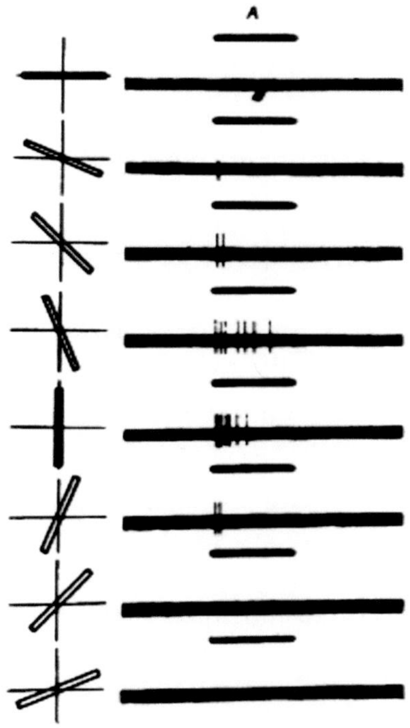

b

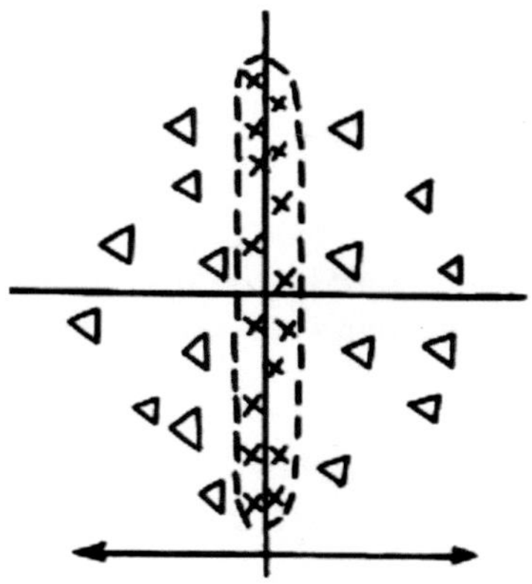

c

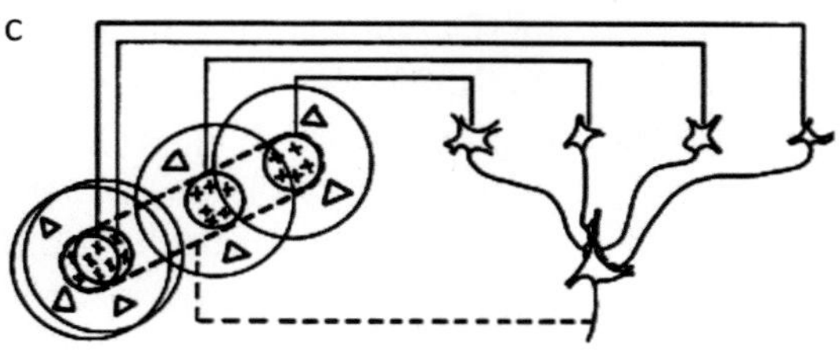

Fig. 4 **a** A bar of light was presented to the receptive field of a simple cell in cat visual cortex. The orientation of the bar was systematically varied between different presentations. The crosses on the left give the centre of the receptive fields, the bar indicates the presented orientation. The corresponding traces to the right show the recordings from the simple cell: The short horizontal bars indicate the time for which the stimulus was on, the thin vertical lines show the spikes. This cell fired more action potentials in response to a bright, vertical bar in the centre of its receptive field (its preferred orientation). The orthogonal orientation (a horizontal bar) is the null orientation for this simple cell, which elicited no spikes. **b** The receptive field plot of the simple cell, from which the orientation responses were obtained. Crosses denote areas where a spot of light excited the cell; triangles denote regions that responded to the spot being turned off. **c** Model proposed by Hubel and Wiesel about how the receptive field of an orientation tuned simple cell is generated. They proposed that the specific pattern of inputs from cells with concentric centre-surround receptive fields provides the simple cell with its selectivity for orientation. If mapped onto the visual field, the ON-centres of the lateral geniculate nucleus (LGN) cells that provide the input would align with the ON-zone of the simple cell. **a, b** Reproduced with permission from Hubel and Wiesel (1959, 1962) J Physiol

There are other cells in the primary visual cortex whose response is more complex. They are also orientation selective but the response is not dependent on bar position within the receptive field and they were – appropriately – termed complex cells (Hubel and Wiesel 1968). Thus, there are even in the primary visual cortex many cells that have increasingly complex response properties for visual stimuli. There are also cells which are sensitive to colour contrast between for instance red and green or to the visual depth of a stimulus.

3.2 The receptive field of a simple cell is constructed by the specific input from neurons with concentric receptive fields

The simple cell provides a good illustration of how visual response properties like orientation selectivity might be constructed in the brain. Hubel and Wiesel (1962) proposed that the receptive field of the orientation tuned simple cell is constructed by a specific pattern of inputs from cells with concentric, antagonistic receptive fields (Fig. 4 c), like the retinal ganglion cells, we have encountered earlier. These inputs would be from cells whose ON-centres respond to the same part of the visual field as the ON-zones of the simple cell to which they are connected.

In the case of the cat, these would be inputs from the relay between retina and visual cortex, the LGN, whose cells have response properties similar to the retinal ganglion cells, from which they receive their input. For the primate, the input to the simple cells would be provided by neurons in the input layer 4C of the primary visual cortex.

That this model is fundamentally correct has been demonstrated by Clay Reid and Jose-Manuel Alonso (1995). They succeeded in simultaneously recording from pairs of neurons in the cat: one cell in the LGN and the other an orientation tuned simple cell in the visual cortex. When these cells had overlapping receptive fields, they showed that the activity of these cells was strongly correlated. Direct connections were very likely when the ON-centre of an LGN cell and the ON-zone of a simple cell overlapped or when an OFF-centre overlapped with an Off-zone. Thus, Reid and Alonso provided direct evidence that there are specific rules that govern the input to the cortical simple cells, as predicted by Hubel and Wiesel.

3.3 Orientation selective cells in primary visual cortex are organised in columns

It is not just the input to simple cells that is governed by specific rules. Orientation tuning in primary visual cortex also illustrates another important point of visual cortical organisation: response properties are often systematically ordered within a visual area.

If one records from neurons in most early visual areas, one finds that neighbouring neurons respond generally to visual stimuli at similar positions in the visual field. But when Hubel and Wiesel (1962) advanced their electrode into primary visual cortex perpendicular to the outer surface, they also found that successive neurons preferred the same stimulus orientation. When the recording electrode advanced at a more oblique angle, the preferred stimulus orientation changed gradually, mostly at regular intervals. They proposed that primary visual cortex contains an ordered map of columns with the same selectivity of bar orientation, such that there are columns for each part of the visual field and each orientation.

About three decades later, such an orientation map in the visual cortex was directly visualized by a technique called optical imaging. Optical imaging records the particular wavelength of light reflected from the surface area of cortex. When neurons are electrically active they consume oxygen and the blood supplying these cells changes from a lighter to a darker red (the intrinsic signal). Tobias Bonhoeffer and Amiram Grinvald (1991) presented gratings of different orientation to an anaesthetized cat and compared the level of activity across a large area of visual cortex. They showed that responses to different orientations are represented in a regular fashion across the visual cortex. Fig. 5 shows an example of such an orientation map from a later paper (Hübener et al. 1997).

It is thought that the regular organisation of response properties in maps and columns provides some of the computing power of the visual cortex. In essence, through regularizing connections between neurons, it might enable the visual cortex to carry out systematically and in parallel the same operations on inputs from different parts of the visual field. As we will see later, the ordered arrangement of cortical visual cells with the same response properties also allows neuroscientists to intervene experimentally in information processing – specifically and effectively.

4. Extrastriate visual areas: from visual representations to perceptual representations

Brain cells in higher, extrastriate visual cortex do not always simply signal the physical properties of a visual stimulus; they can also represent the actual percept. Of course, in many cases, the physical stimulus and the resultant percept are closely related. However, the relationship between physical stimulus and percept can be probed with visual

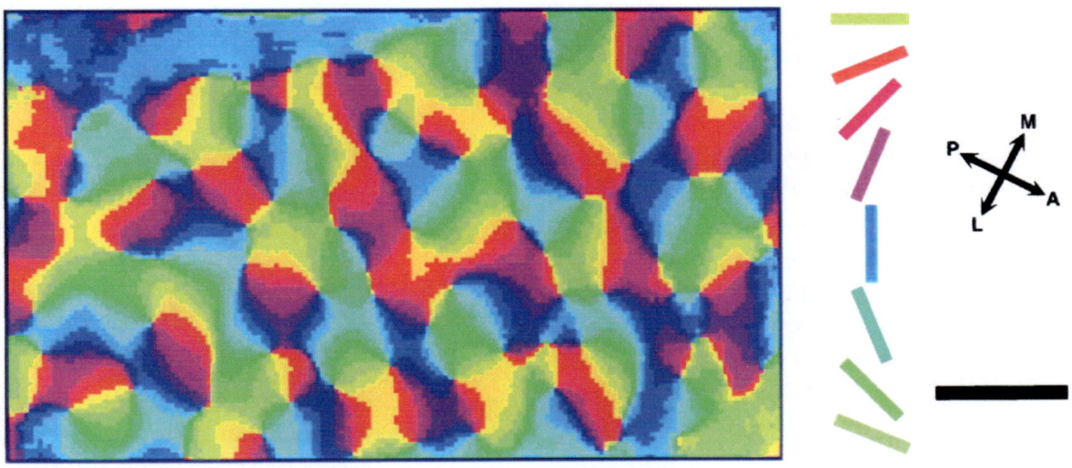

Fig. 5 This is a view onto the visual cortex of the cat. The orientation map in this image has been obtained with optical imaging and provides a view onto the orientation columns in primary visual cortex as viewed from above. Areas of cortex that respond preferentially to different stimulus orientations have been coded with different colours (see key on the right for the respective stimulus orientations). This map indicates that neurons with the same preference for specific orientations cluster together and that neighbouring columns tend to have similar preferred orientations. In particular preferred stimulus orientation changes gradually in a circle around certain points on the map, which are called 'pinwheels'. These pinwheels are singularities from which all colours seem to radiate out. Reproduced with permission from Hübener et al. (1997) J Neurosci

stimuli that can have more than one perceptual interpretation: ambiguous figures, very much like for instance the rabbit-duck illusion (Fig. 6a), allow us to distinguish between neuronal signals that relate to the physical stimulus and to the resultant percept. When this figure is shown to an observer, at some times the observer will report to see the rabbit and at others the duck. Brain cells that signal the percept rather than the unchanging stimulus should have a different activity pattern for different perceptual interpretations. Neurophysiological recordings from single neurons in the behaving monkey allow us to identify such cells whose activity correlates with the reported percept while the actual physical stimulus stays the same.

For instance the activity of neurons in the extrastriate visual area V5/MT (see Fig. 1) has been linked directly to the perception of visual motion and depth. A study by Jon Dodd, Bruce Cumming, Andrew Parker and me showed the relationship between neuronal activity and the perception of bistable structure-from-motion stimuli, which depend on the specific combination of visual depth and direction of motion (Fig. 6b) (Dodd et al. 2001; Krug et al. 2004). This ambiguous figure is made from two transparent, flat sheets of dots moving in opposite directions (Treue et al. 1991). The percept is a robust 3D-cylinder rotating around its central axis – even though there is no depth information to define the 3D structure. The direction of rotation is bistable; when human observers view the cylinder continuously, perception alternates at random with an average duration of about 30s (Krug et al. 2008). We can disambiguate cylinder rotation by putting the left and rightwards moving dots at different depth. If the leftwards moving dots are put in front of the rightwards moving dots, the cylinder rotates clockwise (CW; as viewed from above). With the disambiguated cylinders monkeys can be

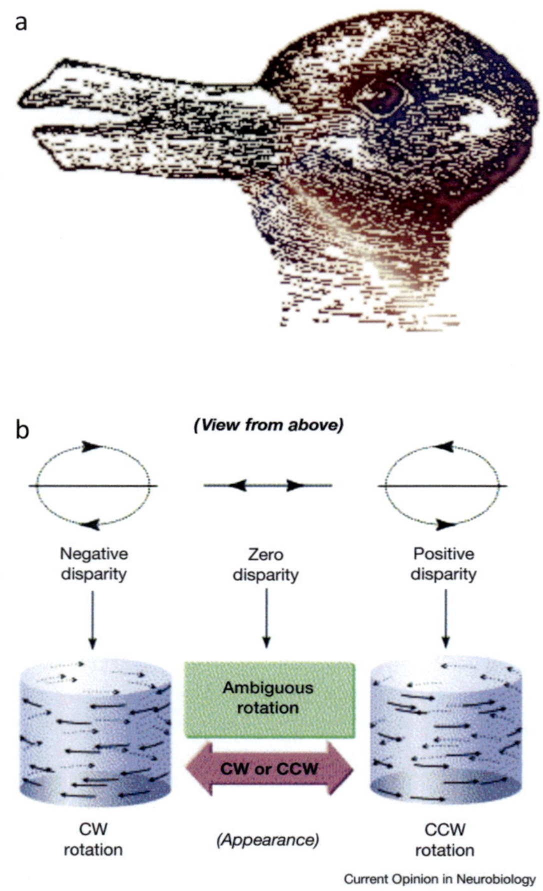

Fig. 6 a In the rabbit-duck figure contradictory cues allow the figure to be seen as either a rabbit or a duck at any given time. This ambiguous figure was described by Joseph Jastrow around 1900 and discussed by Ludwig Wittgenstein in his Philosophical Transactions in the 1950s. **b** The structure-from-motion cylinder consists of two transparent sheets of random dots moving in opposite directions with a sinusoidal velocity profile. The two sets of dots are not separated in depth. The cylinder is alternately perceived as rotating clockwise (CW) or counter-clockwise (CCW) (Treue et al. 1991). The addition of binocular disparity disambiguates the direction of rotation of the cylinder. Reproduced with permission from Parker and Krug (2003) Curr Opin Neurobiol

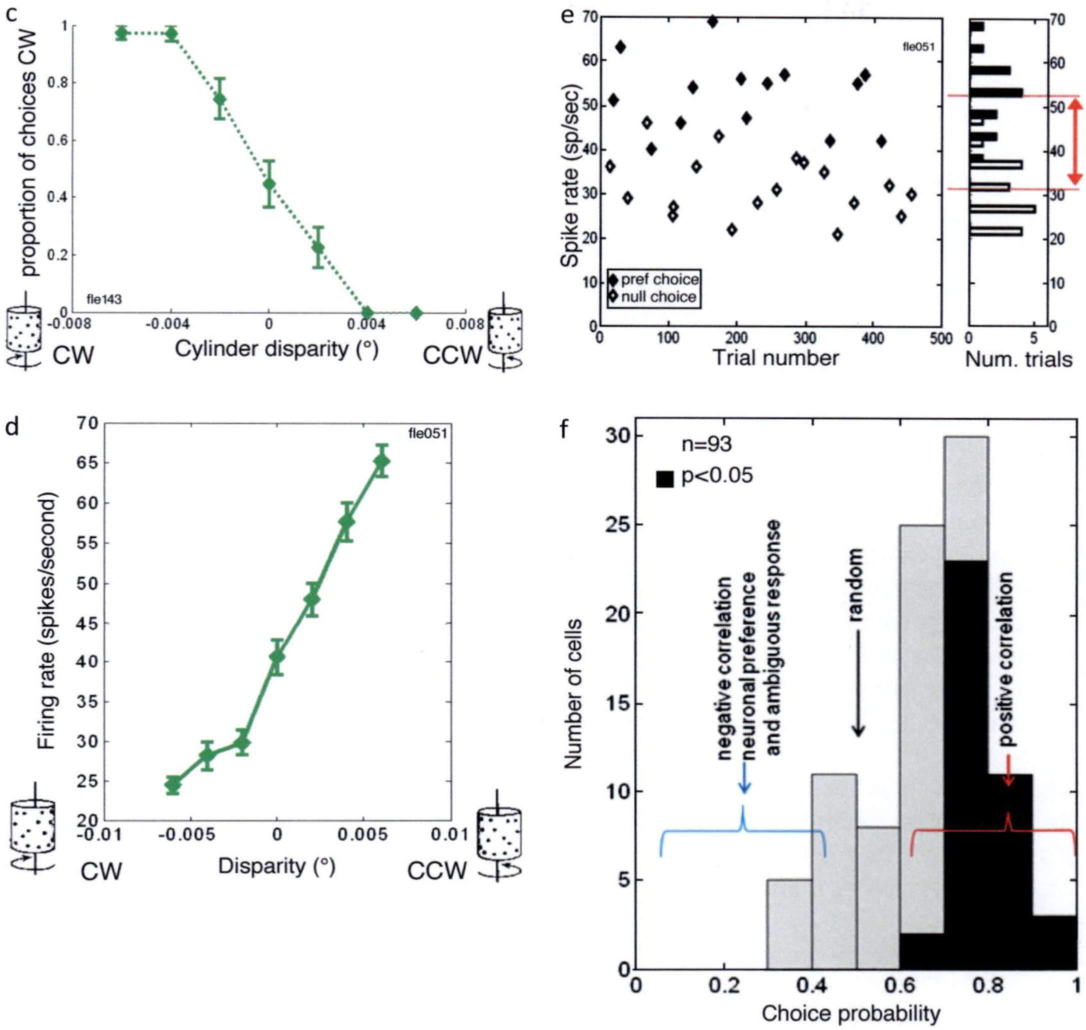

c A monkey's response to the cylinder stimulus. The proportion of CW choices is plotted against the depth signal (disparity) in the cylinder stimulus. For unambiguously CW or CCW rotation, the monkey responds 100% correct; for the ambiguous cylinder (0 disparity), the monkey's choices are about 50% CW and 50% CCW. **d** The response of a V5/MT neuron to the rotating cylinder. The discharge rate of a V5/MT cell is plotted against the depth signal (disparity) in the cylinder stimulus. This neuron prefers CCW rotating cylinders. **e** The graph shows the neuronal responses of one V5/MT neuron when presented with the ambiguous cylinder. Each data point represents the mean discharge rate of one such trial and was coded according to the direction of rotation the monkey chose at the end of the trial (filled symbol: preferred direction of the neuron; open symbol: null direction). The marginal histogram on the right shows that this neuron usually discharged more action potentials on trials when the monkey subsequently chose the preferred direction of rotation of the neuron. **f** This graph summarizes the choice probabilities from 93 V5/MT neurons for the cylinder discrimination task (data from Dodd et al. 2001). The choice probability of 0.67 was significantly different from 0.5 and indicates a significant correlation between neuronal firing and reported percept. Black bars denote choice probabilities that were individually significantly different from 0.5

trained to judge the direction of rotation (Fig. 6c). Monkeys reported the direction of cylinder rotation by making an eye movement to one of two choice targets (one for CW, one for 'counterclockwise' [CCW] rotation) (Dodd et al. 2001; Krug et al. 2004).

In extra-striate visual area V5/MT, which has also been termed the 'visual motion area', most neurons show a preference for a visual stimulus moving in a particular direction and many are sensitive to the depth at which the stimulus is presented (Dubner and Zeki 1971; Maunsell and van Essen 1983). As a consequence, many V5/MT neurons show a specific preference for a particular direction of cylinder rotation when presented with a rotating cylinder. For example when neurons are selective for 'motion to the right' and 'near', they discharge more spikes when a cylinder is presented with a front surface moving to the right, i. e. a cylinder rotating CCW (Fig. 6d). The question is what such a neuron signals when presented with the bistable version of the cylinder. In particular does this neuron fire more spikes when the reported percept of a bistable cylinder is the neuron's preferred rotation (CCW) rather than the non-preferred rotation?

We tested this by recording the activity from single V5/MT neurons while a monkey judged the direction of rotation of a bistable cylinder (Dodd et al. 2001; Krug et al. 2004). Cylinders were matched to the receptive field preferences of the neuron from which recordings were taken. In order to check that the monkey was reporting the direction of rotation accurately, we interleaved trials with unambiguous and ambiguous cylinders. Monkeys were rewarded for correct answers on unambiguous trials and rewarded on ambiguous trials at random on 50% of trials. To test the relationship between neuronal firing and perception, we only analysed the trials on which the ambiguous cylinder was presented. For each cell, we compared the firing rates for trials on which the monkey reported the neuron's preferred direction of rotation with those trials on which the monkey reported the opposite direction of rotation (Fig. 6e). The neuron shown in Fig. 6d,e was generally more active when the monkey subsequently judged the stimulus to rotate in the neuron's preferred direction, even though the displayed cylinders were the same on each trial and did not specify the direction of rotation.

The strength of the correlation between a brain cell's activity and perception can be measured. Simona Celebrini, Kenneth Britten, Bill Newsome, Tony Movshon and Michael Shadlen (Celebrini and Newsome 1994; Britten et al. 1996) first showed that such random fluctuations in neuronal firing rates are seen from trial to trial, even when identical stimuli are displayed, and are systematically related to perceptual decisions. They devised a quantitative measure for this relationship between firing rate and percept: the probability with which an independent observer could predict a monkey's perceptual choice on a given experimental trial based only on the discharge rate of the neuron recorded from and the distribution of discharge rates and choices for that neuron. They termed this the "choice probability". A choice probability of 0.5 signifies a random association between discharge rate and choice: the firing rate predicts with 50% accuracy which of the two directions was chosen. A choice probability of 1 would represent a perfect correlation: one would only have to listen to this one neuron to predict the monkey's subsequent choices about this particular stimulus perfectly. A choice probability of 0 would theoretically mean that the cell fires reliably more for the non-preferred direction. This could be a significant correlation, but would have consequences for the interpretation of the neuronal signals. In practice such responses are rarely observed in neurons.

The choice probability reflects the separation in the distribution of firing rates for the two choices, shown in the marginal histogram in Fig. 6e. The example cell presented in this figure has a very high choice probability, close to 1. Across a population of 93 V5/MT cells, Dodd and colleagues (2001) found an average choice probability of 0.67 (Fig. 6f). This suggests a strong link between the firing of single V5/MT brain cells and the perceived direction of cylinder rotation. On average, by listening to one V5/MT brain cell, an independent observer can predict trial-by-trial with 0.67 accuracy the direction of rotation that the monkey reports to have perceived. Or in other words, if all neurons carried independent information, a monkey would not need many more than three V5/MT cells to account for a cylinder percept.

Neuronal activity in area V5/MT therefore carries information not just about the physical stimulus in the outside world but also about how we perceive this stimulus. For the same task, choice probability was also measured in the primary visual cortex. Although there are cells that are selective for depth and motion, their choice probability is not significantly different from 0.5. Therefore, neurons in the primary visual cortex are not directly linked to visual perception of structure-from-motion cylinders (Grunewald et al. 2002). Thus, choice probabilities allow us to assess the contribution of different groups of neurons to perception (Parker and Newsome 1998). Significant choice probabilities have been measured for other paradigms and in other sensory areas (Gold and Shadlen 2007).

But one caveat must be kept in mind: A significant choice probability, for instance for the perception of cylinder rotation, does not mean that V5/MT neurons directly contribute to perception of all visual stimuli with motion and depth. Other experiments have shown that the very same cortical neurons can have a significant choice probability for the rotating cylinder but might also signal certain visual depth stimuli which the monkey cannot perceive (Krug et al. 2004). The contribution of neurons to perception is specific to the particular task that was assessed experimentally.

5. Extrastriate visual areas: from correlations to causality

In the previous section, we have seen how the activity of single brain cells can be linked to visual perception. But in such experiments, we merely measure correlations. Other techniques are required to establish a causal relationship between the neuronal activity in a visual cortical area and perception. In the awake behaving monkey, researchers have also succeeded in linking small groups of neurons *causally* to perception by using electrical micro-stimulation (Cohen and Newsome 2004). Like for the measurement of choice probabilities, a monkey is reporting the percept of a visual stimulus. But instead of recording the activity at a particular brain site while the monkey makes this judgement, brain cells are activated with small electrical currents with electrical micro-stimulation. The perceptual reports between trials with and without electrical micro-stimulation can be compared.

The first successful electrical micro-stimulation experiment of this kind was carried out also in the extrastriate visual area V5/MT. Daniel Salzman, Kenneth Britten and Bill Newsome (1990) stimulated neurons in area V5/MT while monkeys discriminated the direction of dot motion on a screen. In this experiment, monkeys were required to report – with an eye movement – in which of two opposite directions the dots moved. Dots could be moving 100% coherently in one of the two directions, the direction of motion could be random for each of the dots (0% correlation) or the level of coherence of the dot motion could be somewhere in between (Fig. 7a).

In order for such an electrical stimulation experiment to be successful, a few conditions have to be met. One is that the visual stimulus has to be matched precisely to the receptive field location and size of the electrically stimulated neurons, so that there is a high chance that the monkey should have to 'listen to' the stimulated neurons. Also, neurons with similar receptive field properties have to be clustered together. In the case of this particular experiment, the researchers wanted to stimulate neurons that all preferred the same direction of motion. If for instance half the electrically stimulated neurons were more active for rightward motion and the other half for leftward motion, stimulating these neurons should not alter the perceived direction of motion. Here it is helpful that V5/MT has an ordered representation of motion direction, with columns of neurons that prefer the same direction of motion at the same location of the visual field running perpendicular to the surface of this cortical area (Albright et al. 1984) – not unlike the orientation map in V1 (see section 3). Lastly, to be able to draw firm conclusions, the effect of electrical micro-stimulation on perception should be predictable from the response properties of the neurons at the stimulation site. Ideally, the effect should be positive and not just disrupt perception. Therefore, when the researchers stimulate a column with neurons with a pref-erence for rightward motion, they will expect the monkey to be more likely to report that the dots were moving to the right.

Salzman and colleagues (1990, 1992) inserted a microelectrode into V5/MT. First they recorded the activity of multiple brain cells at the site and characterized their preferred direction of motion. Then, they set up the discrimination in such a way that the monkey had to choose between the preferred direction at this cortical site and its null direction. In half the trials, electrical stimulation was applied through the same microelectrode. Trials were interleaved in a pseudo-random order. The authors noted the proportion of trials, in which the monkey chose the preferred direction (PD), for each motion coherence level they showed in the experiment.

Figure 7b shows the response of the monkey in trials without electrical micro-stimulation (open symbols) and with (filled symbols). When the brain site was electrically stimulated, for each given level of motion coherence the monkey chose more often the preferred direction motion. The shift of the response curve towards preferred direction choices indicates that the monkey acts just as if the visual stimulus would have been a little bit more coherent in this direction on each occasion when electrical stimulation was applied. If the electrical stimulation would have simply disrupted the motion discrimination, the behavioural response curve should have become flatter. Thus, making the cells at the stimulation site in V5/MT more active makes it more likely that the monkey reports the stimulated neurons' preferred direction of motion. This is strong evidence that neuronal activity in V5/MT – or in other areas it projects to – contributes directly to the perception of visual motion.

Thus, electrical micro-stimulation is a powerful tool with which neuroscientists have succeeded to demonstrate that the electrical activity of brain cells in some visual areas can causally affect visual perception. Using the same technique, neurons in V5/MT have also been linked to the perception of coarse visual depth (DeAngelis et al. 1998) and, for instance, neurons in area IT of the monkey have been linked to the perception of faces (Afraz et al. 2006).

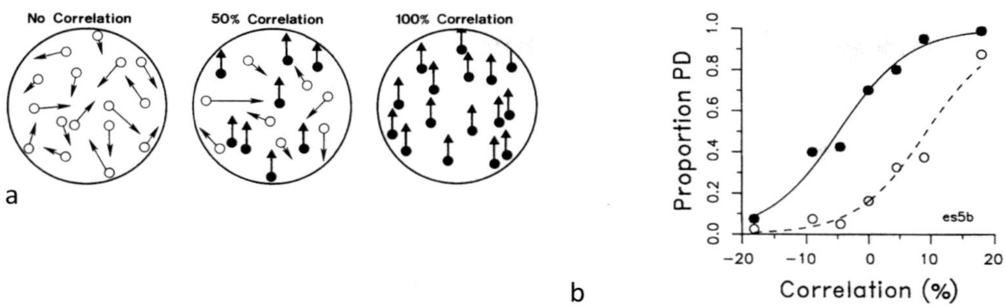

Fig. 7 **a** Diagram of the motion stimulus composed of dots, which was used by Salzman et al. (1990, 1992). The percentage correlation provides the percentage of dots moving coherently in one direction. The 0% correlation stimulus (left) consists of random dots, which are repositioned at a new random location on each computer frame. Dots appear to move in all directions with no overall motion vector. The strongest motion stimulus shows 100% correlation (right) with all dots moving in the same direction. Intermediate level stimuli show only a fraction of dots with a unitary motion signal. In the microstimulation experiment, the coherent element of the motion signal was either directed in the preferred or null direction for the stimulation site. Monkeys had to choose whether dots moved in the preferred or opposite direction. **b** The effect of electrical microstimulation on the monkey's reported percept. This graph plots the percentage of preferred direction (PD) choices by the monkey against the strength of the motion signal. Positive motion strength indicates motion in the preferred direction of the cortical site, negative motion strength indicates motion in the opposite, null direction. Open symbols denote trials without electrical micro-stimulation, filled symbols trials with microstimulation. When electrical microstimulation was applied the monkey was more likely to choose the preferred direction of the stimulated neurons. Reproduced with permission from Salzman et al. (1992) J Neurosci

Conclusion

Vision is a complex problem. However, the processing of visual information is broken down by the nervous system into stages. Each stage of processing is solved in turn. At least some of these stages can be revealed through the investigation of the functional visual representations with neuronal recordings. We have traced different processing stages from the computation of local contrast in the retina to the combination of this information into simple receptive fields for the detection of lines and edges. This is achieved through specific rules that govern connectivity and organisation of neurons in the visual system. At the level of extrastriate visual cortex, neuroscientists have been able to link directly the activity of brain cells to visual perception.

References

Afraz SR, Kiani R, Esteky H (2006) Microstimulation of inferotemporal cortex influences face categorization. Nature 442: 692–695

Albright TD, Desimone R, Gross CG (1984) Columnar organization of directionally selective cells in visual area MT of the macaque. J Neurophysiol 51: 16–31

Bonhoeffer T, Grinvald A (1991) Iso-orientation domains in cat visual cortex are arranged in pinwheel-like patterns. Nature 353: 429–431

Britten KH, Newsome WT, Shadlen MN, Celebrini S, Movshon JA (1996) A relationship between behavioral choice and the visual responses of neurons in macaque MT. Vis Neurosci 13: 87–100

Celebrini S, Newsome WT (1994) Neuronal and psychophysical sensitivity to motion signals in extrastriate area MST of the macaque monkey. J Neurosci 14: 4109–4124

Cohen MR, Newsome WT (2004) What electrical microstimulation has revealed about the neural basis of cognition. Curr Opin Neurobiol 14: 169–177

DeAngelis GC, Cumming BG, Newsome WT (1998) Cortical area MT and the perception of stereoscopic depth. Nature 394: 677–680

Dodd JV, Krug K, Cumming BG, Parker AJ (2001) Perceptually bistable 3D figures evoke high choice probabilities in cortical area MT. J Neurosci 21: 4809–4821

Dubner R, Zeki SM (1971) Response properties and receptive fields of cells in an anatomically defined region of the superior temporal sulcus in the monkey. Brain Res 35: 528–532

Felleman DJ, Van Essen DC (1991) Distributed hierarchical processing in the primate cerebral cortex. Cereb Cortex 1: 1–47

Gold JI, Shadlen MN (2007) The neural basis of decision making. Annu Rev Neurosci 30: 535–574

Grunewald A, Bradley DC, Andersen RA (2002) Neural correlates of structure-from-motion perception in macaque V1 and MT. J Neurosci 22: 6195–6207

Hübener M, Shoham D, Grinvald A, Bonhoeffer T (1997) Spatial relationships among three columnar systems in cat area 17. J Neurosci 17: 9270–9284

Hubel DH, Wiesel TN (1959) Receptive fields of single neurones in the cat's striate cortex. J Physiol 148: 574–591

Hubel DH, Wiesel TN (1962) Receptive fields, binocular interaction and functional architecture in the cat's visual cortex. J Physiol 160: 106–154

Hubel DH, Wiesel TN (1968) Receptive fields and functional architecture of monkey striate cortex. J Physiol 195: 215–243

Krug K, Brunskill E, Scarna A, Goodwin GM, Parker AJ (2008) Perceptual switch rates with ambiguous structure-from-motion figures in bipolar disorder. Proc R Soc Lond B 275: 1839–1848

Krug K, Cumming BG, Parker AJ (2004) Comparing perceptual signals of single V5/MT neurons in two binocular depth tasks. J Neurophysiol 92: 1586–1596

Kuffler SW (1953) Discharge patterns and functional organization of mammalian retina. J Neurophysiol 16: 37–68

Maunsell JH, Van Essen DC (1983) Functional properties of neurons in middle temporal visual area of the macaque monkey. II. Binocular interactions and sensitivity to binocular disparity. J Neurophysiol 49: 1148–1167

Milner AD, Goodale MA (1993) Visual pathways to perception and action. Prog Brain Res 95: 317–337

Parker AJ, Krug K (2003) Neuronal mechanisms for the perception of ambiguous stimuli. Curr Opin Neurobiol 13: 433–439

Parker AJ, Newsome WT (1998) Sense and the single neuron: probing the physiology of perception. Annu Rev Neurosci 21: 227–277

Reid RC, Alonso JM (1995) Specificity of monosynaptic connections from thalamus to visual cortex. Nature 378, 281–284

Sakmann B, Creutzfeldt OD (1969) Scotopic and mesopic light adaptation in the cat's retina. Pflügers Arch 313: 168–185

Salzman CD, Britten KH, Newsome WT (1990) Cortical microstimulation influences perceptual judgments of motion direction. Nature 346: 174–177

Salzman CD, Murasugi CM, Britten KH, Newsome WT (1992) Microstimulation in visual area MT: effects on direction discrimination performance. J Neurosci 12: 2331–2355

Treue S, Husain M, Andersen RA (1991) Human perception of structure from motion. Vision Res 31: 59–75

Ungerleider LG, Mishkin M (1982) Two cortical visual systems. In Ingle DJ, Goodale MA, Mansfield RJW (eds) Analysis of Visual Behavior. MIT Press, Cambridge, MA, pp. 549–586

Van Essen DC, Anderson CH, Felleman DJ (1992) Information processing in the primate visual system: an integrated systems perspective. Science 255: 419–423

What binds it all together? Synchronized oscillatory activity in normal and pathological cognition

4

Wolf Singer

Contents

1. Introduction

It is appropriate to begin with an epistemic caveat, that is motivated by neurobiological evidence. Our perceptions and imaginations as well as our abilities to reason are constrained by the cognitive abilities of our brains – and brains, like all other organs, are the product of an evolutionary process. Hence, our brains have become adapted to the conditions of the mesoscopic world in which life has evolved. It is the world within the scale of millimeters to meters, it is the world where the laws of classical physics are more or less applicable, it is not the world of quantum physics and it is not the world of astrophysics. As a consequence, our cognitive functions have become adjusted to assure survival in this mesoscopic world. Problem solving in this dangerous and poorly predictable world requires the application of pragmatic heuristics and hence cognitive abilities that are with all likelihood not optimized to comprehend the essence behind the perceivable phenomena or the "absolute truth" in the Kantian sense. Evolution did not prepare us to develop intuitions for processes at subatomic or cosmic scales, be-

Wolf Singer
Max-Planck-Institute for Brain Research
Deutschordenstr.46
60528 Frankfurt am Main, Germany
e-mail: singer@mpih-frankfurt.mpg.de

cause they were and are completely irrelevant for our daily struggle to survive.

Even more worrying is the possibility that also the way in which we reason is constrained by adaptation to the narrow range of the world that we can access with our specialized senses (see also Chaper II, 6 by FG Barth). In conclusion, it is very likely that our cognition is constrained and this to an unknown extent. And this may apply not only to primary perception, but also to our way of deriving inferences from observables. If this were true it would pose unsurmountable barriers to our attempts to understand as it would also challenge the consistency of mathematical theories and logical deductions. However, for these very reasons we have no way to know.

2. Current beliefs in neuroscience

We are fairly confident that all cognitive and executive functions that we can observe in human beings including the highest mental activities and consciousness are the result, not the cause of neural interactions. Consequently, they follow neural interactions and do not precede them. Furthermore, we assume that all neural processes follow the known laws of nature. Why do we believe this? Because we can explain the behaviour of organisms of low complexity such as e. g. molluscs or worms by registering the activity of their neurons and establishing causal relations between the spatio-temporal patterns of this activity and the respective behaviour. We have, at present, no reason to postulate any additional unknown forces, laws or modes of interaction in order to explain their behaviour. Furthermore, and I shall review some of the evidence below, neuronal interactions in more evolved brains have only become more complex but continue to obey the same principles. There seems to be no ontological discontinuity in evolution. At present, we need not postulate any additional hitherto unknown ingredients in order to explain how our brains function. Of course, we have to take into consideration that we are embedded in a complex socio-cultural environment and that our brains are shaped not only by the genes but also by the epigenetic influences of this socio-cultural environment – an aspect that has not been considered enough in the recent past – but this concerns the epigenetic modifications of our brains, not the functional principles.

3. Bridges to the humanities

Which, then, are the domains where the neurosciences begin to interact with the humanities? One obvious domain is epistemology. Cognitive neuroscience explores from a third person perspective the mechanisms that mediate our perception and the acquisition of knowledge. The long standing discussions about the objectivity of cognition, the question of how constructive our perceptual processes are, how reliable or idiosyncratic, needs to be reconsidered on the basis of neurobiological data. Another question, to which the neurosciences will have to find an answer is related to the mind-body problem: How can mental phenomena, the immaterial entities such as the qualia of perception and social realities such as belief systems emerge from the material interactions between nerve cells? These phenomena are real and affect our lives but they have another ontological status than the neuronal processes that brought them into this world by allowing *Homo sapiens* to support the evolution of cultures. Yet another question that solicits discussions between neuroscientists and philosophers of the mind is the nature of consciousness. The question of the constitution of the intentional Self is closely related to this issue and so is the conundrum of the existence of free will. If neuronal processes are the basis and cause of all mental phenomena and if brain processes follow the laws of nature, then the principle of causality must hold for neuronal interactions. Even though there is noise and interference, each state of the brain is a necessary consequence of the immediately preceding state and decisions are nothing but special brain states. This notion has far reaching consequences for our self-understanding.

4. Epistemic considerations of cognitive neuroscience

The majority of neuro-biologists consent that perceiving is essentially reconstructing. The sparse sensory signals that we extract from the world are interpreted in a highly idiosyncratic and selective way on the basis of a huge amount of stored *a priori* knowledge about the world. Our brains formulate knowledge based and context dependent expectancies, interpret sensory signals as a function of these inferences and present the result of this reconstruction to the workspace of consciousness. What do we know about the neuronal basis of the *a priori* knowledge required for the interpretation of sensory signals? It is commonly accepted that all the knowledge a brain can have and the rules according to which this knowledge is treated reside in the functional architecture of the brain. This contradicts the often suggested analogy between computers and brains. Computers have processors and separate memories for the programmes and the data. In the brain, there are only neurons and connections. Both the stored knowledge and the programs for the processing of this knowledge reside in the layout of these connections, the polarity of their action – excitatory or inhibitory – and their graded efficacy. The question of the origin of stored information is thus reduced to the question, which processes determine the functional architecture.

5. The origin of implicit knowledge and its role in cognition

The most important determinant of brain architectures is, of course, evolution. Thus, the specific layout of brain architectures can be regarded as a reservoir of evolutionarily acquired, genetically transmitted knowledge about the world. Another determinant of brain architecture and hence a source of knowledge are environmental influences that shape the development of neuronal circuits. There is a long developmental process lasting from birth to the age of 20, during which the brain is susceptible to epigenetic modifications. During this phase neurons still form connections and a large number of initially established connections are destroyed again on the way towards the mature architecture of the brain. This making and breaking of connections is guided by neuronal activity, which is, in turn, influenced by interactions with the environment and sensory experience. Thus, the brain adapts itself to the particular environment in which it evolves. Finally, there is knowledge acquisition by learning that accompanies us life long. This is based on graded changes of the coupling strength of the existing connections between neurons. In the adult brain only few new connections are formed and under normal conditions no breaking of connections occurs (for reviews see: Singer 1995; Hensch and Stryker 2004; Tropea et al. 2008).

In conclusion, evolution can be considered as a cognitive process through which the evolving organisms acquire specific knowledge about the world that they use to organize perception of those features of the world whose consideration facilitates survival and reproduction. These highly adapted, knowledge based cognitive processes are probably not optimized for the understanding of the deeper structure of the world that assures coherence across scales. Knowledge acquired through evolution is stored in the genes, expressed in the functional architecture of the brain and then complemented by developmental imprinting and learning. Together, this *a priori* knowledge determines what and how we perceive. It defines e.g. the sensory categories according to which we assign qualities to our experiences. And these assignments are rather idiosyncratic. We arbitrarily classify electromagnetic radiation with wavelengths between 400 to 700 nanometers as light, because the photoreceptors in the eye are sensitive to this wavelength range. Radiations with slightly longer wavelength stimulate our temperature receptors and we categorize the respective sensations as temperature. A similar arbitrariness of category boundaries is observable in other sensory domains.

A priori knowledge is also required for the definition of the nature of objects (Köhler 1930; Wertheimer 1945). We have clear concepts of what an object is. It must be confined by borders, should be solid and if it moves, all components should move in a coherent way. This definition is appropriate in the mesoscopic world but it does not apply to objects at atomic or subatomic scales. If we had no *a priori* definition of the properties of objects we would not be able to distinguish objects, we would be unable to extract object specific features from the two dimensional brightness distribution, that complex scenes generate on the retina. The principles according to which we associate features to objects reside in the circuitry of the visual system. Similarly, the learning rules according to which we establish associations are implemented by specific molecular mechanisms that have been preserved nearly unchanged since the evolution of primitive nervous systems. A particularly interesting epistemic problem would arise if the same also holds for the way we reason, infer and assign values to certain activity patterns and not others. The respective rules must also reside in the functional architecture of the brain and as mentioned, this architecture is the result of adaptation to the mesoscopic scale and hence to a very special and narrow segment of the world.

6. The conservative nature of evolution and the power of priors

In this context it needs to be emphasized that evolution is a highly conservative process. Once an invention has been made that works it tends to be conserved unless there is a major change in conditions that makes this invention obsolete or maladapted. This is the reason why our nerve cells function in exactly the same way as those of snails, why the same learning rules are implemented and the same mechanisms of signal transduction. Also the development of structures followed a very conservative path. Since the first appearance of the cerebral cortex, the 6-layered sheet of nerve cells that covers the hemispheres of

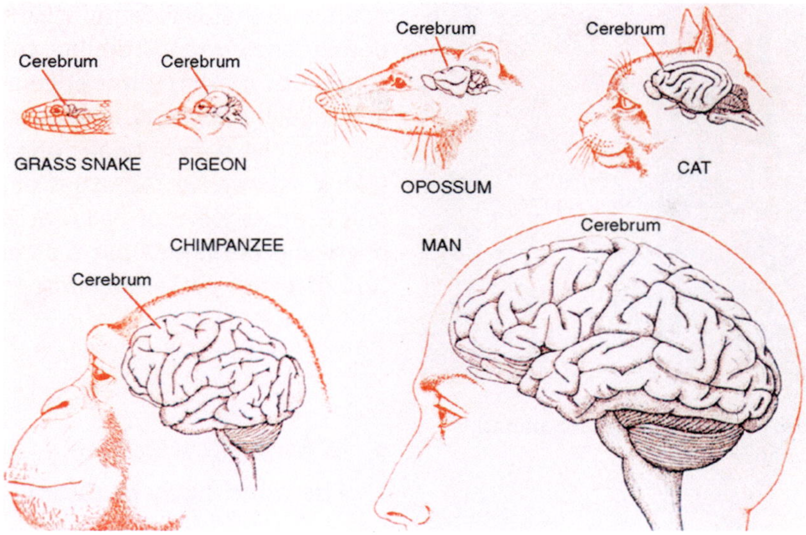

Fig. 1 Comparison of brain sizes between different vertebrate species. Note the dramatic increase of the telencephalon and, in particular, the cerebral cortex in mammals. Note also the striking similarity in the organization of the telencephalon between the great apes (here represented by the chimpanzee) and *Homo sapiens*. The only notable difference is the addition of more cortical substrate in the human brain

the brain, no new structures have emerged. There is just more of the same and this increase of hardware makes all the difference. It is the addition of a few more cortical areas that seems to make the difference between the brain of a human being and that of our nearest neighbours, the great apes.

Apparently it is only this addition of processing substrate and the associated gain of complexity that is responsible for the difference between animals and humans, between species that failed and those that succeeded to promote cultural evolution with all its far reaching consequences. In this context, however, one needs to consider that cultural evolution created a socio-cultural environment of ever increasing complexity that in turn contributes to the epigenetic shaping of brain architectures. Thus, even if the genetically determined layout of brain architectures has changed only little since the beginning of human civilisations, those features that can be modified by epigenetic shaping are likely to have undergone major modifications.

Here are two examples that illustrate how *a priori* knowledge structures our perception (Fig. 2). The object in Fig. 2 is a wooden mould to produce candies. On the left side one sees the front with the concavities and on the right the rear side with the corresponding convex protrusions. In reality, the objects are identical, but one is rotated by 180º. The reason for these very different perceptions is that the brain makes the *a priori* assumption that light comes from above. In this case contours that have the shadow above need to be interpreted as concave and those with the shadow below as convex. Thus, an assumption of which we are not aware of determines what we perceive. Another really striking example is the checkerboard illusion by Adelson shown in Fig. 3. It is hard to believe, but surfaces A and B have exactly the same luminance. They appear so different because the brain sees the shadow that is caused by the cylinder on the right. Even though the amount of light reflected from surfaces A and B and impinging on the retina is exactly the same, the brain inter-

Fig. 2 What we see depends on assumptions about illumination conditions

Fig. 3 Perceived brightness is the result of complex computations that take illumination context into account but we are unaware of these computations

prets the brightness of the two surfaces as different because it infers the following: Given that there is a shadow, surface B must be brighter than surface A which has no shadow on it, in order to reflect the same amount of light. Thus, the brain "computes" the inferred brightness of the surfaces but we are not aware of these computations. We just perceive the result and take it as real, i. e. we see B much brighter than A. One could spend hours with the demonstration of examples which indicate that the brain is generating inferences that we are not aware of, that it is permanently reconstructing the world according to *a priori* knowledge and that we, as perceiving subjects, have to take for granted what the system finally offers us as conscious experience. As expected, this is not only the case with specially designed psychophysical experiments but is an essential feature of all our perceptual processes.

7. The mind-body problem

I had already mentioned that neurobiologists assume that all mental functions including phenomenal awareness and self-consciousness are based on neuronal processes. However, there remains a severe explanatory gap. We encounter extreme difficulties when we attempt to explain how exactly the qualia of our subjective experience emerge from neuronal interactions. We use different reference systems for the description of qualia and neuronal processes, respectively, and lack satisfactory bridging theories between the phenomena defined from first and third person perspectives. Evidence indicates that there is only a minor difference between imagining something and perceiving it in reality. When subjects are asked to imagine for example a rotating wheel, one observes in fMRI (functional magnetic resonance interferometry) scans a characteristic distribution of activities in pre-frontal brain regions that need to be engaged in order to activate the read-out of stored information. In addition visual areas are activated, reflecting the read out of memorized images of rotating wheels. When subjects are then presented with a real image of a rotating wheel, there is only little additional activation which is confined to the primary visual area of the

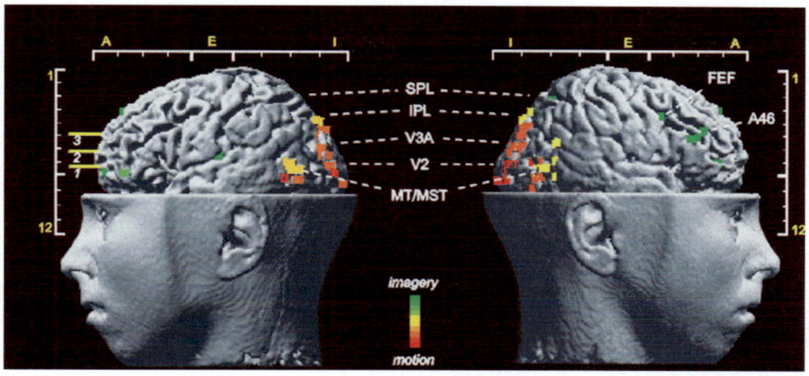

Fig. 4 Brain activation patterns associated with imagery differ only little from those associated with perceiving the imagined object. When subjects merely imagine a moving visual object, all areas indicated by orange, yellow and green pixels are activated in the brain. When the subjects open their eyes, and actually perceive the imagined object, only the areas indicated by dark red pixels are activated in addition. These correspond to very early visual areas close to the retinal input

cerebral cortex where the activity from the eyes is first processed. Thus, the difference between the activity patterns associated with imagination and real perception is minimal.

This is also the reason, why the brain's threshold for hallucinations is so low that sleep deprivation or fever can blur the distinction between imagery and perception. The best examined hallucinatory states are those occurring with schizophrenia. We investigated schizophrenic patients who suffer from verbal hallucinations and registered the activity in the auditory cortex of the speech competent hemisphere (Dierks et al. 1999). Every time when patients reported by a button press in the scanner that they heard voices which they attributed to a real speaker there was activity in the primary auditory cortex. This activity had about the same amplitude as the responses to voices presented by earphones. When healthy subjects imagine silent speech they activate their speech centres, but not the primary auditory cortex. In the patients this self-generated activity in speech centres apparently propagates all the way down to the primary auditory cortex, making it impossible for the subjects to distinguish between self-generated and evoked activity.

8. The conundrum of awareness

How is it possible that we do not only perceive but are in addition aware of the fact that we perceive? How can we explain in neuronal terms phenomenal awareness, the ability to be aware of one's cognitive operations? There are yet no satisfactory answers to that question. However, some hints can be found by considering the evolutionary development of brains. In less evolved brains the paths from sensory to executive areas of the cerebral cortex are short. As evolution proceeds and brains become more and more complex, one observes the addition of new cortical areas, which are no longer connected to the periphery, neither on the executive nor on the receptive side. They receive their information, their input, exclusively from the output of the phylogenetically older areas.

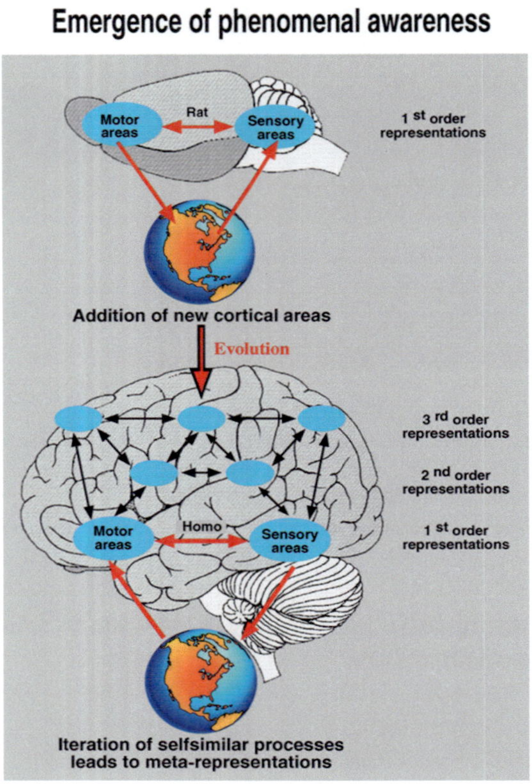

Fig. 5 Schematic illustration of communication pathways implemented by the addition of more cortical areas in more highly evolved brains. These additional areas interpolate more and more regions between primary sensory areas and executive structures in the frontal brain, whereby many of these novel areas no longer maintain connections with input or output structures. It is proposed that the addition of these novel areas provides the option for the generation of higher order representations (meta representations) and excessive polymodal integration which in turn is the prerequisite for abstract and symbolic representations

This process is iterative with more and more areas added that communicate only with other areas. A neuron located in these more recent areas is connected exclusively with other partners in the cerebral cortex. Evidence indicates that all cortical areas, the old and the more recent ones, operate according to the same principles because they have the same intrinsic organization. Thus, the more recent areas process the output of the older areas according to the same principles as these process the signals from the sense organs. This iteration should allow the creation of meta-representations, representations of representations of representations. And this is perhaps the structural basis that allows us to subject to cognitive operations not only signals from the outer world but also those resulting from internal processes, to realize various levels of meta-cognition and to thereby become aware of our own perceptual functions.

A neuron located in these more recent areas is connected exclusively with other partners in the cerebral cortex. Evidence indicates that all cortical areas, the old and the more recent ones, operate according to the same principles because they have the same intrinsic organization. Thus, the more recent areas process the output of the older areas according to the same principles as these process the signals from the sense organs. This iteration should allow the creation of meta-representations, representations of representations. And this is perhaps the structural basis that allows us to subject to cognitive operations not only signals from the outer world but also those resulting from internal processes, to realize various levels of meta-cognition and to thereby become aware of our own perceptual functions.

9. The distributed organization of the brain and wrong intuitions

Another conundrum and challenge to both brain research and our intuitions is the distributed organization of our brains. The neurobiological evidence accumulated over the

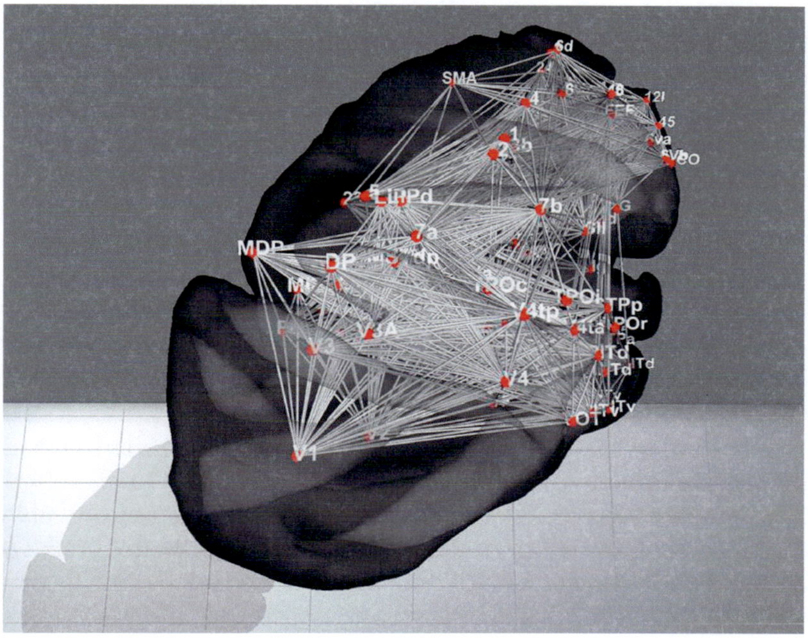

Fig. 6 Wiring diagram of established, mostly reciprocal anatomical connections between visual areas in the occipital, parietal and temporal cortex of the macaque brain and executive cortical regions in the frontal lobe. The red dots correspond to cyto-architectonically defined cortical areas and the white lines to the fibre connections between these areas. A quantitative analysis of this connectivity scheme suggests that it follows the principles of small world networks

last decades has led to radical changes in our views of the brain. In early days intuition and introspection were the major sources of knowledge for the formulation of hypotheses about the organization of the brain. Now we learn that these intuitions are in drastic conflict with the evidence provided by scientific investigations, raising the interesting question why the brain is so agnostic to its own organization. We do not feel our brain, have little intuitive insight into its processes and are surprised to find that it works according to principles that differ substantially from what we thought. Intuition suggests to us that somewhere in the brain there ought to be a convergence centre where all information is coming together to be amenable to coherent interpretations of the world. This would be the site where perception takes place, where an intentional agent is active, where decisions are reached, where plans are developed and where the Self is seated. We assume a hierarchical structure similar to those that we recreate in social and economical systems, – probably not always to our advantage, because they may be maladapted once systems reach a critical degree of complexity.

The reality of our brains as it appears from a third person perspective is very different. The cerebral cortex comprises a large number of different areas that, depending on their input, accomplish different functions but use similar computational algorithms. Thus, the format of exchangeable information is always the same and communication among cortical areas can capitalize on this *lingua franca.* This is a necessary prerequisite for generalization, abstraction, symbolic encoding and last but not least for the constitution of the unity of consciousness. The surprising finding is that the connections linking these areas provide only little evidence for serial processing in strictly hierarchical architectures (see Fig. 6). Rather, the connectivity scheme is dominated by principles of parallelity, reciprocity and distributedness. Thus, neurons located in the visual cortex can talk directly to neurons in the limbic system or in executive areas and most of these interactions are reciprocal. This meshwork of connections is extraordinarily dense and complex but far from random. It is highly structured and has properties of so called "small world networks" (Yu et al. 2008). This architecture is the hardware realization of the programs according to which brains process information and it is also the basis of stored knowledge.

10. Distributed representations

Imagine one perceives a barking dog, touches its fur and judges the dog as friendly. In this case all visual areas will be active and participate in the identification of the dog, the same holds for the tactile areas, which analyze the texture of its fur, the auditory areas which decode the barking and the limbic areas which add the emotional connotations. There is no single locus for the representation of the integrated percept of this dog. Rather, the representation consists of a complex spatio-temporal pattern of neural activity in which myriads of neurons participate that are distributed within and across the various processing areas of the cerebral cortex and subcortical structures. Thus, the brain presents itself as a highly distributed, self-organizing system. It lacks the postulated singular convergence centre, which Descartes actually searched for in cow brains and thought to have found in the pineal organ, because it is impair, occurring only once like the hypophysis. In reality, however, there is no such center, no observer, no coordinator; there is no identifiable seat of the conscious, intentional self. How exactly this distributed system produces the holistic properties that we as observers attribute to the person, the intentional Self, is still largely unknown. Another question is why our intuition is so wrong? My suspicion is that the brain, even though it exploits for information processing the advantages of non-linear dynamics, is tuned to assume that the phenomena to be analyzed are mainly linear because it is only with this premise that predictions can be made on the basis of past experience. However, if the brain assumes the same concerning its own functioning, if it assumes that it is organized like a linear stimulus-response device, it is bound to postulate a mover, because linear systems cannot by themselves produce all the remarkable functions that we observe, they cannot be creative, open towards the future and intentional. But this interpretation is of course a speculation.

11. The binding problem

Self-organizing, distributed and goal directed systems need efficient and flexible mechanisms in order to dynamically coordinate in a context and goal dependent way the many distributed local processes into coherent wholes. One way to bind distributed results is convergence in devoted anatomical circuits. If the messages encoded by units A and B are to be bound it suffices to connect their outputs with a third unit C and then to select appropriate thresholds for unit C, so that C is only active when A and B are active at the same time. In this way relations can be evaluated in fixed, essentially feed forward architectures and encoded by the responses of conjunction specific neurons. The brain makes ample use of this strategy, but because of its rigidity and inflexibility, this strategy can be used only for the encoding of frequently occurring, stereotyped and familiar relations. The alternative is to express relations by dynamic coordination, so that the representation remains distributed but functions as a coherent whole, a strategy called assembly coding. It is a much more flexible coding strategy and shares features with language systems. With 26 letters one can write the world literature, simply by re-combining in a flexible way the letters. With 10^{11} neurons, each having the role of a symbol, and a flexible recombination mechanism, a virtually infinite number of different distributed representations can be formed. The representations of novel objects, of the ever changing constellations of real world conditions and of adaptive motor responses, are, thus, implemented best by dynamically configurated assemblies. However, in assembly coding one needs a mechanism that allows for the flexible and rapid association of ever changing constellations of distributed neurons. A code is required that defines from instance to instance which subset of the myriads of the active neurons actually contributes to a particular representation. As there will always be several coexisting assemblies, an unambiguous signal is needed that tells to the rest of the brain which neurons are actually bound together in an assembly. In essence, neurons supporting assembly codes have to convey two messages in parallel. First they have to signal, whether the feature for which they serve as symbol is present and, second, they have to indicate, in parallel, with which other neurons they are actually collaborating in order to form the coherent whole to which they contribute their feature. There is common agreement that they signal the presence of their symbol, of their feature, by increasing the frequency of their discharges, by becoming active and the more active they become the more reliable is this message. Following a discovery made in our lab in Frankfurt, we pursue the hypothesis that the signature for the *relatedness* of the cells belonging to an assembly is the precise synchronization of the individual discharges (Gray et al. 1989). The required precision is in the range of milliseconds, in order to allow the definition of relations with the necessary temporal resolution. This is required in order to reconfigurate assemblies at a rapid pace. I shall not discuss the technical details that suggest that synchronization is an excellent tag for the definition of relations (for further reading see: Singer 1999; Fries et al. 2007; Uhlhaas et al. 2009). Intuitively, it appears as obvious that events that happen simultaneously are easily bound together. And there are indeed mechanisms that render neurons particularly susceptible to synchronous, i. e. coincident inputs.

12. The role of oscillatory synchronized activity

Since the discovery of stimulus dependent synchronization of oscillatory responses many laboratories have joined the search for its functional implications. A major prerequisite for those studies is to sample simultaneously the responses of at least two neurons, preferably of as many as possible, because otherwise temporal relations cannot be assessed. In this context it is noteworthy that until recently we used to record from only one neuron at the time and related the firing of these isolated cells to stimuli or behaviour in order to identify their functional properties. This precluded analysis of relations and hence the identification of functionally bound assemblies. If one considers the complexity of the system it is obvious that even the multisite recordings have their limits. We are, despite all progress, still at the beginning of understanding the brain processes underlying higher cognitive functions but we seem to know how to proceed. We have decades ahead of us, where we shall need more and more theoretical approaches and new technology, but we know where to go. Since its discovery, synchronization of oscillatory activity has become a candidate mechanism for many different functions. I mentioned already dynamic binding, the flexible definition of relations. However, it also seems to be involved in attentional mechanisms that select activity for further processing out of a sea of competing activities (Fries et al. 1997; Fries et al. 2001a), it appears to serve the read-out of information that is stored in the connectivity (Fries et al. 2001b) and it may also be used to bind different sub-systems together such as sensory and motor systems. Evidence also indicates that it serves the selective routing of signals across the highly interconnected networks of the cerebral cortex (Womelsdorf et al. 2007). The mechanism resembles the tuning of a radio to the frequency of a certain transmitter, thereby allowing brain centres to send a message with high selectivity from point A to point B without spreading it to everybody. Assuring selective routing in a highly connected system is a very difficult task and is apparently solved by synchronization. There are also indications that entrainment into coherent oscillations plays a role in the storage and maintenance of information in short term memory (Tallon-Baudry et al. 2004; Buschman and Miller 2007) and finally large scale synchronization appears to be a prerequisite for signals to have access to conscious processing (Melloni et al. 2007; Gaillard et al. 2009).

Before closing let me briefly turn to one of the limitations of our intuition with respect to assembly coding. It is difficult for us to understand the structure of distributed representations because the essential information is encoded in non-stationary, spatio-temporal relations among the responses of large numbers of neurons rather than in the responses of individual neurons. In principle, one can measure these patterns by registering the activity of these neurons simultaneously. The difficulty is, however, to obtain a comprehensive sample of the relevant cells and to identify the patterns that matter. This identification often requires advanced mathematical methods which, if successful, deliver abstract descriptions of vectors in a high dimensional state space. They are not tangible, difficult to represent in two or three dimensions and consist of lists of numbers or systems of equations. It is not satisfactory to have such explanations for a percept, a thought or a feeling. The problem is, that, in the future, descriptions of neural states will become more and more abstract, more and more mathematical, more and more intuitively implausible.

13. The neuronal correlate of conscious processing

To conclude I want to give one example of a typical experimental approach as it is used in the search of the neuronal substrate of conscious processing. The paradigm contrasts subconscious and conscious processing of stimulus information. Subjects look at a fixation point on a computer screen. Subsequently, a word is presented that is preceded and followed by a noisy mask. With some delay, a word or a figure is presented and subjects are asked to press buttons indicating "same" or "different". The masks are adjusted such that the subjects perceive the first word only occasionally. In trials where they do not see the word they are encouraged to guess and press the buttons as fast as possible. Interestingly, when there was concordance, even if subjects had not seen the first word, the reaction times were shorter. This proves that subjects had processed the stimulus material even though they were not aware of having processed it. During these experiments we obtained EEG (electroencephalogram) recordings and measured the oscillatory responses of the different areas of the cerebral cortex. When we analyzed the power of the oscillations in particular frequency bands, we only observed a delayed increase of theta activity in the conscious condition. However, when we considered not only the amplitude of oscillations, but also the precision of the phase locking across sensors, we noticed shortly after the encoding a brief burst of very precise phase synchronization in the gamma band across widely distributed cortical areas when stimuli were processed consciously. This episode of high coherence was missing in the unconscious condition. The topological distribution of this synchronization phenomenon is wide spread. An extended network of cortical areas engages transiently into very precise synchronous oscillations when stimuli have access to consciousness. This suggests that access to consciousness is associated with very precise synchronization of oscillatory activity in the gamma frequency range across widely distributed cortical areas (Melloni et al. 2007). Thus, the neuronal correlate of conscious processing appears to be a particular dynamic state of a widely distributed network rather than the activation of a special cortical area. Apparently there is no specific area that is responsible for consciousness. This interpretation comes close to Sherrington's statement about consciousness: "Pure conjunction in time without necessarily cerebral conjunction, we mean conjunction in space, lies at the root of the solution of the problem of the unity of mind." It is the synchrony of events that binds processors together and makes that distributed contents are experienced as a coherent whole. Unity is not achieved by converge in space but by coordination in time – and this is not what our intuition assumes.

Outlook

The more we learn about the brain the more abstract will the descriptions become because we shall have to adopt concepts from complexity theory and non-linear dynamics. It is likely that the neuronal correlates of cognitive contents will present themselves as dynamic trajectories of a complex system in a high dimensional space. And these trajectories will never repeat identically because if they did we would have no notion of time. Many of the explanations will further be in conflict with our primary intuitions because we have difficulties to imagine complex non-linear processes. Furthermore, we shall have to consider that the brain is part of a sociocultural network and that some of the phenomena that seem to be so difficult to explain in pure neuronal terms will have their explanation only when considering interactions among brains, if we analyze "brain-networks" rather than individual brains. This widened perspective is probably also necessary to close the explanatory gap between first and third person perspective. We shall have to consider the fact that our brains are the product of their embedding in a complex cultural environment and that many of the constructs that are so difficult to relate to brain processes, such as value systems, consciousness, intentionality and so forth, have only come into the world because brains mirror themselves reciprocally in other brains and assign symbols to the then gained experiences. We are just at the beginning of social brain research but there are already first investigations on phenomena that exist only in the realm of social interactions such as empathy, responsibility, greediness and jealousy.

References

Buschman TJ, Miller EK (2007) Top-down versus bottom-up control of attention in the prefrontal and posterior parietal cortices. Science 315: 1860–1862

Dierks T, Linden DEJ, Jandl M, Formisano E, Goebel R, Lanfermann H, Singer W (1999) Activation of Heschl's gyrus during auditory hallucinations. Neuron 22: 615–621

Fries P, Neuenschwander S, Engel AK, Goebel R, Singer W (2001b) Rapid feature selective neuronal synchronization through correlated latency shifting. Nature Neuroscience 4(2): 194–200

Fries P, Nikolic D, Singer W (2007) The gamma cycle. Trends in Neurosciences 30(7): 309–316

Fries P, Reynolds JH, Rorie AE, Desimone R (2001a) Modulation of oscillatory neuronal synchronization by selective visual attention. Science 291: 1560–1563

Fries P, Roelfsema PR, Engel AK, König P, Singer W (1997) Synchronization of oscillatory responses in visual cortex correlates with perception in interocular rivalry. Proc Natl Acad Sci USA 94: 12699–12704

Gaillard R, Dehaene S, Adam C, Clémenceau S, Hasboun D, Baulac M, Cohen L, Naccache L (2009) Converging intracranial markers of conscious access. PLOS Biology 7(3): e10 00 061

Gray CM, König P, Engel AK, Singer W (1989) Oscillatory responses in cat visual cortex exhibit intercolumnar synchronization which reflects global stimulus properties. Nature 338: 334–337

Hensch TK, Stryker MP (2004) Columnar architecture sculpted by GABA circuits in developing cat visual cortex. Science 303: 1678–1681

Kaiser M, Hilgetag CC (2006) Nonoptimal component placement but short processing paths due to long-distance projections in neural systems. PLoS Computational Biology 2(7): e95

Köhler W (1930) "Gestalt psychology". Bells and Sons, London

Melloni L, Molina C, Pena M, Torres D, Singer W, Rodriguez E (2007) Synchronization of neural activity across cortical areas correlates with conscious perception. J Neurosci 27(11): 2858– 2865

Singer W (1995) Development and plasticity of cortical processing architectures. Science 270: 758–764

Singer W (1999) Neuronal synchrony: A versatile code for the definition of relations? Neuron 24: 49–65

Tallon-Baudry C, Mandon S, Freiwald WA, Kreiter AK (2004) Oscillatory synchrony in the monkey temporal lobe correlates with performance in a visual short-term memory task. Cerebral Cortex 14(7): 713–720

Tropea D, Van Wart A, Sur M (2008) Molecular mechanisms of experience-dependent plasticity in visual cortex. Phil Trans R Soc B 364: 341–355

Uhlhaas PJ, Pipa G, Lima B, Melloni L, Neuenschwander S, Nikolic D, Singer W (2009) Neural synchrony in cortical networks: history, concept and current status. Frontiers in Integrative Neuroscience 3(17): 1–19

Wertheimer M (1945) Productive thinking. Harper, New York

Womelsdorf T, Schoffelen J-M, Oostenveld R, Singer W, Desimone R, Engel AK, Fries P (2007) Modulation of neuronal interactions through neuronal synchronization. Science 316: 1609–1612

Yu S, Huang D, Singer W, Nikolic D (2008) A small world of neuronal synchrony. Cerebral Cortex 18: 2891–2901

II Evolutionary epistemology

Introductory remarks

For biologists it is a fact so well established and accepted that it may appear trivial to mention it: Structures and functions, particularities and peculiarities of organisms are the result of evolution and adapted to survival and reproduction. Man is a product of evolution as well. This has far reaching consequences, which to a large extent affect its perception of the world and what we usually refer to as our human intellectual capabilities. Surprisingly, it still seems to be hard for many a non-biologist to fully acknowledge this, mainly because it is so difficult to give up the thought that man is something basically different from the rest of the animal kingdom. Hopefully, this section of our book will convince all of our readers that taking evolution into account is not just a side interest of secondary importance but has to be an essential and integrative part of all efforts to understand perception and cognition, matter *and* mind.

1. Our knowledge of the world

First a philosopher, Gerhard VOLLMER of Braunschweig University, addresses the old and still pressing problem: How is it that we can know this world? To answer this fundamental question he chooses an approach between biology and philosophy, taking the position of hypothetical realism and highlighting evolutionary epistemology. Evolutionary epistemology indeed explains how the principles used by us to construct our picture of the world came to be in our brains and why cognitive faculties have a strong genetic component. Clearly, these principles are not only dictated by our own senses and external stimuli.

2. An organismal view of sensory perception

In his contribution Friedrich G. BARTH of the University of Vienna then elaborates on another aspect crucial for understanding sensory perception and closely related to evolutionary processes as well. A prominent feature of all biological sensors and sensing is their selectivity for the biologically relevant. Sensory systems are narrow filters not evolved to provide a complete picture of the physical world but instead ignoring most of the potential stimuli actually present. They have not evolved to provide abstract truth but fitness, which is survival and reproduction. From this it follows that the properties of sensory systems are closely matched to the relevant stimulus patterns typical of their species specific environment and guiding their respective species-specific behavior. It also follows that we should not think of man as the culmination of sensory evolution and that there are numerous examples of sensory systems giving animals access to experiences completely alien to us humans.

3. Cognition in insects

Martin GIURFA (University of Tours) reports most exciting learning and cognitive abilities of social insects, mainly the honeybee, which is one of the most intensively studied of all animals. Many a reader will be surprised to learn what can be achieved with a tiny brain so much smaller than ours. For a honeybee visual learning and cognition, which is in the focus of this chapter, is a capacity of fundamental importance in many behavioral contexts like searching for the food provided by flowers and recognizing a partner or enemy. There are many degrees of complexity of visual learning. These range from the simple formation of an associative link to the most remarkable learning of abstract rules applied to unknown visual objects. One of the lessons drawn from *Martin Giurfa's* contribution is that the recent "cognitive revolution" in the biosciences not only inspired research on vertebrates but, with some delay, also research on the perception and learning of honeybees, bumblebees and wasps. This research brought to light surprising capabilities of social insects far beyond elemental associations such as the extraction of the logical structure of what they perceive.

4. Evolution of cognition

Finally, Ludwig HUBER and Anna WILKINSON of the University of Vienna highlight the relevance of a comparative approach of cognition based on evolutionary thought and Darwin's suggestion that apart from the morphological continuity there is also mental continuity across species. The authors trace the historical roots of this young and flourishing field of research. *Huber* and *Wilkinson* emphasize that apart from phylogenetic aspects, questions of adaptation and proximate mechanisms need to be answered in order to fully understand perception and cognitive ("intelligent") behavior. Knowledge about the evolution of our human cognitive capabilities is still sparse. Non-human animals have been found to show very clever behavior which cannot be explained by obvious and simple associations but suggest "higher cognitive" processes to be at work. However, as the authors point out, the neural implementations of the latter are still unknown and therefore it still remains to be demonstrated what the difference between "simple" and "higher" really refers to. Likewise the authors emphasize the importance of not to confuse functional analogy with phylogenetic homology and the danger to "over-intellectualize" by using humans as the sole template instead of following the evolutionary processes which continuously added to and modified pre-existing cognitive systems.

FGB

Between biology and philosophy: our knowledge of the real world

5

Gerhard Vollmer

Contents

Abstract

Cognition takes place in our heads. Using the signals that we receive from our sense organs our brain builds up a picture of the world in to a whole worldview. We construe the world as three-dimensional, as ordered and directed in time, as regular, even structured by laws of nature, and causally connected. With some of our constructions we are successful, with others we fail. The principles by which we construct this world picture are not only dictated by our sense organs or exclusively by external stimuli. How did they come into our heads? This question is answered by *evolutionary epistemology*. We recapitulate its main theses, characterizing it as a naturalistic position and answering three typical objections. We then turn to more recent arguments, concerning language, realism, and the theory of natural selection.

Gerhard Vollmer
Professor-Döllgast-Str.14
86633 Neuburg/Donau
e-mail: g.vollmer@tu-bs.de

1. Introduction: our question

If we have kept our curiosity and our ability to wonder, we must be excited day by day by the new and seemingly endless discoveries of science: from the atomic nucleus to the cell nucleus, from quark to quasar, from white dwarfs to black holes, from the nano engine to the chess computer beating the world chess champion. *How is it that we can know this world?*

In *asking* this we already make several presuppositions. We presuppose that there is (something like) the *world*. In addition, our definite article – *the* world – suggests that there is only *one* such world, i. e. that the object of our cognition is *unique*. We further submit that we can *cognize* this world – maybe not completely, not precisely, not without error, but at least adequately and reliably to a certain degree. If there were no world at all, or if, although it exists, we could not recognize it, then it would not make sense to ask for the "how" and "why" of such cognition: What does not exist needs no explanation. With our question we further presuppose a certain communality in our cognition, some minimal *intersubjectivity*. If our opinions about and our insights into the world were as personal, as subjective, as individual, and therefore as different as our dreams, then we shouldn't dare to speak about cognition. As far, however, as such intersubjectivity prevails – where its scope needs thorough examination – this intersubjectivity is in need of an explanation. This is also implied in our question.

Finally, a question like ours, if meant seriously, presupposes that the answer is *neither trivial nor evidently impossible*. Indeed, why we can know this world is not known to every child, not to the woman on the street, and not to everyday man. Nor is this question totally unanswerable: The answer is no eternal secret, no riddle of the universe. At least, it is not recognized or recognizable as such on first view.

None of these presuppositions is – in view of the history of philosophy – unproblematic. And we have no cogent arguments for our position. Recall Gorgias, sophist in ancient Greece around 450 b.C. with his agnostic three-step declaration: "There is nothing. If there were something, it could not be recognized. And even if it could be recognized, it couldn't be communicated" (After Sextus Empiricus VII 65). Although nobody really accepts Gorgias' pessimistic position, we cannot conclusively refute him. But we may take a position doing justice to common sense thinking, to the approach of empirical science, and to philosophical critique: This position is *hypothetical realism*.

Hypothetical realism is characterized by the following key-concepts:

- existence of a real world independent of consciousness, lawfully structured and connected (called ontological or metaphysical *realism*);
- partial cognizability and intelligibility of this world by perception, thought and intersubjective science (epistemological realism);
- hypothetical (fallible and therefore preliminary) character of all factual knowledge (fallibilism and methodological realism).

Hypothetical realism shares these theses with critical rationalism (which in addition stresses the methodological element of critique). In the context of this position our question makes sense. And in this context we try to answer this question.

2. Our answer: evolutionary epistemology

Thinking and cognition are achievements of the human brain, and this brain originated through organic evolution. Our cognitive structures fit the world (at least partially) because phylogenetically they were formed in adaptation to the real world and because ontogenetically they have to grapple with the environment. The biologist George Gaylord Simpson (1902–1984) makes this point crudely but graphically: "The monkey who did not have a realistic perception of the tree branch he jumped for was soon a dead monkey – and therefore did not become one of our ancestors." (Simpson 1963) Hence we owe the fact that our spatial perception is relatively good to our predecessors living in trees. In this way we may also explain other cognitive achievements.

But why, then, are our cognitive faculties not even better? Why are our sense organs not faster, more sensitive, more encompassing, more reliable? Why is our memory not better? Why is our intuitive reasoning not more correct? For every cognitive faculty of man we find an animal that is superior in that respect. Again the answer is simple enough: Biological adaptation is never ideal or perfect, nor is our cognition. There is no evolutionary premium on perfection, but only on effectiveness. What is decisive for evolutionary success is not pure quality but a defendable cost-benefit relation. It is not essential to find the best possible solution, but only to be better than the competitors. Thus evolutionary epistemology explains not only the achievements of our brain but also its failures. In this context we must think not only of interspecific competition but of intraspecific competition.

That section of the real world to which man is adapted in perception, experience, and action, we call the *mesocosm.* It is a world of medium dimensions: medium distances and time periods, small velocities and forces, low complexity. Our intuition (called our "ratiomorphic apparatus" by Egon Brunswik 1955) is adapted to this world of medium dimensions. Here our intuition is useful, here our spontaneous judgments are reliable, here we feel at home.

Whereas perception and experience are mesocosmically impregnated, scientific cognition may transcend this mesocosm, extending in three directions: to the very small, the very large, and the very complicated. As we know, intuition fails regularly there. Nobody can visualize the conditions of the quantum realm, relativity theory, or deterministic chaos. Yet we must constantly deal with complicated systems. To do so we need working tools and thinking tools, instruction and training. The most important thinking tool is language. Other ladders leading beyond the mesocosm are algorithms, calculi, mathematics, and computers.

3. Evolutionary epistemology as a naturalistic position

Evolutionary epistemology constitutively rests upon organic evolution. This has given evolutionary epistemology its name. The term *evolutionary* is not meant to claim that all epistemological problems can or should be solved by reference to the evolution of the universe, organisms, man, or knowledge. However, it mirrors the idea that the evolutionary origin of our cognitive faculties plays an important role in epistemology, both explanatory and critical. Thus we explicitly deny Ludwig Wittgenstein's claim in his *Tractatus* (1922 § 4.1122) that "Darwin's theory has no more to do with philosophy than any other hypothesis in natural science." To be sure, we must support our denial by arguments, and this is done best by *showing* how evolution pertains to philosophy. It may solve old philosophical problems, pose (and even solve) new problems, or shed new light on problems. Such a triple claim is made by evolutionary epistemology.

It remains unclear, however, how general the concept of evolution is meant here. Do we talk about organic evolution, the evolution of organisms, or do we also talk about the evolution of knowledge, maybe even of science? This ambiguity has been quite confusing. Here we are talking about the biological evolution of cognitive faculties (see also Chapter II, 4 by L Huber and A Wilkinson). Insofar as we are investigating the development of science using evolutionary concepts in general, we prefer to call this enterprise *evolutionary philosophy of science* (Vollmer 1987).

The general orientation of evolutionary epistemology is *naturalistic*. What does that mean? There is talk about naturalism in several areas: theology, philosophy of science, ethics, the arts. In the present context, we take it to be a conception of natural philosophy and anthropology claiming that there are no secrets anywhere in the world. Hence it is distinguished by two traits: by its *claim to universality* and by the *limitation of tools* admitted for describing and explaining the world (Vollmer 1994).

Philosophical naturalism is both a conception and a program. As a program it has at least four parts:

1. It calls for and charts a *cosmic view,* a "worldview".
2. It assigns to *man* a definite place in the universe (which turns out to be rather modest after all).
3. It covers *all* human capacities: language, knowledge, scientific investigation, moral action, aesthetic judgment, even religious faith.
4. Under these premises it calls for and develops in particular
   - a naturalistic anthropology,
   - a naturalistic epistemology,
   - a naturalistic methodology of research,
   - a naturalistic ethics,
   - a naturalistic aesthetics.

There is no naturalistic religion or theology, because every religion has – by definition – some transnatural elements. With respect to epistemological questions, Willard V. O. Quine has formulated such a naturalistic program (Quine 1969), and evolutionary epistemology attempts to carry it out. Occasionally, Quine himself drew on evolutionary arguments: "Natural selection, then, could explain why innate standards of resemblance have been according us and other animals better than random chances in anticipating the course of nature" and "Creatures inveterately wrong in their inductions have a pathetic but praiseworthy tendency to die before reproducing their kind." (Quine 1969). Quite analogously some philosophers have tried to develop evolutionary ethics. Generally, we may speak of an *evolutionary naturalism.* There are even books with this title (Sellars 1922; Ruse 1995).

Finally, evolutionary epistemology is *realistically* oriented. More precisely, it defends a *hypothetical realism,* characterized in section 1.

Numerous objections have been raised against evolutionary epistemology. We look at three of them.

4. Fundamental objections to evolutionary epistemology

4.1 "Does the concept of truth make sense?"

Hypothetical realism makes use of the correspondence theory of truth. By this theory, a proposition is true if what it says corresponds with external reality. But how do we get to know this reality, hence truth? There is no independent access to reality unless for God. We humans cannot take this divine perspective, cannot know the world in itself, and therefore cannot assess truth in the sense of correspondence theory. So goes the objection.

As far as this objection is justified it is directed against all kinds of epistemological realism except perhaps the internal realism proposed by Hilary Putnam (1981) some years ago, which strictly speaking is no realism at all. In fact, we are no gods. But this is not necessary either. The correspondence theory of truth does not supply a criterion of truth, only a definition of it. As epistemologists had to realize after 2,500 years of fruitless search and growing doubts, there are no satisfactorily sufficient criteria for factual truth. What we have are necessary criteria like consistency, corroboration, coherence, or consensus, as exhibited by the different theories of truth. For the definition of truth all these theories rely, in the last analysis, on the correspondence concept. Where this is not the case, the concept of truth is, strictly speaking, superfluous.

We might object that a God's-eye perspective is an undue idealization. However, no theory of truth can do without such idealizations. Internal realism, for instance, regards as true what at the end of all research will be ascertained about the world. What if not this is an idealization? Thus realism and correspondence theory answer this last objection with a *tu quoque* argument: Yes, we make use of an idealization, but all other theories of truth use comparable devices.

4.2 "Is the fit of our cognitive structures ascertainable without a vicious circle?"

Would we not have to know reality *independent* of our cognitive structures? This objection goes beyond the former because what is at stake now is not a definition of truth but our knowledge about reality. Here evolutionary epistemology makes a more ambitious claim.

Let's take an example: Physicists and physiologists disclose to us that our eyes are sensitive in precisely that section of the electromagnetic spectrum where – thanks to the optical window of the terrestrial atmosphere – radiation from the sun can pass through the air and reach the surface of the earth. How could they come to know this? Of course, even physics had to start in the mesocosm, but has definitely left it since. In so doing, physics has objectified its methods as much as its results and its theories. It does not talk about colors anymore but about frequencies, wavelengths, and energies. For the characterization and detection of radiation it does *not* depend on the eye. And it finds electromagnetic radiation of all wavelengths. True, even terms like *wavelength* and *sensitivity* could still be anthropomorphic. However, there is no rational doubt, first, that not all of what in principle could exist does really exist; that, second, with our senses we can process only a section of what there is; and that, third, there is a very good fit between (what we call) daylight and the properties of our eye. We can detect this match without being realists and without an explanation at hand. It is the fit thus established that is interpreted and explained by evolutionary epistemology, not as pure chance, nor as the work of a creator, but as the result of an adaptive process.

Still, one could object that what physicists describe is not the real world but at best a projection, possibly a garbled one, and in the worst case nothing but a wild construction. Certainly we cannot strictly prove the truth, the correctness, the adequacy of our theories. But what on earth can be proven strictly? We cannot even disprove the solipsist claiming or in fact being convinced that there is nothing besides his actual consciousness.

But where proofs are missing, good reasons might still be available. For ontological realism (there is a real world independent of our consciousness) and for epistemological realism (this world can be known at least in part and approximately) there are good arguments. For the suggestion, however, that scientific knowledge is nothing but a wild construction there are *no* good arguments. And it is utterly implausible that we should have adapted over thousands of years to constructs that have been worked out by scientists during the last centuries.

Some constructivists think that organisms are adapted, not to an external world, but to survival. This is nonbiological thinking. If there are no selective demands by the environment, there can be no traits facilitating survival or traits threatening it. In that case any solution will do. But then the concept of adaptation makes no sense at all.

Rounding up: We may sensibly talk of fits and adaptations, we may argue for them, but we cannot prove them. And evolutionary epistemology is happy enough to explain these fits – that is, our cognitive achievements as well as our failures.

4.3 "How can cognitive structures be adapted to an environment that would have to be known to the organism before the organism might adapt to it?"

If this objection were sound there would be no eyes! For how could eyes be adapted to terrestrial illumination if they were necessary *before* any light could be processed? But eyes did originate several times, independently, and, if we follow evolutionary biologists, at least forty times. And most of them are perfectly well adapted to light. How could they originate? The answer is simple: Eyes originated as all things originate in evolution, namely by trial and error elimination, by blind variation and selective retention, by undirected mutations and gene recombinations and preferential reproduction of superior solutions. Nowadays the evolution of the vertebrate eye – and that includes the human eye – can be reconstructed quite well (see also Chapter II, 2 by FG Barth). Similar considerations apply to all other sense organs, to all senses, to all perceptual achievements. There is no reason why they should not be applied to higher cognitive functions, as far as these are genetically conditioned.

For evolutionary origins and explanations, however, it is not essential that an organ be perfect. The intermediate steps are evaluated selectively; they must increase fitness, but it is not necessary that the later function be present and effective from the very beginning. Changes of function are possible and rather common. Here, a trait is built for a function that will be replaced later by another function. From fins arose arms and legs; feathers served at first not for flying but for gliding, catching prey, and keeping warm; the middle-ear ossicles are former jawbones. Since the change of function is not saltatory, it is indispensable that a trait have two or more functions at the same time (Vollmer 1986).

For the eye such intermediate steps and multiple functions are well known because there are so many types of eyes. In other cases we must content ourselves with a scenario – that is, with a sequence of steps for how the appearance of a trait *might* have happened. And sometimes intermediate stages and double functions have yet to be found.

5. A helpful analogy: language

With respect to empirical testing, evolutionary epistemology faces two characteristic difficulties. First, it asserts a strong genetic component for cognitive faculties. In this it sides with the classical nativists (who were mostly rationalists, as were Descartes and Leibniz). For, if at birth the brain is a *tabula rasa*, as John Locke (1706, Book II, Ch. I, § 2) and other strict empiricists have it, if there is no strong innate component, then it remains a mystery how we can ever achieve knowledge. Evolution and genetics could not then be held responsible.

But specifying this supposed innate component is not an easy task. For how do we find out the cognitive inventory of newborn babies? They cannot talk, so we must rely on observations of behavior. But even the behavioral spectrum of newborns is rather limited. And things that a newborn does not master from the outset but only days, months, or years later may always be claimed to be acquired individually, hence to be due not to phylogeny but to ontogeny. Many experiments being, on principle, informative are banned for moral reasons. Thus nobody will intentionally prevent a baby from experiencing color or music merely to find out how it will develop without these stimuli. Therefore, it is very difficult to find conclusive evidence for the genetic preconditions of cognitive achievements.

Now, ethologists find help in comparing species (see also Chapter II, 4 by L Huber and A Wilkinson). Traits occurring in many species, especially if the latter are closely related, are supposed, in the sense of a legitimate working hypothesis, to be innate. Thus it is informative to investigate the cognitive achievements of our kin, the great apes. But they talk even less than human babies, and innate components are again difficult to spot.

Happily there is a fertile analogy to cognition: language. True enough, investigating language acquisition faces obstacles similar to those of investigating cognition. As with cognition, it does not suffice to look at the result – that is, at the different linguistic products or different languages. What is at stake is the *ability* to speak – that is, to learn a language, to use and to form it. Why do we speak? Why can humans do something no other animal can do? Is this due to a biological, hence genetic, hence phylogenetic component? If yes, what does it look like, and how did it arise?

If our language faculty has its origin in organic evolution – for a naturalist there is no doubt about that – then there must have been intermediate steps in this development. Unfortunately such intermediate steps are neither recent nor evidenced by fossils. But the comparison of languages gives at least some cues to innate elements. That is why it is enlightening that the evidence suggesting an innate component of language ability in humans has been strengthened in recent years, and it comes from research on language.

6. Creole languages

Sometimes members of a linguistic group live in an environment where other languages are spoken; they may be merchants, refugees, slaves, or inhabitants of a colony. In such cases they develop, in order to communicate, typical hybrid languages, so-called *pidgin* languages. (The word *pidgin* comes from the Chinese pronunciation of the English word *business.* However, the term is used for all such mixed languages.) Pidgin languages are quite simple in their vocabulary and even more so in their grammar. They do not count as full-blown languages.

Very often the children of such immigrants develop their own languages, called *Creole languages.* Originally the term *Creole* was used to describe the descendants of white Romance-language immigrants in all of South America (white Creoles) or the descendants of black slaves in Brazil (black Creoles). Nowadays, the term applies to all languages developed by immigrants of the second generation, whether on islands or in coastal areas of Middle America, West Africa, the Indian Ocean, or the Pacific region. Creole languages are complex languages whose vocabulary is drawn from totally different "mother tongues," mostly of colonists. Thus Jamaica-Creole rests on English, Guyana-Creole on Dutch, Haiti-Creole on French, Crioulo in West Africa on Portuguese. In their grammar, however, they are quite autonomous, dependent neither on the original language of the immigrants nor on the language of the "host country" (with respect to slavery, the expression *host country* seems inappropriate) or of the colonizers.

Since the colonies are quite separated and have no exchange, or nearly none, the Creole languages must have developed independently of each other. Yet in recent years linguists have discovered that these Creole languages are surprisingly similar in their structure – that is, in morphology and grammar (Bickerton 1983, 1984). How can this be explained?

Many traits common to all humans are explained as being genetically conditioned. If there is an innate language faculty, as claimed for a long time by rationalists and nativists and more recently by Noam Chomsky (1966) and Steven Pinker (1994), then there should be features common to all natural languages. The search for such linguistic universals has not been extremely successful; it has uncovered only very abstract principles. Creole languages, however, share quite concrete traits.

Not conclusive but at least suggestive, therefore, is the conjecture that these shared traits are due to a biological-genetic component. Precisely this is the claim of the leading researcher on Creole languages, Derek Bickerton (1983, 1984). According to him, the innate language component can develop freely only if it is not suppressed by corrections from outside, and just this is the case with Creoles: Their immigrated parents have not mastered the local language yet, and the children usually do not get a formal education. Therefore, the structural similarity of Creole languages of independent origin is evidence for the existence and influence of a strong genetic component in language ability.

7. Two more arguments: infants and deaf-mutes

This conjecture is supported by more recent findings. Children do not master their mother tongue immediately but start with characteristic *mistakes;* they use some kind of an "infant grammar," violating the respective "correct" grammar in many ways – for example, with respect to double negation and interrogative forms. According to Dan Slobin (1985–1997), these infant grammars are strikingly alike. What is more, they have very much in common with Creole languages. This suggests that infant grammar is partly innate. In most cases this innate grammar is overcome by the native tongue learned from outside – except with Creoles.

As much as these findings confirm Chomsky's thesis that there is an innate language acquisition device, they contradict another of Chomsky's conjectures. According to Chomsky all natural languages should fit into the innate linguistic structure (Chomsky 1966). According to Bickerton and Slobin, however, they do *not* fit this innate structure in every respect; that is why children make typical mistakes, and that is why this infant grammar could be detected (Pinker 1994).

New investigations with deaf-mute children point in the same direction. To communicate with each other these children develop an extensive system of signs and gestures. American psychologists have analyzed and compared such sign languages of deaf-mute children from America and Taiwan. They found that although these children had never met before, they gesticulated in stunningly similar ways that they could not have learned from their parents (Goldwin-Meadow and Mylander 1998). This is a kind of involuntary Kaspar-Hauser experiment: Since the deaf-mute children grow up without linguistic stimulation from outside, they have to develop such structures themselves. Here again, a biological-genetic explanation suggests itself.

Suppose such explanations are correct: What does that mean for our cognitive abilities? Language and cognition, though not identical, are closely intertwined: Without language there is no higher cognition, and without cognition language does not make much sense. For Chomsky, language is therefore a kind of probe giving us insight in the organization of mental processes. Thus the evolution of language faculty must have gone along with an evolution of cognitive faculties. If the one part is plausible, so is the other. No wonder then that Chomsky's disciple Steven Pinker wrote not only a book on language, but also one on thinking (Pinker 1997).

8. An argument for realism: the success of theories

In arguing from success we use success as evidence for the quality of a premise. Science is successful as far as it achieves its goals. And a scientific theory is called successful if it promotes our goals. Such successes corroborate the premises made by the respective theory. One fundamental premise of natural science, possibly of all empirical science, is *realism.* How can we argue for realism?

Often enough, it is the success of science that counts as the best argument in favor of realism. According to the early Hilary Putnam (1976), "the typical realist argument against idealism is that it makes the success of science a *miracle.*" In fact, the realist can explain the success of science, whereas the antirealist cannot. For if quarks and quasars really exist then it is no wonder that theories claiming or presupposing their existence are successful. If, however, these objects do not exist at all, how is it that with these theories we make correct predictions and solve so many more problems?

But even the success of science is of course no *proof* of realism. And vice versa:

The fact that idealism, positivism, instrumentalism, or constructivism, cannot explain something does not refute them. Still we may say that realism *explains more.* In theories of empirical science explanatory power is an important trait by which theories are judged. (Other traits are noncircularity, internal and external consistency, testability, and test success.)

True, neither realism nor its counterparts are theories of empirical science. They rather help us to do science and to interpret our results. But if we want to judge metatheories, metaphysical positions, methodological attitudes, and heuristic rules as well, we need criteria on this level, and then explanatory power plays again an important role. And by this criterion realism fares much better. Some philosophers even think that realism is historically *testable,* but they disagree on the question whether it has stood up to the relevant tests.

There is an important objection: Couldn't there be several ways to do justice to the same experiences? Could there not be empirically equivalent theories contradicting each other in their basic premises? Could we not – following radical constructivism or conventionalism – even work with arbitrary theories?

It is not easy to name concrete examples for empirically equivalent theories contradicting each other. Even so, such theories are thinkable. As a matter of principle, for a finite number of experiences or evidences an infinite number of empirically adequate theories may be constructed. Thus, from the success of a theory, we cannot derive its truth. Nor may we infer the truth of realism from its success. Is there a better argument for realism?

9. A better argument for realism: the failure of theories

Failure is the opposite of success. A theory fails if something runs counter to what the theory makes us expect. This applies on the theoretical level – for instance, with predictions, as well as on the practical level – let's say with bridges or tools. What we mean when ascribing success or failure to a theory evidently does not depend on our answer to the question of realism. This independence applies not only to the meaning of the concepts "success" and "failure" but also to the assessment of whether a prediction is confirmed or whether a tool works. There is no danger, then, that the realist will see successes where the antirealist sees none. Nor will the realist want to explain something where the antirealist does not see any problem or any need for an explanation.

Historically, there are more wrecked theories than successful ones. We are not aware of this because we care so little about wrecked theories. And we don't care because, in normal education, there is no time to teach, analyze, and criticize refuted theories. But what makes so many theories fail?

The antirealist has no answer to this question. He may describe the failure: He may say that the set of acknowledged observational statements has turned out to be inconsistent or that his tool did not meet his expectations. But these rewordings do not explain anything. They just say in what sense the theory has failed; they elucidate the failure accepted before. But they don't answer the actual question, they don't explain the failure.

For the realist, the answer is easy enough: A theory fails because it is wrong: The world is not as the theory submits. But to be different the world must not only exist but also have a specific structure that we may hit or miss. Thus realism explains not only the success but also the failure of theories. Even so, there is an asymmetry: For success there are more explanations, even non-realistic ones, but not for failure. The failure of theories is therefore a much better argument for realism, presumably the best one (Vollmer 2003).

10. An argument for natural selection: biodiversity

Natural selection is differential reproduction due to varying fitness. According to evolutionary epistemology, cognitive abilities raise fitness; therefore, selection works for better cognition, at least in cases where such improvements are useful, available, and not too expensive. As far as our cognitive ability is (taken to be) reliable – that is, in the mesocosm – we may explain this reliability by the effect of selection.

The fact that humans survived evolution under competition makes the reverse conjecture plausible, namely that our cognition cannot be too bad. This inverted argument is not altogether compelling; above all, it is not strong enough to exhibit our cognitive faculty as unfailing or to specify some bit of knowledge as certain. However, by this argumentative step we may justify our (limited) trust in our cognitive apparatus.

Obviously evolutionary epistemology makes constitutive use of the theory of evolution, primarily the principle of natural selection. Without natural selection, both argumentative possibilities mentioned above would escape us. For evolutionary epistemology it is therefore relevant whether this important factor of evolution is effective.

What testifies to the effectiveness of natural selection? Usually the multiplicity of species counts as the best argument. Wasn't it the different finches on the Galapagos Islands that aroused in Charles Darwin the idea of natural selection? And if we are told that there exist on earth at least five million, possibly even twenty million, different kinds of organisms (Kaplan 1985), not to count bacteria or viruses, all occupying their own ecological niches, then we are even more easily convinced of the effectiveness of natural selection.

But again there is an objection: Could there not be several, even many, ways to adapt to the same environmental conditions? Could not totally different species occupy the same ecological niche? Is it then not natural selection but mere chance and respective histories that determine which species are formed and populate the earth?

There are in fact arguments supporting this interpretation. We have cases where similar ecological niches are occupied by completely different species: The niche of the great pasture animals is occupied in the savannas of Africa by hoofed animals, in Australia by kangaroos. According to the neutral theory of evolution, developed since 1968 by Motoo Kimura (1924–1994, 1983), many genetic changes follow pure chance processes. From this slow and uniform "genetic clock" we can even determine the age of a species – that is, the time elapsed since it branched off its next relatives. Is organic evolution a mere chance process with natural selection playing a minor role or none at all?

11. A better argument for natural selection: extinction

Again, there is a better argument for the effectiveness of natural selection: the extinction of species. For that we must recall how many species have already died out. Evolutionary biologists take the number of extinct species to be at least one hundred times that of the existing ones (May 1992). Ernst Mayr (1988, p. 43) even guesses that 99.9 percent of all evolutionary lines are extinct, so that the recent ones are surpassed in number a thousand times. Why did so many species die out?

As with individuals it happens occasionally that species go extinct more or less accidentally, by a flood or by the impact of a meteorite. As with individuals we might talk here of situational death. It would be absurd, however, to classify all extinctions under situational death. In contrast to individual aging and dying, there is, as far as we know, no preprogrammed species extinction. Thus we must look for external causes in most cases. Hence we might also ask: What makes organisms, populations, and species fail?

For selectionists the answer is simple: Populations and higher taxonomic units die either because they can no longer adapt to environmental conditions, primarily when these change relatively fast, or because they are displaced by fitter organisms, possibly by superior members of the same species. Both cases demonstrate mechanisms of natural selection.

And how do antiselectionists, such as neutralists, explain species extinction? The answer is, not at all. The reason is not that they have problems with the concept of extinction. Even antiselectionists can state that species become extinct and feel a need to explain it. However, they cannot offer a plausible explanation. The theory of natural selection has more explanatory power than any antiselectionist theory, such as the neutral theory.

And now we may repeat the last section of chapter 9 nearly word for word: Selection theory explains not only the success but also the failure of species. Again, there is a pronounced asymmetry: There are more explanations for success than for failure. The failure of species is therefore a much better argument for the theory of selection, presumably the best one.

12. More arguments: analogy and convergence

It should now be obvious why we switched so abruptly from the epistemological problem of realism to a problem of evolutionary biology, for it enabled us to uncover a far-reaching analogy that we could follow right into its verbal formulations. It is tempting to use this *analogy* as a further argument. Then both conceptions, the realist one and the selectionist one, could support each other. The two arguments do not, however, depend on each other.

A further good analogy with mutual support is furnished by the phenomenon of *convergence*. In the development of science we find a phenomenon we could call convergence of research. There are several kinds of convergence: convergence of measurements, measuring methods, and theories. How do they come about?

Again, the antirealist has no answer, whereas the realist has a ready reply: Research converges because there are real structures that we may uncover and do indeed uncover slowly. This is what the realist rates as scientific progress. Here again the superior explanatory power of realism is remarkable.

Now, convergence is also observed in evolutionary biology. There it refers to similar traits that originated independently, such as the streamlined design of ichthyosaur, shark, tuna, and dolphin. Here the external conditions, especially the need to advance fast in water, have promoted this trait. The effectiveness of natural selection is especially conspicuous here. Again the analogy between the two lines of thought is striking. That the word *convergence* is used in both cases is not essential, but it makes the analogy all the more suggestive.

It is tempting to apply the concept of convergence not only to body traits but also to cognitive achievements. Thus we might say that our different sense organs supply us with a convergent view of the world – for example, when an apple is seen, felt, and tasted. The signals from the sense organs are different but are combined to build an undivided object in perception (see also Chapter I, 4 by W Singer). Similar considerations apply to higher cognitive achievements. This kind of convergence may be interpreted in favor of realism. For only if there are unified outside objects does it pay to reconstruct such objects in our imagination.

Outlook

Evolutionary epistemology has a programmatic character. It shares this character with the theory of biological evolution. This is due, first, to the fact that in both cases the evolution of the systems in question took a very long time: billions of years in the case of biological systems, millions of years in the case of cognitive systems. And it is due, second, to the fact that there are so many different systems to be studied. If we were aiming for completeness we would need descriptions and explanations of *all* relevant systems. Such completeness is neither possible nor necessary. What we really and realistically are after, are satisfying descriptions of and explanations for the main evolutionary steps of cognitive evolution. Of course, since we are humans, for us the evolution of man and his cognitive faculties is most interesting.

Unlike the theory of biological evolution, evolutionary epistemology is not finished when the evolutionary path of cognitive systems can be presented and explained. It also aims at evaluations of the *reliability* of these systems. How far may we trust our cognitive abilities? If we take them as reliable, how can we justify this? If not, why not? May we *improve* our cognition and how? Should we try it by artificial selection, by education, by brain enhancement? Will computers help? Thus, evolutionary epistemology, faithful to its name, combines evolutionary and epistemological approaches. It truly stands "between biology and philosophy".

Obviously evolutionary epistemology is intimately connected with other conceptions without being displaced by them: with realism and naturalism, with cognitive science and evolutionary theory, with the development of science and an evolutionary philosophy of science. This concatenation deserves to be further delineated. Why don't you try?

References

Bickerton D (1983) Creole languages. Sci Am 249 (July): 108–115

Bickerton D (1984) The language bioprogram hypothesis. Behav Brain Sci 7: 173–188

Brunswik E (1955) "Ratiomorphic" models of perception and thinking. Acta Psych 11: 108–109

Chomsky N (1966) Cartesian linguistics: a chapter in the history of rationalist thought. Harper & Row, New York

Goldwin-Meadow S, Mylander C (1998) Spontaneous sign systems created by deaf children in two cultures. Nature 391, no. 6664: 279–280

Gorgias, cited by Sextus Empiricus (ca. 200, 1998) VII 65 ff.

Kaplan R W (1985) On the numbers of extant, extinct, and possible species of organisms. Biol Zbl 104: 647–653

Kimura M (1983) The neutral theory of molecular evolution. Cambridge UP, Cambridge

Locke J (1706) An essay concerning human understanding. Fifth edition

May R M (1992) How many species inhabit the earth? Sci American 267 (October): 42–48

Mayr E (1988) Toward a new philosophy of biology. Harvard UP, Cambridge MA

Pinker S (1994) The language instinct. Morrow, New York

Pinker S (1997) How the mind works. Norton, New York

Putnam H (1976) What is "realism"? Proc Aristotelian Society 76: 177–194, p 177
Putnam H (1981) Reason, truth and history. Cambridge UP, Cambridge
Quine W V O (1969) Epistemology naturalized. In: Ontological relativity and other essays. Columbia University Press, New York, p 90 – Quine: Natural kinds. Ibid., p 126
Ruse M (1995) Evolutionary naturalism. Routledge, London
Sellars R W (1922) Evolutionary naturalism. Open Court, Chicago
Sextus Empiricus (ca. 200, 1998) Adversus mathematicos/Against the mathematicians. Clarendon, Oxford
Simpson G G (1963) Biology and the nature of science. Science 139: 81–88, p 84
Slobin D I (1985–1997) (ed.) The cross-linguistic study of language acquisition. Erlbaum, Mahwah
Vollmer G (1986) Die Unvollständigkeit der Evolutionstheorie. In: Vollmer G (1986) Was können wir wissen? Vol. 2: Die Erkenntnis der Natur. Hirzel, Stuttgart, pp 1–38
Vollmer G (1987) What evolutionary epistemology is not. In: Callebaut W, Pinxten R (eds.) Evolutionary epistemology: a multiparadigm program. Reidel, Dordrecht, pp 203–221
Vollmer G (1994) Was ist Naturalismus? Logos, N. S., 1: 200–219
Vollmer G (2003) Woran scheitern Theorien? In Vollmer G (2003) Wieso können wir die Welt erkennen? Hirzel, Stuttgart 2003, pp 89–120
Wittgenstein L (1922) Tractatus Logico-Philosophicus. Routledge, London

Sensory perception: adaptation to lifestyle and habitat

6

Friedrich G. Barth

Contents

Abstract

Sensory perception is a biological phenomenon and should be treated as such. It has a long history, beginning in the early stages of the evolution of life. Even bacteria depend on information gained about their outside and inside world. Basic mechanisms of sensing have indeed already been invented by the simplest organisms. Among the rich diversity of extant sensory systems are quite a few unknown to man. They give animals access to experiences alien to humans. A prominent characteristic of all sensors and sensing is their selectivity for the biologically relevant and their capacity to ignore most of the potential stimuli actually present. This implies that sensors and sensory systems are windows to the world with an extremely limited but well tuned transparency. They have not evolved to provide abstract truth but rather to achieve fitness (survival and reproduction). The intimate relationship between sensory systems and the habitat and lifestyle of a particular animal species is a logical consequence of this.

Friedrich G. Barth
University of Vienna
Life Sciences
Department of Neurobiology
Althanstr.14
1090 Vienna, Austria
e-mail: friedrich.g.barth@univie.ac.at

1. Introduction

The study of sensors and sensory perception addresses a large range of questions. The search for answers has mainly been driven by a particular interest in our own human perceptions. However, neurobiological findings are often most fascinating where they refer to creatures different from ourselves as humans and living with strange senses in sensory worlds alien to our own.

After a short introduction to some fundamentals this chapter will focus on a few select issues, viewed from a zoologist's point of view, and implying that an understanding of our own brain substantially depends on an understanding of what is going on in animals. There are sensors approaching perfection in the animal kingdom and sensors stimulated by forms of energy humans cannot perceive, such as ultrasound, infrared radiation, and electric fields, to name only a few. The principles of operation in many animal sensors represent interesting alternatives to established ways in which physics is applied to solve a sensory problem. While all these animal sensors are worthy of study in their own right they also shed light on our own human particularities. Like all creatures humans are a product of evolution. No sensory systems came from heaven but rather have evolved by a process of trial and error and by going through simple, progressive stages.

Senses and sensory perception have a *long history*. All forms of life depend on information about their exterior and interior worlds. There is no behavior which would make sense without such information which also provides a basis for motor control. In what follows we focus on the outside world, commonly referred to as the environment. However, it is important to stress that information about the inside world, the invironment, is of crucial importance as well. It is needed to guarantee homeostasis – by controlling body temperature and blood pressure, by keeping the osmotic balance right and by ensuring the proper regulation of metabolic processes. Feedback systems maintaining stable equilibria rely on a corresponding wealth of sensors looking into the organism and controlling its functions. Most sensors respond more strongly and promptly to changes of stimulus conditions than to constant conditions. In other words, organisms are much more interested in knowing about the dynamics of the physical and chemical conditions than their static conditions, which underpins both the relevance of the homeostasis concept and the importance of knowing about changes when executing behavioral programs. Sensory systems make closed-loop real-time control of behavior possible.

In biological sensing the goal *is never abstract truth but fitness*, which entails an organism's capacity to survive and to reproduce, and in more practical terms its ability to orient and to communicate, to move around, to escape from predators and to find prey and a mate.

With all this in mind the wealth of sense organs and sensory capacities found in the animal kingdom is not surprising. Sensory capacities already exist early in the evolution of life, for example, in bacteria, even archaea, which have stretch sensitive cell membrane channels important in keeping the osmotic balance (and internal pressure) right. In unicellular, eukaryotic protozoans like the ciliate *Paramecium*, one already finds highly developed sensory responses to mechanical (like touch) and other (like chemical) stimuli. There are even de- and hyperpolarizing receptor potentials and a sophisticated sensory guidance of oriented swimming behavior that exist despite the absence of any neuron or nervous system (Eckert 1972; Eckert et al. 1972).

Highly effective solutions of sensory problems have thus been found very early in the history of life, and many of the basic mechanisms seem to have been passed on to animal taxa positioned much higher in the phylogenetic tree, including humans. A sensory cell translates the properties of a stimulus into the language of the nervous system. Membrane channels regulating the ion flux across the sensory cell membrane are at the core of *sensory transduction* which describes the processes converting the different forms of stimulus energy to changes of the membrane potential. The membrane potential in turn decides on the occurrence (number and frequency) of nerve impulses sent to the next neuron and / or to the central nervous system. As expected there are variations in the molecular structures and functions which respond to different stimuli like light, sound, touch, and chemicals. However, there are also striking similarities across the animal kingdom and even between different sensory modalities. The diversity of animal sense organs serving the same stimulus modality (form of stimulus energy), such as vision or hearing or chemoreception, is not so much due to the diversity of the transduction processes but to that of the non-nervous structures (e. g. optical apparatus of eyes, sound transforming structures in ears, etc.) first picking up the stimulus from the environment and channelling it to the sensory cell proper, which typically is not directly exposed to the stimulus impinging on the body. The specialist defines these processes as *stimulus transformation* which differs fundamentally from *stimulus transduction* and represents the largest playground and richest soil for biologically applied physics. Here we most readily find proof for the adaptedness of biological sensors to the particular needs of the behavior of a particular species living in a particular habitat. Here a perfect match between sensory filters and the behavioral task is often found in the finest structural and functional details.

Plants have sensory capacities as well. They sense, for example, light, temperature and gravity. However, being autotrophic organisms able to convert solar energy into biomaterials such as sugars and fats by way of photosynthesis and typically living in the same spot for a lifetime, a plant's relation to the environment is much less demanding than that of an animal. *Animals are heterotrophic*, feeding on the energy pre-packaged in the bodies of other organisms. Typically, animals have to run around and search for sources of energy (in the last instance always provided by the plants, the primary producers) and a mate. Their highly developed muscular, sensory, and nervous systems have to be seen in this general context. Interestingly, animals that hardly move at all, such as filter feeding sponges, often look plant like. Their sensory capacities are comparatively limited and some of them appear like plants to the non-specialist and still carry erroneous plant names, like the sea anemone (a coelenterate) or sea lily (an echinoderm).

In the following a few examples will illustrate the intimate relationship between the properties of sensory systems and the environment and lifestyle of animals. The selectivity of sensory systems will first be shown by looking at stimulus transformation way out in the sensory periphery. Subsequently filter properties and selectivities of the brain will be considered. All along the way behavior will be the reference needed to understand the biological significance of particular senses.

2. Sensory periphery – physics, behavior, and ecology

2.1 Color vision

Sense organs are often called the windows of the nervous system to the outside (and inside) world. Although this is true in principle, the main issue here is that the transparency of these windows is very limited. Not only does a sense organ respond to a particular form of energy only – an ear to acoustic but not to electromagnetic waves, and an eye to electromagnetic but not to acoustic waves – but even within its modality it usually selects a small fraction only. To give an example: The wavelengths of light seen by animal eyes are roughly between

300 nm and 800 nm (0.3 to 0.8 µm). This is only a tiny fraction of the wavelengths of extant electromagnetic waves, which cover a range of more than 20 powers of ten, from gamma waves at the short wavelength (high frequency) end to radio waves at the long wavelength (low frequency) end. Even within the small range seen by animals, significant differences in sensitivity are found which can often be related to ecology and behavior.

One of the best known among such cases is the honeybee (*Apis mellifera*, Hymenoptera). The overall width of its spectrum of visible light is the same as in humans, but it is shifted by about 100 nm towards the short wavelength side. This implies that honeybees don't see light that we perceive as red but do see light in the ultraviolet (also known in other insects, fish and birds) which is beyond our short wavelength limit in the violet. Therefore to honeybees the colors of flowers look different from what we see (Fig. 1). For them the petals of a corn poppy flower (*Papaver rhoeas*) which are magnificently red for us would be black were it not for the reflection of ultraviolet radiation which we do not see. For bees and other insects the petals appear ultraviolet, a color alien to us. Similarly, the reflections of the flowers of the marsh marigold (*Caltha palustris*) could not be more yellow to our human perception. However, they reflect both in the yellow and in the ultraviolet range of wavelengths so that the bee sees a color referred to as "bee-purple". Of course we do not know what the color experience of a bee actually is like. However, behavioral and other evidence shows that they can certainly distinguish between different bee colors. The behavioral benefit that ensues from this lies in their ability to distinguish between the flowers of different species. This is the basis for effective pollination of the plants visited and an indispensable part of the "food for pollination deal" between insects and plants. Often flowers bear guide marks to the food (nectar, pollen) which are seen by the insects, because no ultraviolet is reflected by them. Because humans are ultraviolet-blind these guide marks, which are conspicuous to bees, do not exist in our visual experience.

Fig. 1 Color vision. The flower of the corn poppy (*Papaver rhoeas*) shown above, appears red to humans but "ultraviolet" for honey bees and other insects. The flowers of the marsh marigold (*Caltha palustris*) shown below, appear yellow to humans, but to a honey bee they appear in "bee purple". (photographs by author)

To be sure, wavelength and color are two different things. Color is the analyzed wavelength. Although it correlates with the input from sensory cells, it is the product of its integration in the central nervous system. Light of a certain wavelength is colorless. Its subjectively perceived color only exists in our minds and those of the bees and of other animals. However, particular spectral sensitivities of the photoreceptor cells (their wavelength selectivity) are the basis for a particular kind of color vision and the differences such as those between bee and man.

Contradicting our intuition, *the* color vision does not exist. Whether color is seen or not to a large degree depends on the behavioral context. In the bee it is mainly related to the flower visits, but not to visual motion detection, which is a "colorless" operation. Even when visiting flowers, chromatic patterns can only be resolved by the bee at very close range when it already sits on the flower or hovers above it, due to the limited spatial resolution of the bee's compound eye (Hempel 2001, 2002). Other properties of the eye add to the selectivity of the "sensory windows" in the periphery. One of them is the regionalisation and heterogeneity often found in insect (and other) eyes (Kelber 2006). A particular part of the compound eye may be specialized for a particular purpose such as polarization vision (see below) or pursuing a mate, or the distribution of the frequency of the occurrence of photoreceptor cells of a particular spectral type may change between the dorsal and ventral eye.

2.2 Polarization vision

When sunlight is scattered within the atmosphere or reflected from shiny surfaces like those of water or vegetation, it is partially polarised. Such polarised light represents a source of spatial information for many animals. Amongst these is again the honeybee, of which Karl von Frisch (1949) first showed that the polarisation pattern of the sky can be used as a compass cue for orientation, even if only a small patch of sky 10° to 15° wide is seen (von Frisch 1965). Only a small fraction (up to ca. 6 percent) of the bee's compound eye serves the acquisition of compass information from the sky's polarisation pattern (Wehner and Strasser 1985; Wehner and Labhart 2006). The relevant ommatidia look skywards and form the eye's so-called dorsal rim area. In bees the polarisation compass works exclusively with photoreceptor cells which have their peak sensitivity in the ultraviolet; it is therefore monochromatic and color-blind (von Helversen and Edrich 1974).

More recently another case of orientation using polarized light has been found in a spider (Dacke et al.1999). *Drassodes cupreus* (Gnaphosidae) has polarizing filters built into specialized secondary eyes (its posterior median or PM eyes) which look dorsally to the sky but do not form images because they lack a lens (Fig.2). The cornea of these eyes acts like a mere window. The light reflected from the tapetum makes these eyes appear brilliantly blue and is polarized in the direction of the eye's long axis. The built-in polarizer blocks light polarized perpendicular to the eye's long axis. The resulting preference for a particular e-vector orientation in each PM eye and the divergence of the preferred axes of the two PM eyes by 90° form the basis of the polarization analysis of the skylight which is thought to serve the spider when orienting itself home to its burrow after foraging trips. Remarkably, the spider's activity peaks shortly after sunset when the skylight appears particularly strongly polarized to its PM eyes. The main points of the story are the following then. Lacking a lens the stimulus transforming non-nervous optical structures of the PM eyes integrate over a large visual field. Another non-nervous structure, the tapetum, provides the polarizing filter. A comparison of the sensory input from the right and left PM eye will be needed at a higher stage of integration in the central nervous system when using it for polarization navigation (and not mixing up light intensity with direction of polarization). Nevertheless the selectivity and filtering in the periphery is a perfect example of elegantly applied biological physics, which makes the job of the central neurons considerably easier.

Drassodes belongs to a group of spiders with a tapetum of the secondary eyes de-

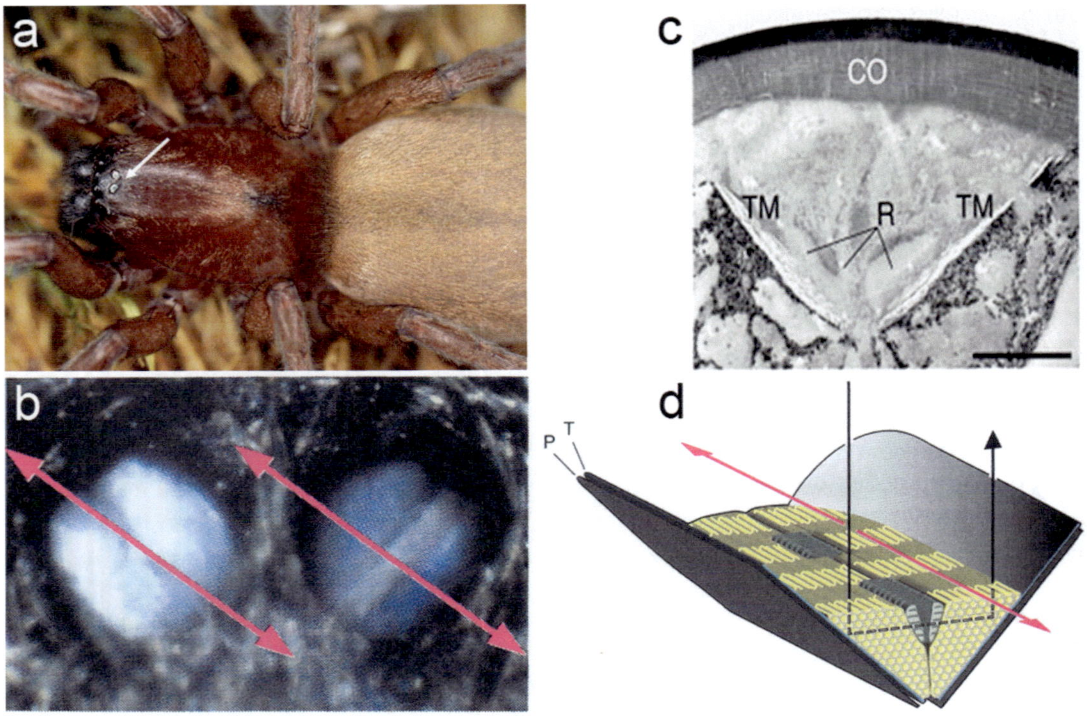

Fig. 2 Polarization vision. **a** The ground spider *Drassodes cupreus* (body length about 1.5 cm). **b** The two posterior median eyes viewed from above under plane-polarized light with the angle of polarization indicated by red arrows. The light reflected from the tapetum is polarized in parallel to each eye's long axis by a polarizer in the eye which is the reason why the right eye remains dark. **c** Section through eye showing its cornea (CO), the two flat mirrors formed by the light reflecting tapetum (TM), and the receptive segments of the visual cells (R). Scale bar 20 µm. **d** Arrangement of the so-called canoe-shaped tapetum (T) (the built-in polarizer) and the visual cells of the retina (yellow and green); black arrow indicates path of light through the eye; red arrow shows direction of polarization parallel to the tubular receptor cell microvilli which implies maximal sensitivity (light absorption). (*a* photograph by Jorgen Lissner; *b, c,* and *d* from Dacke et al. 1999)

scribed as canoe-shaped. The name describes the two flat mirrors forming an angle of 90 to 100° between them which the tapetum consists of (Homann 1928, 1951; Dacke et al. 1999). It is possible that the strange shape of the tapetum serves as polarizing optics in other spiders as well.

2.3 Vibration detection

Most spiders live in a world of vibrations (Barth 2002).They not only use substrate or web vibrations as guides to prey but also actively communicate using self-generated vibratory signals during courtship and male competition. This has been studied in details in the Central American spider genus *Cupiennius* (Ctenidae) (Barth 1997, 1998, 2002). These spiders are wandering spiders which do not live in a web. Instead their vibratory communication with a prospective mate and the detection of prey generated vibrations occur on their dwelling plants (e. g. bromeliads) where they hide in retreats between the leaves during the day and come out to court and prey at night (Fig. 3 a). As in all sensory systems a fundamental problem is to detect and recognize the biologically meaningful signal (vibration) among a lot of background noise. In the given case the background vibrations of the plant due to abiotic sources contain very low frequencies only (below 10 Hz) at the time of the nocturnal activity of *Cupiennius*. Both the courtship and prey generated vibrations (e.g. vibrations produced by a cockroach crawling around on the dwelling plant) contain significantly higher frequency components (male and female courtship signals: main components around 100 Hz and 40 Hz, respectively).

In accordance with this the sensitivity of the vibration sensors on the spider legs is low at low frequencies up to 10 to 30 Hz and increasingly high at higher frequencies, where the vibration amplitude of the tarsus (last segment of the leg) is as low as 10^{-2} to 10^{-3} µm at threshold (Geethabali and Barth 1982; Molina et al. 2009) (Fig. 3b, c). The vibration sensor has prominent high pass characteristics. Although the unusual shape of the threshold curve makes sense from a behavioral and ecological point of view, its origin has for a long time remained enigmatic. Most recently it could be demonstrated

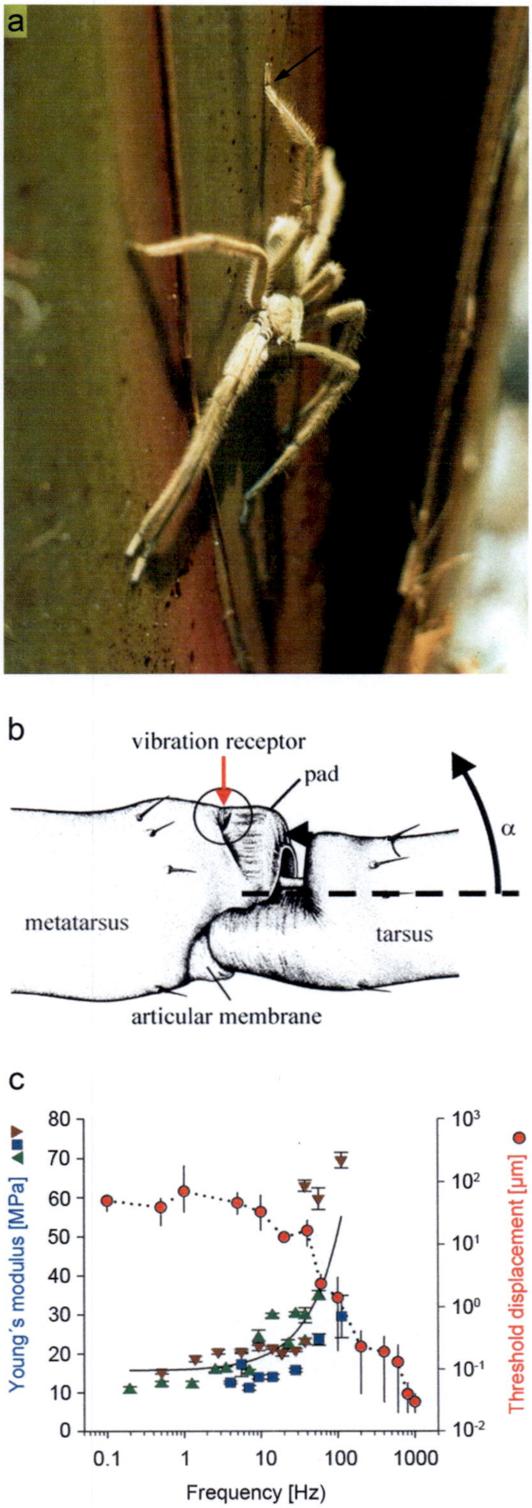

Fig. 3 Spider vibration sensor. **a** A male of the wandering spider *Cupiennius salei* sitting on the pseudostem of a banana plant; arrow points to location of vibration sensor found on all legs. **b** Functional morphology of natural stimulation of the vibration sensitive organ (see red arrow) by the upward movement (see long curved arrow) of the leg's last segment (tarsus), and the subsequent introduction of force (arrowhead) into the cuticular pad which eventually leads to a compression of the vibration sensitive slits making up the organ. **c** Young's modulus (for scaling see left y-axis) of the cuticular pad seen in *b* and tarsal threshold displacement (for scaling see right y-axis) necessary to elicit a nervous response as a function of stimulus frequency. (*a* photograph by author; *b* modified from Barth and Geethabali 1982 with permission, Springer; *c* modified from McConney et al. 2007, with permission, Royal Society)

that the material properties of a tiny cuticular pad in front of the organ perform the function. Again a non-nervous stimulus transmitting structure way out in the sensory periphery forms a highly selective filter enhancing the signal to noise ratio.

The pad works as follows: When the spider's plant is vibrated the tarsus (last segment of the leg) is moved up and down thereby pushing against the distal end of the metatarsus (the second last leg segment). More precisely, it pushes against the cuticular pad in front of the vibration sensor. According to micromechanical measurements using atomic force microscopy (McConney et al. 2007) the pad material has viscoelastic properties. Its stiffness (Young's modulus of elasticity) increases rapidly beyond frequencies of ca.10 to 30 Hz. This implies that vibrations containing higher frequency components are passed on to the sense organ with increasing efficiency (decreasing energy loss). The graph describing the frequency dependence of Young's modulus nicely mirrors the electrophysiologically determined threshold curve (Fig. 3 c). The correlation seems perfect (Fratzl and Barth 2009). In order to see whether it also reflects a causal relationship we made the pad stiffer by cooling it. The predicted increase in sensitivity could indeed be demonstrated even with temperature drops like those occurring naturally (Vogel 2008; Vogel and Barth 2011). Remarkably, this finding nicely links a small cuticular pad on the legs to both behavior and ecology. When the night active spider *Cupiennius* leaves its retreat after sunset, temperatures are lower than during the day. Its vibration sensors are, therefore, more sensitive just at the right time and for the biologically relevant signals.

As in the other cases presented, the biological application of physics seems perfect and reflects the biologically relevant stimulus patterns in surprising detail.

3. Senses and the brain

To summarize, it has become clear by now that sense organs are windows of very limited transparency and that, therefore, the sensory worlds animals (including man) live in are extremely limited compared to physical reality. A major source of the filter properties of sense organs are their auxiliary structures and the processes of stimulus transformation. They contribute substantially to peripheralize information processing, to keep out improper stimulation from the nervous system and to concentrating on what is biologically important. We have also seen that there are many more senses in animals than the five classical ones traditionally attributed to humans. Apart from quantitative differences between the physical properties of stimuli and what we experience, there are also qualitative differences. Electromagnetic waves are experienced as colors, pressure waves as sound; color, sound, odor etc. are all mental constructions which do not exist outside the brain. What we experience as the outside world is to a large extent the experience of ourselves. There is good reason to assume that this situation is also true in non-human animals.

We now enlarge the picture by including a few aspects pertaining to the central nervous system where the information provided by the sense organs is integrated and analyzed, ignored or stored (as memory), and where the decisions are made about what to do or not. It will become even more evident that sensory signals take on a meaning only in the light of behavior and habitat. In the brain there is again alot of weighting and filtering of information. Often there is even a surprisingly influential control of the sense organs by the central nervous system which to a large degree then tells them what to sense and what not.

3.1 Perception as an active process

A few examples will show that sensing is not the result of mere passive accumulation of inevitable sensory stimulation, but that instead it is often a remarkably active process.

The water bug *Notonecta glauca* (Hemiptera) has a regionalized eye with ommatidia specialized for the detection of the polarization of light located ventrally. This is different from the honey bee with its dorsal rim area, but makes sense as well. The water bug does not use the compass information contained in the polarization pattern of the sky above, but detects water bodies below by the light reflected from their surface, even when flying under low light conditions at crepuscular times. Such light, reflected from the air-water interface, is horizontally polarized and the photoreceptor cells ventrally in the compound eye are arranged in a way that leads to maximum sensitivity for just that horizontally polarized light. More precisely, the receptor cells are sensitive to UV light and their microvilli (which contain the photosensitive pigment) optimally oriented for the reception of horizontally polarized light. The biological background is explained below (Schwind 1983).

Although bound to water for most of the time, *Notonecta glauca* flies in search of new habitats. On such dispersal flights it exhibits a remarkable behavior of active sensing and use of physics (Schwind 1983, 1984, 1991). When descending on to the water surface the bug abruptly tilts its long body axis upwards for a moment before diving into it. The tilt is such that the polarization sensitive ventral eye area forms an angle of 53° with the water surface. This behavior turns out to be very clever because as the Scottish physicist David Brewster (1781–1868) revealed, at an angle of incidence of 53° at the air-water interface the light reflected from the surface is maximally linearly polarized (and the angle between the reflected and the refracted light 90°). The water bug seemingly checks the presence of the water surface according to its polarization properties. Unpolarized light does not elicit the bug's tilting response even when it is much brighter. The bug can be fooled, however, by an artificial surface reflecting light like a water surface prompting it to crash. Whereas in *Notonecta* the diving behavior is only elicited by ultraviolet light, it is the blue or green range in other aquatic insects that prompts a plunge reaction (Schwind 1995).

3.2 Three dimensions out of two

There is ample evidence that our human perceptions are to be understood as active processes, with the brain as interpreter. Our perceptions to a large extent are inventions of the brain, although at the same time the brain is at the mercy of the sensory organs. Two examples shall make this point clear.

The first is the perception of two-dimensional pictures in three dimensions as seen in Salvador Dali's painting of his sister Anna-Maria (1925) (Fig. 4). The projections of the more distant windows onto the retina of our eyes obviously are smaller than those of the closer windows. Nevertheless our brain interprets these windows to be of the same size, using logical rules in the unconscious which it has incorporated during evolution and the individual's early experiences. Two-dimensional drawings with lines converging in a way exemplified by Dali's painting, an increasing density of contours and other characteristics make the brain instantly convert the two-dimensional scenery into a three-dimensional one (see also Fig. 5 c). Biologically, the illusion makes more sense than would the physical truth of two-dimensionality, and this is not only true when looking at pieces of art. The subjective size constancy provided by the brain is biologically speaking highly relevant, and may indeed be life

Fig. 4 Three dimensions out of two. Painting by Salvador Dali (1925) showing his sister Anna-Maria. (Salvador Dali "Bildnis meiner Schwester, 1925" / CR Foundation Gala – Salvador Dali/VBK Wien 2010)

saving when, for instance, an enemy is not misinterpreted as tiny and harmless just because of its distance from its potential prey.

The well known Necker cube (Fig. 5 a), named after the Swiss crystallographer Louis Albert Necker, who first presented it as early as 1832, is a line drawing of a cube in isometric perspective with the parallel edges of the cube drawn parallel and so not showing which line is in front or behind, where lines are crossing. The drawing is ambiguous because there are two possible perspectives which flip back and forth. The brain obviously tests both hypothetical perspectives but finds no clue to help it decide for one or the other. Interestingly, the view said to be preferred by the majority is the more common one from above, with the corresponding cube face in front. The Necker cube is often used to support arguments against naïve realism. Indeed, the cube we see is not even there but created by our brain out of 12 lines drawn in two dimensions.

Another well known illusion supporting our main point is the Kanizsa triangle (Fig. 5 b). Again the visual system interprets parts of a picture in a way that seems to make sense, although they are not actually present. Gaetano Kanizsa was an Italian psychologist who presented a "triangle" consisting of illusionary contours in 1955 (Kanizsa 1955). Such illusions are used to support the Gestalt theory of perception. Interestingly, in a similar manner honeybees see "what is not there" under similar conditions (see Chapter II,7 by M Giurfa).

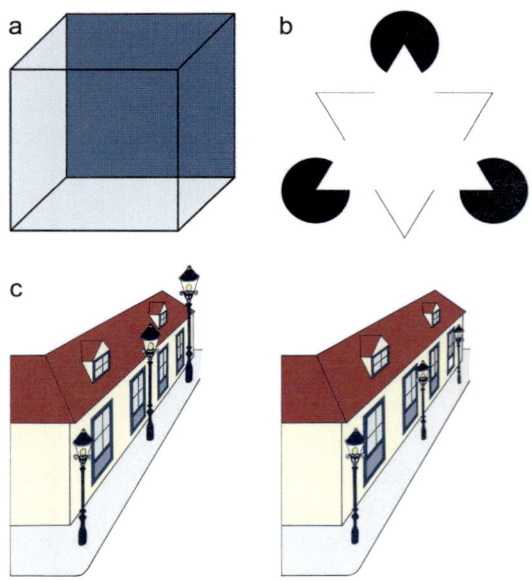

Fig. 5 The brain's active role in perception. **a** Necker cube; **b** Kanizsa triangle; **c** street scene. See text for further explanation. (modified from Barth 1989, with permission Wissenschaftliche Gesellschaft an der Universitaet Frankfurt)

3.3 Efferent control

The water bug is a particularly impressive example of active sensing, though there are many other examples. When we turn and look at something of special interest or focus our attention to the conversation at the neighboring table in a restaurant, this is also active sensing (and at the same time narrowing the sensory field of "view") and telling the sensory systems what to see and hear.

3.3.1 Nervous

The non-specialist may be surprised to hear that the central nervous system not only receives input from the sensory organs but also controls that input. In the vertebrate visual system, for example, there are more efferent than afferent nerve fibers.

Efferent control of sensory organs now seems to be the rule rather than the exception. One of the most striking examples is the circadian change in sensitivity of photoreceptor cells. In the horseshoe crab *Limulus* a circadian oscillator activates efferent fibers in the optic nerve at night. The anatomical and physiological changes induced by them result in an increase of the retina's sensitivity by roughly five orders of magnitude (Batra and Barlow 1990). Adaptations of organisms to planetary periodicities, clock induced daily and seasonal changes, are universal, reflecting the evolution of life under the influences of these temporal changes. As in the case of *Limulus* vision at night is made possible in many other animals by endogenous clock-induced adaptations (Fleissner and Fleissner 2006). As early as 1864 F. Leydig observed rhythmic changes in insect eye structures in constant darkness, indicating that they are endogenous and not induced by changing exposure to light. Among the most intensively studied cases is the scorpion *Androctonus australis*, a night active scorpion (Fleissner and Fleissner 2001, 2006). Its visual sensitivity is also controlled efferently: Due to the activity of efferent neurosecretory fibers its median eye sensitivity changes by three logarithmic units in a day-night rhythm even at constant darkness.

Yet another case is the efferent control of hearing in mammalian ears by the modulation of the basilar membrane motion (Cooper and Guinan 2006). Efferent olivo-cochlear neurons innervate the *outer* hair cells in the cochlea (Fig. 6). Although these cells are not the ones sending information about acoustic stimuli to the brain, they contribute substantially to what the *inner* hair cells tell the brain. Olivo-cochlear neurons control the sensitivity of the peripheral auditory system using the outer hair cells as mechanical effectors whose mechanical activity (contractions) enhances the sound-induced vibrations of the basilar membrane (on which both the outer and inner hair cells sit) in particular frequency ranges. The effect is a remarkable increase of the ear's spectral resolution and, therefore, its ability to distinguish between frequencies. In technical terms one would say that the frequency filter has become much narrower (corresponding to a narrowed spatial range of vibration along the basilar membrane). Latest findings point to the presence of an additional mechanism mediated by the olivo-cochlear neurons which is an inhibition of the mechanical basilar membrane response (Cooper and Guinan 2006).

3.3.2 Hormonal

We primarily associate hormones with the control of metabolism, development and growth, and reproduction, but less with nervous activity. However, there is close interaction between the endocrine and the nervous system. In invertebrates the majority of hormones are indeed neurohormones, secreted by specialized neurons. However,

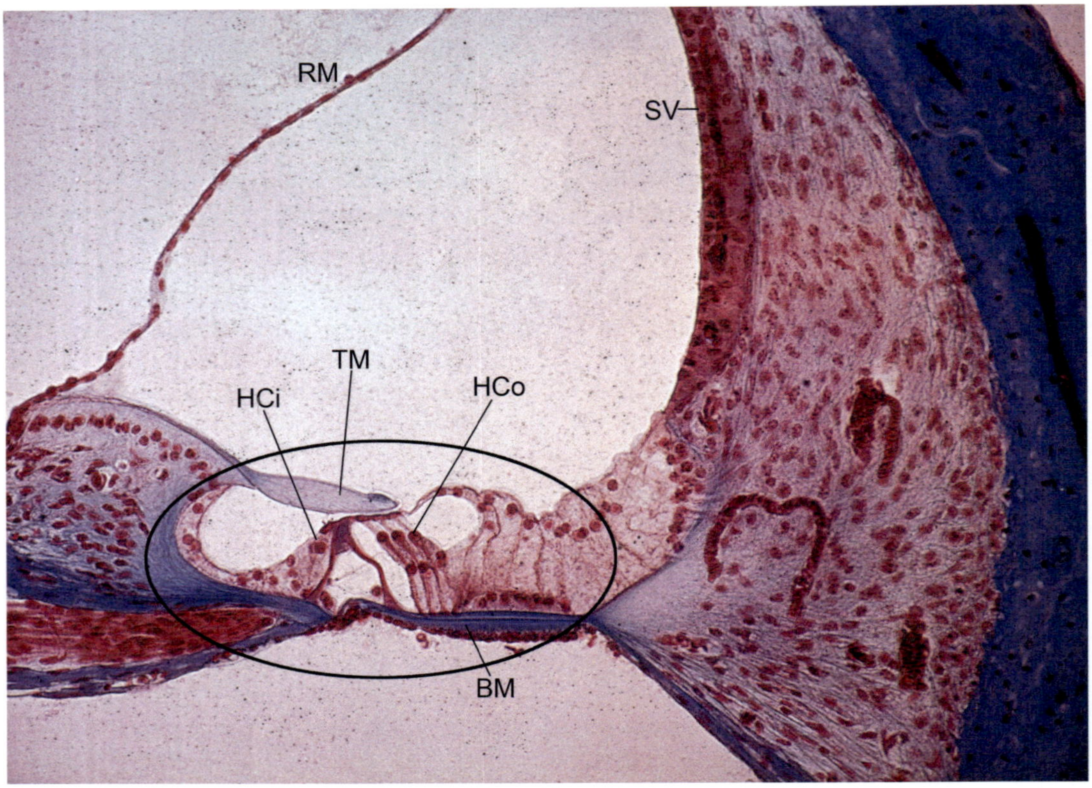

Fig. 6 Mammalian inner ear. Histological section showing details of the sensory apparatus in the cochlea of a guinea pig (*Cavia*). HCo, outer hair cells; HCi, inner hair cells; TM, tectorial membrane; BM, basilar membrane; RM, Reissner membrane; SV, stria vascularis. See text for explanation. (modified from M. P. Kage, W. Nachtigall 1980 Faszination des Lebendigen, Herder Verlag, with permission, M. P. Kage)

here attention shall only be drawn to an additional dependence of sensory input: Hormonally induced state changes in an organism are often paralleled by changes in the responsiveness of sense organs to internal and external stimuli.

Thus changes in olfactory and auditory sensitivity have been known to fluctuate across the female menstrual cycle (Navarrete-Palacios et al. 2003). Olfaction is most sensitive around the time of ovulation and significantly less so during other phases of the menstrual cycle. Similarly, the visual sense changes in color discrimination during the menstrual cycle (Eisner et al. 2004; Giuffre et al. 2007).

Weakly electric fish (Gymnotiformes) have a peculiar sense for the electric fields they produce themselves by their electric organ discharges (EOD). The pulsed self-produced electric fields are used for orientation and communication, and to this end the weakly electric fish have electroreceptor organs widely distributed over their body surface. They detect changes of the field due to objects of either higher or lower conductivity than that of the surrounding water. In the so-called wave-type fish the EOD has a sinusoidal time course and is discharged at a frequency which is under hormonal control and higher in males than in females and juveniles (Fig. 7). By the application of testosterone the discharge frequency can be lowered experimentally. The electrosensors are closely tuned to the fish's own EOD frequency, and this tuning is mediated by steroid hormones which shift the receptor's best frequency according to the EOD discharge frequency (Bass and Hopkins 1985; Zakon 1987, 1998).

3.4 Sensory maps

How is the outside world encoded in sensory signals represented in the brain?

3.4.1 Filtered sensory information is projected onto maps

Of such maps the most familiar is the "somato-sensory homunculus" on the gyrus postcentralis of our cerebral cortex. This map allocates tactile sensitivity (more precisely its spatial resolution) of different skin areas to different brain areas. Areas with high tactile sensitivity like our lips or thumbs correspond to particularly large cortical areas which form kinds of central nervous foveae supporting the concept of filters matched to particular functions. Corresponding maps are found for the other senses like vision and hearing as well.

To qualify as a map the spatial relationship of the position of a stimulus in the sensory space has to be reflected by the position of corresponding neurons on the map. The tactile map then is a topographically ordered, though distorted, representation of the environment, and is called somatotopic. Most likely these maps (termed retinotopic in the case of vision) play a particular role in oriented behavior. However, the map does not always reflect the location of a stimulus source in space. In the acoustic system it is the frequency corresponding to its tonotopic representation in the cochlea. In the barn owl, for example, which uses its acute sense of hearing to localize prey in the dark, the map represents nervously "calculated" values like time of arrival differences or sound intensity differences between the two ears (Konishi 1986, 2003; Wagner et al. 1987). As in the sensory organs the most behaviorally relevant aspects of a stimulus receive particular "attention" and space in the central nervous map (like tactile stimulation of lips and thumbs, echolocation frequencies of bats, relevant time differences in owls etc.).

3.4.2 Different from what a geographical map suggests, the sensory maps laid down in the central nervous system are not always static phenomena.

Change may not only occur depending on the intensity of usage; to use or not to use parts of the receptive field of a sense organ substantially influences the degree of its central nervous representation (Fig. 8) (Jenkins et al. 1990). A map´s position may routinely change depending on behavior and on other stimulus modalities (Sparks 1988). In the monkey (*Macaca*) eye movements induce a shift of midbrain auditory receptive fields of neurons involved in the orientation towards an acoustic stimulus from the left to the right Colliculus superior. This tells us that these sensory maps are organized according to motor coordinates; the auditory neurons use oculomotor coordinates. The absolute position of the goal (a loudspeaker in the experiment) is less important than its position relative to that of the eyes.

3.4.3 An exciting aspect of sensory maps in mammals is their multiplicity

The various senses have not just one but from five to 15 representations in the brain. Each of these multiple maps represents specific invariants of a stimulus, which may also differ in their relationship to behavior and

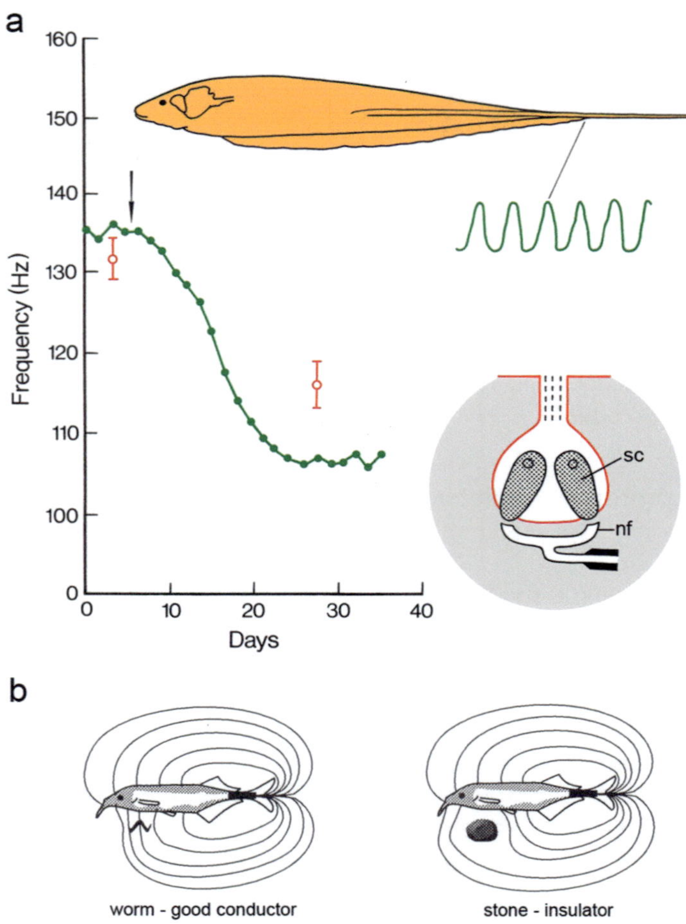

Fig. 7 Efferent control by hormones. **a** Drop of discharge frequency of the electric organ following the application (black arrow) of dihydrotestosterone in a female weakly electric fish (*Sternopygus*). Values in red show simultaneous drop of the frequency to which the electroreceptors are tuned. The oscillation drawn in green below the fish represents the electric discharge of the the electric organ located in the tail region. Inset: the electroreceptor, the so-called tuberous organ, in the skin of the fish; sc, sensory cells; nf, nerve fibers. **b** Schematic drawings showing how an object changes the electric field produced by the fish depending on whether it is a good conductor (left) or an insulator (right). (**a** modified from Zakon 1987, with permission, Elsevier; **b** from von der Emde 2001, with permission, Springer)

thus have different meaning (like one eliciting a shift of one's view, the other affecting emotions). This is best studied in the visual system. Importantly, the stimulus pattern is taken apart and its individual parameters like shape, color and motion are processed separately by the brain. As Semir Zeki (1992) put it: "There is no single central area, to which all other areas would report exclusively, neither in the visual nor in other sensory systems." This type of parallel processing can be nicely demonstrated using non-invasive methods to monitor brain activity and also by studying the often very specific and limited mental deficits of stroke and other patients (Zeki 1992).

The fact that properties of the outside world have parallel representations in the brain, both on the sensory and the motor side, has significant implications. (i) One is the *binding problem*, that is the question of how the representations all come together again and how the holistic nature of our experiences can be explained. One major explanation refers to the temporal synchronization of neurons widely distributed in the brain (see chapter I, 4 by W. Singer). (ii) The other implication is that the older concept of a sequence from sensation to Gestalt to idea is replaced by a model of multiple parallel functions, in which all parts of the brain dealing with the same function cooperate.

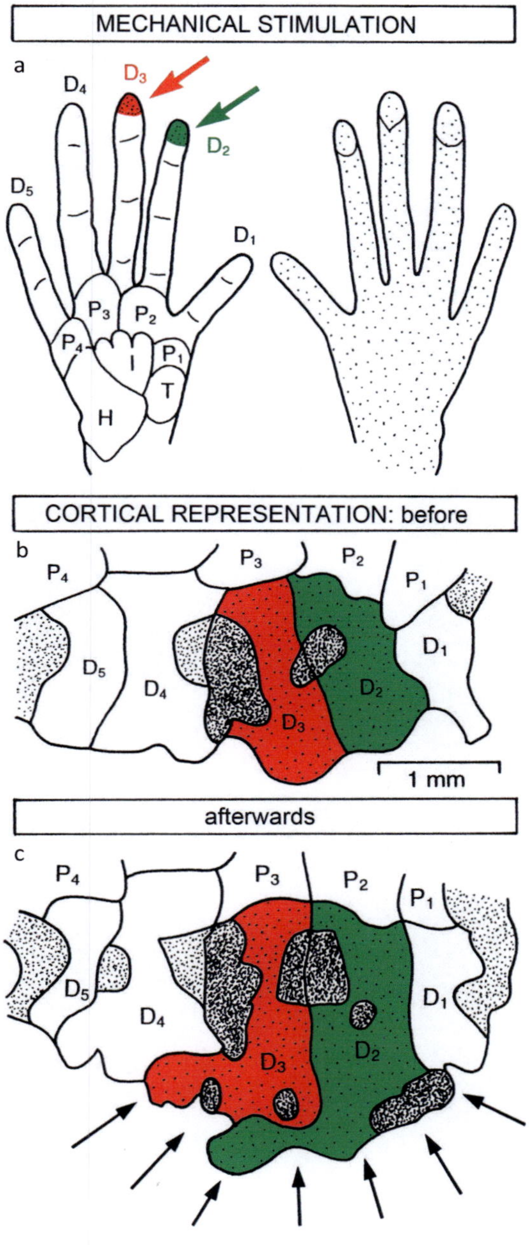

Fig. 8 Sensory maps. **a** Activity-dependent plasticity of cortical representation of an adult owl monkey's hand following continued tactile stimulation of digits D2 and D3. **b** Normal representation of the individual fingers in the somatosensory cortex preceding selective stimulation. **c** After several months of selective stimulation the central nervous representations of the stimulated fingers are enlarged. (modified from Reichert 2000, after Jenkins et al. 1990)

4. Evolution and adaptedness – Darwin's legacy and a few clarifications

Having pointed often to the fit between sensory systems on the one hand and behavior and environment on the other some thoughts on "adaptedness" shall be added in this final section. Adaptedness, an uncommonly used term, focuses on being adapted instead of on the historical process of becoming adapted; it addresses the present role of sensory systems or parts of them and their utility to the individual organism. Throughout this chapter I had in mind process based fitness, convinced that a particular trait of a sensory system can only be classified as an adaptation or adaptedness if its function is identified.

Of course there is no escape from evolutionary thought when we try to understand the emergence of adaptedness and its relation to environmental selection pressures. However, the statement by Theodosius Dob-

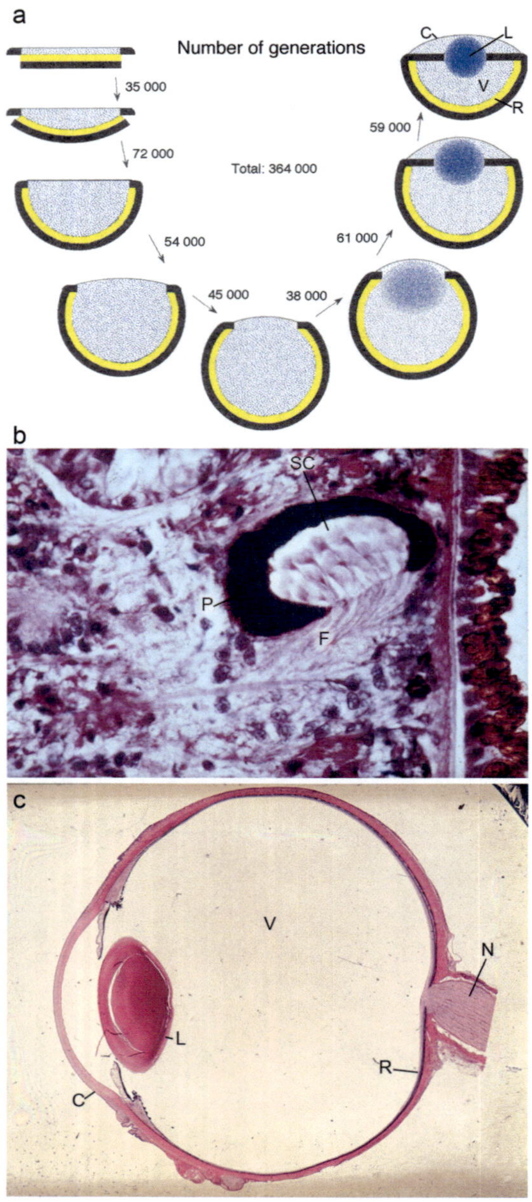

Fig. 9 Evolution and adaptedness. **a** Stages of a modelled sequence of eye evolution starting with a light sensitive flat patch (yellow: photoreceptor cells; blue: protective layer; black: dark pigment). The sequence ends with a focused lens eye characterized by a large invaginated retina (R), a small aperture and a spherical lens (L) with high refractive index and a focal length equal to the distance from the retina. C, cornea; V, vitreous body. The numbers represent generations estimated to be necessary to accomplish the steps between the stages. **b** A simple pigment-cup eye of a flatworm (*Planaria*) located slightly below the body surface. Sensory cells (SC) are seen inside the pigment cup (P) and their axons (F) leaving it. **c** Section through a mammalian camera-type lens eye (mouse). C, cornea; L, lens; V, vitreous body; R, retina; N, optic nerve. (**a** from Land and Nilsson 2002, with permission, Oxford University Press ; **b** and **c** from author)

szansky (1973) that "Nothing in biology makes sense except in the light of evolution," is short sighted and too often quoted carelessly and without reflection. With the same justification one can say that "no aspect of life can be understood without consideration of its physiology; the variants in a population should primarily be understood as physiological experiments" (see also Turner 2007).

In his "Origin of Species" (1859) Charles Darwin was puzzled by the refinement of the human eye. He wrote "To suppose that the eye with all its inimitable continuances could have been formed by natural selection, seems, I freely confess, absurd in the highest degree". How could such impressive design come about without a designer? Following the stringency of logic in his theory of natural selection, however, Darwin concluded (bravely, considering the biological knowledge available at his time) that "Reason tells me that if numerous gradations from a simple and imperfect eye to one complex and perfect can be shown to exist, each grade being useful to its possessor...; if further the eye ever varies and the variations be intended...., and if such variations should be useful.... under changing conditions of life, then the difficulty of believing that a perfected complex eye could be formed by natural selection, though insuperable by our own imagination, should not be considered as subversive of the theory." As it turned out much later it is not subversive at all. Even the habit of disguising lack of knowledge by saying that there were millions and millions of years for a countless number of small steps of improvement needs revision. In their pioneering study Nilsson and Pelger (1994) presented evidence that there need only be 2 000 sequential modifications by one percent in length, width, or protein density to turn a flat patch of light sensitive epithelium into a focussed camera-type eye, provided there is continuous selection for improved spatial vision. Under a few conservative assumptions the authors concluded that the process may be very fast and could indeed have been accomplished in less than 400 000 generations, even with changes from generation to generation of much less than 1 per cent (Fig. 9) and of course without the need for "intelligent design". Assuming one year for the sequence of generations the authors concluded that the formidable vertebrate type of eye may have evolved within less than half a million years, which helps greatly in explaining the rapidity of eye evolution during the Cambrian Explosion (see also Land and Nilsson 2002).

Recent data from Darwin's finches, the true emblems of evolution, underline the surprising speed at which evolution may occur (Grant and Grant 2006). A species of Darwin's finches in a population on one of the islands of the Galapagos archipelago evolved smaller beaks within two decades, this major shift even having occurred within just one generation. When the large ground finch (*Geospiza magnirostris*) arrived on the island of Daphne Major in 1982, the medium ground finch (*Geospiza fortis*), long-time resident on this tiny island, evolved a smaller beak as a result of competition for food. With the smaller beak the finches were more suited to small seeds whereas the much larger newcomer finches with their stronger beaks were mainly interested in the larger seeds. In this way competition among the two species was reduced. In difficult times such as a drought (as in 1977) or unusually rainy weather (as in 1984–85), which are episodes of strong selection, either the large and tough seeds or the small and soft ones are predominant. Under such circumstances the surviving birds and their offspring tend to be bigger with larger and stouter beaks or smaller beaks (better adapted to eat smaller seeds), respectively. Evolutionary biologists refer to this phenomenon as character displacement (Grant and Grant 2006). Obviously natural selection not only operates over vast periods of time but sometimes the evolutionary course of diversification can be directly observed within one or two generations only. The story of the finches also tells us that the identification of an adaptation relies on the identification of function which in turn needs the determination of the specific selection pressures at work. These do not work single-sided, but also depend on the animal's interaction with them. There is a wise saying that "Structure without function is a corpse, function without structure is a ghost." (Vogel and Wainwright1969). The two have to be brought together by integrative biology to include knowledge of the intact and behaving organism.

References

Barth FG (1989) Vom Sinn der Sinne. Sinnesorgane zwischen Umwelt und Verhalten. Sitzungsber Wiss Ges Univ Frankfurt am Main, Franz Steiner, Wiesbaden, 1–38

Barth FG (1997) Vibratory communication in spiders: Adaptation and compromise at many levels. In: Lehrer M (ed) Orientation and communication in arthropods. Birkhäuser, pp 247– 272

Barth FG (1998) The vibrational sense of spiders. In: Hoy RR, Popper AN, Fay RR (eds) Springer Handbook of Auditory Research. Comparative hearing: Insects. Springer, New York, pp 228–278

Barth FG (2002) A spider's world. Senses and behavior. Springer, Berlin Heidelberg

Barth FG, Geethabali (1982) Spider vibration receptors. Threshold curves of individual slits in the metatarsal lyriform organ. J Comp Physiol A 148: 175–185

Bass AH, Hopkins CDC (1985) Hormonal control of sex differences in the electric organ discharge (EOD) of mormyrid fishes. J Comp Physiol A 156: 587–604

Barlow RB Jr, Chamberlain SC, Levinson JZ (1989) Circadian rhythms in the invertebrate retina. In: Stavenga D, Hardie R (eds) Facets of vision. Springer, Berlin Heidelberg New York, pp 257–280

Batra R, Barlow RB (1990) Efferent control of temporal response properties of the *Limulus* lateral eye. J Gen Physiol 95: 229–244

Cooper NP, Guinan JJ Jr (2006) Efferent-mediated control of basilar membrane motion. J Physiol 576/1: 49–54

Dacke M, Nilsson D-E, Warrant EJ, Blest AD, Land MF, O'Carroll DC (1999) Built-in polarizers form part of a compass organ in spiders. Nature 401: 470–473

Darwin CR (1859) On the origin of species by means of natural selection or the preservation of favoured races in the struggle for life. Murray, London

Dobshanzky T (1970) Genetics of the evolutionary process. Columbia University Press, New York

Dobshanzky T (1973) Nothing in biology makes sense except in the light of evolution. The American Biology Teacher 35: 125–129

Eckert R (1972) Bioelectric control of ciliary activity. Science 176 (4034): 473–481

Eckert R, Naitoh Y, Friedman K (1972) Sensory mechanisms in *Paramecium*. I. Two components of the electric response to mechanical stimulation of the anterior surface. J Exp Biol 56: 683- 694

Eisner A, Burke SN, Toomey MD (2004) Visual sensitivity across the menstrual cycle. Vis Neurosci 21(4): 513–531

Fleissner G, Fleissner G (2001) Neuronal organization of circadian systems. In: Polis G, Brownell P (eds) Scorpion biology and research. Oxford University Press, New York, pp 107–137

Fleissner G, Fleissner G (2006) Endogenous control of visual adaptation in invertebrates. In: Warrant E, Nilsson D-E (eds) Invertebrate vision. Cambridge University Press, Cambridge UK, pp 127–166

Fratzl P, Barth FG (2009) Biomaterial systems for mechanosensing and actuation. Nature 462: 442–448

Frisch K von (1949) Die Polarisation des Himmelslichtes als orientierender Faktor bei den Tänzen der Bienen. Experientia 5: 142–148

Frisch K von (1965) Tanzsprache und Orientierung der Bienen. Springer, Berlin Heidelberg New York

Giuffre G, DiRosa L,Fiorino F (2007) Changes in color discrimination during the menstrual cycle. Ophthalmologica 221: 47–50

Grant PR, Grant BR (2006) Evolution of character displacement in Darwin finches. Science 313, 5784: 224–226

Helversen O von, Edrich W (1974) Der Polarisationsempfänger im Bienenauge: ein Ultraviolettrezeptor. J Comp Physiol 94: 33–47

Hempel de Ibarra N, Giurfa M, Vorobyev M (2001) Detection of coloured patterns by honeybees through chromatic and achromatic cues. J Comp Physiol A 187: 215–224

Hempel de Ibarra N, Giurfa M, Vorobyev M (2002) Discrimination of coloured patterns by honeybees through chromatic and achromatic cues. J Comp Physiol A 188: 503–512

Homann H (1928) Beiträge zur Physiologie der Spinnenaugen. I Untersuchungsmethoden. II Das Sehvermögen der Salticiden. Z vergl Physiol 7: 201–269

Homann H (1951) Die Nebenaugen der Araneen. Zool Jb Anat 71: 56–144

Jenkins WM, Merzenich MM, Ochs MT, Allard T, Guic-Robles E (1990) Functional reorganization of primary somatosensory cortex in adult owl monkeys after behaviorally controlled tactile stimulation. J Neurophysiol 63 (1): 82–104

Kage MP, Nachtigall W (1980) Faszination des Lebendigen. Herder Verlag, Freiburg

Kanizsa G(1955) Margini quasi-percettivi in campi con stimolazione omogena. Revista di psicologia 49 (1): 7–30

Kelber A (2006) Invertebrate colour vision. In: Warrant E, Nilsson D-E (eds) Invertebrate vision. Cambridge University Press, Cambridge UK, pp. 250–290

Konishi M (1986) Centrally synthesized maps of sensory space. TINS 9: 163–168

Konishi M (2003) Coding of auditory space. Ann Rev Neurosci 26: 31–55

Land MF, Nilsson D-E (2002) Animal eyes. Oxford University Press, Oxford

Leydig F (1864) Das Auge der Gliederthiere. Verlag H Lauppsche Buchhandlung, Tübingen

McConney ME, Schaber CF, Julian MD, Barth FG, Tsukruk VV (2007) Viscoelastic nanoscale properties of cuticle contribute to the high-pass properties of spider vibration receptor (*Cupiennius salei* Keys.). J R Soc Interface 4: 1135–1143

Molina J, Schaber CF, Barth FG (2009) In search of differences between the two types of sensory cells innervating spider slit sensilla. J Comp Physiol A 195 (11): 1031–1041

Navarrete-Palacios E, Hudson R, Reyes-Guerrero G, Guevara-Guzmán R (2003) Lower olfactory threshold during the ovulatory phase of the menstrual cycle. Biol Psychol 63: 269–279

Necker LA (1832) Observations on some remarkable phenomenon which occurs in viewing a figure of a crystal of geometrical solid. London and Edinburgh Phil Mag J Sci 3: 329–337

Nilsson D-E, Pelger S (1994) A pessimistic estimate of the time for an eye to evolve. Proc R Soc Lond B 256: 53–58

Reichert (2000) Neurobiologie. Thieme, Stuttgart

Schwind R (1983) A polarization-sensitive response of the flying water bug *Notonecta glauca* to UV light. J Comp Physiol A 150: 87–91

Schwind R (1984) Evidence for true polarization vision based on two-channel analyzer system in the eye of the water bug, *Notonecta glauca*. J Comp Physiol A 154: 53–57

Schwind R (1991) Polarization vision in water insects and insects living on a moist substrate. J Comp Physiol A 169: 531–540

Schwind R (1995) Spectral regions in which aquatic insects see reflected polarized light. J Comp Physiol A 177: 439–448

Sparks DL (1988) Neural cartography: sensory and motor maps in the superior colliculus. Brain Behav Evol 31: 49–56

Turner JS (2007) The tinkerer's accomplice. How design emerges from life itself. Harvard University Press, Cambridge Mass

Vogel E, Barth FG (2011) Spider-Vibration receptor: efferent control and effect of temperature (in prep.)

Vogel S, Wainwright SA (1969) A functional bestiary. Addison-Wesley, Reading UK

Wagner H, Takahashi T, Konishi M (1987) Representation of interaural time difference in the central nucleus of the barn owl's inferior colliculus. J Neurosci 7: 3105–3116

von der Emde G (2001) Electric fields and electroreception: How electrosensory fish perceive their environment. In: Barth FG, Schmid A (eds) Ecology of sensing. Springer, Berlin Heidelberg, pp. 313–329

Wehner R, Strasser S (1985) The POL area of the honeybee's eye: behavioural evidence. Physiol Entomol 10: 337–349

Wehner R, Labhart T (2006) Polarisation vision. In: Warrant E, Nilsson D-E (eds) Invertebrate vision. Cambridge Universiy Press, Cambridge UK, pp. 291–348

Zakon HH (1987) Hormone-mediated plasticity in the electrosensory system of weakly electric fish. TINS 10/10: 416–420

Zakon HH (1998) The effect of steroid hormones on electrical activity of excitable cells. Trends in Neurosci 21: 202–207

Zeki S (1992) The visual image in mind and brain. Sci Am Sept 1992: 69–76

Visual learning in social insects: from simple associations to higher-order problem solving

7

Martin Giurfa

Contents

Abstract

Visual learning allows the acquisition of new environmental information, which in turn allows adaptive responses when viewing already experienced events again. This capacity is crucial in contexts such as search for food, partner recognition, navigation and defense against potential enemies. It admits different levels of complexity, from simple associative link formation between a visual stimulus (e. g. a specific color) and the consequence of it (e. g. reward or punishment), to more sophisticated performances such as categorization of objects (e. g. animal vs. non-animal) or apprehending abstract rules applicable to unknown visual objects (e. g. "larger than" or "on top of"). In principle, mastering categories and rules allows flexible responses beyond simple forms of learning. Not surprisingly, higher-order forms of visual learning have been mainly studied in vertebrates with larger brains, while the study of simple visual learning has been restricted to animals with small brains such as insects. However, this dichotomy has recently changed, as research on visual learning in social insects (mainly bees and wasps) has yielded surprising results in terms of the sophistication of the tasks that can be mastered. In parallel, the accessibility and small size of insect brains have allowed the characterization of some neural mechanisms of visual learning. Here I review a spectrum of visual learning forms in social insects, from color and pattern learning, visual attention, and top-down image recognition to

Martin Giurfa
Université Paul Sabatier – Toulouse III
Centre de Recherches sur la Cognition Animale CNRS
118 Route de Narbonne
31062 Toulouse cedex 4, France
e-mail: giurfa@cict.fr

inter-individual recognition, conditional discrimination, category learning and rule extraction. I discuss the necessity and sufficiency of simple associations to account for complex visual learning and profit from the extensive knowledge on brain organization in insects to discuss neural mechanisms underlying these visual performances.

1. Introduction

Visual learning refers to an individual's capacity to acquire experience-based information pertaining to visual stimuli so that adaptive responses can be produced when viewing such stimuli again. This capacity, which is present in almost all living animals capable of seeing, intervenes in contexts as diverse as the search for food, partner recognition, navigation and orientation, and defense against potential enemies. It admits different levels of complexity as it varies from the establishment of a simple associative link connecting a visual stimulus (e. g. a specific color) and the consequences that result (e. g. a reward or a punishment), to more sophisticated performances such as learning to categorize distinct objects (e. g. animal vs. non-animal) or apprehending abstract rules applicable to unknown visual objects (e. g. "larger than," "on top of" or "inside of).

The first situation, the establishment of unequivocal, unambiguous links between a visual target and its outcome, constitutes a case of elemental learning. What is learned for a color is valid only for that color and not for different ones. Red light means not crossing the street while green light means that crossing is allowed. In this context, however, blue light, does not mean anything. In contrast, learning about categories and rules constitutes a case of non-elemental learning, which is not based on one stimulus – one consequential relationship, as an appropriate response can be then transferred to novel, unknown stimuli for which the subject has no personal experience. We are able to categorize a dodo as a bird, or as an animal, even if we have never observed one of these extinct creatures. Similarly, if trained to classify objects based on a size rule (e. g. 'larger than') we would be able to respond appropriately to unknown objects just by considering their size relative to each other. In these cases, the subject's response is flexible and relatively independent of the physical nature of the stimuli considered.

Social Hymenoptera, particularly bees (*Apis* sp. and *Bombus* sp.) and wasps (several genera), which are at the center of this article, are interesting models for the study of visual learning because in the context of their natural behavior they have to solve a diversity of visual problems of varying complexity. For instance, these insects learn and memorize the local cues characterizing the places of interest, which are essentially the hive and the food sources (Menzel 1985; Menzel et al. 1993; Zeil et al. 1996; Fauria et al. 2000, 2002). Honeybees, and to a minor extent bumblebees too, are 'flower constant'. They forage on a particular flower species as long as it offers profitable nectar and/or pollen reward (Grant 1951; Heinrich 1979; Chittka et al. 1999). This capacity is partly based on visual cues provided by flowers such as colors or patterns. Learning and memorizing the visual cues of the exploited flower through their association with nectar and/or pollen reward is what allows a bee forager tracking a particular species in the field (von Frisch 1965; Menzel 1985). Similarly, learning abilities for landmark constellations and for celestial cues used in navigation (azimuthal position of the sun, polarized light pattern of the blue sky) ensure a safe return to the nest and enhance foraging efficiency (Collett and Collett 2002; Collett et al. 2003).

Visual capacities are highly developed in social Hymenoptera, especially in those species that fly freely. Bees and wasps see the world in color (Beier and Menzel 1972; Menzel and Backhaus 1991; Chittka et al. 1992; Campan and Lehrer 2002; Dyer and Chittka 2004; Lehrer and Campan 2004, 2005; Lotto and Chittka 2005), perceive shapes and patterns (Wehner 1981; Srinivasan 1994; Giurfa and Lehrer 2001; Lehrer and Campan 2004, 2005; Dyer et al. 2005) and resolve movements with a high temporal resolution (Srinivasan et al. 1999). They implement these capacities in navigational journeys that may lead them several kilometers away from their nest to which they nevertheless return after each successful foraging trip. The complexity and richness of their natural life is therefore appealing in terms of the opportunities it offers for the study of visual learning and memory. Such an appeal would be, however, useless if these phenomena were not amenable to controlled laboratory conditions. Indeed, one of the reasons why bees and wasps constitute an attractive model for the study of visual learning resides precisely in the existence of controlled experimental methods for the study of this capacity at the individual level.

2. Visual conditioning of bees

Visual conditioning of freely-flying honeybees (von Frisch 1914) has allowed the uncovering of the perceptual capabilities of these insects and has been used to this end for more than nine decades. The common experimental protocol exploits the fact that free-flying honeybees learn visual cues such as colors, shapes and patterns, depth and motion contrast, among others (von Frisch 1914; Wehner 1981; Giurfa and Menzel 1997; Lehrer 1997; Giurfa and Lehrer 2001) when these are presented together with a reward of sucrose solution (Fig. 1a). Each bee is individually marked by means of a color spot on its thorax or abdomen so that performances of individuals can be recorded. In general the marked bee is displaced by the experimenter towards a site (called experimental place) where it is rewarded with sucrose solution in order to promote its regular return. Such pre-training is performed without presenting the training stimuli in order to avoid uncontrolled learning. When the bee starts visiting the experimental place actively (i. e., without being displaced by the experimenter), the training stimuli are presented and the choice of the appropriate visual target rewarded with sucrose solution.

Bees have to be trained and tested *individually* to achieve a precise control of the experience made by each subject. It is also important to control the distance at which a choice is made because visual orientation and choice are mediated by different visual cues at different distances or angles subtended by the target (Giurfa et al. 1996; Giurfa and Menzel 1997; Giurfa and Lehrer 2001). The time between visits to the experimental place has also to be recorded as it reflects the appetitive motivation of the bee (Núñez 1982) and thus its motivation to learn. Bees coming irregularly to the experimental place are not highly motivated and their performances are therefore unreliable.

Several actions can be used to quantify the bees' choice in these experiments. Contacts (i. e. flights towards a target that end with a contact of the bee's antennae or legs with the stimulus surface) and landings on a given visual target are usually recorded. The associations built in this context can be either classical, operant or both, i. e. they may link a visual stimulus (conditioned stimulus

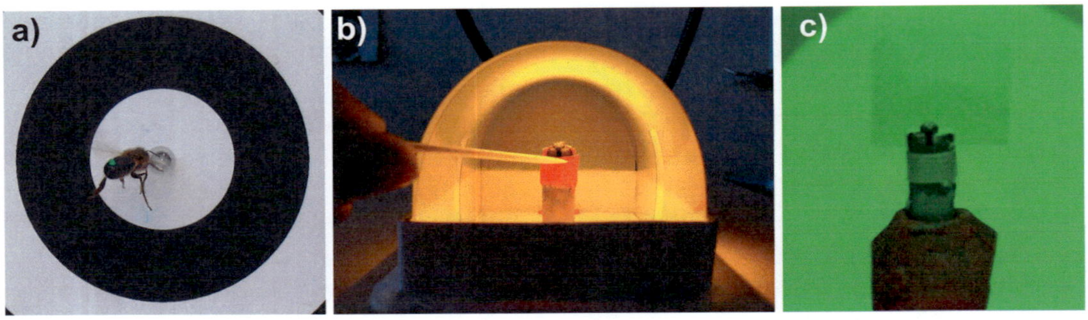

Fig. 1 Experimental protocols for the study of visual learning and perception in honeybees. **a** *Visual appetitive conditioning of free-flying bees*. A bee marked with a colored spot on the abdomen is trained to collect sugar solution in the middle of a ring pattern. **b, c** *Color conditioning of the proboscis extension reflex in harnessed honeybees.* **b** A hungry honeybee immobilized in a tube extends its proboscis when its antennae are touched with a drop of sucrose solution. **c** Pairing a chromatic (green) illumination with such a reward leads to the establishment of a color – reward association that makes the bee extend the proboscis when presented with the chromatic illumination alone (see Hori et al. 2006)

or CS) and sucrose reward (unconditioned stimulus or US), the response of the animal (e. g. landing) and the US, or both, respectively. The experimental framework is nevertheless mainly operant as the bee's behavior determines whether it obtains the sucrose reinforcement or not. The same experimental method has been successfully adapted to train other freely flying social insects to a variety of visual targets (bumblebees: Dyer and Chittka 2004; Lotto and Chittka 2005; Dyer et al. 2005; solitary bees: Campan and Lehrer 2002; Menzel et al. 1988; stingless bees: Menzel et al. 1988; wasps: Beier and Menzel 1972; Lehrer and Campan 2004, 2006).

Visual conditioning of freely-flying bees presents a fundamental problem for researchers interested in the mechanistic basis of visual learning as it does not allow the study of visual learning at the cellular level. Because bees fly freely during the experiment, to simultaneously study the neural activity in visual centers in the brain so far has remained impossible. Recently, however, a protocol for visual conditioning of harnessed bees has been developed (Hori et al. 2006, 2007). It is based on pioneer studies by Kuwabara (1957) and consists in training a harnessed bee to extend its proboscis to colors (Hori et al. 2006) or visual motion cues (Hori et al. 2007) paired with sucrose solution (Fig. 1b). Hungry bees reflexively extend the proboscis when their antennae are touched with sucrose solution, the equivalent of a nectar reward. In this protocol, colors or patterns are paired with a sucrose reward to create a Pavlovian association in which the visual stimuli are the conditioned stimuli (CS) and sucrose is the unconditioned stimulus (US). With some effort (learning takes two days and acquisition levels remain relatively low at 40%), bees learn the visual task only if their antennae have been previously cut. The reasons for this apparent interference of the antennae on visual learning remain unknown but it may simply be that cutting the antennae affects the general motivation of the bee so that the sucrose reward is not as attractive as expected by the experimenter (de Brito Sanchez et al. 2008). Improving this protocol is a priority for future research on visual learning as it will allow a combination of behavioral quantification with access to the nervous system. As in honeybee olfactory learning (see Giurfa 2007 for review), the fact that the bee is immobilized while it learns allows access to the nervous system and the study of it by the application of invasive methods.

3. Attentional and experience-dependent modulation of visual learning

The first study of bee learning and memory that used controlled protocols for characterizing individual learning and memory, employed colors as rewarding stimuli (Menzel 1967). Freely flying honeybees were trained to choose a rewarded monochromatic light and then presented with the rewarded light versus an alternative color on a horizontal plane in dual choice situations. This early study reported learning curves for different wavelengths and showed that, under the given experimental conditions, bees learned all the wavelengths after only a few learning trials. Some wavelengths, particularly 413 nm, were learned faster than others, requiring only one to three acquisition trials (Menzel 1967; but see below). This finding argued in favor of innate biases in color learning, probably reflecting the particular biological relevance of the color signals that are learned faster (Menzel 1985). Indeed, in their first foraging flight color-naïve honeybees prefer those colors that experienced bees learn faster (Giurfa et al. 1995), and preliminary findings indicate that these very colors may correspond to floral colors that are strongly associated with a profitable nectar reward (Giurfa et al. 1995).

Visual learning, as studied in these color conditioning experiments, is elemental as bees are merely presented with a single color target paired with sucrose solution. It was intended to be a fast form of learning (Menzel 1967; see above), compared, for instance, to learning of visual patterns which usually takes longer (20 or more trials). Recent studies on bumblebee and honeybee color learning (Dyer and Chittka 2004; Giurfa 2004) have, nevertheless, introduced a new twist to these conclusions. It was long thought that what an animal sees and visually learns is constrained by its perceptual machinery with little or no room for experience-dependent modulations of perception. However, studies of honeybees (Giurfa 2004; Dyer and Neumeyer 2005) and of bumblebees (Dyer and Chittka 2004) have shown that this idea is incorrect: In some cases, learning one and the same color may need few trials only but in other cases it may take more than twenty trials (Fig. 2a). The critical feature is *how* the bees learn the task. For instance, *absolute conditioning*, in which a subject is trained with a single color rewarded with sugar water, in general yields fast learning. *Differential conditioning*, in which the same subject has to learn to discriminate between a rewarded from a non-rewarded color, takes more trials, even if the rewarded color is the same as in absolute conditioning. When these animals are asked to discriminate between colors in a test, their performance differs dramatically. While bees trained in differential conditioning can discriminate between colors that are very similar (Fig. 2c), bees trained in absolute conditioning cannot discriminate between the same pair of colors (Fig. 2b; Giurfa 2004). Interestingly, similar results were obtained for bumblebees (Dyer and Chittka 2004) and for ants trained to discriminate between colors in a Y-maze (Camlitepe and Aksoy 2010). Differential conditioning promoted fine color discrimination while absolute conditioning did not.

Comparable results were obtained in a study on pattern learning and discrimination by honeybees (Giurfa et al. 1999). While differential conditioning results in a visual recognition strategy that uses the cues present

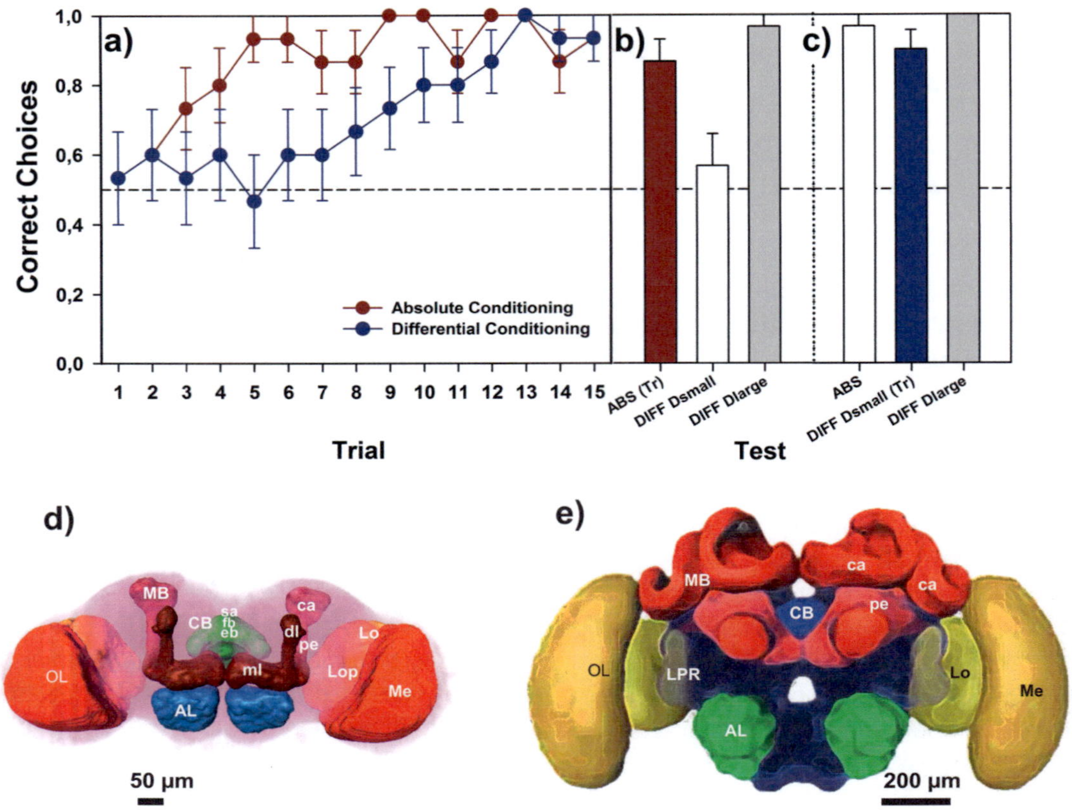

Fig. 2 a, b, c Attention-like processes in honeybees. Performance of the free-flying bees trained with colors under absolute (one color rewarded) and differential conditioning (one color rewarded vs. a color non-rewarded). Adapted from Giurfa (2004) **a *Acquisition*** along 15 trials (mean ±S. E.; n = 15 bees for each curve). Red circles: absolute conditioning; blue circles: differential conditioning. **b *Performance in the tests of the group trained in absolute conditioning.*** *Red Bar:* test presenting the trained situation (ABSTr), i. e. the single color that was previously rewarded. *White Bar*: test presenting a novel differential situation (DIFF D_{small}), i. e. the color that was previously rewarded vs. a new color that was very similar to the trained one. *Grey Bar:* test presenting a novel differential situation (DIFF D_{large}), i. e. the color that was previously rewarded vs. a new color that was very different from the trained one. **c *Performance in the tests of the group trained in differential conditioning***. *Blue Bar:* test presenting the trained situation (DIFF D_{small}Tr), i. e. the previously rewarded and the non-rewarded colors which were very similar. *White Bar:* test presenting onlyt the previously rewarded color (ABS). *Grey Bar:* Test presenting the previously rewarded color vs. a novel color very different from the rewarded one (DIFF D_{large}). **d *Three-dimensional reconstruction of a fruit fly brain***; AL: antennal lobe; OL: optic lobe; Me: medulla; Lo: lobula; Lop: lobula plate; MB: mushroom body; Ca: calyx; dl: dorsal lobe; ml: medial lobe; pe: peduncle; CB: central body; eb: ellipsoid body; fb: frontal bridge; fb: fan-shaped body; sa: superior arch. **e *Three-dimensional reconstruction of a honeybee brain***; LPR: lateral protocerebrum

in the whole pattern, absolute conditioning results in a recognition strategy that restricts cue sampling mainly to the lower half of the pattern. In other words, bees recognize a pattern differently, depending on the kind of learning implicit to the conditioning task. In both cases (color and pattern learning), however, differential conditioning increases the demands imposed to the perceptual system of the bees. They not only have to go where a rewarded stimulus is presented (absolute conditioning) but to discriminate it from a non-rewarding alternative (differential conditioning). The difference in performance suggests, therefore, that attentional processes are involved. In differential conditioning the bee has to focus on the difference and not on the mere presence of a visual target, thus making learning slower. In any case, the result contradicts the idea that the difference between two colors is a fixed, immutable property constrained by the visual machinery.

At the time at which Menzel first characterized color learning (1967, 1968; see above) studies on pattern perception were performed by Wehner (1967, 1971, 1972) and others (e. g. Anderson 1972), continuing the tradition started by von Frisch's students (Hertz 1929 a,b, 1933; Wolf 1933, 1934; Wolf and Zerrahn-Wolf 1935). In contrast to Menzel's work, these studies did not focus on learning but on the perceptual capabilities of bees confronted with pattern discrimination tasks. Certainly, visual conditioning was also used in these and in later works on pattern perception (for review see Wehner 1981; Srinivasan 1994; Lehrer 1997; Srinivasan and Zhang 1997) but there was no quantification of acquisition curves and/or a characterization of pattern memory. This tradition was continued in the 1970s, 80s and even 90s when visual learning was used mainly as a tool to answer questions of visual perception and discrimination close to the feeding place. The questions asked by these works focused on visual spatial resolution, shape discrimination, orientation detection, movement perception and distance estimation based on image movement on the retina (the so-called parallax), among others. An accurate control of the bees' individual experience prior to the tests to which they were subjected, would have been desirable. As mentioned above, in pattern vision, like in color vision, what a bee sees depends on its previous visual experience and on possible attentional processes. Zhang and Srinivasan (1994) showed, for instance, that the previous visual experience of a bee can speed up the analysis of the retinal image when a familiar object or scene is encountered. They first attempted to train bees to distinguish between a ring and a disk when each shape was presented as a textured figure placed a few centimetres in front of a similarly textured background (Fig. 3 a). In principle the figures were detectable through the relative motion of the figure outlines, whose distance differed from that of the background when bees flew towards the targets. Despite intensive training, the bees were incapable of learning the difference between a ring and a disk (Fig. 3 a), a discrimination task that usually poses no problems when the bees see the stimuli as plain (non-textured) shapes. Zhang and Srinivasan (1994) then trained a group of bees to solve the 'easy' problem, presenting a plain black disk and ring positioned a few cm in front of a white background (Fig. 3 b). As expected, the bees easily learned the task. They were then confronted with the 'difficult' problem of learning the textured disk versus the ring and this time they immediately solved the discrimination task (Fig. 3 c). Thus, pre-training with plain stimuli primed the pattern recognition system in such a way that it was able to detect shapes that otherwise could not be distinguished. Again, it may be that such pre-training triggers attentional processes that allow a better focus on the targets that have to be discriminated between.

Thus, color and pattern vision studies in bees have introduced the idea that visual capabilities are modulated by attentional and experience-dependent processes. This idea has been explicitly studied in honeybees trained to choose a colored disc ('target') among a varying number of differently color-

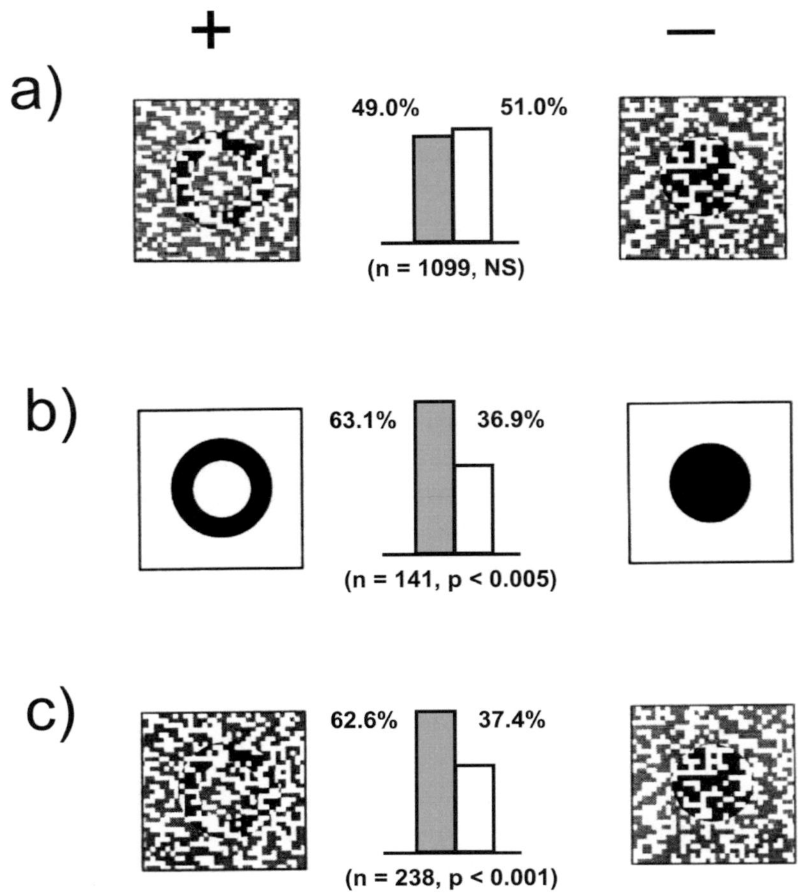

Fig. 3 Prior experience enhances pattern discrimination in honeybees (adapted from Zhang and Srinivasan 2004). **a** Bees were trained in a dual-choice Y-maze to distinguish between a ring (rewarded) and a disk (non-rewarded) when each shape was presented as a textured figure placed a few cm in front of a similarly textured background. Despite intensive training, the bees were incapable of learning the difference between a ring and a disk (n: number of choices; the percentages correspond to the choice of stimuli presented). **b** When these stimuli were presented as plain (non-textured) shapes, few cm in front of a white background, the bees could, as expected, easily learn the task. **c** They were then confronted with the difficult problem of learning the textured disk vs. the ring and this time they solved the discrimination challenge. Pre-training with the plain stimuli may trigger attentional processes that allow a better focus on the targets whose discrimination is difficult

ed discs ('distractors') (Spaethe et al. 2006). Accuracy and decision time were measured as a function of distractor number and color. For all color combinations, decision time increased and accuracy decreased with increasing distractor number, whereas performance increased when more targets were present. These findings are characteristic of a serial search in primates, when stimuli are examined sequentially, thus indicating that at the behavioral level, the strategies implemented by bees converge with those of animals in which attention is commonly studied (Spaethe et al. 2006).

In addition, other results have shown that the visual strategies used by bees to accomplish visual discrimination may be affected by the amount of experience accumulated at the moment of a test (Giurfa et al. 2003; Stach and Giurfa 2005). Quantifying individual experience and reporting visual acquisition curves is therefore crucial even for the simplest visual discrimination experiment that involves training to a single visual target.

Uncovering how attentional processes and learning modulate visual perception constitutes an unexplored and promising research field. The existence of attentional processes in insect brains is not far-fetched and recent research has been able to locate such processes in identified structures of the insect brain. In the fruit fly *Drosophila melanogaster*, attention can be demonstrated and characterized at the physiological level (van Swinderen and Greenspan 2003). A tethered fruit fly within a circular arena and tracking a visual object (a vertical black bar) moving at a constant frequency around it, exhibits anticipatory behavior consistent with attention for the bar it tracks. The neural correlate of such an anticipatory tracking is a transient increase in a 20–30 Hz local field-potential recorded in a region of the brain called the medial protocerebrum (Fig. 2d; van Swinderen and Greenspan 2003). The neural response is not only anticipatory, but also selective to the stimulus presented, increased by novelty and salience and reduced when the fly is in a sleep-like state (van Swinderen and Greenspan 2003). Moreover, the use of mutants showed that a subset of neurons of the mushroom bodies, which are a higher-order structure of the insect brain (Fig. 2d, e), are required for both the tracking response and the 20–30Hz response. Mutants in which the output of these neurons was suppressed did not retain either of these characteristics, thus showing that the neural basis of attentional processes can be located in a specific set of neurons of the fruit fly brain (van Swinderen and Greenspan 2003). This result is consistent with the finding that mushroom bodies are required for choice behavior in *Drosophila* facing contradictory visual cues (Tang and Guo 2001). In this case, a tethered fly flying stationarily is trained in a circular arena in which one kind of visual stimulus (say, a T pattern) represents a permitted flight direction, while another kind of visual stimulus (say, an inverted T pattern) represents a forbidden flight direction associated with an unpleasant heat beam on the thorax. Tang and Guo (2001) conditioned flies to choose one of two directions in response to color and shape cues; after the training, the flies were tested using contradictory cues. Wild-type flies made a discrete choice that switched from one alternative to the other as the relative salience of color and shape cues gradually changed. However, this ability was greatly diminished in mutant flies with miniature mushroom bodies or in flies with chemically ablated mushroom bodies. Obviously, mushroom bodies mediate the assessment of the relative saliency of conflicting visual cues (Tang and Guo 2001, Xi et al. 2008) and are also involved in improving the extraction of visual cues after pre-training in *Drosophila* (Peng et al. 2007). The mushroom bodies of hymenopterans may play similar roles (Fig. 1e), favoring attention processes and better problem solving and discrimination in the visual domain.

Yet, visual learning and the neural circuits mediating it are still poorly understood in the fruit fly. The mushroom bodies, which are the main site for olfactory memories, are not directly involved in visual learning. In *Drosophila*, different from hymenopterans, there is no direct input from the visual areas of the brain to these structures (Wolf et al. 1998). Recent studies succeeded in precisely identifying the neuronal substrates of two forms of visual memory in the *Drosophila* brain, outside the mushroom bodies (Liu et al. 2006). Memory for pattern elevation and orientation, two parameters whose variations can be easily distinguished by flies, were retraced to different regions of the central complex, a structure in the centre of the insect brain (Fig. 2d). Two neuronal layers of the central complex are required for visual discrimination based on pattern elevation and orientation, respectively. Because in all cases only visual short-term memory was studied, the localization of visual long-term memory remains an open question.

In bees and wasps, the localization of visual memories may differ from that in *Drosophila*. In contrast to the fruit fly, the visual areas of the hymenopteran brain do provide direct input to the mushroom bodies (Ehmer and Gronenberg 2002). The mushroom bodies may well, therefore, be the seat of visual memories in addition to the central complex (Fig. 2d).

4. Complex forms of visual learning (that may not be so complex)

Only in the 1990s, when researchers became interested in the existence of cognitive processing in insects and the honeybee, was the model chosen to address most of the questions asked at that time. The delay with respect to the general "cognitive revolution" which already flourished at the end of the 1970s and the early 1980s (Miller 2003) was due to the reluctance to view invertebrates, and therefore insects, as cognitive organisms (by cognitive we understand here as 'capable of non-elemental', higher-order forms of learning). For instance, the main idea about visual pattern learning, which is still sometimes defended, was that insects can only view isolated spots, blobs and bars without having the capacity to integrate them in a given configuration (Horridge 2000, 2003, 2006, 2009). Even a basic capacity of recognition such as generalization, the ability to respond to stimuli that despite being different are perceptually similar to a trained target (Spence 1937, Shepard 1987), was and still is considered by some researchers as being too high-level for a honeybee (Horridge 2006, 2009). Yet, dozens of works had already shown that honeybees generalize their choice of visual patterns to novel figures that have some similarity with those that have been trained (e.g. Wehner 1971, Anderson 1972). This denial of generalization capacities is consistent with the preconception that bees and insects in general have limited plasticity and should instead be viewed as reflex machines or robots reacting to specific features in the environment to which they are attuned.

In the last decade, however, researchers have shown that bees are not robots but exhibit visual learning capabilities that to date had only been attributed to various vertebrates. Some of these capacities are surprising and may be viewed as non-elemental. However, alternatively one might argue that it is possible to explain them as being based on simple, elemental associations. The experiments reviewed in the next section were not conceived to address these opposite views, so we are currently unable to determine whether the fascinating performances observed are forms of elemental or higher-order learning.

4.1 Visually-based recognition of individuals in wasps

The capacity to recognize the distinctive identity of individuals has long been dismissed in social insects because of the cognitive requirements that such performance may impose in colonies with thousands of individuals. For instance, Wilson (1971) stated that *"insect societies are, for the most part, impersonal* [...]. *The sheer size of the colonies and the short life of the members make it inefficient, if not impossible, to establish individual bonds"*. Not all social insects live, however, in huge overcrowded societies. Small colonies of bumble bees, wasps and some ant species are based on dominance hierarchies where individual recognition may be crucial for responding appropriately to a conspecific. Indeed, recent studies have shown that queens of the ant *Pachycondyla villosa* recognize each other using olfactory cuticular cues (d'Ettorre and Heinze 2005). In the visual domain, studies on the paper wasp *Polistes fuscatus* have shown that individual recognition is achieved through learning the yellow-black patterns on the wasp faces and/or abdomens (Tibbetts 2002) (Fig. 4). More variable patterns with larger black components were found to be carried by individuals ranking higher in the nest hierarchy. Altering these facial and/or abdominal color patterns induced aggression against such animals, irrespective of whether their patterns were made to signal higher or lower ranking. These results, therefore, are in favor of the capacity to learn visual features allowing the recognition of individual wasps (Tibbetts and Dale 2004). This capacity should not necessarily be a surprise with regard to its cognitive implications. Each individual facial mask would have an unambiguous significance in terms of its ranking in the social structure (i. e. mask A $\rightarrow\alpha$ individual; mask B $\rightarrow\beta$ individual, etc.) Thus, wasps would learn a series of elemental associations between mask patterns and social ranking. Given the small size of colonies in which five to ten individuals can coexist, storing several memories, one for each individual, seems to be a plausible possibility. If this were the case, a fundamental research goal would now be to characterize the storage capacity of the visual memory and its accuracy in relation to colony size (number of individuals).

4.2 Learning by observation in bumblebees

According to recent studies on bumblebees (Leadbeater and Chittka 2005, 2007; Worden and Papaj 2005), these insects copy the learnt foraging preferences of other bees by observing their choices of visual rewarded targets. Bumblebees, *Bombus terrestris,* are influenced by other conspecifics when sampling unfamiliar flowers. They land on unknown flowers where other bees are already present (Leadbeater and Chittka 2005). This behavior is exhibited even when naïve bees are separated from experienced foragers by a transparent screen such that they can neither sample the flowers by themselves nor interact with their foraging conspecifics (Worden and Papaj 2005). Similarly, naïve

Fig. 4 Portraits of nine *Polistes dominulus* wasp foundresses showing the diversity of facial patterns (from Tibbets and Dale 2004)

bumblebees would abandon an unrewarding flower species and switch to a more rewarding alternative more quickly when accompanied by experienced foragers (Leadbeater and Chittka 2007).

As surprising as this performance may appear, it can be accounted for by an elemental form of associative learning called second-order conditioning (Pavlov 1927), which involves two connected associations. In this scenario, an animal first learns an association between a conditioned stimulus (CS) and an unconditioned stimulus (US). It then experiences a pairing between a new conditioned stimulus CS2 and CS1 so that CS2 becomes meaningful through its association with CS1, and indirectly with the US. How would this apply to the observational learning of bumblebees? One could propose that naïve bumblebees first associate the presence of a conspecific (CS1) with reward (US) simply by foraging close to experienced foragers. Subsequently, the observation of a conspecific landing on a given color (CS2) may allow establishing an association between color (CS2) and conspecific (CS1) (Leadbeater and Chittka 2007). The hypothesis of connected elemental links is supported by the fact that honeybees can learn 2nd-order associations while searching for food. They learn to connect both two odors (Odor 1 + Sucrose Reward; Odor 2 + Odor 1; Bittermann et al. 1983) and one odor and one color (Odor + Sucrose Reward; Color + Odor; Grossman 1971).

4.3 'Symbolic matching to sample' and other forms of conditional discrimination in bees and wasps

'Symbolic matching to sample' is a term used to describe an experimental situation in which the correct response to a problem depends on a specific background or condition. In other words, animals have to learn, for instance, that in condition A response C is correct while in condition B response D is correct. This is why symbolic matching to sample is considered to be a form of *conditional discrimination* because a given stimulus, the sample (also called the 'occasion setter'), sets the condition for the next choice. Using this design, Zhang et al. (1999) trained honeybees to fly through a compound Y-maze consisting of a series of interconnected cylinders. The first cylinder carried the sample stimulus (e. g. a vertical or a horizontal black-and-white grating). The second and third cylinders each had two exits. Each exit presented a visual stimulus so that the bee had to choose between them. In the second cylinder, bees had to choose between a blue and a green square whereas in the third cylinder they had to choose between a radially sectored and a ring pattern. Correct sequences of choices were '*Vertical – Green – Ring*' and '*Horizontal – Blue – Radial*'. Only after making a succession of correct choices both in the second and in the third cylinder a bee could reach the feeder with sucrose solution. The bees learned to master these successive associations between different kinds of visual cues (Zhang et al. 1999). This finding was also extended to other sensory modalities; the same principle applied when visual cues were combined with odors in a similar symbolic matching protocol (Srinivasan et al. 1998).

Conditional learning admits other variants that, depending on the number of occasion setters and discriminations involved, have received different names. For instance, another form of conditional discrimination involving two occasion setters is the so-called *transwitching problem*. In this experiment an animal is trained differentially using two stimuli, A and B, and two different occasion setters, C and D. When C is available, stimulus A is rewarded but not stimulus B (A+ vs. B-), while the opposite is the case (A- vs. B+) when D is available. The transwitching problem is also considered a form of contextual learning because the occasion setters C1 and C2 can be viewed as contexts determining the appropriateness of each choice. Bumblebees have been trained in a transwitching problem to choose a 45° grating and to avoid a 135° grating to reach a feeder, and to do the opposite to reach their nest (Fauria et al. 2002) (Fig. 5). Here, the nest and the feeder provide the appropriate contexts defining what has to be chosen. Bumblebees can also learn that an annular or a radial disc must be chosen, depending on the disc's association with a 45° or a 135° grating either at the feeder or the nest entrance: At the nest, access was allowed by the combinations 45° + radial disc and 135° + annular disc, but not by the combinations 45° + annular disc and 135° + radial disc; at the feeder, the opposite applied (Fauria et al. 2000). In both cases, the potentially competing visuo-motor associations were insulated from each other by being set in different contexts.

Solving this kind of problem can be viewed as a form of non-elemental learning and thus as a sophisticated form of cognitive visual

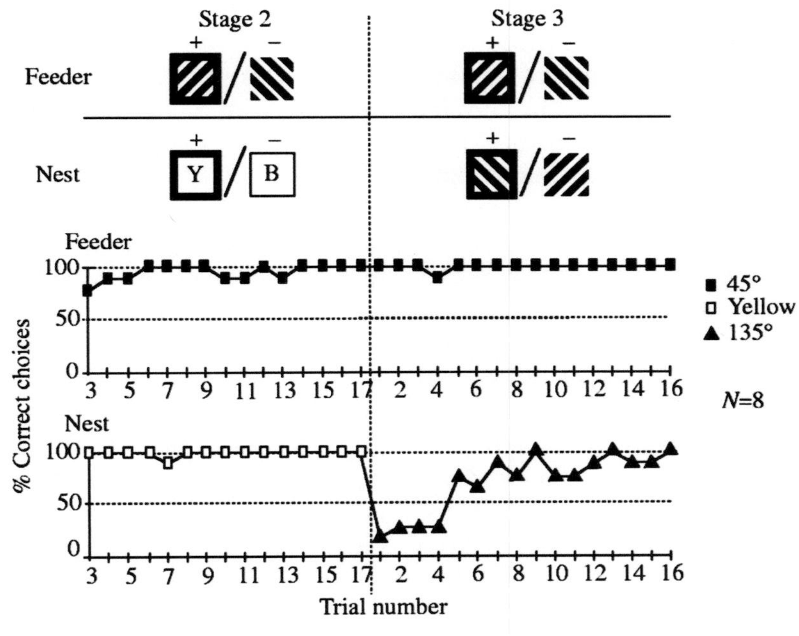

Fig. 5 Performance of bumblebees in a transwitching problem in which they had to choose between two gratings, which have different outcomes depending on the site (feeding place or nest entrance) in which they are presented (adapted from Fauria et al. 2002). **Top:** Stimuli used to train the bumblebees in phases 2 and 3 of the experiment; B: blue; Y: yellow. The stimuli framed in bold are those that allowed animals to access either the food (grating oriented at 45°) or the nest (yellow in phase 2, and grating oriented at 135° in phase 3). While phase 2 poses a simple, non-ambiguous problem to the bees, phase 3 constitutes a case of transwitching as the grating oriented at 135° (and not at 45°) becomes positive at the nest entrance (i. e. allows accessing the nest). **Bottom:** Percentage of correct choices of groups of eight bees plotted against trial number for phases 2 and 3. The curves show the performance with 45° *versus* 135° gratings at the feeder and yellow *versus* blue at the nest in phase 2, and with 45° *versus* 135° at the feeder and 135° *versus* 45° at the nest in phase 3. In phase 2, bees were assisted in trials 1 and 2, so these trials are not plotted. Changing the stimuli from phase 2 to 3 affected performance at the nest but left intact that at the feeder. At the end of phase 3, the bees mastered both problems simultaneously

processing. Indeed, as for other forms of conditional discrimination, one could describe this protocol as CA+, CB- (if C then A but not B), and DA-, DB+ (if D then B but not A). Each stimulus, A, B, C, and D, is rewarded as often as it is non-rewarded so that solutions cannot be based on the mere consequence of A, B, C or D. A higher-order solution would then be to learn the outcome of each particular configuration CA, CB, DA, DB. However, an alternative explanation could argue that what the insects do is to establish hierarchically simple associations like those underlying 2nd-order conditioning (see above). Indeed, one could imagine that bees learn to associate a radial disc with sucrose reward and that they then learn to associate a 45° grating with the radial disc. This is a relatively simple strategy probably used by bees for navigational purposes (Zhang et al. 1996) when they are confronted with successions of different landmarks en route to the goal.

A critical factor determining the application of one strategy or the other may, therefore, be the temporal order of stimulus presentation. If these are presented *serially*, learning chains of simple associations could be primed while *simultaneous* presentation of stimuli may prime learning of configurations and their specific consequences. An example of the latter is the case of honeybees trained to solve a *biconditional discrimination* AC+, BD+, AD-, BC- in which all four stimuli were presented simultaneously, and were as often rewarded as non-rewarded (Schubert et al. 2002). Four different gratings combining one color (yellow or violet = A or B) with one orientation (horizontal or vertical = C or D) were used in such a way that bees had to learn that, for instance, yellow-horizontal (AC) and violet-vertical (BD) were rewarded while yellow-vertical (AD) and violet-horizontal (BC) were non-rewarded. Bees learned to choose the rewarded stimuli despite the fact that colors and orientations were ambiguous when considered alone. They thus learned the configurations and not the specific outcome of each element (Schubert et al. 2002).

The last example shows that it is possible to distinguish between elemental and non-elemental visual learning. However, leaving this exception aside, despite their sophistication the visual performances presented in this section can be accounted for by elemental associations. Further research should determine to what extent visual performances, which appear complex to us, are indeed based on higher-order or on simple associations.

5. Non-elemental visual learning

A higher level of complexity is reached when animals respond in an adaptive manner to novel stimuli *that they have never encountered before and that do not predict a specific outcome per se based on the animals' past experience*. Such a positive transfer of learning (Robertson 2001) is therefore different from elemental forms of learning, which link known stimuli or actions to specific rewards (or punishments). In the previous section, symbolic matching to sample, for instance, does not pose this kind of problem to a bee: Horizontal leads always to Blue which in turn leads always to Radial which in turn always leads to a reward (Zhang et al. 1999). In the cases considered in this section, the insects' response seems to reflect a *rule* guiding the animal's behavior. Examples are relational rules such as 'on top of' or 'larger than' which can be applied irrespective of the physical similarity of the stimuli presented or the previous 'knowledge' that the animal has of these stimuli.

5.1 Categorization of visual stimuli in honeybees

Positive transfer of learning is a distinctive characteristic of categorization performances. Visual categorization refers to the classification of visual stimuli into defined functional groups (Harnard 1987). It can be defined as the ability to group distinguishable objects or events on the basis of a common feature or set of features, and therefore to respond similarly to them (Troje et al. 1999; Delius et al. 2000; Zentall et al. 2002). Categorization deals, therefore, with the extraction of defining features from objects of the subject's environment. A typical categorization experiment trains an animal to extract the basic attributes of a category and then tests it with novel stimuli that were never encountered before and that may or may not present the attributes of the category learned. If the animal chooses the novel stimuli based on these attributes it classifies them as belonging to the category and therefore exhibits a positive transfer of learning. Categorization differs between generalizations because generalization implies a gradual decrease in responses along a perceptual dimension while categories have abrupt borders. In other words, if we are trained to search for a specific fruit of a given red color, we may respond less and less (we generalize less) to fruits whose hue varies progressively from the known red to different red hues. However, if the task we learned is to search for red objects in general and not for blue ones, we may respond equally to very different reds. We assign the different reds to the same category, even though we can distinguish them well. As long as we identify the presence of red, we respond positively. However, if the hue is changed to an extent which makes us decide that we are out of the category, we will stop responding abruptly.

According to several recent studies, free-flying honeybees trained to discriminate between different patterns and shapes are indeed able to visually categorize. For instance, van Hateren et al. (1990) trained bees to discriminate between two given gratings presented vertically and in other orientations (e. g. 45° vs. 135°) by rewarding the choice of only one of these gratings with sucrose solution. Each bee was trained with a changing succession of pairs of different gratings, one of which was always rewarded and the other not. Despite the difference in pattern quality, all the rewarded patterns had the same edge orientation and all the non- rewarded patterns had a common orientation as well (perpendicular to the rewarded one). Under these circumstances, the bees had to extract and learn the orientation common to all rewarded patterns to solve the task. This was the only cue predicting reward delivery. In the tests, bees were presented with novel patterns, which they had never been exposed to before. These patterns were all non-rewarded but had the same stripe orientations as the rewarding and non-rewarding patterns employed during the training. In such transfer tests, bees chose the appropriate orientation despite the novelty of the structural details of the stimuli. Thus, bees could categorize visual stimuli on the basis of their global orientation.

Bees can also categorize visual patterns based on their bilateral symmetry. When trained with a succession of changing patterns to discriminate bilateral symmetry from asymmetry, they learn to extract this information from very different figures and indeed transfer it to novel symmetrical and asymmetrical patterns (Giurfa et al. 1996). Similar conclusions apply to other visual features such as radial symmetry, concentric pattern organization and pattern disruption (see Benard et al. 2006for review) and even photographs belonging to a given class (e. g. radial flower, landscape, plant stem) (Zhang et al. 2004).

How could bees classify different photographs of radial flowers appropriately if these vary in color, size, outline, etc.? An explanation was provided by Stach et al. (2004) who expanded the demonstration that bees can categorize visual stimuli based on their global orientation. The authors showed that different coexisting orientations can be considered at a time, and can be integrated into a global stimulus representation that is the basis for the category (Stach et al. 2004). Thus, a radial flower would in fact be the conjunction of five or more radiating edges. Besides focusing on a single orientation, honeybees were shown to assemble different features to build a generic pattern representation, which could be used to respond appropriately to novel stimuli sharing such a basic layout. Honeybees trained with a series of complex patterns sharing a common layout comprising four edge orientations remembered these orientations simultaneously in their appropriate positions, and transferred their response to novel stimuli that preserved the trained layout (Fig. 6). Honeybees also transferred their response to patterns with fewer correct orientations, depending on their match with the trained layout. Obviously honeybees extract regularities in their visual environment and establish correspondences among correlated features such that they generate a large set of object descriptions from a finite set of elements.

This capacity can also explain the recent claim that honeybees can learn to recognize human faces (Dyer et al. 2005). Bees were rewarded with sugar water to choose a photograph of a person's face and to distinguish it from other persons' photographs. They were indeed able to do that. Does this mean that bees do recognize human faces? Not really. For the bees rewarded on the photographs these were just strange flowers. The more interesting question is which information contained in the photographs was used to recognize the correct stimulus. This question was recently tackled by a work that studied whether bees can bind the features of a face-like stimulus (two dots in the upper part as the eyes, a vertical line below as the nose, and a horizontal line in the lower part as the mouth) and recognize faces using this basic configuration (Avargues et al. 2010). Bees did indeed distinguish between different variants of the face-like stimuli, thus showing that they discriminate between these options, but they grouped the stimuli and therefore reacted similarly to faces if trained to do so. Stimuli made of the same elements (two dots, a vertical and a horizontal line) but not preserving the configuration of a face were not recognized as positive. This shows that bees learn that the rewarded stimulus consists of a series of elements arranged in a specific spatial configuration. Interestingly, the recognition of face-like stimuli by bees is flexible enough to transfer the choice towards real photographs which obviously preserve the basic configuration learned (Avargues et al. 2010). Furthermore, when trained with real faces, bees can learn to recognize novel views of a face by interpolating between or 'averaging' views they have experienced (Dyer and Vuong 2008).

In any case, honeybees show a positive transfer of learning from a trained to a novel set of stimuli, and their performances are consistent with the definition of categorization. Visual stimulus categorization is not, therefore, a privilege of certain vertebrates. At the same time this finding may not be surprising as it admits (again) an interpretation based on elemental learning. To understand this interpretation, the possible neural mechanisms underlying categorization have to be considered. If we admit that visual stimuli are categorized on the basis of specific features such as orientation, the neural implementation of category recognition could be relatively simple. The feature(s) allowing stimulus classification would then have to activate specific neuronal detectors

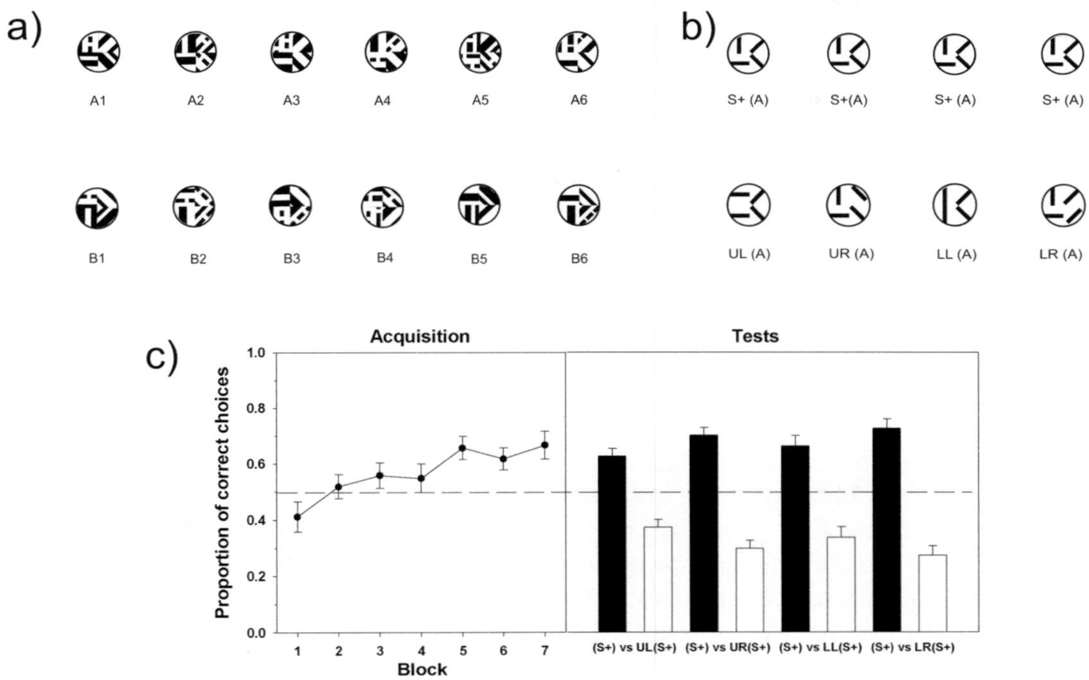

Fig. 6 Categorization of visual patterns based on sets of multiple features (adapted from Stach et al. 2004). **a** Training stimuli used in Stach et al.'s experiments (2004). Bees were trained to discriminate A from B patterns during a random succession of A vs. B patterns. A patterns (A1 to A6) differed from each other but shared a common layout of orientations in the four quadrants. B patterns (B1 to B6) shared a common layout perpendicular to that of A patterns. **b** Test stimuli used to determine whether bees extract the simplified layout of four bars from the rewarded A patterns. S+, simplified layout of the rewarded A patterns; UL, upper-left bar rotated; UR, upper-right bar rotated; LL, lower-left bar rotated; LR, lower-right bar rotated. **c** *Left panel*: acquisition curve showing the pooled performance of bees rewarded on A and B patterns. The proportion of correct choices along seven blocks of six consecutive visits is shown. Bees learned to discriminate between the rewarding patterns (A or B and improved significantly their correct choices during training. *Right panel*: proportion of correct choices in the tests with the novel patterns. Bees always preferred the simplified layout of the training patterns previously rewarded (S+) to any variant in which one bar was rotated

in the optic lobes, the visual areas of the bee brain. Examples of such feature detectors are orientation detectors whose selectivity for stimulus orientation was already characterized electrophysiologically by recording from neurons in the honeybee optic lobes (Yang and Maddess 1997). Thus, responding behaviorally to different gratings with a common orientation of, say, 60°, is simple because they all will elicit activity in the same set of neural orientation detectors despite their different structural quality. In the case of category learning, the activation of an additional neural element is needed. Such an element would be a 'reward neuron' whose activity substitutes for the sucrose reward. A neuron with these properties has been identified in the honeybee brain. It is called $VUM_{m\times1}$ (from 'ventral unpaired median' neuron located in the maxillar neuromere 1; Hammer 1993). $VUM_{m\times1}$ mediates olfactory learning in the honeybee as it contacts the olfactory circuit at its key processing stages in the brain. In other words, when an odor activates the olfactory circuit, concomitant sucrose stimulation activates VUM_{mx1} thus providing the basis for neural coincidence between odor and reward. The branching of VUM_{mx1} makes it specific for the olfactory circuit and thus for olfactory learning (Hammer 1993). Other VUM neurons whose function is still unknown are present in the bee brain (Schroter et al. 2006). One or a few of them could be the neural basis of reward in associative visual learning. Category learning could thus be reduced to the progressive reinforcement of a neural circuit relating visual-coding and reward-coding neurons to each other, similar to the behavior observed in simple associative (e. g. Pavlovian) conditioning. Even if categorization is viewed as a non-elemental learning form because it involves the positive transfer of learning, it may simply rely on elemental links between conditioned and unconditioned stimuli.

An even simpler alternative may account for the categorization task described. The mechanism just explained could be viewed as a form of supervised learning, in which a visual network is instructed by the external signal of the reinforcement neuron to respond to the right combination of features. Recent modeling work on the vertebrate visual system has shown that visual networks can learn to extract the distinctive features of a category without any kind of supervision (Masquelier et al. 2007). The model relies on the fact that neurons strongly activated fire first, a mechanism that encodes image information, and on the presence of the so-called spike timing dependent plasticity (STDP), which is a rule that concentrates high synaptic weights on afferents that systematically fire early. When a network with these properties is repeatedly presented with natural images belonging to a given category, neurons become selective to the features that are predictive of the category, while their latencies decrease. Thus, those features that are both salient and consistently present in the images are highly informative and enable robust object recognition (Masquelier et al. 2007). Testing whether similar neural mechanisms underlie object categorization in the insect visual system would be a fascinating task.

5.2 Rule learning in honeybees

In rule learning, positive transfer occurs independently of the physical nature of the stimuli considered. The animal learns relations between objects and not the objects themselves. Typical examples are the so-called rules of *sameness* and *difference*. They are demonstrated through the protocols of delayed matching to sample (DMTS) and delayed non-matching to sample (DNMTS), respectively. In DMTS, animals are presented with a sample and then with a set of stimuli, one of which is identical to the sample. Choice of this stimulus is rewarded while choice of the different stimuli is not. Since the sample is regularly changed, animals must learn the sameness rule, i. e. '*always choose what is shown to you (the sample), independent of what else is shown to you*'. In DNMTS, the animal has to learn the opposite, i. e. '*always choose the opposite of what is shown to you (the sample)*'. Honeybees foraging in a Y-maze learn both rules (Giurfa et al. 2001). They were trained in a DMTS experiment in which they were presented with a changing non-rewarded sample (i. e. one of two different color disks or one of two different black-and-white gratings, vertical or horizontal) at the entrance of a maze (Fig. 7). The bees were rewarded only if they chose the stimulus identical to the sample once within the maze. Bees trained with colors and presented in transfer tests with black-and-white gratings that they had not experienced before solved the problem and chose the grating identical to the sample at the entrance of the maze. Similarly, bees trained

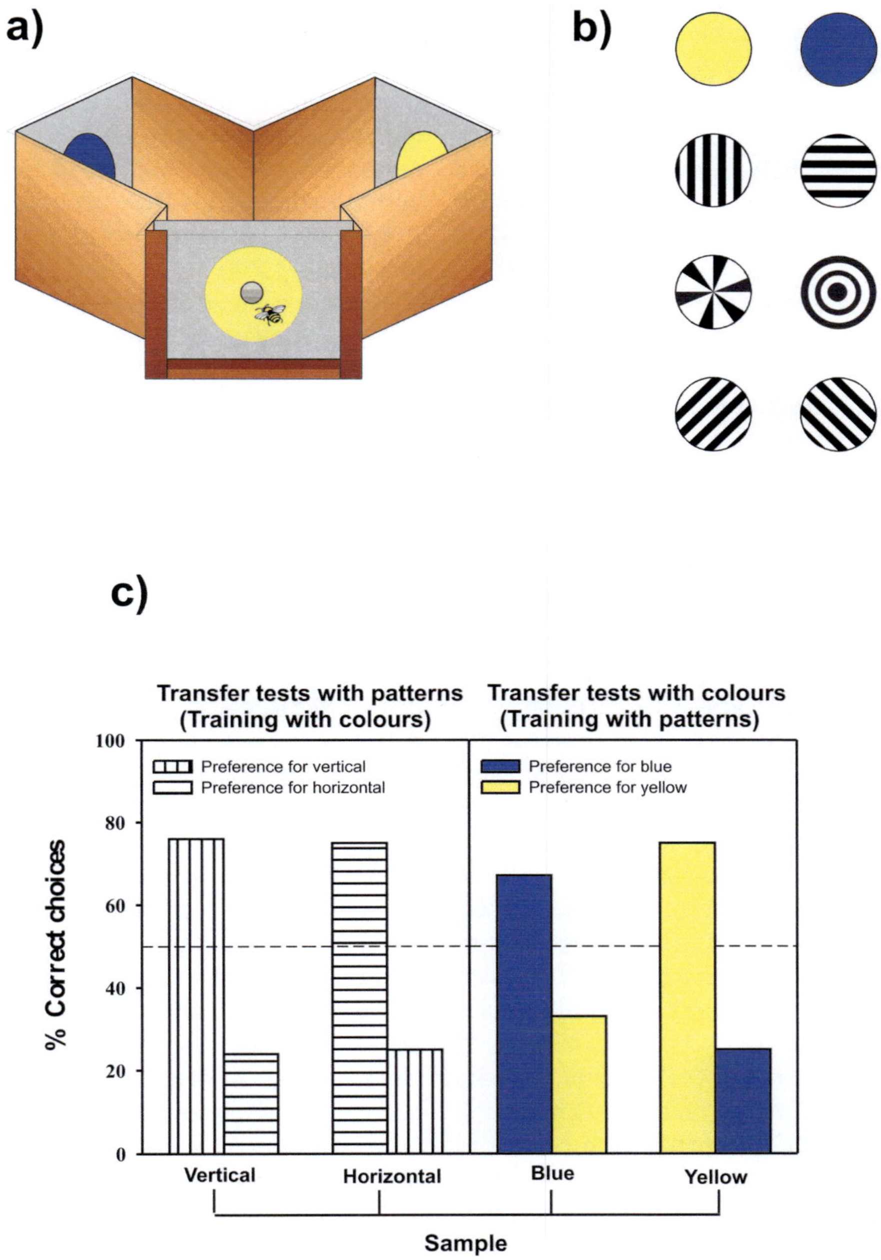

Fig. 7 Rule learning in honeybees (adapted from Giurfa et al. 2001). **a** Honeybees were trained to collect sugar solution in a Y-maze. **b** A series of different patterns was used to train a rule of sameness. A sample was shown at the maze entrance and bees had to match their choice to the sample once they entered the maze. When the bees mastered the trained discrimination, they were presented with novel simuli. **c** Transfer tests with novel stimuli. In Experiment 1, bees trained with the colors were tested with the gratings. In Experiment 2, bees trained with the gratings were tested with the colors. In both cases the bees chose the novel stimuli corresponding to the sample shown at the entrance of the maze, although they had no experience with such test stimuli

with the gratings and tested with colors in transfer tests also solved the problem and chose the novel color corresponding to that of the sample grating at the maze entrance. Transfer was not limited to different types of visual stimuli (pattern vs. color), but could also operate between drastically different sensory modalities such as olfaction and vision (Giurfa et al. 2001). Bees also mastered a DNMTS task, thus showing that they learn a rule of difference between stimuli as well (Giurfa et al. 2001). These results document that bees learn rules relating to stimuli in their environment.

The capacity of honeybees to solve a DMTS task has recently been verified and studied with respect to the working memory underlying it (Zhang et al. 2005). It was found that the working memory for the sample underlying the solving of DMTS lasts for approximately 5 s (Zhang et al. 2005). This length of time coincides with the duration of other visual and olfactory short-term memories characterized in simpler forms of associative learning in honeybees (Menzel 1999; see above). Moreover, bees trained in a DMTS task can learn to pay attention to one of two different samples presented successively in a flight tunnel (either to the first or to the second) and can transfer the learnt relevance of the sequence to novel samples (Zhang et al. 2005).

Despite the honeybees' evident capacity to solve relational problems such as the DMTS or the DNMTS tasks, such capacities are not unlimited. In some cases, biological constraints may impede the solving of a particular problem for which the extraction of a rule is necessary. It is therefore interesting to focus on a different example of rule learning which bees could not master, the *transitive inference problem* (Benard and Giurfa 2004). In this problem, animals have to learn a transitive rule, i. e. A > B, B > C, then A > C. Preference for A over C in this context can be explained as the result of two strategies: 1) deductive reasoning (Fersen et al. 1990) in which the experimental subjects construct and manipulate avlinear representation of the implicit hierarchy A > B > C; or 2) responding as a function of the effective number of reinforced and non-reinforced experiences with the stimuli (A is always reinforced while C is always non-reinforced) (Terrace and McGonigle 1994).

To determine whether bees learn a transitive rule, they were trained using five different visual stimuli A, B, C, D, and E in a multiple discrimination task: A+ vs. B–, B+ vs. C–, C+ vs. D–, D+ vs. E– (Benard and Giurfa 2004). Training involved the overlap of adjacent premise pairs (A > B, B > C, C > D, D > E), which underlie a linear hierarchy A > B > C > D > E. After training, the bees were tested by presenting B vs. D, a non-adjacent pair of stimuli that were never explicitly trained together. In theory, B and D have equivalent associative strengths because they are, in principle, equally associated with reinforcement or its absence during training. Thus, if the bees were guided by the associative strength of the stimulus, they should choose randomly between B and D. If, however, the bees used a transitive rule, they should prefer B to D. Honeybees learned the premise pairs as long as these were trained as uninterrupted, consecutive blocks of trials (Benard and Giurfa 2004). However, if shorter and interspersed blocks of trials were used, such that the bees had to master all pairs practically simultaneously, performance collapsed and the bees did not learn the premise pairs. The bees' choice was significantly influenced by their experience with the last pair of stimuli (D+ vs. E–) such that they preferred D and avoided E. In the tests, no preference for B to D was found. Although this finding agrees with the idea of an evaluation of stimuli according to their associative strength (see above), during training the bees more often visited B when it was rewarding than they visited D. Therefore a preference for B should have been expected if only the associative strength were guiding the bees' choices. From the experimental results it was concluded that bees do not establish transitive inferences between stimuli but rather guide their choices by the joint action of a recency effect (preference of the last rewarded stimulus, D) and by an evaluation of the associative strength of the stimuli (in which case preference for B should be evident). As the former supports choice of D while the latter supports choice of B, the equally frequent choice of B and of D in the tests could be explained (Benard and Giurfa 2004). In any case, memory constraints (simultaneous mastering of the different premise pairs was not possible and the predominance of the last excitatory memory over previous memories) impeded the learning of the transitive rule. Recently, Cheng and Wignall (2006) demonstrated that the failure to master several consecutive visual discrimination tasks is due to response competition occurring when animals are tested. This may explain why bees in the transitive inference protocol were unable to master the successive short blocks of training with different premise pairs.

Conclusion

Almost one hundred years of research on visual learning in bees and other social Hymenoptera, starting with Karl von Frisch's (1914) first demonstrations of color and pattern learning in bees, have yielded an impressive amount of information about how honeybees, bumblebees and wasps see the world and learn about visual cues in their environment. New discoveries in this field, inspired by the 'cognitive revolution,' which had a delayed influence on studies of insect perception and learning, have shown that besides simple forms of visual learning whose relevance can be easily conceived in the life of a bee (e.g. associating a flower color and nectar reward), social Hymenoptera also master complex forms of visual learning, ranging from conditional discriminations and observational learning to rule learning. Visual learning capabilities, therefore, allow the extraction of the logical structure of the perceptual world of insects. Such capabilities reach different levels of complexity but are not unlimited. Although the cognitive capabilities of bees and wasps may surprise due to their sophistication, limitations related to natural life seem inescapable. For instance, in the case of wasps learning facial mask patterns of conspecifics, one could imagine that memory has specific size constraints related to the small sizes of the colonies in which such wasps live. In other words, interindividual recognition is certainly possible but probably has limitations in terms of the number of individuals that can be learned and remembered. Similarly, mastering simultaneously several different associations would be facilitated if these are organized *serially or hierarchically* in chains of associations that can mediate successful navigation in a complex environment. But if these associations have to be mastered *simultaneously* at the same place, learning them would be probably difficult given the bees' biological specialization as a serial forager. In this case, learning configurations of stimuli may be more adaptive than learning each component separately.

Considering that social bees and wasps exhibit such complex forms of visual learning, the question arises, which kind of limitation do these insects present as models for the study of the mechanisms of these phenomena? So far the main limitation resides in the impossibility of addressing questions related to the cellular and molecular level. This is especially obvious when considering the achievements reached in *Drosophila*. The learning protocols used have the advantage of not restraining the animals' movements so that the behaviors recorded reflect the potential of the mini brains of social Hymenoptera. However, the protocols so far used are limiting because in a flying bee no access to the brain is so far possible. As mentioned above, new protocols in which bees learn color – reward and motion cues– reward associations under restrained conditions (Hori et al. 2006, 2007), are promising because they allow access to the neural circuits involved in these learning forms (Giurfa 2007). The critical question would then be to what extent experimental constraints limit the expression of forms of visual learning more complex than elemental associations. Why should bees and wasps continue to be attractive for research into visual cognition despite this technical limitation? The answer is simple: because the sophisticated visual performances that they exhibit, reviewed in this article, have not been found so far in the fruit fly. Future research should benefit from an analysis comparing the visual performances and mechanisms of bees and flies. The historic burden of not having a window open to the neural and molecular basis of visual learning (irrespective of the level of complexity considered), as is the case in bees, has to be overcome.

Acknowledgments

I am grateful for the support received from the French National Research Agency (ANR; Project APICOLOR), the French Research Council (CNRS) and the University Paul Sabatier. Thanks are also due to Aurore Avarguès-Weber, Julie Benard, and Silke Stach for fruitful discussions and productive collaborations.

References

Anderson AM (1972) The ability of honey bees to generalize visual stimuli. In: Wehner R (ed) Information processing in the visual systems of arthropods. Springer, Berlin, pp 207–212

Avarguès-Weber A, Portelli G, Benard J, Dyer A, Giurfa M (2010) Configural processing enables discrimination and categorization of face-like stimuli in honeybees. J Exp Biol 213: 593–601

Beier W, Menzel R (1972) Untersuchungen über den Farbensinn der deutschen Wespe (*Paravespula germanica* F., Hymenoptera, Vespidae): verhaltensphysiologischer Nachweis des Farbensehens. Zool Jb Physiol 76: 441–454

Benard J, Giurfa M (2004) A test of transitive inferences in free-flying honeybees: unsuccessful performance due to memory constraints. Learn Mem 11: 328–336

Benard J, Stach S, Giurfa M (2006) Categorization of visual stimuli in the honeybee *Apis mellifera*. Anim Cogn 9: 257–270

Bitterman ME, Menzel R, Fietz A, Schäfer S (1983) Classical conditioning of proboscis extension in honeybees (*Apis mellifera*). J Comp Psychol 97: 107–119

Camlitepe Y, Aksoy V. (2010) First evidence of fine colour discrimination ability in ants (Hymenoptera, Formicidae). J Exp Biol 213: 72–77

Campan R, Lehrer M (2002) Discrimination of closed shapes by two species of bee, *Apis mellifera* and *Megachile rotundata*. J Exp Biol 205: 559–572

Cheng K, Wignall AE (2006) Honeybees (*Apis mellifera*) holding on to memories: response competition causes retroactive interference effects. Anim Cogn 9: 141–150

Chittka L, Thomson JD, Waser NM (1999) Flower constancy, insect psychology, and plant evolution. Naturwissenschaften 86: 361–377

Chittka L, Beier W, Hertel H, Steinmann E, Menzel R (1992) Opponent colour coding is a universal strategy to evaluate the photoreceptor inputs in Hymenoptera. J Comp Physiol A 170: 545–563

Collett TS, Collett M (2002) Memory use in insect visual navigation. Nat Rev Neurosci 3: 542–552

Collett TS, Graham P, Durier V (2003) Route learning by insects. Curr Opin Neurobiol 13: 718–725

D'Ettorre P, Heinze J (2005) Individual recognition in ant queens. Curr Biol 15: 2170–2174

de Brito Sanchez MG, Chen C, Li J, Liu F, Gauthier M, Giurfa M (2008) Behavioral studies on tarsal gustation in honeybees: sucrose responsiveness and sucrose-mediated olfactory conditioning. J Comp Physiol A 194: 861–869

Delius JD, Jitsumori M, Siemann M (2000) Stimulus equivalences through discrimination reversals. In: Heyes C, Huber L (eds) The evolution of cognition. MIT Press, Cambridge, Massachusetts, pp 103–122

Dyer AG, Chittka L (2004) Fine colour discrimination requires differential conditioning in bumblebees. Naturwissenschaften 91: 224–227

Dyer AG, Neumeyer C (2005) Simultaneous and successive colour discrimination in the honeybee (*Apis mellifera*). J Comp Physiol A 191: 547–557

Dyer AG, Neumeyer C, Chittka L (2005) Honeybee (*Apis mellifera*) vision can discriminate between and recognise images of human faces. J Exp Biol 208: 4709–4714

Dyer AG, Rosa MGP, Reser DH (2008) Honeybees can recognise images of complex natural scenes for use as potential landmarks. J Exp Biol 211: 1180–1186

Ehmer B, Gronenberg W (2002) Segregation of visual input to the mushroom bodies in the honeybee (*Apis mellifera*). J Comp Neurol 451: 362–373

Fauria K, Colborn M, Collett TS (2000) The binding of visual patterns in bumblebees. Curr Biol 10: 935–938

Fauria K, Dale K, Colborn M, Collett TS (2002) Learning speed and contextual isolation in bumblebees. J Exp Biol 205: 1009–1018

Fersen L v, Wynne CDL, Delius JD (1990) Deductive reasoning in pigeons. Naturwissenschaften 77: 548–549

Frisch K v (1914) Der Farbensinn und Formensinn der Biene. Zool Jb Physiol 37: 1–238

Frisch K v (1965) *Tanzsprache und Orientierung der Bienen*. Springer Verlag Berlin-Heidelberg-New York.

Giurfa M (2004) Conditioning procedure and color discrimination in the honeybee *Apis mellifera*. Naturwissenschaften 91(5):228–231

Giurfa M (2007) Behavioral and neural analysis of associative learning in the honeybee: a taste from the magic well. J Comp Physiol A 193: 801–824

Giurfa M, Lehrer M (2001) Honeybee vision and floral displays: from detection to close-up recognition. In: L Chittka, JD Thomson (eds) Cognitive ecology of pollination. Cambridge University Press, Cambridge, pp 61–82

Giurfa M, Menzel R (1997) Insect visual perception: complex abilities of simple nervous systems. Curr Opin Neurobiol 7: 505–513

Giurfa M, Eichmann B, Menzel R (1996) Symmetry perception in an insect. Nature 382: 458–461

Giurfa M, Núñez JA, Chittka L, Menzel R (1995). Colour preferences of flower-naive honeybees. J Comp Physiol A 177: 247–259

Giurfa M, Vorobyev M, Kevan P, Menzel R (1996) Detection of coloured stimuli by honeybees: minimum visual angles and receptor specific contrasts. J Comp Physiol A 178: 699–709

Giurfa M, Zhang S, Jenett A, Menzel R, Srinivasan MV (2001) The concepts of 'sameness' and 'difference' in an insect. Nature 410: 930–933

Giurfa M, Schubert M, Reisenman C, Gerber B, Lachnit H (2003) The effect of cumulative experience on the use of elemental and configural visual discrimination strategies in honeybees. Behav Brain Res 145: 161–169

Grant V (1951) The fertilization of flowers. Sci Amer 12: 1–6

Grossmann KE (1971) Belohnungsverzögerung beim Erlernen einer Farbe an einer künstlichen Futterstelle durch Honigbienen. Z Tierpsychol 29: 28–41

Hammer M (1993) An identified neuron mediates the unconditioned stimulus in associative olfactory learning in honeybees. Nature 366: 59–63

Harnard S (1987) *Categorical perception. The groundwork of cognition*. Cambridge University Press, Cambridge

Hateren JH v, Srinivasan MV, Wait PB (1990) Pattern recognition in bees: orientation discrimination. J Comp Physiol A 197: 649–654

Heinrich B (1979) "Majoring" and "minoring" by foraging bumblebees, *Bombus vagans*: an experimental analysis. Ecology 60: 245–255

Hertz M (1929a) Die Organisation des optischen Feldes bei der Biene I. Z vergl Physiol 8: 693–748

Hertz M (1929b) Die Organisation des optischen Feldes bei der Biene II. Z vergl Physiol 11: 107–145

Hertz M (1933) Über figurale Intensitäten und Qualitäten in der optischen Wahrnehmung der Biene. Biol Zb 53: 11–40

Hori S, Takeuchi H, Arikawa K, Kinoshita M, Ichikawa N, Sasaki M, Kubo T (2006) Associative visual learning, color discrimination, and chromatic adaptation in the harnessed honeybee *Apis mellifera* L. J Comp Physiol A 192: 691–700

Hori S, Takeuchi H, Kubo T (2007) Associative learning and discrimination of motion cues in the harnessed honeybee *Apis mellifera* L. J Comp Physiol A 193: 825–833

Horridge A (2000) Seven experiments on pattern vision of the honeybee, with a model. Vision Res 40: 2589–2603

Horridge A (2003) The effect of complexity on the discrimination of oriented bars by the honeybee (*Apis mellifera*). J Comp Physiol A 189: 703–714

Horridge A (2006) Visual discriminations of spokes, sectors, and circles by the honeybee (*Apis mellifera*). J Insect Physiol 52: 984–1003

Kuwabara M (1957) Bildung des bedingten Reflexes von Pavlovs Typus bei der Honigbiene, *Apis mellifica*. J Fac Sci Hokkaido Univ Ser VI Zool 13: 458–464

Leadbeater E, Chittka L (2005) A new mode of information transfer in foraging bumblebees? Curr Biol 15:R447-R448

Leadbeater E, Chittka L (2007) The dynamics of social learning in an insect model, the bumblebee (*Bombus terrestris*). Behav Ecol Sociobiol 61: 1789–1796

Lehrer M (1997) Honeybee's visual orientation at the feeding site. In: Lehrer M (ed) Orientation and communication in arthropods. Birkhäuser, Basel, pp 115–144

Lehrer M, Campan R (2004) Shape discrimination by wasps (*Paravespula germanica*) at the food source: generalization among various types of contrast. J Comp Physiol A 190: 651–663

Lehrer M, Campan R (2005) Generalization of convex shapes by bees: what are shapes made of? J Exp Biol 208: 3233–3247

Liu G, Seiler H, Wen A, Zars T, Ito K, Wolf R, Heisenberg M, Liu L (2006) Distinct memory traces for two visual features in the *Drosophila* brain. Nature 439: 551–556

Lotto RB, Chittka L (2005) Seeing the light: Illumination as a contextual cue to color choice behavior in bumblebees. Proc Nat Acad Sci USA 102: 3852–3856

Masquelier T, Thorpe SJ (2007) Unsupervised learning of visual features through spike timing dependent plasticity. PLoS Comput Biol 3(2):e31

Menzel R (1967) Untersuchungen zum Erlernen von Spektralfarben durch die Honigbiene (*Apis mellifica*). Z vergl Physiol 56: 22–62

Menzel R (1968) Das Gedächtnis der Honigbiene für Spektralfarben. I. Kurzzeitiges und langzeitiges Behalten. Z vergl Physiol 60: 82–102

Menzel R (1985) Learning in honey bees in an ecological and behavioral context. In: Hölldobler B, Lindauer M (eds) *Experimental Behavioral Ecology and Sociobiology*. Fischer, Stuttgart, pp 55–74

Menzel R (1999) Memory dynamics in the honeybee. J Comp Physiol A 185: 323–340

Menzel R, Backhaus W (1991) Colour vision in insects. In: Gouras P (ed) *Vision and visual dysfunction. The perception of colour*. MacMillan Press, London, pp 262–288

Menzel R, Greggers U, Hammer M (1993) Functional organization of appetitive learning and memory in a generalist pollinator, the honey bee. In: Papaj D, Lewis AC (eds) *Insect Learning: Ecological and Evolutionary Perspectives.* Chapman and Hall, New York, pp 79–125

Menzel R, Steinmann E, de Souza J, Backhaus W (1988) Spectral sensitivity of photoreceptors and colour vision in the solitary bee, *Osmia rufa.* J Exp Biol 136: 35–52

Menzel R, Geiger K, Müller U, Joerges J, Chittka L (1998) Bees travel novel homeward routes by integrating separately acquired vector memories. Anim Behav 55: 139 -152

Miller GA (2003) The cognitive revolution: a historical perspective. Trends Cognit Sci 7: 141–144

Núñez JA (1982) Honeybee foraging strategies at a food source in relation to its distance from the hive and the rate of sugar flow. J Apicult Res 21: 139–150

Pavlov IP (1927) *Lectures on conditioned reflexes*. International publishers, New York

Robertson I (2001) *Problem Solving*. Psychology Press, Hove

Schroter U, Malun D, Menzel R (2007) Innervation pattern of suboesophageal ventral unpaired median neurones in the honeybee brain. Cell Tissue Res 327: 647–667

Schubert M, Giurfa M, Francucci S, Lachnit H (2002) Nonelemental visual learning in honeybees. Anim Behav 64: 175–184

Schwaerzel M, Müller U (2006) Dynamic memory networks: dissecting molecular mechanisms underlying associative memory in the temporal domain. Cell Mol Life Sci 63: 989–998

Shepard RN (1987) Towards a universal law of generalisation for psychological science. Science 237: 1317–1323

Spaethe J, Tautz J, Chittka L (2006) Do honeybees detect colour targets using serial or parallel visual search? J Exp Biol 209: 987–993

Spence KW (1937) The differential response to stimuli varying within a single dimension. Psychol Rev 44: 430–444

Srinivasan MV (1994) Pattern recognition in the honeybee: recent progress. J Insect Physiol 40: 183–194

Srinivasan MV, Zhang SW, Zhu H (1998). Honeybees link sights to smells. Nature 396: 637–638

Srinivasan MV, Poteser M, Kral K (1999) Motion detection in insect orientation and navigation. Vision Res 39: 2749–2766

Stach S, Giurfa M (2005) The influence of training length on generalization of visual feature assemblies in honeybees. Behav Brain Res 161: 8–17

Stach S, Benard J, Giurfa M (2004) Local-feature assembling in visual pattern recognition and generalization in honeybees. Nature 429: 758–761

Swinderen Bv, Greenspan RJ (2003) Salience modulates 20–30 Hz brain activity in *Drosophila*. Nature Neurosci 6: 579–586

Tang S, Guo A (2001) Choice behavior of *Drosophila* facing contradictory visual cues. Science 294: 1543–1547

Terrace HS, McGonigle B (1994) Memory and representation of serial order by children, monkeys and pigeons. Curr Dir Psychol Sci 3: 180–185

Tibbets EA (2002) Visual signals of individual identity in the wasp *Polistes fuscatus*. Proc Biol Sci 269: 1423–1428

Tibbets EA, Dale J (2004) A socially enforced signal of quality in a paper wasp. Nature 432: 218–222

Troje F, Huber L, Loidolt M, Aust U, Fieder M (1999) Categorical learning in pigeons: the role of texture and shape in complex static stimuli. Vis Res 39: 353–366

Wehner R (1967) Pattern recognition in bees. Nature 215: 1244–1248

Wehner R (1971) The generalization of directional visual stimuli in the honeybee *Apis mellifera*. J Insect Physiol 17: 1579–1591

Wehner R (1972) Dorsoventral asymmetry in the visual field of the bee, *Apis mellifera*. J Comp Physiol A 77: 256–277

Wehner R (1981) Spatial vision in arthropods. In: Autrum H (ed) *Invertebrate Visual Centers and Behavior. Handbook of Sensory Physiology VII/6C.* Springer, Berlin, pp 287–616

Wilson EO (1971) *The Insect Societies*. Belknap Press, Cambridge, pp 548

Wittstock S, Menzel R (1994) Color learning and memory in honey bees are not affected by protein synthesis inhibition. Behav Neural Biol 62: 224–229

Wolf E (1933) Critical frequency of flicker as a function of intensity of illumination for the eye of the bee. J Gen Physiol 17: 7–19

Wolf E (1934) Das Verhalten der Biene gegenüber flimmernden Feldern und bewegten Objekten. Z vergl Physiol 20: 151–161

Wolf E, Zerrahn-Wolf G (1935) The effect of light intensity, area, and flicker frequency on the visual reactions of the honey bee. J Gen Physiol 18: 853–863

Wolf R, Wittig T, Liu L, Wustmann G, Eyding D, Heisenberg M (1998) *Drosophila* mushroom bodies are dispensable for visual, tactile, and motor learning. Learn Mem 5: 166–78

Worden BD, Papaj DR (2005) Flower choice copying in bumblebees. Biol Lett 1: 504–507

Wüstenberg D, Gerber B, Menzel R (1998) Long- but not medium-term retention of olfactory memories in honeybees is impaired by actinomycin D and anisomycin. Eur J Neurosci 10: 2742–2745

Yang EC, Maddess T (1997) Orientation-sensitive neurons in the brain of the honey bee (*Apis mellifera*). J Insect Physiol 43: 329–336

Zentall TR, Galizio M, Critchfield TS (2002) Categorization, concept learning and behavior analysis: an introduction. J Exp Anal Behav 78: 237–248

Zhang S, Srinivasan MV (1994) Prior experience enhances pattern discrimination in insect vision. Nature 368: 330–333

Zhang SW, Bartsch K, Srinivasan MV (1996) Maze learning by honeybee. Neurobiol Learn Mem 66: 267–282

Zhang SW, Lehrer M, Srinivasan MV (1999) Honeybee memory: navigation by associative grouping and recall of visual stimuli. Neurobiol Learn Mem 72: 180–201

Zhang SW, Srinivasan MV, Zhu H, Wong J (2004) Grouping of visual objects by honeybees. J Exp Biol 207: 3289–3298

Zhang S, Bock F, Si A, Tautz J, Srinivasan M (2005) Visual working memory in decision making by honey bees. Proc Natl Acad Sci USA 102: 5250– 5255

Zeil J, Kelber A, Voss R (1996) Structure and function of learning flights in ground-nesting bees and wasps. J Exp Biol 199: 245–252

Evolution of cognition: a comparative approach

8

Ludwig Huber and Anna Wilkinson

Contents

Abstract

Evolutionary biology has uncovered remarkable anatomical and physiological similarities among animal species (including humans). The most parsimonious assumption is that they may also share some cognitive and behavioural traits. This idea was put forward by Charles Darwin, who suggested that humans are not separated from the animal kingdom in terms of cognition, but rather, that there is mental continuity across species. This hypothesis has now transformed into the flourishing field of Cognitive Biology, which examines the cognitive abilities of non-human animals from a comparative perspective. In this chapter we describe the historical roots of the field and discuss why an evolutionary account alone is insufficient for a full understanding of cognition in animals. We emphasize the importance of studying a behaviour at four different levels: phylogeny (history), adaptation (ultimate function), proximate mechanisms (neurosciences) and development (ontogeny) before it can be entirely understood. Finally we review a selected subset of recent results and discuss their implications for our understanding of the evolution of cognition.

Ludwig Huber
Vetmeduni Vienna
Comparative Cognition Research
Messerli Research Institute
Veterinärplatz 1
1210 Vienna, Austria
e-mail: ludwig.huber@vetmeduni.ac.at

1. Introduction

In traditional Western thinking, humans are considered superior to the rest of the animal kingdom; they were thought to possess qualities and capabilities that are either unique or greatly exceed those found in other species. However, as science advanced throughout the 19th century, it became increasingly accepted that, at least in terms of anatomy and physiology, humans are simply another representative of the animal kingdom. Despite this obvious insight many philosophers continued to believe that in terms of cognitive abilities, humans were different: Only humans have consciousness, can argue, think and plan, only humans are responsible for their behavior and have morality.

Charles Darwin himself originally doubted that natural selection could explain the evolution of human cognition, because he could not see how natural selection was responsible for the stepwise progression to complex organs such as the eye or the brain. The problem for him and his contemporaries was to understand how complex organs could have evolved if their proper function depends on the interaction of mutually adjusted forms of each of its components. In his notebook he wrote: "*The soul by the consent of all is superadded. Animals not got it*". Later Darwin abandoned his initial skepticism and in *The Descent of Man* (1871) he argued against the idea that the difference between human and non-human animals, which is so small with respect to morphology, is fundamental when it comes to mental prowess.

Although biologists nowadays are inclined to accept Darwin's intuition, that where there has been evolution there must also be continuity, it is by no means self-evident and requires strong empirical support. But where might this support come from? At the brink of a new millennium, how close are we to an understanding of the mind? An answer to the question "What does it mean to become human?"[2] requires understanding of the origins and evolution of human cognitive faculties.

To explain human cognition as the result of evolution through natural selection is tempting. However, Richard Lewontin (1998), one of the most influential evolutionary biologists of the 20th century, acknowledged the difficulty of such an effort. Despite the existence of a vast and highly developed mathematical theory of evolutionary processes we still know little about the evolution of our cognitive capabilities. Is our situation really hopeless then? Are there any fields of research that have made substantial progress in uncovering cognitive processes in an evolutionary context? Or are we confined to the difficult and sometimes frustrating enterprise of 'Evolutionary Psychology' (the study of the evolution of cognition in hominids)? We believe not. In this paper we will outline an alternative, broader enterprise: Biology of cognition. We will first describe the historical roots of this field and then discuss why an evolutionary account alone is insufficient. Finally we will review a selected subset of recent results and discuss their implications for our understanding of the evolution of cognition.

2. Historical roots of a biology of cognition

Before Charles Darwin published the "Origins", Herbert Spencer (1855) laid the foundation for evolutionary psychology, emphasizing that "*Mind can be understood only by showing how mind is evolved*". With this claim he contributed to 'naturalizing the mind'. This is opposed to the Cartesian view that mind is substantially distinct from the rest of nature and to theories that link mental properties with the possession of an immaterial soul or spirit.

A specific version of this kind of naturalism, called *Evolutionary Epistemology* (Campbell

[1] The title of a special initiative within the 6th EU Programme for Research

1974), subscribes to the idea that cognition should be considered a product of biological evolution (Riedl 1984).

2.1 Evolutionary epistemology

Despite noteworthy differences in focus and emphasis among its adherents, the ultimate aim of Evolutionary Epistemology (EE) is to explain *"man's status as a product of biological and social evolution"* (Campbell 1974, p. 413). Biological evolution is regarded as a precondition of the cognitive, cultural, and social behavior that an organism, group or species can portray. Therefore, as a kind of satellite theory of evolutionary theory, EE provides a framework for the study of the origin, evolution and mechanisms of all cognitive capacities of all biological organisms. These emerge from the perceptual processes of animals and lead to human-specific cognitive products such as science, culture and language.

EE emphasizes the role of natural selection in two ways. In the first, selection is the generator and maintainer of the reliability of our senses and cognitive mechanisms, creates the "fit" between those mechanisms and the real world. In the second, trial and error learning and the evolution of scientific theories are construed as selection processes (Bradie 1986). In this chapter we will focus mainly on the first.

Darwin was, and remains, the 'prime mover' behind this naturalistic movement. The capacities of humans for reasoning, knowledge and belief are the products of natural evolution. The underlying cognitive structures and processes therefore could and should be analyzed from a biological viewpoint. Their original function is to make survival and reproduction of an organism more likely. Thus the phylogenetic evolution of knowledge depends on the degree to which its carrier survives natural selection through its environment. This is the classical, adaptationist approach to EE.

One of its most influential representatives was Konrad Lorenz (1977). For him, cognition first and foremost was a phenotypic trait, which all organisms show to some extent. Like all behavioral traits, it is partly determined by genetics and its structure and function reflects adaptation, an ongoing, non-teleological process of interaction between the organism and its environment. A cognitive trait tending to increase fitness in a given population should therefore become more common over time, and a trait that tends to prevent its carriers from passing on their genes should show up less and less frequently.

Thus, through adaptation, there is a correspondence between our sensory and perceptual representation of the world and the world itself, or between organism and environment. This of course is not a one-to-one correspondence; our image of a tree is not the real tree, but because our cognitive apparatus is adapted to the world, there is a *partial isomorphism* between the two (Lorenz 1977).

In the remainder of this chapter we will expand upon one branch of contemporary cognitive science, which is under-represented in the ongoing (popular) debate about the evolution of cognition. We start with the notion that a naturalist position with regard to the mind does not require the assumption of mental phenomena in nonhuman organisms. *"Skepticism about animal minds is one of a number of pieces of the Cartesian legacy whose defeat would go some way toward vindicating naturalism"* (Allen and Bekoff 1997, p. 12). New life can be breathed into EE through three related endeavors. The first, to distinguish between the different cognitive processes that organisms from all major kingdoms of life can display. The second, to find out how these various cognitive structures and processes are distributed across the animal kingdom. And finally, to examine the products of cognition within an evolutionary framework.

2.2 How to study animal cognition properly

The study of cognitive processes in animals has emerged as a major field of research. One reason for this is the synthesis of comparative psychology and ethology. Though both scientific areas have traditionally been concerned with cognitive issues, the integration into a coherent field of research did not happen until the 1960s (for two recent reviews, see Bekoff et al. 2002, Wasserman and Zentall 2006).

In comparative psychology behaviorism developed as a scientific alliance of cerebral physiology (Pavlov 1927) and learning theory (Thorndike 1898), entirely confining itself to the study of now-and-forever causation (mechanism). Despite using the adjective 'comparative' to label their discipline, these psychologists studied a very small sample of species only (e.g. rats, pigeons and rhesus monkeys) and in sterile laboratory settings. Instead of considering natural behavior, artificial tasks were created to guarantee experimental rigor and straightforward analysis. Ethological work was considered virtually worthless and contaminated by anecdotal and anthropomorphic errors.

In contrast, ethologists focused on phylogenetic comparisons of species-specific behavior patterns in closely related groups to identify homologies (defined by common descent). They also worked on ecological comparisons based on the form and function of behavior (analogies, defined by convergent selection pressures). Lorenz believed that homologous traits of behavior depend on conserved genetic programs and therefore considered the evolutionary development of behavior patterns inevitable. However, Lehrman (1953) argued that individual development is not only a simple inevitable unfolding of some inner (genetic) potential, but a highly variable process resulting from a cascade of complex interactions between genetic information, the developing features of the individual, and the environment.[2]

2.3 Cognitive ethology

The 1960s and 1970s saw two 'scientific revolutions' in the study of animal behavior: the cognitive revolution in psychology, and the behavioral ecology and sociobiology revolution in ethology. However, most psychologists remained ignorant about the behavior of animals in their natural environment, and ethologists remained skeptical about the study of advanced cognitive processes.

In 1976, Donald Griffin published a highly influential book (*The Question of Animal Awareness*) in which he integrated psychology and ethology. With this and several follow-up books he sought to facilitate the study of taboo subjects like consciousness, choice, planning, intentions, and other mental processes in non-human species. Griffin and his followers called themselves 'cognitive ethologists' to stress their strong relationship with (classical) ethology. However they viewed the application of cognitive science to ethology as conceptually richer than Lorenzian constructs.

By the time cognitive ethology emerged, many of the old distinctions between subdisciplines in the research of animal behavior had fallen away. A major event in the resolu-

2 The behaviorists also recognized this evolutionary shaping of learning that leads to an increase of the fitness of the learner, but their biological turn (termed 'constraints on learning') happened quite late and in complete ignorance of the writings of Lorenz and other European ethologists

tion of the problems was Tinbergen's seminal paper *'On aims and methods of ethology'* (1963), in which he emphasized the necessity of understanding behavior at four different levels: phylogeny (history), adaptation (ultimate function), proximate mechanisms (neuroscience) and development (ontogeny). This integrative, multilevel approach has proven to be a progressive and successful research program.

3. Animal cognition: the core assumptions

Animal cognition addresses phenomena as different as spatial navigation in invertebrates, birds and mammals; the perception and production of auditory signals; and the comprehension and use of symbolic communication systems. This diversity signals both the strength and weakness of the cognitive approach which can appear fragmented, and lacking a central definition. Many different labels are used to describe it: animal cognition, comparative cognition, cognitive ecology, cognitive ethology, etc. We therefore have to ask what the core assumptions of animal cognition research are.

In the contrast to non-cognitive approaches, most cognitive researchers postulate processes that organize stored and actual sensory information into coherent internal representations of external events, which allow the animal to respond appropriately to stimuli when experiencing them in novel combinations or contexts. The central structural assumption of the cognitive research program is that organisms possess some type of internal representation of the external world. Examples include categorical or abstract concepts, cognitive maps, episodic-like memories and intentional plans. These internal representations are related to the external world in two ways: a coding process on the input side and a relationship to behavior on the output side. From a neurobiological point of view, the term cognition designates brain functions that exclude primary sensory and motor functions, autonomic brain functions, reflexes and reflex-like stereotyped behavior. Cognition thus includes such diverse functions as perception, learning, memory, expectation, and planning, whether accompanied by consciousness or not. From this it follows that cognition is not necessarily restricted to human beings and cognition research needs proper comparison between species based on sound evolutionary questions.

3.1 Comparing cognitive traits

Let us start with the fundamental functional question. What are animal minds (including the human mind) *for*? More specifically: Is it possible to make a general statement about the kinds of selective pressures and advantages that are responsible for the evolution of cognitive mechanisms? Why has the expensive and delicate biological machinery underlying mental life evolved? This question should be approached not only in terms of the continuities between human mental capacities and cognition in nonhuman animals (Darwin's primary question), but also in recognition of continuities between cognition and a wider class of "protocognitive" phenomena. Cognitive mechanisms serve to control behavior; they comprise a subset of a class of mechanisms that enable organisms to adapt to changing environments.

Ethologists, like Lorenz, emphasized that learning is only one of several ways to cope with an unpredictable environment. If behavior is the means of adjusting the organism to external and internal environmental changes, then the behavior must be sufficiently plastic in the face of unexpected change. Thus organisms must operate efficiently both under known conditions, and under new conditions with a minimum of disruption. The solution is a tradeoff between 'fast and frugal heuristics' and behavioral flexibility.

What kinds of cognitive processes have evolved in the history of life? Unfortunately, the phylogenetic investigation of the major transitions in cognition termed 'psychophylogenesis' by Huber (2000) has been underrepresented compared to *ecological*, *microevolutionary*, or *proto-evolutionary* concerns (Timberlake 1993). Even worse, at least four different proto-evolutionary, anthropocentric approaches, have dominated traditional approaches to cognition. These traditions have been concerned with a) whether animals comprehend what they are doing, b) how the various species can be sorted according to their performance on standard intelligence tests, c) whether (or to what degree) animals do what we ourselves do, and d) whether or not it is possible to 'shape' intelligent performance in animals or to create artificial cognitive systems. In behavioristic psychology the focus was mainly on the assessment of (functional) similarity, rather than asking questions about how and when these abilities came into existence (Hodos and Campbell 1969). There has been little concern about specific phylogenetic relations between species or about the selection pressures that may have produced them.

In biology, the *ecological* approach has focused on describing the evolution of cognition by tracing the path of specialized adaptations to the demands of the environment. Behavioral ecologists have concentrated on closely related species that fill divergent niches and distant relatives that fill convergent niches (Shettleworth 1998). They commonly suppose that cognitive faculties (e.g. recovering food) are the result of selective forces exerted by specific environmental circumstances (e.g. periods of meager food supply). However, evolution is seldom as straightforward as this. The adequate metaphor is to view evolution as a 'tinkerer', building innovations by modifying existing structures, rather than by designing them from scratch. Unfortunately, the reconstruction of the sequence of cognitive breakthroughs lies outside the research program of most traditional ecological learning theorists or behavioral ecologists.

The crucial question, whether the ultimate function of a behavior can explain the underlying mechanism, cannot be answered using the ecological approach. By taking a comparative approach one can look at examples of enhanced function in search for clues to what specific brain areas do. Nevertheless, functional considerations in an evolutionary sense cannot fully explain the neural mechanisms of behavior in general, and of learning and memory in particular (Bolhuis and Macphail 2001). Evolutionary study has two major problems: to elucidate the course that evolution must be assumed to have taken, and to unravel its dynamics. So what are the key principles of evolutionary change?

3.2 Evolutionary questions

Evolution is considered as a change in the inherited traits of a population from one generation to the next. These changes are caused by a combination of three main processes: variation (differences between genotypes underlying the traits, mainly caused by sexual reproduction and mutation), reproduction (heritable traits are propagated between generations), and selection (individuals with advantageous traits are more likely to survive and reproduce).

In the animal kingdom there is not one brain, one sense, one problem, and one solution. At every level of organization, variation prevails. Of great value to understanding how novel features emerge is the study of the relationships between individual development and phenotypic change during evolution (the new science of 'Evo-Devo'; see Müller 2007, for a recent review). The inescapable difficulty is to track the process of innovation and variation (resulting in the evolution of new traits) on the one hand, and that of stabilization and conservatism (resulting in traits remaining stable across time) on the other. Why have many solutions to the same problem evolved in some cases, and in other cases only one? Why did some structures not disappear, despite their obvious redundancy? Why do the receptor cells in the retina face backwards in vertebrates but not in many invertebrates?

3.3 A systemic viewpoint

Understanding the main classes of evolutionary processes – innovation and constraint – and their interrelationship, requires the adoption of a systemic viewpoint. Generally, the maintenance of equlilibrium is the most fundamental characteristic of all self-reproducing systems. Riedl (1995) distinguished two important principles: (1) *correspondence* with the conditions external to the system (the organism), and (2) *coherence* with the internal conditions of the system. The processes contributing to the latter task are mostly obscure to the external observer, and were underestimated by behaviorists who focused on the processes that lead to observable changes in behavior. Consequently, the processes involved in supporting adaptation to external circumstances were considered one-sided, requiring the animal to 'track' environmental changes, with information being fed into the organism from outside and thereafter stored either by genetic processes during phylogeny or by memory processes during ontogeny.

Systemic approaches have challenged this view by suggesting that external influences are only a kind of 'trigger' or 'selector' of internal modifications, and depend on the limited potentials of the organism. Growing from this notion, it has been emphasized that cognition is not only the representation of a pregiven world by a pregiven mind but rather the construction of a world and a mind based on the history of actions that a being performs in its world (Maturana and Varela 1979).

Learning must thus be viewed as a process that occurs in accordance with those processes that operate as control mechanisms to keep the system in coherence and correspondence. These conservative processes are the result of an earlier process of natural selection that has led to a high degree of functional coupling between the organism and its environment. This high degree of adaptation is responsible for the fact that only a limited amount of information is acquired during the process of ontogenetic learning. Learning constraints are the result of past selection events which now exert their influence through genetic shaping of those structures of the brain that subserve learning and memory. It is no coincidence that current best estimates suggest that more than half of the genetic information in humans is directed towards the construction of the central nervous system. However, gene products are a minute fraction of the total number of behavioral determinants. Another small fraction is caused by environmental factors. Most importantly, however, the vast majority of deterministic factors reside in the multitude of, as yet unpredictable, interactions between genetic and environmental factors (Greenspan 2001). Understanding exactly how the information contained in our genes is translated into the structures of our brains will be one of the great growth areas of twenty-first-century science.

4. Examples of cognitive abilities of non-human animals

Throughout evolution, animal brains, both big and small, have solved recurring problems as their owners orient through complex environments, find food, escape predators, reproduce etc. The most frequently adopted assumption is that small brains apply simple and rigid solutions, while larger brains are more flexible and creative. However, recent research provides considerable evidence against such a view. In the case of insects, for example, we find high flexibility, and cross-communication between different behavioral routines. Thus, it is appropriate to conceptualize insect behavior from a cognitive perspective. We devote this final section to a short review of recent findings in the perceptual, physical and social domain of cognition in non-human animals. (see also Chapter II, 4 by M Giurfa)

4.1 Perceptual cognition

Categorization is probably the most fundamental component of cognition. It allows animals to treat similar but non-identical stimuli as equivalent, by sorting them into their proper categories and reacting to them similarly. The brains of many animals solve this task in a natural, effortless manner and with an efficiency that is difficult to reproduce in computational models and artificial systems.

What is the functional value of categorization? Considering the vast amount of potential information arriving at the perceptual systems and the limited number of behavioral output patterns possible in some non-human animals, categorization may be understood as an adequate solution to this 'information bottleneck'. A drastic reduction of information is considered a fundamental principle of cognitive economics and can be found widely among species. The evolutionary pressures to minimize signal processing requirements are self-evident. How is this accomplished?

To behave appropriately an animal needs a description of those stimuli that are predictors of significant consequences. Somehow the animal must select particular aspects of experience and use them as a basis for judgments. Different occasions may dictate different descriptions, thus categories are neither purely perceptual nor merely functional, but both. Categorization problems are co-determined by the physical variation of the stimuli and the consequences for behavior. Evidence suggests that natural selection has equipped animals with adaptations for dealing with these problems.

Animals may represent (encode) classes of stimuli in (at least) six rather distinct ways: (i) pictorially as arrays of features or elements defined in their own absolute values; (ii) associatively as collections of objects or events signaling the same consequence or follow-up event; (iii) functionally as collections of items with the same inherent function; (iv) abstractly as relations between two or more objects or events; (v) analogically, as relationship between two or more other relationships, and (vi) symbolically as relations to other classes of stimuli (Huber in press).

Although it is tempting to think that 'lower' animals are confined to the first two or three mechanisms of this list, while 'higher' animals, like primates have progressed to the more advanced levels, comparative research has challenged this view. As birds are highly mobile animals, vision is of particular importance to them. This fact is reflected in the huge relative size of their eyes and the large portion of their brain that is dedicated to visual processing. On an absolute scale, however, bird brains are quite small. The brain of a pigeon (*Columba livia*) is about

1/1000 the size of our own. Nevertheless, pigeons are able to solve many of the perceptual problems that we can. In our lab we have demonstrated that pigeons can categorize highly complex visual stimuli (reviewed in Huber and Aust 2006). Furthermore, they can form concepts like 'food', 'tool' or 'enemy', mental constructs and abstract representations.

Memorization, feature learning, and acquired equivalence are sufficient to account for many instances of categorization. The underlying perceptual and associative processes seem to be widely distributed across the animal kingdom. There is evidence that some corvids and parrots, as well as dolphins, some monkeys and apes, show excellent transfer to wholly novel stimuli after being trained with relational classes. Pigeons can also do this, but require a much larger stimulus set to show convincing transfer. Finally, the learning and understanding of higher-order relations between relations seems to require explicit symbol training. Monkeys and pigeons have so far failed to acquire symbolic referential meaning. Only a few chimpanzees and a grey parrot (*Psittacus erithacus*), all previously language trained (e. g. lexigrams, sign language, vocalizations), have shown this ability (reviewed in Bekoff et al. 2002). Though essentially related, this ability is not equivalent to the abilities of humans, who use relational concepts as an integral part of language.

Recent findings of surprisingly advanced cognitive abilities in invertebrates have posed a huge problem for the search for causal links between brain size and intelligence. Although the brain of a honeybee contains only 960 000 neurons (approximately hundred and fourteen thousand times less than our human brain) and its volume is only 1 mm^3, it supports impressive behavioral capabilities (review in Menzel and Giurfa 2006; Giurfa 2007). Honeybees are equipped with sophisticated sensory systems and have well developed learning and memory capacities, whose essential mechanisms do not differ drastically from those of vertebrates. Applying a cognitive, neuroethological approach, researchers have shown that these animals exhibit complex, non-elemental forms of learning, such as contextual learning, categorization and learning of abstract rules. In particular, they can learn colors and shapes, can generalize abstract features, such as figure orientation and symmetry. They also readily learn cross-modal associations, for example between the scent and color of flowers showing more flexible abilities than is seen in many mammalian and avian species (see also Chapter II, 4 by M Giurfa).

4.2 Physical / technical cognition

The ability to act on information flexibly is one of the cornerstones of intelligent behavior. A particularly informative example, tool-oriented behavior has been used to determine to what extent non-human animals understand means-end relations and object affordances, and have specific motor skills for these tasks. Further, planning with foresight, goal-directed problem solving and causal inference have also been a focus of this research. These abilities may not be restricted to humans or other tool-using animals but may also be found in animals that show high levels of curiosity, object exploration and manipulation.

When animals interact with non-animate objects, they may gain an understanding of the 'folk physics' of time, space and causality. For many decades, researchers investigating tool use in non-human animals suggested that the mechanisms underlying this behavior were simply trial-and-error exploration and fortuitous discovery of solutions to problems reinforced through conditioning. We believe that this view underestimates the cognitive abilities of many species. Using the kea (*Nestor notabilis*), a New Zealand parrot, as a model species, we have studied causal understanding, insightful behavior, object permanence, the understanding of tertiary relations, and other forms of sensory-motor intelligence (Huber and Gajdon 2006). By using a truly Tinbergian 4-whys approach, we have examined how these capacities develop during ontogeny. Through hand-raising animals we have studied cognitive development in both the physical and the social domain. Further, our integration of results from laboratory experiments and field observations of keas have revealed surprising cognitive capacities in both domains.

In one experiment we investigated how captive keas face a novel problem, the 'string-pulling' task, where food objects are suspended from a long string which the animal has to pull up to gain access to the food (Werdenich and Huber 2006). In this popular test for means-end comprehension we found that the majority of keas completed the first trial within a few seconds, showing only goal-directed behavior and executing the solution in a manner that could not be improved upon in nine further trials. We interpreted the performance of the keas as insightful in the sense that they were sensitive of the relevant functional properties of the task and thereby produced a new, adaptive response without trial-and-error learning. This was confirmed in various two-string discrimination tests differing in the attachments and the spatial relationship of the strings (e. g., crossed or slanted strings).

Another benchmark test for investigating the understanding of spatial relationships between objects is the 'support problem' (Fig. 1), a paradigm that has previously only been studied in primates. It has been considered a special case of a weak physical connection between two objects, afforded by the weight of one object resting on the other (i. e., the 'support'). Keas performed better than tamarins when they were confronted with a choice between two support devices (wooden slats), one of which consisted of a choice between a reward resting on it and the other slightly next to it, another between a continuous and a disrupted support (Auersperg et al. 2009). These tasks allow an initial assessment of the animals' ability to appreciate the distinction between physical connection and mere contact. Interestingly, the keas performed better than chimpanzees in some tasks in which the perceptual connection of the food to the support was altered. Obviously, they pulled a support as a 'means' to reach a desired 'end'.

Together, these findings contribute to the ongoing debate regarding the distribution of higher cognitive skills in the animal kingdom by showing high levels of sensorimotor intelligence in animals that do not use tools. Several new findings have challenged the theory behind the evolution of tool use, which had been seen as one of the crucial cognitive distinctions between man ('the toolmaker') and the rest of the living universe. When Goodall (1986) reported that chimpanzees use sticks to retrieve termites from their nests, and later Boesch and Boesch (1990) investigated the nut-cracking behavior of this species, the 'right' order was maintained. This allowed the claim that only man's closest relatives are able to use 'proto-tools' (McGrew 1992) and also more recently, complex tools (Carvalho et al. 2008).

A greater challenge to this '*scala naturae*' approach to tool use behavior came with the exciting findings of tool using monkeys and birds. Like chimpanzees, wild bearded capuchin monkeys (*Cebus libidinosus*) use stones both as cracking tools and as anvils, thereby also showing a combination of different elements. When faced with stones differing in the functional features of friability and weight, they choose, transport, and use the more effective stone to crack nuts (Visalberghi et al. 2009). Moreover, when

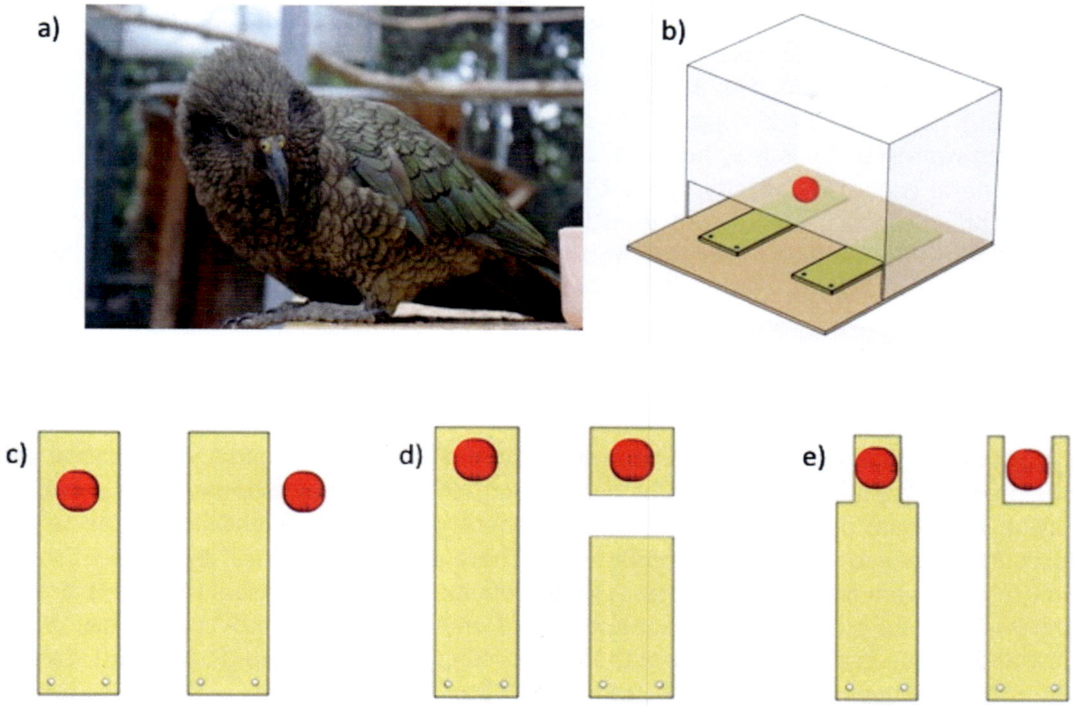

Fig. 1 Keas, the alpine parrots of New Zealand, are spontaneously able to apply the concepts of support and connectivity when confronted with a choice between two wooden slats. **a** a young kea; **b** a sketch of the experimental apparatus; **c** two equal rewards (peanuts) at the same distance from the kea, one on the slat, one next to the other slat; **d** both slats baited but one is interrupted by a gap; **e** the reward appeared less contained by the slat of the correct option than the incorrect one. (Drawing by A. Auersperg)

weight cannot be judged by visual attributes, capuchins act to gain information to guide their selection, thereby outperforming their captive counterparts (Schrauf et al. 2008).

New Caledonian crows (*Corvus moneduloides*) have become famous for their highly sophisticated and diverse tool manufacture and use, both in the wild (Hunt 1996) and in captivity (Weir et al. 2002). In contrast to woodpecker finches (*Cactospiza pallida*), which also show some manipulation of sticks to retrieve food (insect larvae) from tree holes, it has been claimed that these birds have come close to early human tool making by not only modifying an impressive range of stick and leaf tools with some selection and trimming, but even fine sculpting. This evolution of 'crafting' is technologically important because it allows the open-ended development of tools (McGrew 1992).

A common assumption about the origin of the monkey/ape shift in intelligence is a selection pressure based on increased efficiency in foraging behavior. The evidence from New Caledonian crows asks for a broader view. The bending of wire by a captive New Caledonian crow to make hooks from straight pieces in order to retrieve food (Weir et al. 2002) has previously confirmed the view that tool use had evolved in this species as a result of cognitive specialization and ecological pressures. However, rooks (*Corvus frugilegus*) have now been shown to be capable of insightful problem solving related to sophisticated tool use. They spontaneously modify and use a variety of tools, shaping hooks out of wire, and applying a series of tools in a particular sequence to gain a reward (so-called meta-tool use; Bird and Emery 2009). These findings suggest that the ability to represent tools may be the result of a general cognitive capacity rather than an adaptive specialization. They question the relationship between physical intelligence and tool use in the wild.

4.3 Social cognition

Among all possible domains in which cognition provides the adaptive fine-tuning of behavior, the social domain has become the most interesting one. Social cognition not only refers to general learning processes that are employed for group behavior, but also to new and unique strategies to control the relationships between con-specifics and to transmit information between individuals and generations.

Humphrey (1976) suggested that the physical environment did not present the kind of challenges which would lead to the evolution of a flexible, intelligent mind, but that the social environment did. He proposed that it was the intricate social interactions of primates, their ability to recognize individuals, track their relationships and deceive one another, which occupied their time and substantial brainpower. This has been named the 'Social Intelligence Hypothesis' (SIH). The social environment is ever changing and largely unpredictable, particularly in those societies in which individuals have a history of interactions with other individuals. Humphrey's proposal has blossomed with the support of positive data, and mutated into the 'Machiavellian Intelligence Hypothesis' (Byrne and Whiten 1988) or 'Social Brain Hypothesis' (Dunbar and Shultz 2007), which focus on the manipulative and deceptive aspects of social life and the relationship between sociality and brainpower. From its inception, the social intelligence hypothesis was discussed primarily with respect to the evolution of primate intelligence, however it rapidly became clear that other social animals, such as dolphins, hyenas, elephants, corvids and parrots, also demonstrated many of the biological, ecological and behavioral preconditions for intelligence (de Waal and Tyack 2003, Emery et al. 2008).

A particularly interesting phenomenon is the relation between social learning and culture. The last two decades have seen a virtual explosion in research investigating the role of social interactions in the development of animals' behavioral repertoires. There is recent empirical evidence of social influences on food choice, tool use, patterns of movement, predator avoidance, mate choice, and courtship (Galef and Laland 2005).

One experimental focus is on the mechanisms that control learning through observing a conspecific and their contribution to the transmission of innovations. Experiments have been conducted with a wide range of animals to investigate their ability to imitate, i. e., to learn part of a demonstrated behavior. Here again we have witnessed a paradigm shift, starting with a long-held belief that only humans can imitate, then allowing great apes to show this skill, and only in the last decade acknowledging that in addition other taxa are living in an imitative universe (Heyes and Galef 1996).

A recurring question in imitation research is what exactly is copied and what information the observer uses from the observed action. In many situations, instead of precisely copying actions of others, it is more useful to understand the goal of the demonstrator's actions, and only to copy those elements that are relevant to the task or preferred by the observer. Furthermore, animals may learn through observation how the environment works, by learning about the affordances of objects or causal relationships between them. Experimental evidence of the latter ability has been found in keas that learned to dismantle several locking devices through observation, but did not copy the demonstrated actions (Huber et al. 2001).

So-called demonstrator-consistent responding implies that the subjects copied some part of the observed actions. Interestingly, the largest body of evidence comes from birds; budgerigars (*Melopsittacus undulates*), European starlings (*Sturnus vulgaris*), pigeons and Japanese quail (*Coturnix japonica*) have provided evidence for action imitation (see Zentall 2004 for a review). In our laboratory we have studied the imitative abilities of New World monkeys, common marmosets (*Callithrix jacchus*). The most stringent test of whether animals can *learn* a new movement by observation involves the demonstration of at least one action that is unlikely to be performed by animals without having had the opportunity to witness its performance (Bugnyar and Huber 1997; Voelkl and Huber 2000). From a neuropsychological point of view, it is interesting to consider how it is possible to transform visual information into matching motor acts (Brass and Heyes 2005). Here, the important issue is not the learning of a complex skill, but determining the copying fidelity of animals at different levels of behavioral organization. Recent studies with marmosets have provided evidence of very precise copying of movements, challenging current theories of imitation in terms of associative learning, human-specific adaptations and mirror neurons (Voelkl and Huber 2007). Surprising findings from archer fish (*Toxotes chatareus*) suggest that precise movement copying is widespread in the animal kingdom. These fish seem capable of learning by observation from a skilled conspecific how to anticipate the path of a moving prey item when hunting (knocking down with a precisely aimed shot of water) aerial prey (Schuster et al. 2006).

Experiments with dogs (*Canis familiaris*) have provided results similar to those gained from the great apes in respect to the selectivity and goal-directedness of imitation (Huber et al. 2009). Like apes, dogs are not particularly sensitive to details of actions, but mostly achieve a functional fit. Their actions seem to be goal directed and object bound, and shortcuts revealed that they are often driven by efficiency. They show superior performance with object manipulations in comparison to body-oriented movements. This finding is not only congruent with those for great apes, but also with those for children.

Can dogs evaluate the efficiency of an action and the constraints for its execution when observing another dog perform? Or would they copy blindly? A recent experiment provided convincing evidence for selective re-enactment resulting from inference about efficiency and circumstantial constraints (Range et al. 2007). Two groups of dogs watched a demonstrator dog pulling a rod using a paw instead of the mouth, which is usually preferred. In the first group, using the 'inefficient' paw action was justified by the mouth being occupied carrying a ball, whereas in the second group no constraints could explain the demonstrator's choice. In a subsequent test, the observers of both groups were encouraged to manipulate the apparatus to get the food by themselves. Only in the second (mouth-free) group the majority of dogs used the non-preferred paw action, thereby showing clear evidence of imitation (Fig. 2). In contrast, most of the dogs of the first (mouth-occupied) group used the mouth to manipulate the rod, thus showing emulation instead of imitation. Emulation is defined as copying the goals or the end results of an observed action, rather than the action itself. It appears that the observer dog has taken the perspective of the model dog and thereby evaluated the effectiveness of the action in relation to the possible ones. Consequently,

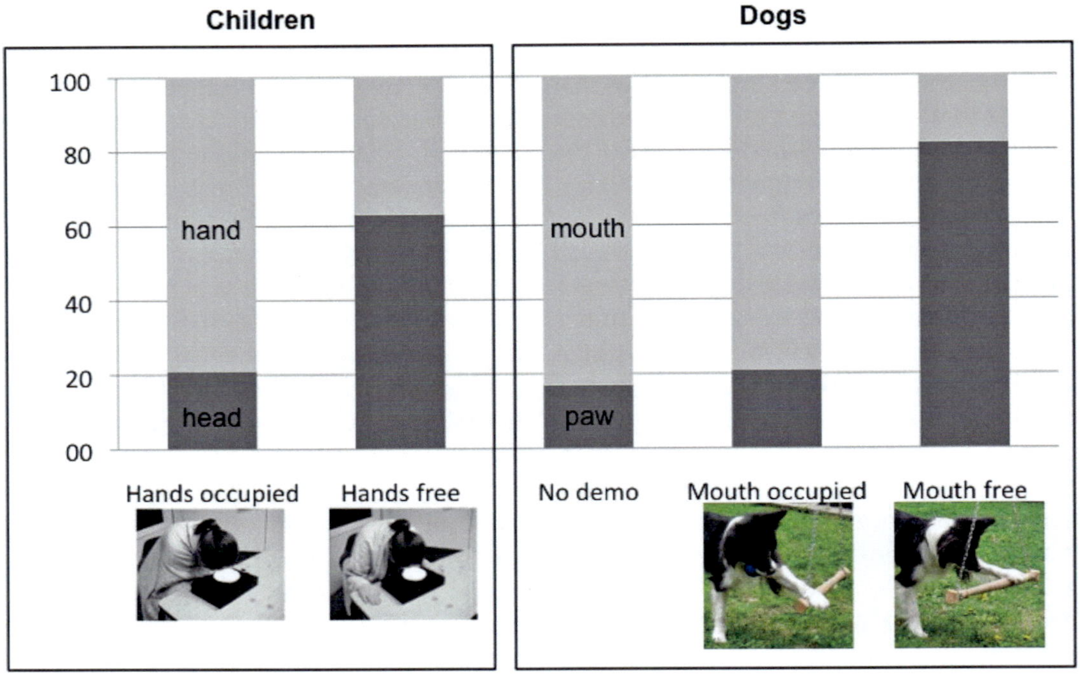

Fig. 2 Dogs, like human children, take into account the circumstances of a conspecific demonstrator, who acts in an unusual manner. The left diagram illustrates the results of Gergely et al. (2002), the right diagram illustrates the results of Range et al. (2007). The dark bars show the proportion of subjects using the demonstrated action (head or paw, respectively), the light bars show the proportion of subjects using their preferred action (hand or mouth, respectively) in the first trial

dogs seem able to imitate selectively based on an inferential, evaluative process.

This selective way of imitation reflects action-understanding not only in terms of comprehension (evaluating the model's action) but also of anticipation (choosing another unseen method), and thus has invoked vivid discussion about the underlying cognitive mechanisms (e. g. Brass et al. 2007). Human infants also "rationalize" the model's inefficient action by using a more efficient method instead of copying it. However, they only do this if the demonstration is accompanied by ostensive-communicative cues, that is, if the human demonstrator looked at them and talked to them during the demonstration (Gergely et al. 2002).

Finally, exciting social learning reports have come from bees and reptiles. Honeybees can use symbolic communication for cross-species talk as a result of social learning between two species, which supports the view that the honeybee dance language has a learned component (Su et al. 2008). Even more exciting from an evolutionary point of view is the fact that the solitary red-footed tortoise (*Geochelone carbonaria*) has been found to be able to use social cues in two entirely different experiments. In a task that examined whether the tortoises would respond to the gaze direction of a conspecific we found that if they observe another animal look up into the air they will also look in the same direction (Wilkinson et al. 2010 a). A second even more exciting finding showed that this species can learn to solve a detour task by observing the actions of a conspecific (Wilkinson et al. 2010 b). This

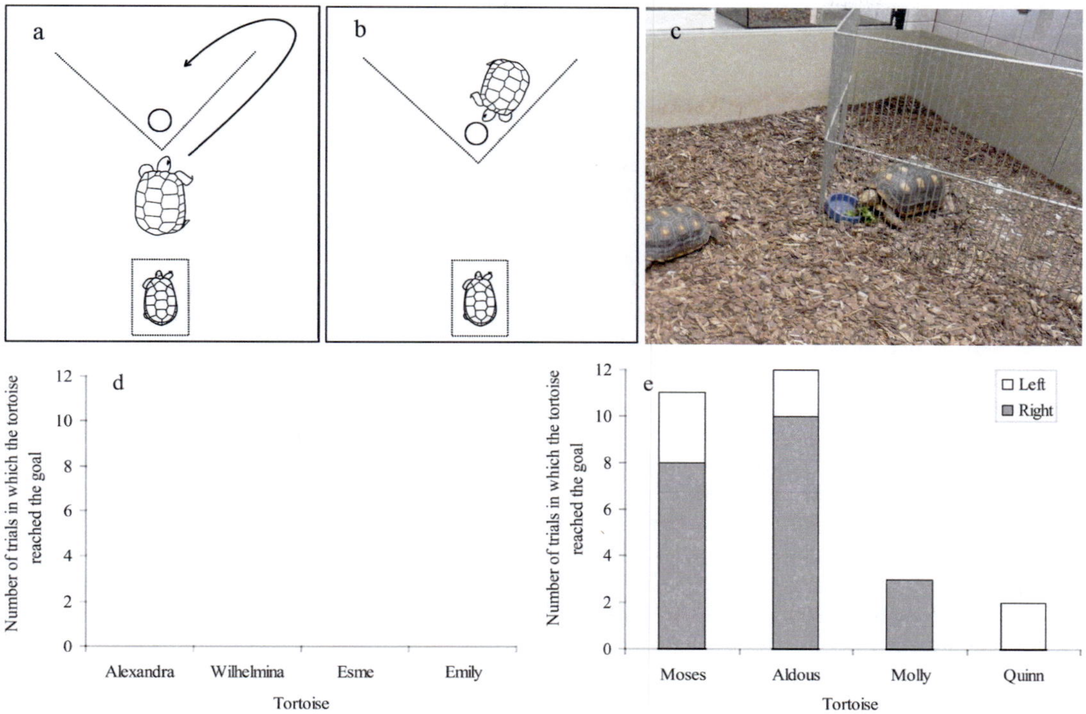

Fig. 3 The solitary red-footed tortoise is able to learn to reach an otherwise inaccessible goal by observing the actions of a conspecific. **a** and **b** represent the experimental setup, the arrow indicates the direction in which the demonstrator tortoise moved; **c** a photograph of the observer tortoise (cage removed for display purposes) watching the demonstrator eat; **d** and **e** the experimental results; **d** the number of trials in which the tortoises of the non-observer group completed the detour; none were successful; **e** represents the same data for the observer group

challenges all current theories about the evolution of social learning, as these are based on the core assumption that social learning is an adaptive specialization for social living. It rather suggests that an animal's ability to learn socially or to use social cues may be a simple reflection of its general ability to learn.

Conclusion

For the last few decades animal research has provided increasing evidence of 'intelligent' behavior in non-human animals, showing that they have the capacity to adaptively modify their behavior in various ways. The degree of flexibility accomplished is far higher than would be possible if it were based purely on the similarity between events stored in memory. Of course, the underlying abstract representations may not be domain-general, they may not extend far from the here and now, and they may not be part of a conscious evaluation for and against beliefs. Although linguistic abilities are of great advantage for inferential reasoning, the abilities demonstrated by animals in exclusion studies (knowing by excluding an alternative; e. g. Aust et al. 2008, Schloegl et al. 2009), and, in tactical social maneuvers (Heinrich and Bugnyar 2007) support the view of 'Popperian animals', that is animals which are rational in the sense of being able to conduct trial and error operations in their head. Further research may unravel not only which animal species share the most advanced cognitive abilities in the perceptual, technical and social domain, but also what kind of representations and what kind of core mechanisms account for the behavior that we like to call cognitive (or intelligent or rational). The most difficult question here is whether advanced forms of associative mechanisms at the subpersonal level may be sufficient. So far, non-human animals have demonstrated surprisingly clever behavior, for which obvious associative explanations are not available. These forms of animal cognition have been explained in terms of causal understanding, declarative memories and intentionality. However, it is premature to contrast associative and 'higher cognitive' processes, because we still do not know the neuronal implementation of the latter. Often, the similarity between the way non-human animals solve tricky problems – such as manufacturing tools, competing with group mates, and making provisions for the future – and the way Western-European thinkers would solve such problems, is used as a guide to describe and explain the underlying processes. Although the comparison with humans may have heuristic values, our knowledge of how evolution works should warn us not to confuse (functional) analogies and true (genealogical) homologies (Heyes and Huber 2000).

Evolution has not used humans as templates against which non-human abilities are designed. Instead, cognitive evolution is part of biological evolution, which is characterized by constraints and innovations (Huber 2000). It proceeds by adding to, or tinkering with, pre-existing cognitive systems. This means that we must be careful to avoid over-intellectualizing. Instead of reinventing Descartian dichotomies we should try to disentangle continuities and discontinuities, which may cut through taxonomies as well as through species and individuals.

Finally, we hope to have shown convincingly that the kind of research described in this paper, which we call "Biology of Cognition", has a high potential to foster intellectual collaboration between philosophers and natural scientists in order to evaluate debates about animal minds and cognition that too easily come down to a confrontation between people who argue that many animals have rich cognitive lives and methodological or ontological behaviorists who view animals as mere stimulus-response machines. We favor pluralism and 'non-speciecist' approaches.

References

Allen C, Bekoff M (1997) Species of mind. MIT Press, Cambridge, Mass

Auersperg AMI, Gajdon GK, Huber L (2009) Kea (*Nestor notabilis*) consider spatial relationships between objects in the support problem. Biol Lett 5: 455–458

Aust U, Range F, Steurer M, Huber L (2008) Inferential reasoning by exclusion in pigeons, dogs, and humans. Anim Cog 11: 587–597

Bekoff M, Allen C, Burghardt GM (2002) The cognitive animal. MIT Press, Cambridge, Mass.

Bird CD, Emery NJ (2009) Insightful problem solving and creative tool modification by captive non-tool-using rooks. PNAS 106: 10 370–10 375

Boesch C, Boesch H (1990) Tool use and tool making in wild chimpanzees. Folia Primatol 54: 86–99

Bolhuis JJ, Macphail EM (2001) A critique of the neuroecology of learning and memory. TICS 5: 426–433

Bradie M (1986) Assessing evolutionary epistemology. Biol Phil 1: 401–459

Brass M, Schmitt R, Spengler S, Gergely G (2007) Investigating action understanding: inferential processes versus motor simulation. Curr Biol 17: 2117–2121

Brass M, Heyes CM (2005) Imitation: is cognitive neuroscience solving the correspondence problem? TICS 9: 489–495

Bugnyar T, Huber L (1997) Push or pull: an experimental study on imitation in marmosets. Anim Behav 54: 817–831

Byrne RW, Whiten A (1988) Machiavellian intelligence. Oxford University Press, Oxford

Campbell DT (1974) Evolutionary epistomology. In: Schilpp PA (ed) The philosophy of Karl R. Popper, Vol. 1. Open Court, La Salle, pp 413–463

Carvalho S, Cunha E, Sousa C, Matsuzawa T (2008) Chaines operatoires and resource-exploitation strategies in chimpanzee (*Pan troglodytes*) nut cracking. J Hum Evol 55: 148–163

Darwin C (1871) The descent of man and selection in relation to sex. Murray, London

de Waal F, Tyack P (2003) Animal social complexity. Harvard University Press, Cambridge, Mass.

Dunbar RI, Shultz S (2007) Evolution in the social brain. Science 317: 1344–1347

Emery N, Clayton N, Frith CD (2008) Social intelligence: from brain to culture. Oxford University Press, Oxford

Galef BGJ, Laland KN (2005) Social learning in animals: empirical studies and theoretical models. BioScience 55: 489–499

Giurfa M (2007) Behavioral and neural analysis of associative learning in the honeybee: a taste from the magic well. J Comp Physiol A 193: 801– 824

Gergely G, Bekkering H, Kiraly I (2002) Rational imitation in preverbal infants. Nature 415: 755

Goodall J (1986) The chimpanzees of Gombe. Harvard University Press, Cambridge, Mass.

Greenspan RJ (2001) The flexible genome. Nat Rev Genet 2: 383–387

Griffin DR (1976) The question of animal awareness. Rockefeller University Press, New York

Heinrich B, Bugnyar T (2007) Just how smart are ravens? Sci Am 296: 64–71

Heyes CM, Galef BGJ (1996) Social learning in animals: the roots of culture. Academic Press, San Diego

Heyes CM, Huber L (2000) The evolution of cognition. MIT Press, Cambridge, Mass.

Hodos W, Campbell CBG (1969) Why there is no theory in comparative psychology. Psychol Rev 76: 337–350

Huber L (2000) Psychophylogenesis: innovations and limitations in the evolution of cognition. In: Heyes C, Huber L (eds) The evolution of cognition. MIT Press, Cambridge, Mass., pp 23–41

Huber L (2010) Categories and concepts: language-related competences in non-linguistic species. In: Breed MD, Moore J (eds) Encyclopedia of animal behavior. Elsevier/Academic Press, Oxford

Huber L, Aust U (2006) A modified feature theory as an account of pigeon visual categorization. In: Wasserman EA, Zentall TR (eds) Comparative cognition: experimental explorations of animal intelligence. Oxford University Press, New York, pp 325–342

Huber L, Gajdon GK (2006) Technical intelligence in animals: the kea model. Anim Cogn V9: 295–305

Huber L, Range F, Voelkl B, Szucsich A, Viranyi Z, Miklosi A (2009) The evolution of imitation: what do the capacities of nonhuman animals tell us about the mechanisms of imitation? Phil Trans R Soc 364: 2299–2309

Huber L, Rechberger S, Taborsky M (2001) Social learning affects object exploration and manipulation in keas, *Nestor notabilis*. Anim Behav 62: 945–954

Humphrey NK (1976) The social function of intellect. In: Bateson PPG, Hinde RA (eds) Growing points in ethology. Cambridge University Press, Cambridge, pp 303–317

Hunt GR (1996) Manufacture and use of hook-tools by New Caledonian crows. Nature 379: 249–251

Lehrman DS (1953) A critique of Konrad Lorenz's theory of instinctive behavior. Q Rev Biol 28: 337–363

Lewontin RC (1998) The evolution of cognition: questions we will never answer. In: Scarborough D, Sternberg S (eds) An invitation to cognitive science (Vol 4). MIT Press, Cambridge, Mass., pp 107–132

Lorenz K (1977) Behind the mirror. A search for a natural history of human knowledge. Methuen, London

Maturana H, Varela F (1979) Autopoiesis and cognition. Reidel Publishers, Dordrecht

McGrew WC (1992) Chimpanzee material culture: implications for human evolution. Cambridge University Press, Cambridge

Menzel R, Giurfa M (2006) Dimensions of cognition in an insect, the honeybee. Behav Cogn Neurosci Rev 5: 24–40

Müller GB (2007) Evo–devo: extending the evolutionary synthesis. Nat Rev Genet 8: 943–949

Pavlov IP (1927) Conditioned reflexes. Oxford University Press, Oxford

Range F, Viranyi Z, Huber L (2007) Selective imitation in domestic dogs. Curr Biol 17: 1–5

Riedl R (1984) Biology of knowledge. The evolutionary basis of reason. Wiley, Chichester, New York

Riedl R (1995) Deficiencies of adaptation in human reason: A constructivistic extension of evolutionary epistemology. Evol Cogn 1: 27–37

Schloegl C, Dierks A, Gajdon GK, Huber L, Kotrschal K, Bugnyar T (2009) What you see is what you get? Exclusion performances in ravens and keas. PLoS ONE 4: e6368

Schrauf C, Huber L, Visalberghi E (2008) Do capuchin monkeys use weight to select hammer tools? Anim Cogn 11: 413–422

Schuster S, Wohl S, Griebsch M, Klostermeier I (2006) Animal cognition: How archer fish learn to down rapidly moving targets. Curr Biol 16: 378–383

Shettleworth S (1998) Cognition, evolution and behavior. Oxford University Press, Oxford

Spencer H (1855) Principles of psychology. Longman, London

Su S, Cai F, Si A, Zhang S, Tautz J, Chen S (2008) East learns from West: Asiatic honeybees can understand dance language of European honeybees. PLoS ONE 3:e2365

Thorndike EL (1898) Animal intelligence: An experimental study of the associative processes in animals. Psychol Rev Monogr 2 (Whole No. 8)

Timberlake W (1993) Animal behavior: a continuing synthesis. Ann Rev Psychol 44: 675–708

Tinbergen N (1963) On aims and methods of ethology. Z Tierpsych 20: 410–433

Visalberghi E, Addessi E, Truppa V, Spagnoletti N, Ottoni E, Izar P, Fragaszy D (2009) Selection of effective stone tools by wild bearded capuchin monkeys. Curr Biol 19: 213–217

Voelkl B, Huber L (2000) True imitation in marmosets. Anim Behav 60: 195–202

Voelkl B, Huber L (2007) Imitation as faithful copying of a novel technique in marmoset monkeys. PLoS ONE July:e611

Wasserman EA, Zentall TR (2006) Comparative cognition: experimental explorations of animal intelligence. Oxford University Press, New York

Weir AS, Chappel J, Kacelnik A (2002) Shaping of hooks in New Caledonian crows. Science 297: 981

Werdenich D, Huber L (2006) A case of quick problem solving in birds: string-pulling in keas (*Nestor notabilis*). Anim Behav 71: 855–863

Wilkinson A, Mandl I, Bugnyar T, Huber L (2010 a) Gaze following in the red-footed tortoise (*Geochelone carbonaria*): Anim Cogn 13: 765–769

Wilkinson A, Kuenstner K, Mueller J, Huber L (2010 b) Social learning in a non-social reptile (Geochelone carbonaria) Biol Lett. 6: 614–616

Zentall T (2004) Action imitation in birds. Learn & Behav 32: 15–23

Perception, art, and illusion III

Introductory remarks

Section III of the present volume brings together five chapters dealing with quite different aspects of perception, art, and illusion.

1. Psychophysics of color

Although the "color-opponency" theory of color vision (Jameson and Hurvich 1997; Hurvich 1981, 1997; cf Byrne and Hilbert 1997) provides a reliable model of how color is processed and represented within the human brain, the subjective aspect of our experience of color still is a major question. What is it like to see the colors and what is the qualitative character of our subjective experience of colors?

In their contribution *"Understanding color vision, with comments on mind and matter"*, Christoph von CAMPENHAUSEN and Jürgen SCHRAMME consider the relationship between the mental and the physical aspects of color vision unknowable – "a problem transcending human understanding" as they say, referring to Emil Du Bois-Reymond. Nevertheless it is worth mentioning current research in this field, which indeed provides a number of interesting hypotheses. According to the philosophical position of eliminative materialism, the human visual color qualia and human opponent-cell coding triplets are one and the same thing (Churchland 2005). Embracing the position of nonreductive physicalism, this relation can also be described as supervenience, where the subjective color qualia supervene upon the neuronal coding vector (Chalmers 1996). According to Donald Davidson's theory of "anomalous monism" of the mental, another form of nonreductive physicalism, one can ask whether "lawlike" or "nomological" connections between subjective sensations of colors and their underlying neuronal processes are possible or not. Davidson denies the possibility of psychophysical laws (Davidson 1970). In the last Section of this volume, Gerhard Roth (Chapter VI, 19) provides an excellent overview of the current debates on the "mind-body problem". Pleading for the position of nonreductive physicalism, Roth suggests that most of the mind-body positions are in fact testable and a solution of the problem is indeed within reach.

Based on current advances in research on both mental and physical aspects of color vision the chapter by *von Campenhausen* and *Schramme* addresses specific observations rather than general issues. The authors emphasize that the processes of transduction of stimuli to sensory excitation represent an early step critical with regard to perception. Therefore, they subdivide the term "matter" into "physics of the stimuli" on the one hand and "eye and brain" on the other, in order to clarify why the transition from the physical aspects of color vision to the mental domain of color is possible, whereas subjective color sensations cannot be traced back to their physical causes. Their chapter argues for the importance of science for an understanding of the mind by drawing from specific results of psychophysical experiments. At the same time *von Campenhausen* and *Schramme* criticize philosophical theories of a more general nature.

Interestingly, mental and physical color systems (see also Chapter V, 17 by John Gage) reveal common structures in spite of being derived from aesthetical introspection and objective science, respectively. As the authors point out, reductionism does allow the prediction of color sensations on the basis of the stimuli but this does not work in the other direction, starting with the sensations and asking for their physical basis (because different stimuli may generate the same sensation). Obviously, colors are presented to the mind in a format, which cannot be translated back to the colors' physical origin. In the struggle for life the biological meaning of stimuli must have been more important than the physical aspects of sensory perception. According to *von Campenhausen* and *Schramme* this is one of the reasons for the evolution of the reduced complexity of the mental format. The mind was the original stage of all human experience. Today its restrictions can be understood by means of psychophysics.

2. The double reality of pictures

In his contribution *"Pictures as strange objects of perception"*, Richard GREGORY emphasizes that looking at pictures evokes "other worlds". Pictures indeed have a double reality: They are objects in their own right that we can touch like any other object, and at the same time they are windows into different space and time for the beholder. *Richard Gregory* draws our attention to the astonishing fact that we see people and mountains, ships and flowers, from blobs of paint on paper or canvas. He calls these the "other worlds" of pictures and stresses that seeing what pictures represent strongly depends on knowledge derived from the primary experience with touchable objects and emotional relationships with other subjects, which allow us to appreciate portraits (Gregory 1987).

Although the first artists were cave painters natural selection had produced pictures much earlier, birds' eyes on butterfly wings being an example. In regard to perception

Gregory distinguishes between innate "cues" and learned "clues" which represent innate knowledge and knowledge learned by individual experience, respectively. Perception both depends on and provides knowledge. These issues are dramatised in trompe l'oeil pictures. They deserve special interest in regard to the question of how pictures are perceived and potentially confused with "reality". According to *Gregory* trompe l'oeil paintings show the most impressive use of cues and clues. The more realistic a painting the more dramatic are its illusions, especially those associated with moving around it or with the trompe l'oeil dome of a cathedral.

For the present volume *Gregory* also provides a table classifying the phenomena of illusions as "kinds" and "causes". This table differs from its previous tentative version (Gregory 1987) by leaving out the optical causes of illusions. According to *Gregory*, optical phenomena like the rainbow should not be classified as illusions of vision because they are already fully explained by the physical concepts of optics. In his classification for the present volume, *Gregory* instead focuses on causes concerning physiological and psychological rules, knowledge as well as understanding of the objects.

An intriguing question put forward by *Gregory* is why our conceptual understanding of a picture rarely destroys the perceptual illusion. Although illusions are recognized as such and fully explained, we continue to experience them: "Although one may know intellectually that one is seeing an illusion, one still *sees* the illusion. This difference between perceiving and conceiving, which applies to all robust illusions, may be because perception has to work very fast for our survival into the next second or so. [...] This remarkable speed could not be achieved if perception drew upon all our knowledge." (Gregory 2001, p. 22)

In a previous publication, *Richard Gregory* touched on a similar phenomenon. Remarkably, behaviour does not always correspond to the illusion experienced (Gregory 1987) which is related to the presence of two complementary systems in human vision (Goodale and Milner 1992; Milner and Goodale 1995). Vision for perception is a complex experience of the present surroundings helping to plan behaviour. Vision for action is older and it is simpler for actions not requiring consciousness (such as reaching for and grasping objects) (Jacob 2005). It was suggested that the two visual systems are based on two different pathways in the primate visual system, the ventral and the dorsal pathway. Despite the interactions between the two systems, each stream uses visual information about objects and events in a different way (Goodale and Milner 1992). According to *Gregory* only the ventral system (vision for perception) seems to be affected by cognitive illusions, whereas reaching out for and grasping objects (vision for action) should not be affected by visual illusions.

3. Psychostimulant action

Among our conscious perceptual experiences there are altered states of consciousness induced by drugs. In 1954 Aldous Huxley took mescaline for the first time and described its effects as an opening of the "doors of perception". According to his report objects appeared colourful and fantastic and the world seemed to be perfect like in a description of mystical experiences (Masters and Houston 2000; Grunenberg 2005).

Over the centuries, the alteration of consciousness by the use of drugs has been common in many societies during religious practices. Mild psychostimulants served the enhancement of alertness, the reduction of anxiety, or far more dramatic purposes in case of addiction. As brains use their own specific chemical molecules to induce changes in their states, artificial psychoactive drugs widely affect consciousness by interacting with the natural chemical systems already present (Whitlock 2001; Pace-Schott and Hobson 2007).

René WEISSENSTEINER, Thomas STEINKELLNER, Andreas JURIK, Simon BULLING, Walter SANDTNER, Oliver KUDLACEK, Michael FREISSMUTH, Gerhard ECKER, and Harald SITTE provide an update of the action of psychostimulant drugs such as amphetamines and cocaine targeting the catecholaminergic system.

Although these substances have been marketed and used by clinicians for many years they became branded as illicit drugs owing to their abuse potential. In their contribution *"Towards an understanding of the psychostimulant action of amphetamine and cocaine"*, the authors discuss recent advances in the understanding of the molecular mechanisms underlying the psychostimulant activity of these substances, their abuse potential and current employment in clinical practice.

Cocaine and amphetamine affect sensory perception by targeting the neurotransmitter sodium symporters (NSS) at the synapses between neurons. They both increase the concentration of the neurotransmitter in the synaptic cleft but do this in different ways which are explained in some detail by the authors.

While these overall actions have long been appreciated, the underlying mechanistic details have been surprisingly difficult to understand. The advent of a crystal structure of a prokaryotic NSS protein and the concomitant development of homology models for eukaryotic NSS family members finally generated novel insights into the structure-function relationships of this clinically relevant class of transporters. These as well as remaining questions are detailed by the authors. Structural biology will play an important role in fully revealing the molecular basis of the psychopharmacological action of amphetamines and cocaine and their respective congeners. Especially, molecular dynamics simulations will allow to dynamically picture the possible conformational changes within this class of transport proteins. Likewise, homology modelling and docking of ligands will help in drug discovery and design.

As we learn from the authors of this chapter, for the treatment of cocaine addiction antibodies against cocaine have been tested in a clinical trial without big success. Small chemical compounds, refined by rational approaches and guided by homology modeling and docking studies may finally succeed. Hence, the authors hope that the effects of drug use like disturbances and enhancement of sensory perception can finally be understood on a molecular basis.

4. Music and speech

In their contribution *"Music, speech and the empirical nature of perception"*, Daniel BOWLING and Dale PURVES hold that percepts are generated in a wholly empirical manner. In vision, the phenomenology of brightness, color, form and motion percepts has provided ways of exploring this idea in quantitative terms. Likewise, the phenomenology of tone perception in music is said to provide a way to examine this conceptual framework.

Bowling and *Purves* discuss how associations between the sound of music and speech can lead to the emotional impact of music. They ask whether the constraints and preferences found in music can be understood in terms of the physical characteristics of speech. Like other sensory capabilities, our ability to perceive tonal sound stimuli has presumably evolved because it provided information on natural and biologically relevant stimulus sources. Although a variety of tonal sounds are present in the human auditory environment, the major source of tonal periodic stimuli in our auditory environment is the vocalization of other humans. Accordingly, our appreciation of tonal sounds has arisen primarily from the benefits of conspecific information.

From this the authors conclude that the structure and function of the tonal sounds produced by the human vocal apparatus may provide the key to understanding how and why we perceive tonality in music the way that we do. *Bowling* and *Purves* indeed suggest that associations made between the spectra of speech uttered in different emotional states and the spectra of thirds and sixths in major and minor music are the basis for the different emotional effects of these different tone collections in music. Their results support the more general idea that tonality can be understood in terms of human vocalization which is also considered a biological basis for musical aesthetics.

5. Synaesthesia

In his contribution *"Synaesthesia and synergy in art. Gustav Mahler's 'Symphony No. 2 in C minor' as an example of interactive music visualization"*, Johannes DEUTSCH deals with a striking phenomenon: the stimulation of one sensory modality induced by an experience in another sensory modality. Most of the synaesthetic people hear colors, but they may also see smells or experience a similar interaction between other senses. This strange phenomenon can be involuntary or intentionally induced by artists through a synthesis of image or color on the one hand and music on the other. In other words, synaesthesia may be a neurological condition and characterized by anomalous sensory perception or emerge from a longstanding tradition of synaesthetic œuvres in the history and theory of art and culture.

The author follows the history of scientific research on this phenomenon starting with Gustav Theodor Fechner (1876), Sir Francis Galton (1883), and Alfred Binet (1892) and includes current research of Richard Cytowic, Vilayanur S. Ramachandran, Simon Baron-Cohen, Hinderk Emrich, and Julia Simner in his review. The most intensively studied synaesthetic relation is that between auditory and visual perception. However, corresponding connections may involve other senses (like olfaction and taste) as well.

Although some artists refer to their involuntary perceptual experience of synaesthesia, in the history and theory of art and culture, in art synaesthesia is conceived mainly as the intentional effort to evoke sensory experiences in the audience mainly through the synergy of music and colors (unidirectional – like in clinical synaesthesia – from the music to the colors). *Johannes Deutsch* explores the long history of aesthetics connecting colors to music, starting with the Pythagoreans, through Leonardo da Vinci, the French "audition colorée", Alexander Scriabin, Wassily Kandinsky, the "Sonchromatoschope" of Alexander László, and concluding with the most recent efforts in contemporary art and design.

Up to now, a major problem remained for both the theoreticians of culture and art and the artists themselves: how precisely can the simultaneity of synaesthetic perceptions, typically experienced in the neurological condition of syaesthesia, be achieved in art?

The author, both an artist and art theoretician, shows how he reached the goal not only in avant-garde projects, but also in operas like Wagner's "Rheingold" and Schumann's "Manfred". He subsequently elaborates on how he visualized Mahler's "Symphony No. 2 in C minor", also known as "Resurrection". Regarding the technical realization, Johannes Deutsch adopted a complex computer cluster to generate the real-time visualization of the symphony. The cluster simultaneously processed and implemented any modulations picked up from the orchestra's 56 instruments which were equipped with microphones and individual soundtracks.

Johannes Deutsch concludes his chapter addressing some major theoretical questions concerning the frequency of processes such as cross sensory coupling, the prevalence of fixed paired pattern of association as well as their relevance for art. **P G-D**

References

Byrne A, Hilbert D R (eds) (1997) Readings on color. Vol 1: The philosophy of color. Vol 2: The science of color. MIT Press, Cambridge, Mass

Chalmers D (1996) The conscious mind. Oxford University Press, Oxford

Churchland P M (2005) Chimerical colours: some novel predictions from cognitive neuroscience. In: Brook A, Akins K (eds) Cognition and the brain. The philosophy and neuroscience movement. Cambridge University Press, Cambridge, New York, pp 309–335

Davidson D (1970) Mental events. In: Davidson D (1980) Essay on actions and events. Clarendon Press, Oxford, pp 207–225

Goodale M A, Milner A D (1992) Separate visual pathway for perception and action. Trends in neurosciences 15: 20–25

Gregory R L (1987) Illusions. In: Gregory R L (ed) The Oxford companion to the mind. Oxford University Press, Oxford

Gregory R L (2001) Illusions. In: Underwood G (ed) Oxford guide to the mind. Oxford University Press, Oxford, pp 15–22

Grunenberg C (ed) (2005) Summer of love. Psychedelische Kunst der 60er Jahre. Hatje Cantz Verlag, Ostfildern-Ruit

Hurvich L M (1981) Color vision. Sinauer, Sunderland, Mass

Hurvich L M (1997) Chromatic and achromatic response functions. In: Byrne A, Hilbert D R (eds) Readings on color. Vol 2: The science of color. MIT Press, Cambridge, Mass, pp 67–91

Jacob P (2005) Grasping and perceiving objects. In: Brook A, Akins K (eds) Cognition and the brain. The philosophy and neuroscience movement. Cambridge University Press, Cambridge, New York, pp 241–283

Jameson D, Hurvich L M (1997) Essay concerning color constancy. In: Byrne A, Hilbert D R (eds) Readings on color. Vol 2: The science of color. MIT Press, Cambridge, Mass, pp 177–198

Masters R, Houston S (2000). The varieties of psychedelic experience. The classic guide to the effects of LSD on the human psyche. Vermont, Rochester

Milner A D, Goodale M A (1995) The visual brain in action. Oxford University Press, Oxford

Pace-Schott E F, Hobson A J (2007) Altered states of consciousness: drug induced states. In: Velmans, M, Schneider, S (eds) The Blackwell companion to consciousness. Blackwell, Oxford, pp 141–153

Whitlock F A (2001) Addiction. In: Underwood G (ed) The Oxford guide to the mind. Oxford University Press, Oxford, pp 149–152

Understanding color vision, with comments on mind and matter

9

Christoph v. Campenhausen and Jürgen Schramme

Contents

Abstract

Much is known about the mental and physical aspects of color vision. Color vision, therefore, is a paradigm well suited for the discussion of the relationship between mind and matter. The aim of the present chapter is to support the proposition that mental affairs cannot be adequately understood if their neurobiological aspects are neglected. Although it is possible to focus on fundamental problems of general relevance when discussing mind and matter, this chapter will deal with specific observations rather than general issues. The possibility of generalizations derived from empirical results is always limited. Provided the conditions under which these observations were made can be confirmed, specific results are reliable. The relationship of mind and matter often comes under the headline 'mind-body' or 'mind-brain-problem' to facilitate the discussion of general aspects. With regard to perception, a critical early step is the transduction of the stimuli to sensory excitation. Subsequently, the term 'matter′ will be subdivided here into 'physics of the stimuli′ on the one hand, and 'eye and brain,′ on the other. This will help to clarify why the transition from the physical aspects of color vision to the mental domain of color is possible, whereas subjective color sensations cannot be traced back to their physical causes.

Christoph v. Campenhausen
Johannes Gutenberg-University Mainz
Institute of Zoology III-Neurobiology
Kleinmannweg 2 – SBII
55099 Mainz, Germany
e-mail:campenha@uni-mainz.de

1. Introduction

In the complete absence of neurobiological knowledge, as was perhaps the case in earlier periods of human evolution, the mental domain may have been the only stage of perception, imagination and communication. Today neurobiological knowledge does exist. The question may be raised, therefore, as to whether this influences the understanding of mental experiences. There are, of course, philosophical traditions which regard neurobiology as irrelevant with respect to the mental domain, as indicated in this statement: "At present brain science does not offer any theory on the connection of mind and brain, consciousness and nervous system which may be tested empirically" (Küng 2005). This chapter does not discuss such philosophical ideas. Nor does it aim to postulate or deny a categorical boundary separating the domains of mind and matter. The purpose of the chapter is more modest. It will report some specific results of psychophysical research on color vision. Each case will be considered to determine whether the examples are of relevance in understanding the relationship between mind and matter.

The concept and terminology will be taken from Emil Du Bois-Reymond (1818–1896). For Du Bois-Reymond (Campenhausen von 1981, 1994) the relationship between mind and matter was a mystery that could never be explained, rather like the quadrature of the circle. He considered the relation of mind to matter as a riddle which transcended human understanding. His comment was "ignoramus, ignorabimus" (we do not know and we will not know anything about it) (Du Bois-Reymond 1872, 1880).

Two obviously erroneous conclusions from this comment must be avoided. (a) Because the important question of how mind and brain work together may never be answered does not mean that brain research will never be successful. According to Du Bois-Reymond, the increasing collection of evidence of the interdependence of mind and matter sufficiently rewards brain science. (b) Du Bois-Reymond was not a dualist putting mind and matter in categorically different domains. He clearly considered himself a psychophysical monist. In his opinion, the Leibniz paradigm of two clocks representing mind and body was misconceived. In this dualistic model, the synchrony of the clocks indicates the psychophysical connection. Synchrony may be caused either by mechanical coupling, continuous adjustment or harmony pre-stabilized by the creator. All these proposals were irrelevant to Du Bois Reymond. He replaced the two clocks by a single one representing both mind and body and thus denied the need for a philosophical explanation of the dualism of mind and matter.

But the "ignoramus, ignorabimus" still has to be justified, though the explanation is very simple. It is derived from the methods of research, whereby, all results are predetermined by the methods applied. The mind cannot be found in the brain with microscopes or electrodes. One has to talk with a person to meet the mind. Thus, the categorical difference between mind and matter can no longer be treated as an ontological problem. Recognizing the special restrictions defined by the methods in each specific research project is more instructive than insinuating a basic philosophical problem concerning mind and matter in general. Illustrations to this approach will be presented in this chapter.

2. Speculations versus facts in color vision

The renowned painter Josef Albers (1888–1976), who also taught color theory to artists, formulated the opinions of many people when he stated the following: "If one says ′Red′ (the name of a color) and there are 50 people listening, it can be expected that there will be 50 reds in their minds. And one can be sure that these reds will be very different" (Albers 1963). He even remained doubtful as to whether or not different people looking at the same colored object have the same color sensation. It is, of course, perfectly true that there is no trick available to experience what another subject perceives. As a consequence, the opposite speculation may also be true: everyone looking at the same object will have exactly the same color sensation. The ambiguity of these speculations can be avoided in more realistic situations of color vision, which also provide access to their physical aspects.

An appropriate example is taken from the field of color blindness. There are human individuals who cannot discriminate between the colors of the sky and a flower, which to most people appear blue and pink, respectively. Obviously, the color sensations of these subjects looking at pink and blue patches on paper, for instance, will differ from what the majority perceives. This mental defect was discovered by Johann Wolfgang von Goethe (1749–1832), who correctly assumed that the subject suffered from the inability to generate certain color sensations. He named the defect Akyanoblebsie (no blue in vision) since he assumed that the mental capability for blue sensations was lacking. But, obviously, the confusion of pink and blue could have been due to the absence of either blue or red sensations. The explanation is once more ambiguous.

Meanwhile, the origin of this type of color blindness has been traced to the lack of a single well-known gene present in any cell of color-capable subjects. The mental deficiency, therefore, can be predicted from the molecular structure of the human genome. This can be investigated with any cell of the subject, taken from a drop of blood, for example. Wolfgang Jaeger (1917–1995) examined a great-grandson of Goethe's subject, who had inherited the color blindness. We now know that Goethe provided the correct interpretation in general (one kind of color sensation is lacking), but he chose the wrong color (not blue but red sensations were absent) (Jaeger 1989a, b). This type of color blindness is called Protanopia today.

Goethe could have avoided the wrong interpretation if he had demonstrated the light spectrum to his subject. Then he might have discovered that the subject did not see the red end of the spectrum as readily as the human majority, thus proving that the sensation red rather than blue was lacking.

The origin of the defect causing Protanopia is localised in the sensory cells of the eyes known as cones (Fig. 1a). In the cones the stimuli (electromagnetic radiation) are transduced to sensory excitation. There are three types of cones equipped with different visual pigments (rhodopsins). The l-cones respond to radiation of longer wavelengths λ, compared to the m- and s-cones (middle and short λ)(Fig 1b). Excitation of l-cones causes red sensations. Since the l-cones do not work in the protanopic eye, no message from these cones will reach the brain. The protanope′s brain only receives the messages of the remaining two cone types. In the surrounding scenery, the protanope finds fewer discernible colors than the majority of humans with three types of cones in their eyes. Many colors which are easily discriminated by the majority appear identical to the protanope. Recently, it was shown that injecting the missing genes into the eyes

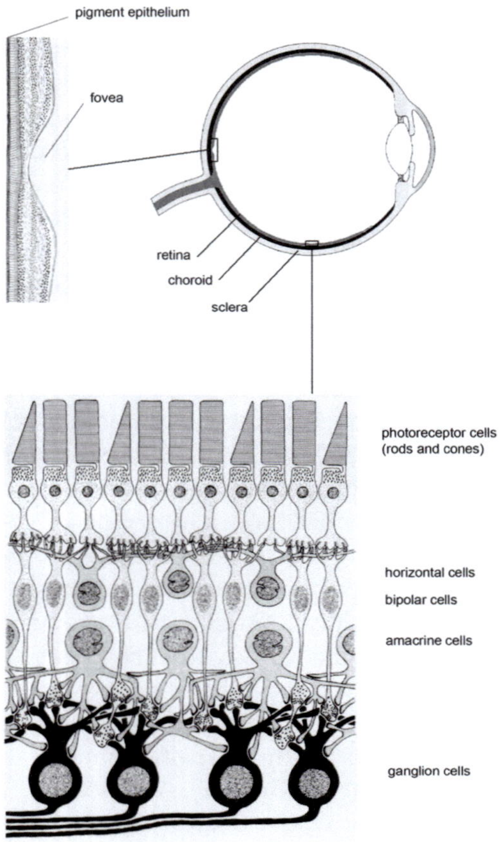

a

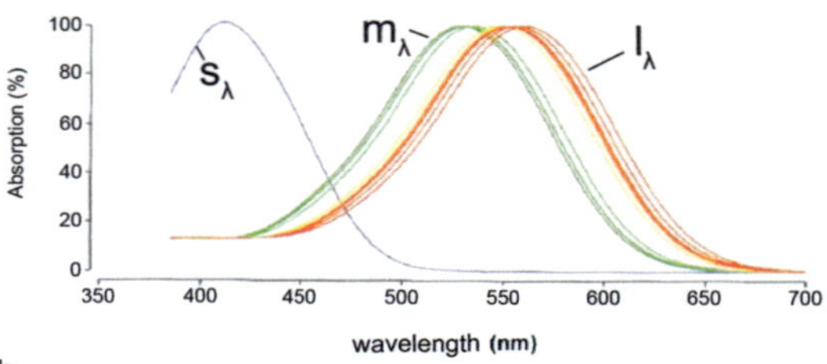

b

Fig. 1 **a** Horizontal cross-section of the human left eye and a part of the retina with fovea centralis enlarged (left), and enlarged once more (below). **b** Spectral absorption functions of the cones s. m and l (named according to the wavelength region of maximal absorption short, middle and long, normalized to their maxima) (Campenhausen von 2001, 2007)

of color-blind monkeys established full color vision (Mancuso K et al. 2009).

These results confirm the prediction: considering the physical aspects of color opens up additional perspectives about the mental aspects of color vision. Former ambiguities can be settled. Perceptual abnormalities such as color blindness can be classified and explained. Medical therapy of color blindness can be predicted. But such results do not change our attitude towards the different character of mind and matter.

3. Proof of presence and absence of colors in the mind

It is worth noting that, just like other people, protanopes happily talk about the colors they perceive. Unfortunately, however, nobody can participate in the subjective color experiences of other people. Knowing that protanopes see fewer colors in the scenery around them, the question arises as to whether they can imagine those colors which they never see. Do protanopes dream of colors they cannot see with their eyes open?

There is an interesting case of a protanopic boy who reported the wonderful colors he used to perceive when he rubbed his eyes before falling asleep in the evening (Hebenstreit M, personal communication). Pressure on the eyes is known to be a stimulus, which, though inadequate, can generate vivid light sensations. The color capability of the protanopic boy is restricted due to the lack of l cones. But the nerve cells being stimulated inadequately may still have the capability of generating all color sensations. This hypothesis was confirmed experimentally using still another trick to generate color sensations.

Pattern-induced flicker colors (PIFCs) were generated by means of a variety of Benham's Top (Fig. 2a). Looking at the spinning disc, the circling arcs fuse, forming rings of different colors on the disc. The outer circle appears reddish, the middle ring green, and the inner bluish, if the disc rotates at the adequate frequency (Campenhausen von, Schramme 1995). The circling black and white pattern is imaged on the retina in the eyes of the observer. The cones under the rings will receive the periodically modulated light-dark stimuli, a, b and c shown in Fig. 2b. All rings obviously reflect stimuli of exactly the same spectral composition. The colors, therefore, cannot be caused by different stimulation of the cones. The stimulus time courses of the inner and the outer ring are even identical, but they appear in different colors on the spinning disc. The time course of the periodical modulation and phase relationship between the stimuli (a, b, c) and (d) at adjacent retinal areas determine the PIFCs. The origin of the PIFCs has been localised in neural networks interconnecting neighboring areas within the retinae of the eyes.

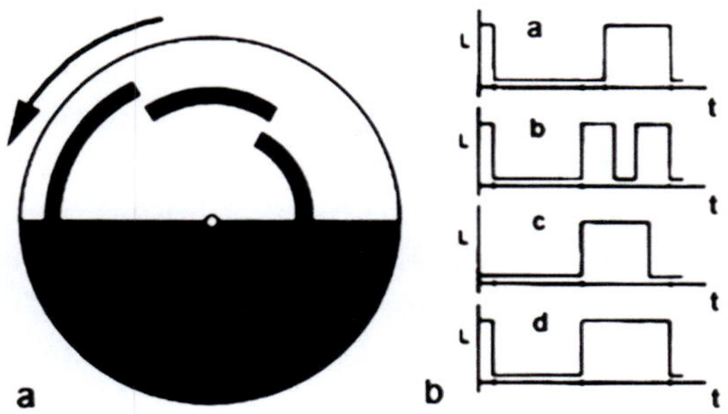

Fig. 2 **a** Disc similar to Benham´s top to demonstrate pattern-induced flicker colors (PIFCs). **b** Periodical black and white stimuli reflected by the outer, middle and inner ring (a, b, c) of the spinning disc and by the areas between the rings d

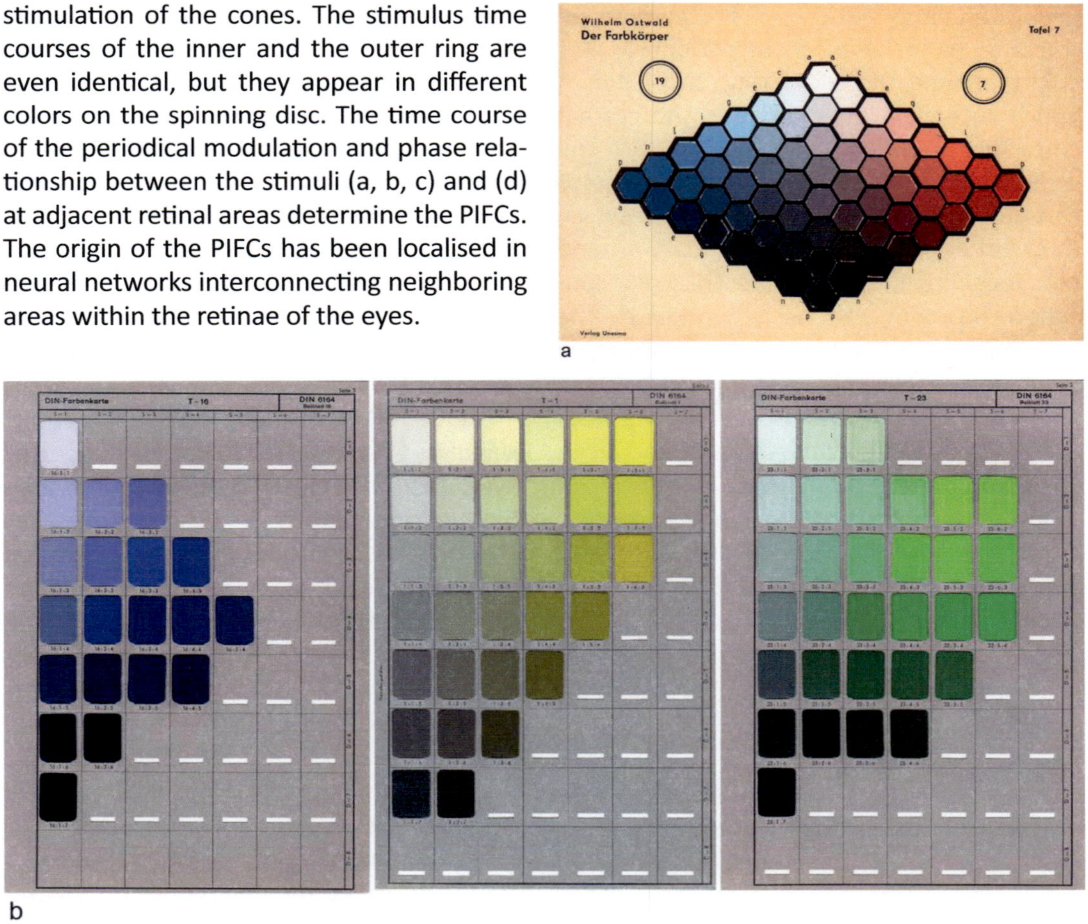

Fig. 3 Color charts taken **a** from the color atlas of Wilhelm Ostwald (Ostwald 1925) and **b** from the DIN-color atlas (DIN 1961)

The important message is: there are different ways of producing color sensations. If one way in the protanopic eye is blocked, there may be others bringing the missing color sensation to the mind. Whereas pressure to the eyes is difficult to reproduce, the stimuli and color sensations of the PIFC are well defined and can therefore be tested.

The PIFCs can be matched with color sensations generated by adequate stimulation of the eye. For this purpose, color chips matching PIFCs can be selected from a collection, typically a color atlas (Fig. 3). Color-capable people have no problem with these matchings. Their color sensations can be generated along both ways. Protanopes, on the other hand, have difficulty finding a color chip matching the outer ring, the color of which is reddish. But they find the correct chips for the green and bluish PIFCs of the other rings. The outer ring provokes a color sensation in the protanope´s mind, which does not exist for him in the scenery around him, including the color atlas. This color sensation, however, can be generated by the spinning disc. It is clear that there are more colors available in the mind of protanopes than expected considering the reduced color discrimination capabilities (Campenhausen von, Schramme 1995).

Knowledge of the physical aspects of color vision was necessary to establish this result concerning the mental capabilities of other people.

4. Mental and scientific color systems

4.1 Color appearance systems

Colors may be arranged according to the mental criterion of resemblance. The results are color charts (Fig. 3), a collection of which may establish a color atlas. No physical or physiological knowledge is needed to design color charts. Color charts are useful in many applications. Manufacturers and merchants of paints and their customers can refer to a certain color on the chart to agree on the color of the product to be purchased. In this way, the color chip bridges the gap between mind and matter. There are, of course, many color collections available for this purpose (Silvestrini et al. 1998).

Would it be feasible to construct a collection of all the colors a human being might possibly experience? A color arrangement fulfilling this claim is called color appearance system. The color sphere of Philipp Otto Runge (1777–1810) is a prominent example (Fig. 4a). The colors are ordered according to their resemblance. There is a white and a black pole connected by the grey scale running from black to white (Fig. 4b). The brilliant colors of the color circle make up the equator. Colors of minor saturation are placed in the space within the sphere according to their similarity to colors at neighboring loci. In this three-dimensional system a locus may be found for any color one can think of.

About 100 years later, the great chemist Wilhelm Ostwald (1853–1932), who was also a good painter of landscapes, developed his color appearance system (Fig. 4c), which was successfully applied in business, science, and art. The colors are ordered according to their similarity, as in the Runge sphere, but the distances between the color loci are slightly different. The color chart of Fig. 3a is a sectional plane through the Ostwald color system.

In the most ideal color appearance system imaginable, the distances between the different color loci would correspond exactly to the resemblance of the colors: the greater the similarity, the smaller the distances. An improvement in regard to this claim was introduced by Alfred Munsell (1858–1918), who also started as a painter (Fig. 4d). In the Munsell system, the equator forming the color circle is tilted in comparison to the Ostwald system. This geometrical arrangement reduces the distance between yellow and the white pole and brings blue closer to the black pole. This improvement was retained in all color appearance systems designed later, for instance the DIN-System (Fig.3b) and the color appearance system of the Optical Society of America (OSA system).

Surprisingly, this arrangement is already found in the color appearance system described in the textbook on physiological optics by Franciscus Aguilonius (1567–1617) (Aguilonius 1613). He carefully described the order of colors according to their similarity in the book, which was illustrated by Peter Paul Rubens (1567–1617). The great painter transferred these ideas to a geometrical drawing, in which similarity of colors is coded in distances along lines connecting the color loci (Fig. 5a): The greater the resemblance, the shorter the distance. This geometrical construction can be transformed topologically, maintaining the lengths of the lines (Fig. 5b-d). The resulting color appearance system (Fig.5d) is again three-dimensional, and the circle with the colors of greatest saturation has the oblique orientation familiar in most modern color appearance systems. The idea behind Rubens' drawing was not understood until 2001 (Campenhausen von et al. 2001).

The structure of all color appearance systems – despite being based exclusively on mental judgement – appears to follow certain rules which have not changed for over four centuries. They all are three-dimensional and demonstrate that the multiplicity of colors is somehow limited. The three-dimensional systems provide loci for any color one can think of. The earliest and the most advanced color appearance systems are very similar.

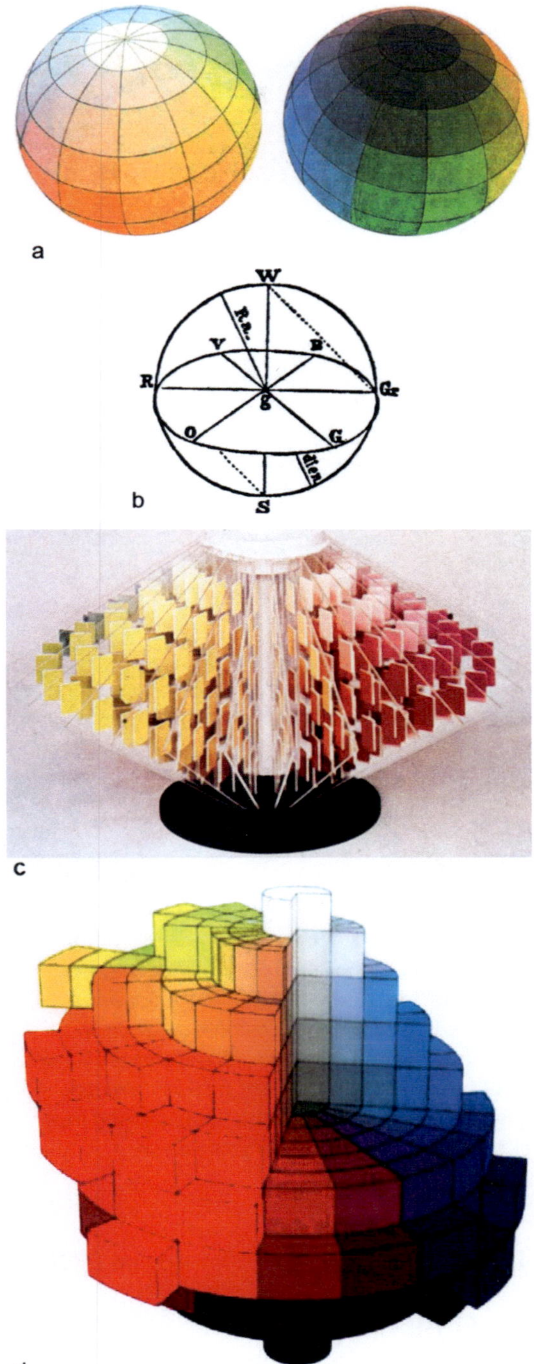

Fig. 4 **a** Color sphere of Philipp Otto Runge (view on the white and black pole) (Runge 1810) and **b** design of the Runge sphere (*W*hite, *S*: Black, *R*ed, *V*iolet, *B*lue, *G*reen, *G*:Yellow, *O*range) **c** Ostwald and **d** Munsell Color System

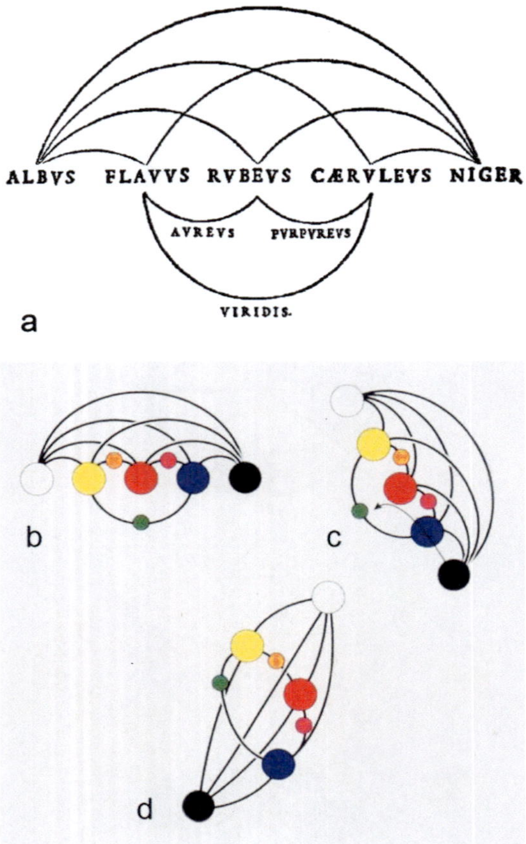

Fig. 5 **a** Color system according to Aguilonius (Aguilonius 1613, p. 40). Albus: white, Flavus: yellow, Rubeus: red, Caeruleus: blue, Niger: black, Aureus: gold, Purpureus: purple, Viridis: green. (**b-d**) Transformation of Aguilonius´ flat color system into a three- dimensional color body. **b** Connecting arcs between yellow and red as well as between red and blue are turned upwards. **c** Black is moved through the heart-shaped inner circle downwards and **d** its completion

4.2 The physiological color system (PCS)

The Physiological Color System (PCS) is derived exclusively from physiological and physical data (Campenhausen von and Schramme 2003). The physics of the stimulus will be explained with reference to the color circle (Fig. 6a). This particular color circle was composed without much care by selecting colored papers from a collection which happened to be available. The spectral stimuli of the colored papers can be measured by an instrument recording the spectral power distribution ($P_\lambda \beta_\lambda$) of the electromagnetic radiation reflected by the color papers. For each color paper, the energy S, M, and L absorbed in the cones s, m, l can be calculated (Fig. 6b), where s_λ, m_λ, and l_λ are the spectral absorption functions of the rhodopsins in the three types of cone (Fig. 1b). With S, M, and L the first physiological step towards color perception is described quantitatively. S, M, and L are different for each color paper because of the different spectral power distributions of the stimuli emitted by the color papers. The three quantities S, M, and L can be plotted in a three-dimensional diagram (Fig. 6c). Each point in this receptor absorption space represents the absorption values in the three types of cones (receptors) when the subject looks at one of the color papers.

As one can see, the points produced by the stimuli of the color circle once more form something similar to a circle in the receptor absorption space. This is a formal coincidence connecting the mental color arrangement and the sensory excitation. If illumination of the paper with the color circle is shut off completely, stimuli will no longer be re-

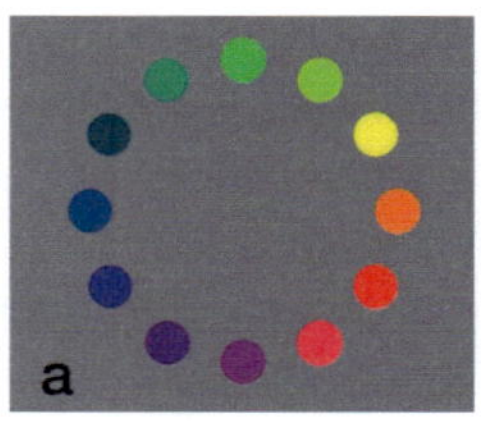

b

$$S = \int P_\lambda \beta_\lambda s_\lambda \, d\lambda$$
$$M = \int P_\lambda \beta_\lambda m_\lambda \, d\lambda$$
$$L = \int P_\lambda \beta_\lambda l_\lambda \, d\lambda$$

Fig. 6 **a** Color circle. **b** Calculation of absorption values S, M, L in the three kinds of rhodopsin of the cones with the absorption functions s_λ, m_λ l_λ (Fig. 1b). The term $P_\lambda \beta_\lambda$ stands for the stimuli (spectral power distribution of the radiation) emitted by the color papers of the color circle

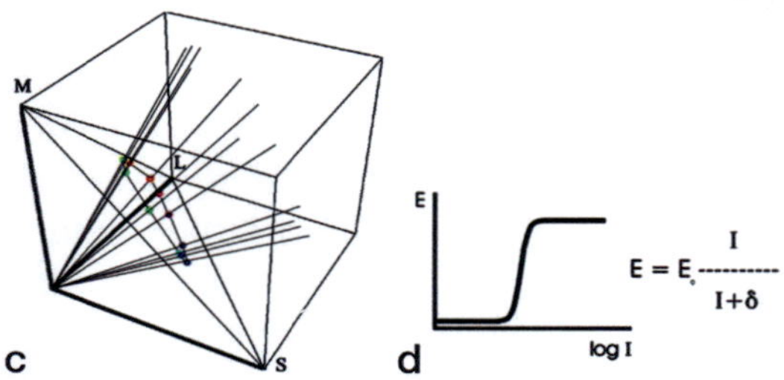

Fig. 6 **c** Receptor absorption space. The points represent the absorptions S, M, L for the different color papers. **d** Calculation of the excitatory output E of the cones as a function of the stimulus input I, given by the absorptions S, M, L.
e Receptor excitation space with the PCS. Excitations E_S, E_M, E_L calculated for the three cones with stimulus strength increased from zero to infinity

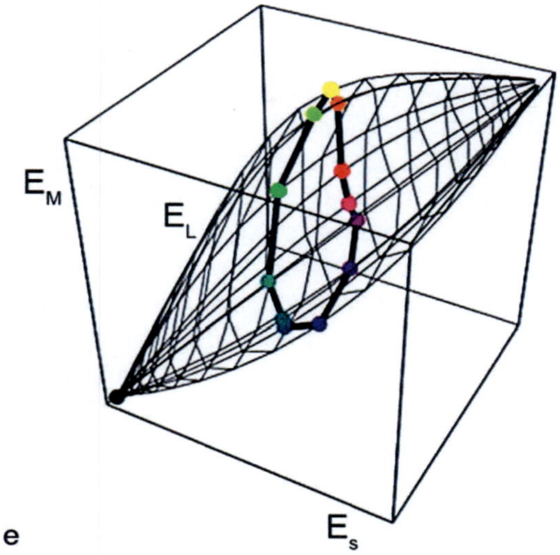

flected. Hence, all the points coding the absorption will disappear in the origin of the receptor absorption space. This is another formal coincidence connecting the black pole of the color appearance systems with the origin of the receptor absorption space. An even more instructive analogy between the mental and the physiological domain can be observed if the excitatory output of the cones is plotted in the diagram instead of the input (Fig. 6e). The excitatory output of the cones is the message sent to the brain.

The sensory excitation of the cones has been studied by electrophysiological methods. The transformation of the inputs (I, which is S, M or L) to output (E, which is E_S, E_M, E_L) can be calculated, using the mathematical formula (Fig. 6d), where δ is a constant. The sigmoid function shows correctly that excitation E of the cones does not grow infinitely with increasing stimulus strength I, but approaches a maximal value instead. The calculated excitations E_S, E_M, and E_L can be plotted in a three-dimensional diagram, which is called receptor excitation space (Fig. 6e).

Each dot of the receptor excitation space codes the excitations of the three cone types if a subject is looking at one of the color papers of the color circle (Fig. 6a). If illumination of the color circle is increased from zero to infinity, the points in the receptor absorption space will move along the lines from the origin to the point of maximal excitation. The resulting figure is the physiological color system (PCS) shown in Fig. 6e. The PCS is derived from physical data exclusively, but its formal construction resembles the color appearance systems, which is a mental achievement. This can be easily seen by comparing Fig.6e and Fig.5d. The point of maximal excitation corresponds to the white pole of the color appearance systems. This is a further coincidence of the mental and physical aspects of color vision.

4.3 The common basis of mental and scientific color systems in the brain

The mental color systems designed independently by artists over the last 400 years have a common structure (Figs 4a, c, d, 5d), which is shared with the PCS (Fig. 6e). The simplest explanation for this structural coincidence would be that there is a common basis for the mental and physical aspects of color vision in the brain. This is by no means a wild hypothesis but rather the simplest explanation available. In addition, this explanation is confirmed by the neurophysiological investigations of the cerebral cortex, especially its area V4. If this part of the brain is destroyed in the right half of the cerebral cortex, the left-hand side of the scenery around the subject appears grey in grey, whereas colors are readily perceived on the right-hand side. If V4 does not work on both sides, the whole scenery around the subject appears grey in grey. Studies of single neurons in V4 overwhelmingly show that the basic structure of both the mental and physical color systems is present in the brain (Zeki 1993).

In spite of the strong correlations of the organisation of colors in the mind and in the brain, Du Bois-Reymond's "ignoramus, ignorabimus" remains justified. Personal experiences of color have to be distinguished clearly from the physiological facts concerning excitation of sensory and neural cells. Before interpreting the different character of mind and brain more closely, another difference will be described here, namely the structural difference of the color system in the mind and the domain of stimuli.

5. Comparing the complexities of the mental and the stimulus domain in color vision

Whereas mental experiences of color sensations are closely correlated with certain processes in the eyes and brain, the physical stimuli producing color sensations are a good deal more complex. To avoid confusion: it is not the complications of inadequate stimulation like pressure to the eyes, PIFCs, drugs or electric stimulation which are considered here. The domain of the adequate stimuli alone is much more complex than the whole of all colors to be perceived. Isaac Newton (1643–1727) already knew that each color sensation can be evoked by a multitude of physically different stimuli. Popular examples are the complementary pairs of color stimuli which, if observed separately, may appear yellow and blue, but white if superimposed. Another pair may be blue-green and red and yield the same white if mixed additively. The single stimuli and their combinations are physically very different in these cases, but the white sensations are indistinguishable. Obviously, more physically different color stimuli exist than color sensations.

In the case of color vision, entrance to the mental domain is possible for scientists investigating the stimuli which cause excitations leading to certain color sensations. If the electromagnetic spectral power distribution of a stimulus is known exactly, the resulting color sensation can be predicted by calculation of the locus of the PCS using the equations of Figs. 6b and d. The color sensation can be taken from the corresponding color locus of a color appearance system. Because of the greater complexity of the material basis, however, the exit from the mental domain to the stimuli is closed. In this case also, the scientist cannot determine which of the many possible stimuli might have caused the color sensation.

Reductionism is a philosophical term often used to criticize the attempt to explain mental experiences on the basis of the factual findings of neuroscience. The belief prevails that this is impossible because the mental domain is assumed to be of greater complexity than the stimulus domain, which is considered more primitive. The message from color vision is different. There are more different stimuli than colors to be distinguished. The complexity is greater in the stimulus domain than in mental color systems. Prediction of color sensations on the basis of physical stimuli is possible and, in fact, a reductionistic operation from the stimulus domain of higher complexity to color sensations, which are of lower complexity. The colors in the mind, however, cannot be traced to their stimuli because of the greater complexity of the stimulus domain.

Nobody is aware of his liver, brain, or genes unless he is instructed. Fortunately, we do not feel the excitation of all sensory and nerve cells, neither do we perceive the highly complex stimuli. A visual system which records all this can be constructed. But such an enormous amount of information would be of little value in the struggle for survival. We prefer simple solutions which can be applied to our problems without much thought. You never look under the hood of a working car. In the same way, our brain presents to the mind what has been useful under the conditions of natural selection.

In spite of being caused by physical stimuli, colors cannot be retranslated into their material origin. The mental format of colors is a perceptual meta-language of reduced complexity compared to the domain of the stimuli. The color must, therefore, have a code of its own. Studies of colors as mental phenomena take place in the confined domain of subjective experiences, and nature did not provide a way out of this confinement as the mental aspect of colors remains private. Du Bois-Reymond correctly stated that crossing the border between mind and matter is a problem transcending human understanding. The explanation of the problem itself, however, is rather simple.

6. Why colors at all?

Colors are enjoyable natural gifts, but they are also useful. Changes in the color of a person's face indicate changes in mood. In fruits, colors provide information about the degree of ripeness. Colors provide cues for additional information in many situations and may improve the visual contrast. No scientific education is required to agree with these notions.

Additional ideas explaining color vision have been derived from the scientific investigations of natural illumination. On the planet earth the spectral composition of the natural light varies with the weather, the geographical location, and the time of day. If there were no means to compensate for these variations of illumination, a white paper would appear blue at noon and orange or even red in the morning and evening. Scattering is the main cause of the diurnal variation of the illuminant spectrum. The shorter the wavelength, the more radiation is scattered in the atmosphere. The sky is blue because of the scattered light of short wavelength. At noon, the sun beams cross the atmosphere more or less vertically, and the length of the path through the atmosphere is short. When the sun is close to the horizon, the length of the path through the atmosphere is much longer and the effect of scattering stronger. The sun is red in the morning and the evening due to the greater portion of short-wave radiation lost due to the scatter.

Reliable perception of objects under the varying conditions of illumination is achieved by compensatory adaptation of the cones. If long wavelengths prevail in the morning and evening, the cone channels sensitive to long wavelength radiation reduce their sensitivity. Thanks to this compensatory adaptation, the white paper retains the same appearance despite reflecting relatively more radiation of long wavelengths. More than one type of sensory receptors is needed for both detection and compensation of spectral variations of the daylight spectrum.

Fortunately, everyone can experience what happens if vision has to be accomplished with just one type of sensory cell in the eye. By means of tightly fitting goggles of very low transmittance (Fig. 7) stimulus energy can be lowered under the threshold of excitability of the cones. Subjects wearing these goggles see nothing but darkness for several minutes. Thereafter, receptors of a fourth type, the rods (Fig. 1a) become active. Rods are inactive in daylight but take over at low levels of stimulation. There is only one type of rod. In rod vision the world appears grey in grey like black and white photography. Wearing these goggles at noon, when short-wave radiation prevails in natural light, strawberries appear dark if ripe, but bright if unripe. In morning and evening light, when the illuminant spectrum is shifted to long wavelengths, ripe strawberries appear brighter than unripe ones. Without cones, even the lightness of objects cannot be trusted. Animals living on strawberries would be unable to visually discriminate good fruits from indigestible ones. They would die out due to permanent gastric trouble.

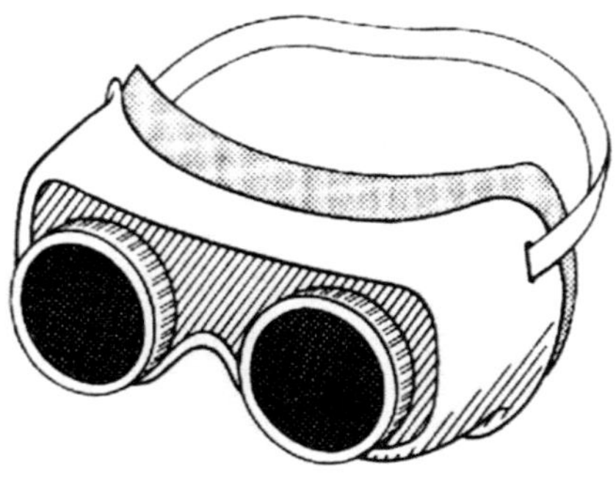

Fig. 7 Tightly fitting goggles of very low transmittance for rod vision

Critical experiments with the goggles (Campenhausen von 1986, Campenhausen von and Tausch 1993, Schneider and Campenhausen von 1998) show that the lack of lightness constancy is a serious problem. Sensory capabilities are useless if they are not reliable. Under the conditions of our planet, the natural variations of illumination have to be compensated for. This is possible only by means of cones which are sensitive in different spectral ranges (Fig. 1b). The original benefit of having three cone types was not color vision. Constancy of lightness perception must have been the primary purpose of the cones in evolution. The ability to discriminate between different spectral compositions of stimuli and to perceive colors was an additional natural gift, which improved visual capabilities.

Detailed knowledge of the physical aspects of the daylight spectrum and vision was required to understand the indispensable natural gift of lightness and color constancy.

Conclusion

The relationship of the mental and scientific aspects of color vision poses a problem transcending human understanding, according to Emil Du Bois-Reymond. This chapter has illustrated the relevance of science for the understanding of the mental domain by means of very specific psychophysical results, as opposed to philosophical doctrines of a general nature. Various examples taken from the psychophysics of color vision have been presented to demonstrate that it is possible to explain why the mental and physical domains must be different in order to serve biological needs. It has been shown that vague speculations about the color experiences of other persons may be settled (2. + 3.) by means of scientific research. Mental and scientific color systems are described which reveal common structures in spite of being derived from esthetical introspection and objective science, respectively (4.). The difficulties of crossing the border between the mental and the physical domain is to be expected considering the higher level of complexity of the stimuli generating color sensations, compared with the whole of all possible colors in the mind (5.). Reductionism is possible in predicting color sensations on the basis of the stimuli (from higher to lower complexity) but not in the other direction, starting with color sensations and asking for their physical basis (because there are many stimuli generating the same color sensation). Therefore, colors must be presented to the mind in a format, which cannot be translated back to their physical origin (6.). In the evolution of mind and body, information of relevance in the struggle for life was of importance rather than the physical aspects of sensory perception. This is another reason for the development of a mental format of reduced complexity. The mind was the original stage of all human experiences. Today its restrictions can be understood by means of psychophysics. Considering evolution, it emerges that the sensory and nerve cells providing color vision serve a more important purpose. They compensate for the variation of the daylight spectrum on the planet earth to create visual constancy (7.). This natural gift is unknown to most people because of its nearly perfect performance, which conceals the natural variations of illumination.

References

Aguilonius F (1613) Francisci Aguilonii e Societate Iesu Opticorum Libri Sex Philosophis juxta ac Mathematicis utiles. Officina Plantiniana, Antwerpen

Albers J (1963) Interaction of color. Yale University Press, New Haven

DIN-Farbenatlas (1961) DIN-Farbenkarten. Beuth-Vertrieb, Berlin

Du Bois-Reymond E (1872) Über die Grenzen des Naturerkennens. In: Du Bois-Reymond E (1886) Reden. Leipzig, 105–140

Du Bois-Reymond E (1880) Die sieben Welträtsel. In: Du Bois-Reymond E (1886) Reden. Bd. 2. Leipzig, 65–98

Campenhausen C von (1981) Elektrophysiologie und physiologische Modellvorstellungen bei Emil Du Bois-Reymond. In: Mann G (ed.) Naturwissen und Erkenntnis im 19. Jahrhundert: Emil Du Bois-Reymond. Gerstenberg, Hildesheim 79–104

Campenhausen C von (1986) Photoreceptors, lightness constancy and color vision. Naturwissenschaften 73: 674

Campenhausen C von, Tausch A H (1993) Absence of lightness constancy as a deficit of monochromatic vision. In: Drum B (d.) Color vision deficiencies XII (Doc ophth proc series 57) Kluwer, Dordrecht 197–202

Campenhausen C von (1994) Zufall und Notwendigkeit bei der Einführung früher elektrophysiologischer Begriffe und Konzepte durch Emil du Bois-Reymond. In: Meier W, Zoglauer T (eds.) Technomorphe Organismuskonzepte. Modellübertragungen zwischen Biologie und Technik. Frommann-Holzboog, Stuttgart, 269–268

Campenhausen C von, Schramme J (1995) 100 years of Benham's top in colour science. Perception 24: 695–717

Campenhausen C von, Pfaff C, Schramme J (2001) Three-dimensional interpretation of the color system of Aguilonius/Rubens 1613. COLOR research and application, Supplement Volume 26: 17–19

Campenhausen C von (2001) Eine Vorform der dreidimensionalen Farbensysteme in dem von Rubens illustrierten Lehrbuch des Franciscus Aguilonius (1613). Med hist J 36: 267–307. English translation in: Fehr M, Wurmfeld S (2004) Seeing red. On non-objective painting and color theory. Salon Verlag, Cologne, 192–230

Campenhausen C von, Schramme J (2003) Some properties of the physiological colour system. In: Mollon J D, Pokorny J, Knoblauch K (eds.) Normal and defective colour vision. Oxford University Press, Oxford 289–295

Campenhausen Christoph von (2007) Von der Psychophysik der Farben. Bildhafte Farbsysteme von Malern und Physikern. In: Elsner N (ed.) Bilderwelten. Vom farbigen Abglanz der Natur. Wallstein, Göttingen, 47–65

Jaeger W (1989a) Endgültige Abklärung der Untersuchungsbefunde Goethes an Farbenblinden. Klin Mbl Augenklinik195, 6: 382–387

Jaeger W (1989b) Goethes Untersuchungen an Farbenblinden 'Eine sehr enge Pforte, um in das Allerheiligste der Farbenlehre zu dringen'. Jb Heidelberger Akad Wiss für 1989. Heidelberg 1990, 103–112

Mancuso K, Hauswirth W W, Li Q, Conor T B, Kuchenbecker J A, Mauck M C, Neitz J, Neitz M (2009) Gene therapy for red-green colour blindness in adult primates. Nature 461: 784–787

Küng H (2005) Der Anfang aller Dinge. Naturwissenschaft und Religion. Piper, München

Ostwald W (1925) Der Farbenkörper und seine Anwendung zur Herstellung farbiger Harmonien. 12 Tafeln und Text. Unesma, Leipzig

Runge P O (1810) Die Farbenkugel oder die Konstruktion des Verhältnisses aller Mischungen der Farben zueinander und ihrer vollständigen Affinität, mit angehängtem Versuch einer Ableitung der Harmonie in der Zusammenstellung der Farben. Perthes, Hamburg

Schneider N, Campenhausen C von (1998) Color and lightness constancy in different perceptual tasks. Biol Cybern 79: 445–455

Silvestrini N, Fischer E P, Stromer K (1998) Farbsysteme in Kunst und Wissenschaft. DuMont, Köln

Zeki S (ed.) (1993) A vision of the brain. Whiley, New York

Pictures as strange objects of perception

10

Richard L. Gregory †

Contents

Abstract

Pictures have a double reality – objects in their own right and representing quite different objects in a different space and time. We may call these Other Worlds of pictures.

The first artists were cave painters; though there were far earlier pictures of bird's eyes on butterfly wings, produced by natural selection. For perception I distinguish 'cues' from 'clues'. This is from innate knowledge, to knowledge learned by individual experience. Knowledge is key to perception – as perceptions depend on and also provide knowledge.

Trompe l'oeil is the most dramatic use of cues and clues; but the more realistic a painting the more dramatic are its illusions, especially associated with moving around the painting, or (most dramatic) a trompe l'oeil dome of a cathedral. There is an attempt to classify phenomena of illusions.

Richard L. Gregory

† deceased May 17, 2010. CBE, FRS, FRSE

Emeritus Professor and Senior Research Fellow, University of Bristol

Department of Experimental Psychology

12a Priory Road, Clifton

Bristol BS8 1TU, UK

1. Introduction

Rather than looking at art history we look here at the biological origins of pictures – with associated phenomena of illusions. It is hoped that this may provide a context for thinking about the history of art.

Any picture is remarkable as it has a double reality. Like any ordinary object it is in our world of Here and Now, and miraculously a picture is also in Another World in a different space and time. It is indeed miraculous that we see entirely different objects, such as people, in a blob of paint.

Fig. 1 Bird's eyes 'painted' on butterfly wings by natural selection. This is pre-human art, designed to attract and warn

Other Worlds of pictures was the invention of cave artists of, perhaps, half a million years ago. Pictures are as magical to us now as they must have been to the first artists. Actually these were not quite the first pictures, for pictures have a much earlier biological origin deep in evolution – bird's eyes painted on butterfly wings, by Natural Selection millions of years ago (Fig. 1). These are remarkably realistic, with their eyespot reflections, serving as false though effective warnings to birds. As we also see them as eyes, these first pictures suggest that bird vision (and birds are descended from dinosaurs) is quite similar to ours.

2. Cues and clues

Vision works from innate *cues* triggering reflexes and inherited behaviour patterns, learned by the genetic code, and in 'higher' animals and man from sophisticated *clues,* based on individual knowledge learned by brains. The ideal example is the inferences of the Victorian fictional detective Sherlock Holmes from small clues, depending on his wide knowledge including brands of tobacco and varieties of human behaviour. Although this suggested distinction between cues and clues may not be generally accepted, it seems useful for thinking about origins of perception in evolution and from individual experience.

3. Trompe l'oeil

Pictures present patterns that are accepted as familiar cues and clues for recognising objects and events in the normal world. In art, depicted cues and clues are most effectively presented in trompe l'oeil paintings. We may ask . . .

3.1 Why is trompe l'oeil so effective?

We have said that everyday perception works from innate cues and learned clues for recognising objects, with their sizes and where they are in surrounding space, for interactive behaviour in a largely hostile world. Trompe l'oeil presents these same cues and clues but pictured as patterns on flat surfaces, generally with marked perspective and painted shadows suggesting depth – though in a flat picture this is illusory (Fig. 2). It is interesting that photographs seem to be less effective than the best trompe l'oeil paintings. This is presumably because the artist is free to use exaggerated though fictional cues and clues, based on experience of real objects and events. As a picture is physically very different from the objects it represents it can evoke bizarre illusions, especially with movement.

Fig. 2 Though a flat pattern, this still life evokes three-dimensional objects from painted cues and clues normally from the world of objects (Julian Merrow-Smith, Still Life with Autumn Fruits, 2010)

3.2 Moving illusions of trompe l'oeil

Visual cues and clues derive from interacting with objects, and they work well for interactive behaviour with normal objects. But strange distortions occur when the viewer moves around a trompe l'oeil picture, which, though flat, represents objects in three-dimensional space. As the observer moves around, the picture's flat images do not transform at all like retinal images of normal objects, including those portrayed in the picture. This absence, or very different change at the eyes with movement, produces remarkable dynamic distortions that defy description.

The bizarre movements and distortions of a large trompe l'oeil dome, as one moves under it, are truly wonderful. For example, the painted ceiling of the church of Sant' Ignazio in Rome or the false dome of the cathedral in Arezzo in Tuscany are dramatic and tell us a lot about how eyes and brains work, and sometimes get things wrong.

The founder of the modern science of vision, Hermann von Helmholtz (1821–1894), suggested a General Principle for explaining many visual phenomena, which comes to our aid here. We may state Helmholtz's Principle simply as: *Perception attributes objects to images.* This explains many illusions, as being false attributions. Seeing pictures as other objects depends on attributing more or less familiar objects to patterns of pigments. It is curious that we see a picture both as a picture and, at the same time, attribute other (often fictional) objects to it. So we see both the picture and what it represents. This is quite different for retinal images, as we do not see into our own eyes.

Helmholtz's Principle applies also to movement. When one moves around a portrait, with its eyes staring out at the viewer, *its eyes seem to follow one's every movement*. This is because, in normal life, eyes that continue to stare at one must be rotating to follow one's movements. We attribute this rotation of normal eyes to the fixed and flat but apparently three-dimensional eyes of a picture. The more realistic the picture, the more powerful is this illusory movement of its eyes.

Returning to the false dome: like the painted eyes, a true dome could only keep an unchanging image in our eyes as we move under it if (impossibly) it rotates to follow our movements. So this is what we see, although it is impossible and has zero probability! But there is more to these illusions of movement with large architectural trompe l'oeil, such as a false dome, for they undergo remarkable static distortions when we have moved away from the 'correct' position for viewing them. This is because these retinal images have not changed as they would for a three-dimensional object, such as a truly hollow dome. There is a continually up-dated expectation of shape changes according to the three-dimensional shape the object is *seen* to have – though this is inappropriate for a picture, as the picture is truly flat. As these pictures are *seen* in depth though they are flat, so there are discrepancies between what is expected and what happens at the eyes. The difference between the expected and the optical changes of the retinal image gives these static and dynamic distortions.

We attribute a face even to the sketchiest portrait, making the task of the artist easier than might be thought. Indeed, except for trompe l'oeil, the viewer contributes at least as much as the artist for seeing a picture. So we see 'faces in the fire' and the 'man in the moon' even from random patterns. There is a continuum in art from inkblots to trompe l'oeil, with wonderful revealing phenomena along the way (Fig. 3).

4. Classifying illusions

Vision evolved over millions of years – though not of course originally for pictures – which we see through processes that evolved for interacting with dangerous and rewarding objects. Vision's perception from a distance – and so power to predict – was immensely useful for survival. Pictures, though not immediately useful, depend on these life-death lessons from dangers and rewards of normal objects. Pictures are a fortunate gift for humans, bought by ancestral disasters.

Fig. 3 Seeing scenes and people in ink blots shows the creative dynamics of vision. www.inkblottestwallpaper.com/INK-BLOT TEST 01

With the risks of any dangerous speculation, key stages of evolution of vision seem to be from: passive *Receptions* of stimuli, for behaviour by reflexes and tropisms from cues, to active constructive *Perceptions* from visual clues, from knowledge of objects and how they behave. Finally, at least human brains developed *Conceptions* for thinking and communicating abstract ideas with symbols. Paintings, in their strange Other Worlds, involve all three.

The linking theme is increasing use of knowledge, from *receptions* to *perceptions* to *conceptions*. As misapplied knowledge can be deeply misleading it is not surprising if errors (phenomena of illusions) become richer and more interesting through this development from control by stimuli to perceptions as guesses – hypotheses of what is out there.

These speculations suggest that it might be interesting to classify phenomena of illusions. Here is an attempt, by kinds and causes, which will take us to pictures (Table 1). The idea is that there are very general rules for perception (here called 'sideways rules', rather like grammar in language) and top-down knowledge of objects and their characteristics and uses, and finally conceptual understanding. As classifying has been so important through the history of science, we may find it useful here for understanding the phenomena of illusions and the wonder of pictures.

Conclusion

Art works by using 'cues' and 'clues' important for survival in a dangerous world, where objects are more than patterns in the eyes. Art is not only cultural, it has its roots in the biological evolution of perception by which brains represent the world of objects for intelligent predictive behaviour.

References

Gregory RL (1997) Eye and brain: The psychology of seeing (5th edn.). Oxford University Press, Oxford, UK

Gregory RL (2009) Seeing through illusions (1st edn.). Oxford University Press, Oxford, UK

Ebert-Schifferer S, Staiti P, Wheelock A K, Singer W (2003) Deceptions and illusions:Five centuries of Trompe L'Oeil paintings (1st edn.), National Gallery of Art, Washington DC

Gombrich E (1950) The story of art (1st edn.). Phaidon, London, UK

Websites for trompe l'oeil dome:
www. museumplanet. com/tour.php/venice/l/9; http://commons.wikimedia.org/wiki/File:Sant' Ignazio_-_painted_dome_-_antmoose.jpg

Table 1

Kinds and Causes of Illusions

CAUSES

KINDS	Receptions → Side-ways Rules	Perceptions	Top-down Knowledge →	Conceptions Understanding
NOTHING	**Blindness** Long-term total blindness has no sensation – like behind one's head.	**Perceptions as hypotheses** Perception is *attributing objects to images*, forming predictive hypotheses. **Inappropriate rules** Produce illusions with normal physiology.	**Agnosia** *Lack of visual knowledge* Failure to recognise even familiar objects (generally from brain damage.)	**Ignorance** Without understanding, the world looks like a conjuring trick.
INSTABILITY	**Jazzing** Op. Art. McKay Rays (Repeated lines stimulate on-off cells with eye tremor). ***Shifting contours*** Ouchi illusion. (Lack of 'border-locking'?)	**Grouping** Random dot patterns group and regroup by Gestalt Rules: closure, contiguity, common fate etc.	**Constancy Scaling** 'Constancies' partially compensate changes of size and shape to make the visual world stable. Inappropriate Constancy scaling produces distortions.	**Brain changing its mind** Conceptions are constructed from memory and imagination, by following cognitive rules and guided by probabilities.
CON-FOUNDED AMBIGUITY	**Sensory differences** Limited by neural noise, and overlapping of response curves. ***Colour*** Red+green yellow light looks the same as monochromatic yellow, (as the R and G pigments overlap, giving yellow when nearly equal).	**Object differences** When their images are the same, different objects must look the same. *Ames Room* Has the same retinal image as a normal room – so must look the same.	**Classifying** Different kinds of objects are often confounded when not familiar or understood.	**Explanations** Depend on analogies and paradigms. As science develops, understanding and appearances separate and may conflict.
FLIPPING AMBIGUITY	**Epilepsy** **Spontaneous brain activity** Neural nets are dynamic, and can be physically unstable. Migraine Visual brain disturbances associated with headache.	**Figure-ground** The most basic decision is whether there is an object present. This is seen dramatically in flipping figure-ground ambiguity, when the brain cannot make up its mind.	**Alternative Perceptions** Perceptions flip to alternatives when the brain can't make up its mind. E. g. *Necker Cube, Duck-Rabbit, young woman/grandmother.* The *Hollow Face illusion* Probabilities normally give stability, but can mislead – a concave hollow facemask looks convex.	**Collapsing wave packets** Objects do not 'flip' to other objects, except in quantum physics. Measurements or perceptions are supposed to 'collapse' many possibilities into particular realities. This creation of physical reality by perception has been ascribed to consciousness. This is totally mysterious.

DISTORTION	**Signal errors** *Contrast:* brightness, colour illusions. *Lateral inhibition* gives only small spatial distortions. *Delay* gives Pulfrich Pendulum phenomenon *After effects* of continuous motion, tilt, curvature, spatial frequency, colour, etc. may serve to calibrate the senses,	**Cognitive distortions** "Geometrical or ' perspective' distortions: Müller-Lyer; Ponzo; Hering; Poggendorff; Horizontal-Vertical; Harvest Moon etc. On the Misapplied Scaling theory, Depth cues miss-set size-scaling – features signaled as more a distant being expanded. Scaling can also be set 'downwards'.	**Anticipation** Prediction is essential for cognitive perception, but can mislead.	**Reference truths** An object cannot itself be distorted, but may differ from accepted references. Thus a ruler is bent, or too long or too short by reference to some other ruler, accepted as 'true'. Reference to non-illusions is essential for measuring illusions; though illusions can show up as internal inconsistencies.
PARADOX	**Conflicting cues** Parallel channels can disagree when adapted differently. Then perception may reject a discrepant channel or combine aberrant signals or cues into paradox.	**Conflicting clues** The Penrose *Impossible Triangle* can exist as a three-dimensional object, yet appear impossible from certain positions when the sides meet, in the retinal image, though are separated in depth. The false assumption of two-dimensions generates the paradox.	**Conflicting knowledge** Magritte's painting of the back of a man's head – appearing in the mirror instead of his face. This disturbs, as it goes counter to one's visual knowledge of reflections.	**Cognition and physics** Although the brain is a physical system, its 'virtual realities' of perception are not limited by physics; making it possible to imagine and see impossibilities.
FICTION	**Spurious signals** After-images appear as objects that would give the same images, so seen in external space.	**Grouping** Random dots group with the Gestalt Laws – similarity, common fate etc. – into common object features.	**Phantasms** *Faces-in-the-fire; man-in-the-moon; inkblots.* These show the creative dynamics of perception, when alternative hypotheses are evoked from minimal evidence.	**Perceptions and Conceptions** Are both indirectly related to reality, and subject to illusions and delusions.
OTHER WORLDS OF PICTURES	**Cues** Visual cues trigger neural systems, to signal edges, shapes, colours, movement. Artists are free to combine cues in new ways to produce the Other World of pictures.	**Clues** Visual clues are evidence for objects, based on probabilities from previous experience.	**Object knowledge** Confers non-optical properties to images, such as hardness, and warning of dangers and promises of rewards, to personalities of faces. Pictures work by attributing more or less familiar objects to patterns of pigments.	**Pictures** Pictures evoke Other Worlds, of a different space and time from our world.

Towards an understanding of the psychostimulant action of amphetamine and cocaine

11

René Weissensteiner*, Thomas Steinkellner**, Andreas Jurik*, Simon Bulling**, Walter Sandtner**, Oliver Kudlacek**, Michael Freissmuth**, Gerhard F. Ecker * and Harald H. Sitte**

Contents

Abstract

Cocaine and amphetamine are psychostimulant drugs that are illicitly used; they affect sensory perception by targeting the neurotransmitter:sodium symporters (NSS) at the synapses between neurons. They both increase the concentration of the neurotransmitter in the synaptic cleft but by different means.

The physiological role of NSS is the reuptake of their endogenous substrate. For this task, they exploit the pre-existing sodium-gradient across the cellular membrane that is maintained by the activity of the sodium:potassium pump. This reuptake process terminates synaptic transmission because the neurotransmitter is removed from the synaptic cleft – and its action on pre- and postsynaptic receptor molecules is stopped.

Amphetamines induce the reverse operation of distinct NSS family members, whereas cocaine merely inhibits the same transporters and thereby blocks the reuptake of neurotransmitter. These effects, although completely different in molecular mechanism, lead to an increase in the synaptic concentration of non-exocytotically released neurotransmitters. While these actions have long been appreciated, the underlying mechanistic details have been surprisingly difficult to understand. The advent of a crystal structure of a prokaryotic NSS protein and the concomitant development of homology models for eukaryotic NSS family members generated novel insights into the structure-function relationships

René Weissensteiner
University of Vienna
Life Sciences
Department of Medical / Pharmaceutical Chemistry
Althanstr.14
1090 Vienna, Austria
e-mail: rene.weissensteiner@univie.ac.at

Harald H. Sitte
Medical University of Vienna
Center of Biomolecular Medicine and Pharmacology
Institute of Pharmacology
Waehringerstr.13a
1090 Vienna, Austria
e-mail: harald.sitte@meduniwien.ac.at

Pharmacoinformatics Research Group
* University of Vienna, Department of Medicinal Chemistry, Emerging Field Pharmacoinformatics
** Medical University of Vienna, Center of Biomolecular Medicine and Pharmacology, Institute of Pharmacology, Währinger Str. 13a, A-1090 Vienna, Austria

of this clinically relevant class of transporters. Ultimately, we hope to understand the effects of amphetamines and cocaine on a molecular level to elucidate their profound effects on sensory perception.

1. Introduction

Among the many psychostimulant drugs illicitly used by mankind, amphetamines and cocaine are of particular importance (Amara and Sonders 1998). In part, these drugs play a role as recreational drugs, in part, they are abused as *psychostimulants*. These drugs, in part, exert remarkable effects on sensory perception in that they influence the perception of colours (e. g. mescaline) or decrease the urge for rest (e. g. methylene-dioximethamphetamine, better known as "ecstasy"). Clinically, cocaine elicits hyperactivity, suppresses sleep and fatigue resulting in insomnia. Cocaine also induces improved performance both psychologically and physically – the latter is the reason why cocaine is blacklisted as a drug used for doping in sports. Administration of higher cocaine doses leads most commonly to euphoria and also to hallucinations and the development of fear conditions. The clinical profile of amphetamine closely resembles the actions of cocaine with notable additional effects: it enhances the power of concentration and, in general, the well-being of the subject in conjunction with improved self-confidence. The administration of higher doses of amphetamines leads to psychosis and hallucinations. In addition, some individual amphetamine congeners may lead to additional disturbances of sensory perception, for instance mescaline (also see below) that leads to an intensified perception of colours (for review, see Seiden et al. 1993).

Interestingly, both cocaine and amphetamines target the same protein, i. e. the neurotransmitter transporters that mediate the reuptake of monoamines. This is the basis for the physiological function of neurotransmitter transporters that act to re-accumulate neurotransmitter into the presynaptic specialization by using the sodium gradient across the cell membrane as the driving force (Rudnick and Clark 1993). Therefore, this transport process constitutes an efficient and fast way to retrieve released neurotransmitter from the surroundings of the neuron. It is one of the mechanisms that terminate synaptic transmission. Other mechanisms include (i) diffusion of the neurotransmitter out of the synaptic cleft or (ii) enzymatic degradation (Iversen 1971). The *neurotransmitter:sodium symporters* (Saier Jr et al. 2006) comprise the largest family among all transporters using electrochemical gradients to drive the accumulation of substrates (Nelson 1998). They include transporters for *monoamines* (i. e. *dopamine*, DAT, *norepinephrine*, NET, and *serotonin*, SERT), amino acids (i. e. the transporter for γ-amino butyric acid, GABA, thus the GABA-transporter, henceforth GAT, glycine, proline and taurine), as well as osmolytes (i. e. betaine and creatine) and the so-called "orphan" transporters with currently unknown substrate specificity (Masson et al. 1999; Saier Jr et al. 2006).

The neurotransmitter:sodium symporter family is of obvious medical relevance since a number of different disorders can be efficiently treated by inhibiting transporters. Here, primarily mental diseases like depression are treated by inhibition of SERT and NET or epilepsy where inhibition of GAT1 by tiagabine is a sufficient add-on (Iversen 2000).

The transport mechanism itself is still an enigmatic process although there is a wealth of data on ionic requirements, stoichiometry, regulation, physiological and pharmacological properties of the re-uptake process mediated by neurotransmitter:sodium symporters (Fig.

1; for review see Torres et al. 2003b). Experiments have clarified that these transport proteins consist of 12 hydrophobic transmembrane helices (Amara and Kuhar 1993), studies employing antibodies (Bruss et al. 1995), electron microscopic immunocytochemistry (Nirenberg et al. 1997) and functional assays using thiol-reactive reagents (Chen and Reith 2000; Sen et al. 2005).

A breakthrough in the biology of neurotransmitter:sodium symporters was the *molecular cloning* of GAT1 by Kanner and colleagues (Guastella et al. 1990). Subsequently, the sequence of the first monoamine transporter, NET, was unravelled by Susan Amara and her colleagues (Pacholczyk et al. 1991), followed by the SERT (Blakely et al. 1991) and the DAT (Giros et al. 1991). The knowledge of the primary sequence of the neurotransmitter transporters initiated the possibility to change it by mutagenesis approaches. Almost two decades of molecular cloning ensued and helped to characterize the portions of the individual neurotransmitter:sodium symporter members important to afford binding and translocation of substrate (Torres et al. 2003b). Until recently, educated guesses have been based on the results of these mutagenesis studies and observations employing chimeric proteins. However, the interest in the transport process has been filled with life again because several crystallographic snapshots have become available that allow the re-examination of early educated guesses: the x-ray crystal structures of the bacterial transporter LeuT from *Aquifex aeolicus* has been published in several different conformations. LeuT is a homolog to family members of the neurotransmitter:sodium symporter family (Protein Data Bank accession code 2A65; Yamashita et al. 2005; Singh et al. 2008; Singh 2008). In spite of the low overall homology (some 20–25% residues in LeuT and mammalian neurotransmitter:sodium symporters are identical), there is a much higher conservation around the binding site (i.e., higher than 50% identity; Singh 2008). Hence, homology modelling has fuelled a situation that clearly allows for making more educated guesses than those relying solely on site-directed mutagenesis. Importantly, the different crystallographic results describe a sequence of events (Singh 2008) that is consistent with a basic model that has been conceptualized some 40 years ago by Jar-

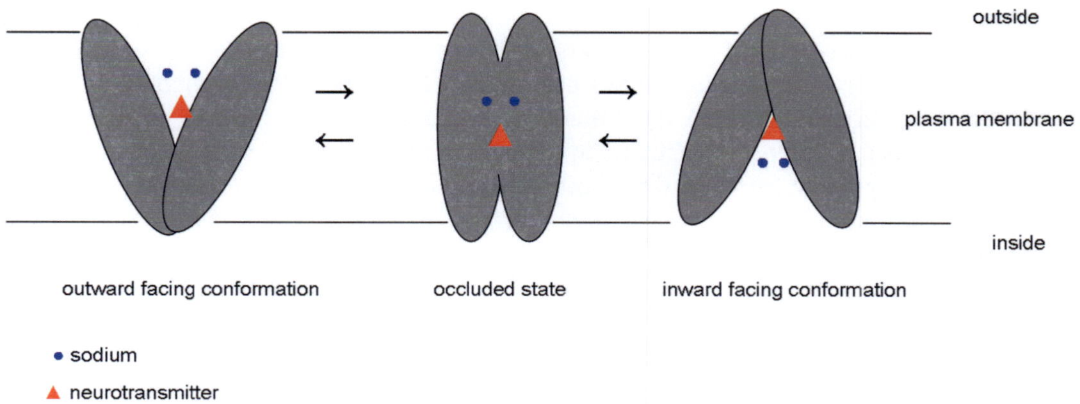

Fig. 1 Alternating access model: A mechanism to describe the activity of the bacterial transporters $LeuT_{Aa}$ and Glt_{Ph} in the cell membrane that involves binding of substrates and ions to an open-to-outside conformation, isomerization to an occluded state, and subsequent isomerization to an open-to-the-inside conformation

detzky (1966) and termed the "*alternating access model*" (Fig. 1). The hypothesis predicted a transition state in which the protein changes from outward- to inward-facing conformations via an occluded state. The various crystal structures (Singh 2008) support this scheme and thus have proven very fruitful in guiding attempts to elucidate the molecular mechanism of transport (Shi et al. 2008; Forrest et al. 2008).

The *quaternary structure* of neurotransmitter:sodium symporters has been addressed by a number of different biochemical methods (Jess et al. 1996; Kilic and Rudnick 2000; Hastrup et al. 2001) and microscopic techniques in living cells (Schmid et al. 2001; Sorkina et al. 2003; Just et al. 2004; Egana et al. 2009) including neurons expressing transporters in their natural environment (Egana et al. 2009; Fjorback et al. 2009); they all report that neurotransmitter:sodium symporter family members form *constitutive oligomers* (Sitte and Freissmuth 2003; Sitte et al. 2004). At present, the functional consequence for this thermodynamically complex structural arrangement is not clarified. Most importantly, the transport process is not affected in mutants that have been rendered monomeric (Scholze et al. 2002a). Recently, it has been shown that oligomer formation was evolutionarily supported to efficiently export transporters from the endoplasmic reticulum (Farhan et al. 2006). While cocaine acts on the transporter merely as a blocker of its activity, the quaternary arrangement is required to allow the action of amphetamine and its congeners on DAT, SERT and NET. In the present chapter, we will focus on the molecular mechanism underlying the psychostimulant activity of cocaine and amphetamine.

Both compounds exert their effects by competing with the substrate of the transporter for the substrate binding site. However, they differ in the way the transporter handles them. While cocaine is a true uptake inhibitor, amphetamine is a substrate readily transported into the cells. There is no rule without exceptions; methylphenidate, an amphetamine derivative also blocks the uptake without being transported.

2. Amphetamine and its congeners

The structural requirements that define a substance to be a member of the amphetamine family are (1) an unsubstituted phenyl ring, (2) a two-carbon side chain between the phenyl ring and nitrogen, (3) an α-methyl group and (4) a primary amino group (Biel and Bopp 1978). Not all substances considered to be amphetamines also classify according to these strict rules (Fig. 2); however, they are all related to phenethylamine.

Naturally occurring amphetamines have been used by mankind ever since owing to their psychostimulant effects. The pharmacologically active ingredients of the plant *Ephedra sinica* (ephedrine) or of *Catha edulis* (of the family of the *Celastraceae* that produces cathinone, norpseudoephedrine/cathine also named 'Kath') have not only been used to elicit psychostimulant effects but also as medication. For instance, ephedrine has been widely administered in the treatment of asthma and upper respiratory infections. On the other hand, *over-the-counter* ephedrine was used as appetite suppressant and abused by athletes to enhance their performance. Ephedrine has been removed from the market in the United-States in 2004 after a series of deaths among athletes but still is the most common source for the production of methamphetamine by amateurs. Cathinone is widely used in the region close to the horn of Africa but also in Yemen, where it is also cultivated in larger scale. Another naturally occur-

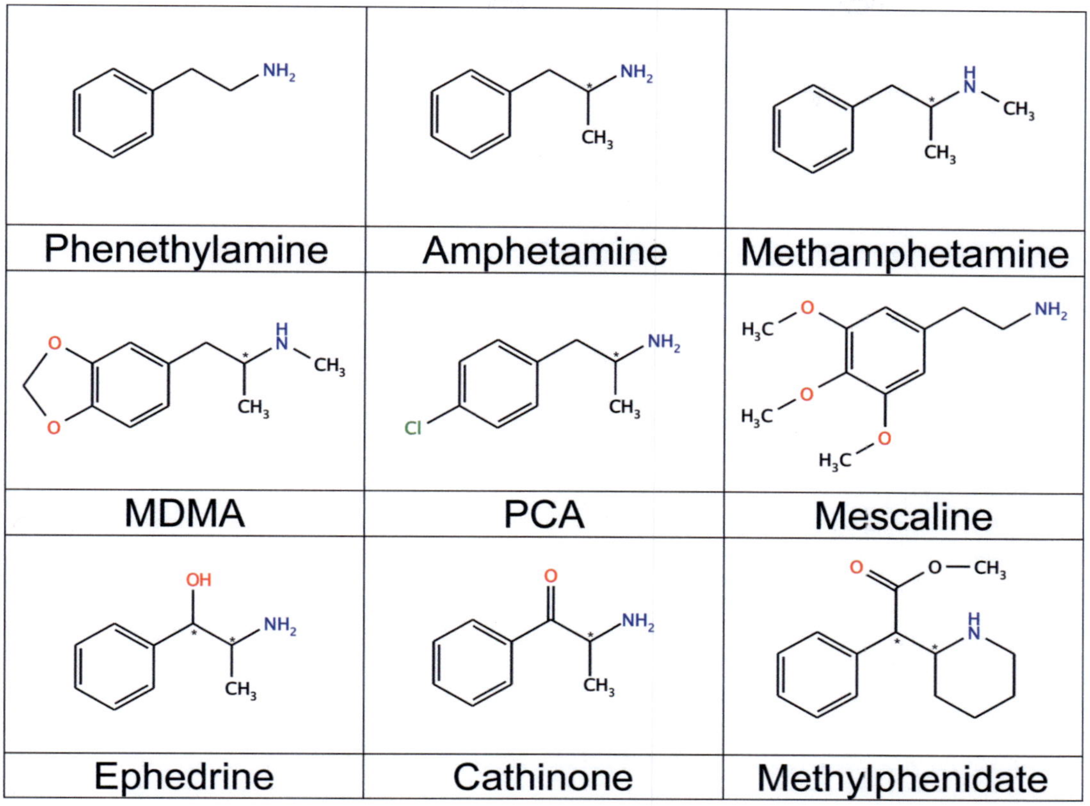

Fig. 2 Amphetamine and congeners: Several different amphetamines and congeners are shown (the chemical names in parentheses represent the IUPAC names). Starting from the "mother compound" phenethylamine (2-phenylethanamine) over amphetamine (street name: speed; D-form: (2S)-1-phenyl-propan-2-amine), methamphetamine (street name: crank or crystal meth; D-form: (2S)-N-methyl-1-phenylpropan-2-amine), MDMA or methylenedioxymethamphetamine (street name: ecstasy; 1-(1,3-benzodioxol-5-yl)-N-methylpropan-2-amine), PCA = p-Chloroamphetamine (1-(4-chlorophenyl) propan-2-amine), to naturally occurring amphetamines like the extract of the peyote cactus mescaline (2-(3,4,5-trimethoxyphenyl)ethanamine), ephedrine ((1R,2S)-2-(methylamino)-1-phenylpropan-1-ol), cathinone (also known as norephedrone, street name kath; (2S)-2-amino-1-phenylpropan-1-one) to methylphenidate (methyl 2-phenyl-2-piperidin-2-ylacetate) that is not transported in contrast to virtually all other amphetamines

ring amphetamine-like substance is Mescaline that is produced in the peyote cactus *Lophophora williamsii.* Mescaline was used extensively by the Aztecs and mentioned in a historic description on South-American plants (Sulzer et al. 2005). There are a number of synthetic amphetamines including D-amphetamine which was the first amphetamine marketed for clinical use, and methylphenidate. Both compounds are administered in patients suffering from attention deficit hyperactivity disorder. However, the mechanisms by which the two exert their effects markedly differ. Whereas D-amphetamine is a substrate of monoamine transporters and clearly induces reverse transport (Sitte et al. 1998), methylphenidate is a blocker and elevates extracellular monoamine concentrations by inhibiting the reuptake of previously released monoamines (Mazei-Robinson and Blakely 2006). Importantly, the notion that methylphenidate acts as an inhibitor is the exception of the rule: normally, amphetamines are transported and induce reverse transport or carrier-mediated release (Sulzer et al. 2005). In their seminal publication, Hertting and Axelrod (Axelrod et al. 1961) observed the uptake-inhibiting effect of amphetamines, which is similar to the effect of uptake blockers. However, these authors already insinuated that this could also be seen as simple competitive inhibition and that amphetamine could possibly play a different role regarding neurotransmitter transporters (Axelrod et al. 1961).

In subsequent experiments, reverse transport (Fig. 3) was addressed using the effect of tyramine on noradrenaline-containing neurons which was absent in denervated organs (Fleckenstein and Burn 1953) and in reserpine-treated animals (Burn and Rand 1958). These and related experiments, where sympathetic nerve endings in organ preparations were challenged with amphetamine or its congeners (Axelrod et al. 1961; Furchgott et al. 1963; Glowinski and Axelrod 1965; Ross and Renyi 1966;), paved the way for a concept, in which the release of noradrenaline (or dopamine) by reverse transport was the crucial step to understand their mode of action. This concept was extensively verified in brain slices and synaptosomes (Gobbi et al. 1997). However, all these preparations contain presynaptic nerve terminals with the complete machinery for synaptic vesicle exocytosis. The same is true for platelets and membranes prepared thereof, which contain SERT and the exocytotic machinery of platelets (Fishkes and Rudnick 1982; Rudnick and Wall 1992). Because of the simultaneous presence of vesicular and membrane transporters – and hence of different pools of substrates – it is difficult to dissect the contribution of each substrate pool to reverse transport.

With the advent of the cloned neurotransmitter transporters and their coding DNA's,

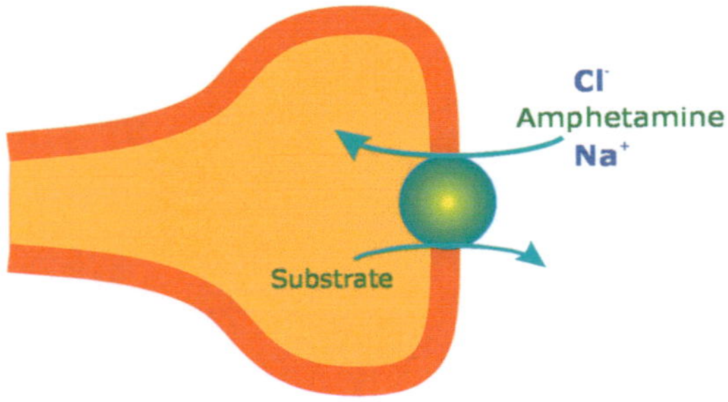

Fig. 3 Schematic representation of a synaptic button expressing a neurotransmitter transporter targeted by amphetamine and co-substrates that trigger the transporter-mediated release of cytosolic substrate, the neurotransmitter

respectively, the way was free to examine individual transporter properties in heterologous expression systems (e. g. cell lines, nowadays also neurons). This approach definitely led to a more refined picture since there were no confounding effects of vesicular storage mechanisms and of the exocytotic release machinery. Clearly, amphetamine and its derivatives induce reverse transport and thereby reproduce all the characteristic features known from native systems (Eshleman et al. 1994; Wall et al. 1995; Pifl et al. 1995; Pifl and Singer 1999; Pifl et al. 1999). Furthermore, observations in DAT-knockout mice (Jones et al. 1998) provided a formal proof that monoamine transporters are the principle site of action of amphetamine and its congeners.

Not only amphetamines are capable of inducing reverse transport; rather, all substrates of neurotransmitter:sodium symporter family members can, in principle, reverse the transport direction of these transporters (Eshleman et al. 1994; Pifl et al. 1995; Cinquanta et al. 1997; Sitte et al. 1998; Scholze et al. 2000; Sitte et al. 2001; Gobbi et al. 2002, 2008). However, under physiological conditions natural substrates do not act as releasers because futile cycling would ensue and efficient removal of neurotransmitter from the synaptic cleft made impossible. In contrast, the biological actions of amphetamines and other releasers are contingent on transport reversal and there isn't any evidence that they elicit a biological action other than that in vivo. In addition, if transmembrane ion gradients are changed, reverse transport is triggered. This is true for both, lowering the extracellular sodium concentration (Pifl et al. 1995, 1997; Pifl and Singer 1999) and raising the extracellular potassium concentration (Scholze et al. 2000, 2002b). Thus, there are two different types of reverse transport: (i) binding of releasers to the transporter protein and (ii) changes of the ion composition of the extra- or intracellular fluid as triggers. The latter situation differs from the former by the fact that releasers need not be present on the extracellular side. Obviously, in either case, a releasable pool of substrate must be accessible on the cytoplasmic side.

Initial models to explain the reverse operation of neurotransmitter:sodium symporters were based on the "alternating access hypothesis" that explained the inward transport of substrates from the extracellular side by a substrate-induced conformational change in a sodium- and chloride-dependent manner. Following this hypothesis the concept of reverse transport as a "revolving door" emerged, where reverse transport of any intracellular substrate was driven by and coupled to the inward transport of any extracellular substrate. The fact that both, forward and reverse transport is accomplished by the very same transporter molecule was inferred from the substrate selectivity and the blockage by specific inhibitors. For example, cocaine blocked both substrate uptake and amphetamine-induced substrate efflux. This revolving door model was termed *"facilitated exchange diffusion model"* (Fischer and Cho 1979; Trendelenburg et al. 1987; Bönisch and Trendelenburg 1989).

Inward transport is the favoured mode and outward transport of substrate is per se a rare event (because of a low intracellular sodium concentration). In case the protein transports larger amounts of substrates, reverse transport is more likely to occur (e.g., because intracellular sodium builds up). It can therefore be conceived that the half-maximal effect for substrate-induced substrate release is higher than the affinity for inward transport (Scholze et al. 2000). Likewise, the maximal velocity of outward transport only reaches <20% of V_{max} for inward transport (Sitte et al. 2001, 2002).

The switch from outward- to inward-facing conformation enhances the availability of inward-facing transporters by all transported substrates. Hence, the probability of the binding and subsequent translocation of cytoplasmic substrates to the cell exterior is enhanced. Fischer and Cho (1979) inferred that the tight coupling between substrate uptake and reverse transport must be expected: "If uptake of the releasing amine is the first step in the chemical release process, then it must follow that the initial rate of release of [^{3}H]dopamine induced by an indirectly acting sympathomimetic amine must be related to the initial rate of uptake of the releasing agent." This "facilitated exchange diffusion model" was supported by numerous publications on reverse transport examined in different neurotransmitter:sodium symporter family members (Hilgemann and Lu 1999; Lu and Hilgemann 1999a,b; Roux and Supplisson 2000; Wang et al. 2003).

Under control conditions, one does not expect that transporter-mediated reverse transport is likely to occur. In cells containing vesicles like neurons and platelets, the free cytosolic concentration of substrate is kept low because neurotransmitters are sequestered into these synaptic vesicles by the action of specialized vesicular transporters. The intracellular concentration of dopamine has been estimated to about 2 µM (Ewing et al. 1992). Since the affinity for outward transport is by far higher than that for inward transport (by three orders of magnitude; Sitte et al. 2001), this concentration is far too small to initiate reverse transport. Jones and colleagues (1999) measured reverse transport of dopamine by cyclic voltammetry in mouse striatal slices. However, an elevation of the intracellular dopamine concentration by inhibition of the vesicular monoamine transporter did not produce DAT-mediated reverse transport.

Following the arguments above, we propose two different possibilities how efflux is driven: (i) Inward facing conformations are favoured, if the Na^+-concentration is increased on the intracellular side. Likewise, reducing the external Na^+ concentration ablates the inward driving force and, in principle, precludes the accumulation of the outward facing conformation. This statement is supported by the following arguments: inhibitors of NET, DAT and SERT (e. g., cocaine analogues, tricyclic antidepressants) bind preferentially to the outward facing conformations (Singh 2008; Singh et al. 2008); high-affinity binding is dependent on Na^+ (Humphreys et al. 1994) and proceeds in an ordered fashion with Na^+ binding prior to the inhibitor (Korkhov et al. 2006). Furthermore, currents elicited by amphetamine are also consistent with the interpretation that high concentrations of Na^+ drive the transporter into the outward conformation (Erreger et al. 2008). Manipulations of the Na^+-gradient result in the predicted change in influx and efflux rates: ouabain inhibits the Na^+/K^+-ATPase, thereby enhancing the intracellular Na^+ concentration. This has long been known to reduce substrate uptake and to facilitate efflux (Bönisch 1986; Trendelenburg et al. 1987; Wolfel and Graefe 1992; Sitte et al. 1998; Jones et al. 1999) and has also been shown using heterologous expression systems, i. e. in the absence of vesicular transporters (Sitte et al. 2000, 2002; Scholze et al. 2000). Khoshbouei and colleagues (2003) correlated a rise in the cytoplasmic Na^+-concentration and the increase in DAT-mediated dopamine efflux. Taken together, these results indicate that substrate efflux can be triggered by an increase in the intracellular Na^+ concentration. (ii) Amphetamines and other releasers must elicit a conformational change that drives reverse transport.

Neurotransmitter:sodium symporter-family members carry a current which is intricately linked to the actions of amphetamines. Binding of amphetamine to the transporter appears to favour a so-called *channel mode*, i. e., a conducting pore may be stabilized and support efflux. Transient elevations of extracellular dopamine can be observed by amperometry in DAT-expressing cells and these "bursts" were only detected after amphetamine application but not of dopamine application (Kahlig et al. 2005). The spikes are consistent with fast, burst-like release events of neurotransmitter through a channel. Amperometry is limited in its sensitivity (and the amount of released detectable substrate), thus, it is difficult to discern the relative proportion of dopamine that is released by bursts and by counter-transport. Kahlig and colleagues estimated a ratio of 9:1 in favour of continuous efflux (2005); this estimate suggests that amphetamine does allow the transporter to cycle efficiently through outward and inward facing conformations and that the channel-like mode is a comparably rare event. However, amphetamine action on DAT *per se* depolarizes the cell (Meinild et al. 2004; Carvelli et al. 2004). The *uncoupled conductance* is postulated to supply enough Na+ to trigger reverse transport: based on the "alternating access model", Na^+ accumulates intracellularly and favours the transporters in their inward facing conformation. Thus, the transporter are more prone to support substrate efflux: binding of substrate to the inward-facing conformation of the transporter becomes a more likely event as the intracellular Na^+

concentration rises. Thereafter, the transporters return to the outward-facing conformation in a substrate-loaded state and translocate substrate from the cell interior to the extracellular space.

2.1 What are the differences between physiological substrates and amphetamines?

A question that remains unanswered by the arguments laid out above is: What are the differences between physiological substrates and amphetamines that finally lead to their dissimilar properties? The apparent affinity in substrate and amphetamine uptake is very similar (being in the low micromolar range; Sitte et al. 1998; Seidel et al. 2005). The number of transported substrate molecules differs. Since amphetamines are difficult to measure precisely owing to their notorious lipophilic nature (which causes rapid back-diffusion), there are only a few systematic investigations in this area: HEK293 cells stably expressing the human DAT differed in substrate uptake rates (dopamine, tyramine and the two isomers of amphetamine) (by more than a factor of 20 (Sitte et al. 1998). This has also been observed by other groups (Bönisch 1984; Zaczek et al. 1991). In contrast, superfusion experiments performed in parallel sister cultures revealed that D-amphetamine induced transporter-mediated release with the highest potency and efficacy. In these studies, the released substrate was the metabolically stable MPP^+ (Langston et al. 1984; Javitch et al. 1985); this obviates any confounding effects due to the action of amphetamines on monoamine-oxidases (MAO-A or MAO-B, that are not seen with physiological substrates such as dopamine and serotonin; see below). Furthermore, patch-clamp experiments carried out in parallel showed that the rank order of potency for the induction of inwardly directed current was comparable to that obtained in superfusion experiments (Sitte et al. 1998). Importantly, when currents were consecutively assessed in the same DAT-expressing cells (Sitte et al. 1998), amphetamine induced a 25% higher inwardly-directed uncoupled conductance compared to dopamine. Hence, the propensity of amphetamines to induce larger currents suggests that they are more effective in triggering the channel mode in neurotransmitter transporters. This may account for some if not all of the differences between amphetamines and physiological substrates.

However, the action of amphetamines also requires transporter phosphorylation: inhibition of protein kinase C or of *Ca^{2+}-calmodulin-dependent kinase-2* (CaMK2) blunts the action of amphetamines but does not completely eliminate substrate-induced efflux (in other words: exchange diffusion) (Kantor and Gnegy 1998; Kantor et al. 2001; Gnegy 2003; Seidel et al. 2005; Fog et al. 2006). These observations can also be recapitulated upon genetic ablation of *protein kinase Cβ* (Chen et al. 2009).

2.2 Amphetamine induces activation of protein kinases

The membrane-impermeant Ca^{2+}-chelator BAPTA reduces amphetamine-induced dopamine efflux (Gnegy et al. 2004). Hence, this supported the notion that the action of amphetamines is – in part – contingent on protein kinase C-mediated (Kantor and Gnegy 1998) and Ca^{2+}/calmodulin-dependent phosphorylation as shown for DAT (Kantor et al. 1999; Fog et al. 2006), NET (Kantor et al. 2001) and SERT (Seidel et al. 2005). However, the direct activation of protein kinase-C by amphetamines (Giambalvo 1992a,b) and of Ca^{2+}/calmodulin-dependent protein kinase has not been shown so far. The stimulation is dependent on intracellular Ca^{2+} (Giambalvo 2003). Nevertheless, the pleiotropic actions of kinases hint to a number of different mechanisms that affect transport reversal, e. g. phosphorylation-dependent inhibition of Na^+/K^+-ATPase (Kazanietz et al. 2001). Activators of PKC such as 12-O-tetra-decanoyl phorbol-13-acetate (TPA), diacylglycerol and arachidonic acid enhance DAT-mediated efflux of dopamine (Pozzan et al. 1984; Davis and Patrick 1990; L'hirondel et al. 1995). Nevertheless, there is evidence for a direct action of PKC-isoforms on transporters: truncation of the 22 N-terminal residues of DAT eliminates the ability of DAT to undergo PKC-dependent phosphorylation (Granas et al. 2003) and suppresses reverse transport induced by amphetamine (Khoshbouei et al. 2004). Foster et al. (Foster et al. 2002) detected DAT phosphorylation after stimulation of PKC phorbol esters. The N-terminus of DAT can be phosporylated by PKC (Fog et al. 2006; Gorentla et al. 2009). Nevertheless, amphetamine-induced reverse transport is blunted after mutational replacement of the 5 serine residues in the N-terminal segment of DAT by alanine (Khoshbouei et al. 2004). From these mutagenesis experiments, the authors concluded that amino terminal phosphorylation by PKC shifts the DAT from a "reluctant" to a "willing" state for efflux (Robertson et al. 2009). This was further underscored by mutation of the same 5 serine residues to aspartate and thereby restoring the efflux property of the transporter (Khoshbouei et al. 2004).

PKC's are not the only intracellular kinases postulated to phosphorylate neurotransmitter transporters. As mentioned above, CamK2α phosphorylates DAT (but not NET or SERT) and the mechanistic details are understood. First, the enzyme engages the carboxyl terminus of DAT and subsequently phosphorylates the amino terminus (Fog et al. 2006). However, it is unclear to date whether the two kinases exert their effects independently from each other or in strictly hierarchical order. Furthermore, other kinases (e. g., Moron et al. 2003; Zhu et al. 2005; Ramamoorthy et al. 2007) impinge on neurotransmitter transporters. This causes changes in trafficking (Moron et al. 2003) and/or velocity of substrate influx (Zhu et al. 2005). It is not clear whether these kinases also affect amphetamine-induced outflow but it may be worth considering the following possibility: amphetamines are used to treat *attention-deficit hyperactivity disorder* (ADHD; Mazei-Robinson and Blakely 2006). The effectiveness of amphetamine is counterintuitive, because amphetamines elicit psychostimulant effects (an action which is sought in their illicit consumption). However, a (rare) mutation in the dopamine transporter allows for rationalizing the paradoxical effect of amphetamines in ADHD: a substitution of alanine 559 by valine (found in an affected family) results in a transporter which is targeted to the cell surface and mediates dopamine uptake in a manner indistinguishable from the wild type. However, DAT-A559V has augmented substrate basal efflux, which is blocked by amphetamine (and methylphenidate) (Mazei-Robison et al. 2008). It is conceivable that in other individuals affected by ADHD, it is not the transporter that is mutated but the defect is in a regulatory pathway which switches the transporter into a mode analogous to that adopted by DAT-A559V.

2.3 The oligomer-based counter-transport model

Neurotransmitter transporters are organized in an oligomeric quaternary structure (Kilic and Rudnick 2000; Schmid et al. 2001; Kocabas et al. 2003; Sorkina et al. 2003; Torres et al. 2003 a). Oligomerization has been recognized as an important prerequisite for exit from the endoplasmic reticulum (Farhan et al. 2004, 2007, 2008). However, we posit that oligomerization is also relevant to understand the mechanism of action of amphetamines. We refer to this model as "oligomer-based counter-transport model", in which the releaser is taken up by one moiety and transporter-mediated substrate efflux is achieved by the neighbouring subunit (Fig. 4). There are several arguments that support this concept.

(i) A single transporter moiety cannot be both permeation pathway for amphetamine from out to in and for substrate from in to out which would lead to steric hindrance. At least there is no supporting experimental evidence for such a notion; furthermore, the permeation pathway within the crystal structure of LeuT is by far too narrow to afford the concomitant translocation. Hence, we assume that amphetamine supports an adjacent transporter moiety within the oligomeric complex to adopt a channel-like conformation for the efflux of intracellular substrate. Given these premises, amphetamine will induce maximum substrate release at a concentration where it occupies about 50% of the available transporters in a transporter oligomer. A further increase of amphetamine concentration is predicted to cause a decline in substrate release, because progressive occupation of transporters with inward moving amphetamine reduces the number of those available for outward transport of substrate resulting in a bell-shaped concentration-response curve (as shown by Seidel et al. 2005).

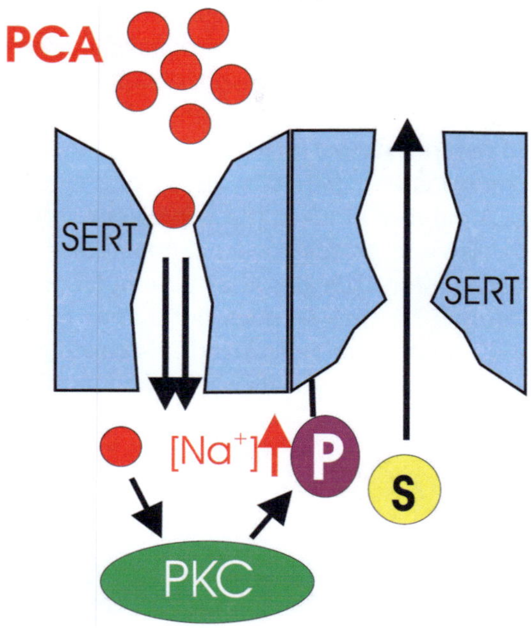

Fig. 4 Oligomer-based counter-transport model: Schematic illustration of the oligomer-based counter-transport model (described in more detail in Seidel et al. 2005); as an example, the effect of low concentrations of amphetamine (pCA = para-chloroamphetamine) inside the cell is shown: only a fraction of SERT moieties is occupied by pCA in the oligomeric complex (shown as a dimer). Occupancy by pCA precludes phosphorylation by PKC (Ramamoorthy and Blakely 1999). The other SERT moieties in the oligomeric complex that have not been occupied by pCA are subject to phosphorylation and thereby primed for outward transport of neurotransmitter

(ii) The statement presented above has been tested by using a concatemeric transporter-fusion protein composed of a SERT moiety and a GABA-transporter-1 (GAT1) moiety. In this SERT-GAT1 concatemer, SERT-selective amphetamines have been employed (pCA; MDMA) and elicited [^{3}H]GABA efflux (Seidel et al. 2005). This effect demonstrates that – in the concatemer – GAT1 can sense the action of amphetamines on the adjacent SERT. It is likely that the signal that is sensed is the Na^+ influx in the vicinity of the inner vestibule, which is conceivable given the fact that Na^+ accumulates close to the plasma membrane (Saunders et al. 2000). This finding is most parsimoniously explained by the oligomer-based counter-transport model (see Seidel et al. 2005). Nevertheless, it is impossible to formally rule out other mechanisms, e. g. amphetamine-induced changes in second messenger-triggered cascades or alterations of the membrane potential propagated through ion channels. But, based on our current understanding of the biological actions of amphetamine, these alternative possibilities cannot readily account for driving the GAT1-moiety of the concatemer into the reverse-transport conformation. (iii) It is worth noting that pCA-induced efflux in this concatemer is still sensitive to protein kinase C-inhibition. This observation indicates that accumulation of internal Na^+ per se does not suffice to explain amphetamine-induced efflux and again stresses the importance of amphetamine-induced protein kinase activation (see above).

3. Cocaine

Cocaine has been consumed by humans for more than 1,000 years. Cocaine (benzoylmethylecgonine; Fig. 5) is derived from the green leaves of coca (*Erythroxylon coca*, a South American shrub of the family of *Erythroxylaceae*). The psychostimulant effects of cocaine have been regarded as 'divine' in the northern Andes of Peru and Bolivia and therefore played a role in religious ceremonies. It was not only the supernatural background of cocaine's effects that also included distinct curing abilities ascribed but also the alleviation of ordinary shortcomings like hunger and thirst. Needless to say that mankind quickly found out the effects cocaine elicits regarding the stamina required for daily labor and life at great altitudes. The use of cocaine by messengers of the Aztecs resulted in an amazing reduction of the time to delivery of the messages; this has been among the first descriptions of cocaine by Spaniards in the mid-sixteenth century.

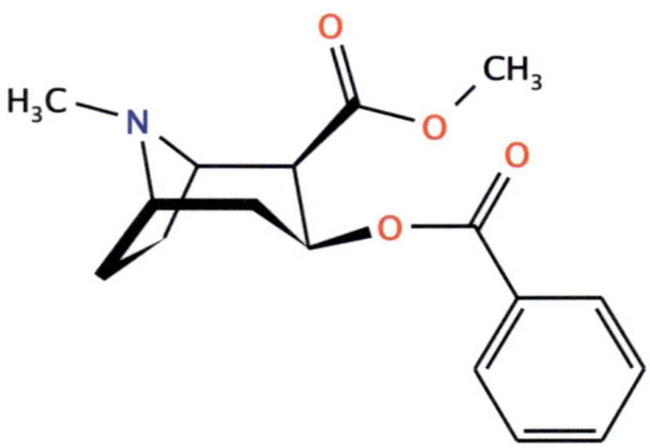

Fig. 5 Cocaine: The figure shows the chemical structure of cocaine (IUPAC name: methyl (1R,2R,3S,5S)-3-benzoyloxy-8-methyl-8-azabicyclo[3.2.1]octane-2-carboxylate)

3.1 The molecular mechanism of the cocaine action

As already outlined in the general introduction to this chapter, the sympathomimetic effects of cocaine can be explained in a much easier way than those of amphetamine and its congeners. While nearly all amphetamines are substrates of certain neurotransmitter:sodium symporter-family members, cocaine is a non-transportable inhibitor of these proteins. Thus, the transporter substrates remain for a longer period of time and in higher concentrations in the synaptic cleft upon cocaine inhibition of substrate reuptake. Cocaine targets to and blocks substrate uptake by DAT, NET and SERT with similar potency. The DAT has been in the focus among these three different transporters because dopamine is the point of convergence of all neurobiological pathways that lead to addiction (Nestler 2005; Kalivas and Volkow 2005). More than one clinical study points to the fact that the DAT is the main biological target of the effects underlying cocaine's abuse liability (Kuhar et al. 1991). Cocaine binding at the DAT inhibits dopamine reuptake and leads to an increase in extracellular dopamine concentrations (Javitch et al. 1984). The final proof of principle was the lack of any psychostimulant effect after administration of cocaine in mice lacking the DAT (Giros et al. 1996). However, the molecular mechanism of the cocaine action was still enigmatic at that time although the answer to the question seems to be quite simple, being "just" reuptake inhibition. Any structural prediction was rendered much easier after the crystal structure of the "bacterial cousin" of the neurotransmitter:sodium symporter family had been reported by the Gouaux group (Yamashita et al. 2005). Subsequently, homology modelling was stimulated and directly used to compare the crystal structure of LeuT with the known molecular determinants obtained in previous mutagenesis experiments (Beuming et al. 2006). When properly aligned, LeuT and other prokaryotic neurotransmitter:sodium symporter proteins can be safely used to infer structural and functional determinants of their eukaryotic homologs. However, the alignment has to be accurate. This is especially important in the case of low sequence homology: in the case of LeuT, the homology with its eukaryotic counterparts such as DAT, SERT, or NET is only 20, 21, and 24%, respectively. Some of the important regions within TMs 1, 3, 6 and 8 have been recognized as regions where conservation approaches ~50%; almost all of these parts surround the putative binding sites conserved over the eu- and prokaryotic neurotransmitter:sodium symporter proteins.

Importantly, the residues lining the substrate binding sites that have been reported in the pre-crystal-era closely overlapped with the predictions from homology modelling (SERT, for example: Rudnick 2006; Jorgensen et al. 2007). The introduction of the crystal structure of LeuT started a new era for the building of homology models. Especially, the availability of different conformational states fuelled ideas used to create models that predict possible interactions with inhibitors and substrates (Huang and Zhan 2007; Indarte et al. 2008); furthermore, the models also allowed to test the old predictions. Modelling is a powerful tool but it may be tricky to avoid steric clashes and possible false interpretations of the electron density maps by the crystallographers. As Beuming and colleagues pointed out: "Analysis of the conservation of the binding site residues in the context of the complete alignment leads to a striking dissociation of the contact residues into those that are conserved from those that are not conserved between subgroups" (Beuming et al. 2006). It is obvious that a homology modelling study without experimental data does not prove the model. Therefore, Beuming and colleagues embarked on a survey on the binding site for cocaine in the DAT, directed by their previously established homology model (Beuming et al. 2006) and validated by a combination of mutagenesis analysis and intramolecular cross-linking strategies (Beuming et al. 2008). The intriguing result of their study was that the binding site of cocaine overlaps with the binding site for dopamine (Beuming et al. 2008; Fig. 6).

Conclusion and outlook

Homology modelling has shown that both amphetamine and cocaine bind in a pocket of neurotransmitter:sodium symporter family members that overlaps with the substrate binding site. Thus, in the case of cocaine, the mode of action seems to be unravelled. However, we still do neither know exactly how the conformational changes of the transporters are initiated nor have the exact binding modes of substrates/inhibitors and/or co-substrates been elucidated to date. For the amphetamines, we do know that they are transported as substrates; we know that the quarternary structure may influence their principal mode of pharmacological action – triggering reverse transport – but we do not exactly know whether dimeric or even tetrameric transporters need a specific cooperativity to make reverse transport happen. Also, the structural basis for the currents that flow through the

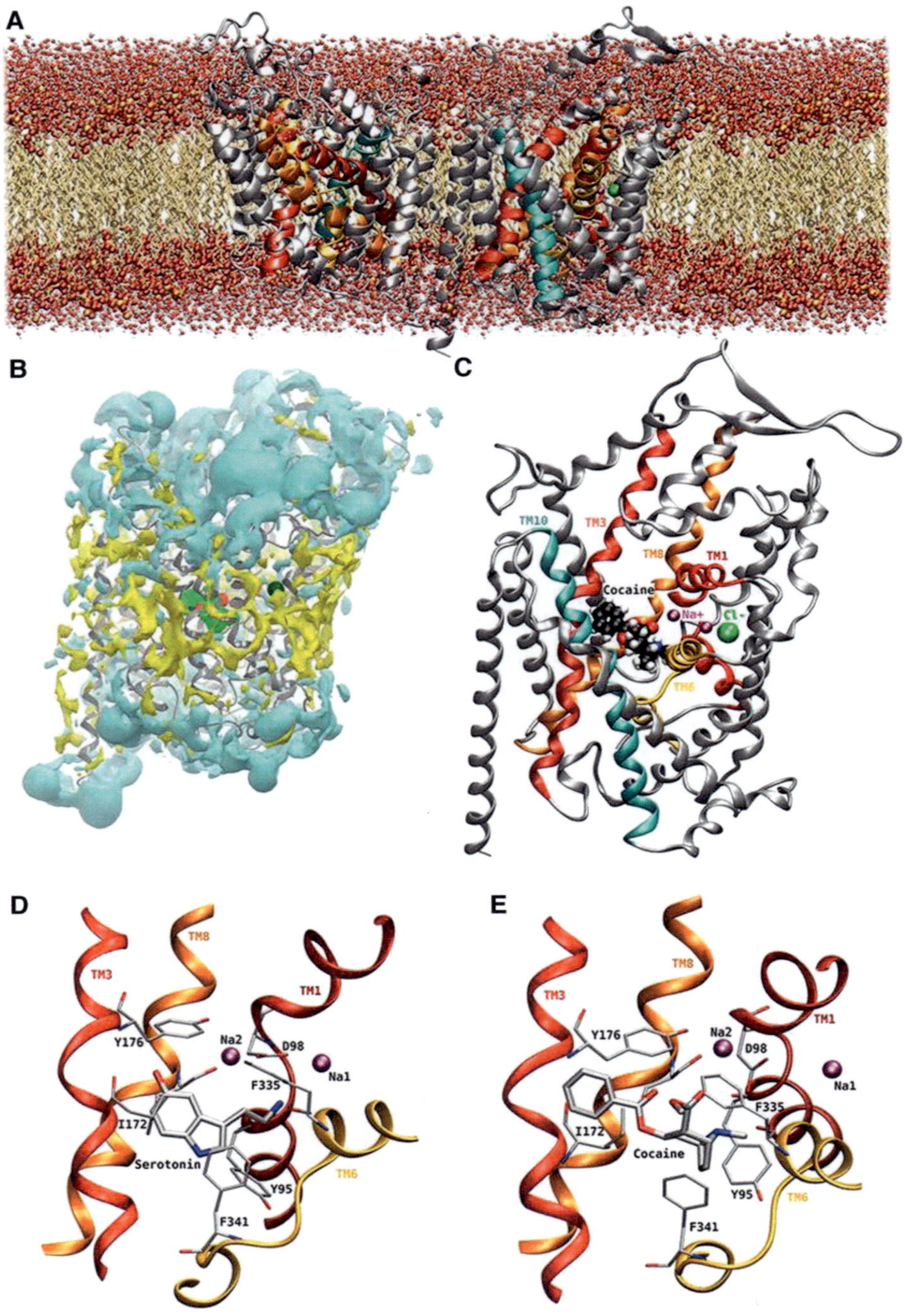

Fig. 6 The natural habitat of transporters and the binding of neurotransmitter and cocaine.
A Symbolic representation of a phospholipid bilayer that accommodates the homology model of dimeric human serotonin transporter. The model was created with software MOE 2007.09 based on the crystal structure of LeuT (pdb-code 2A65) published by Yamashita et al. (2005). The alignment is according to Beuming et al. (2006). Software VMD 1.8.7 was used to create the membrane. **B** The lipophilic surface of the human SERT. Light blue marks the hydrophilic regions of the protein on the intracellular (bottom) and extracellular (top) side. Lipophilic regions are marked in yellow. They are supposed to be buried within

the membrane. This representation was created with software Maestro Version 9.0. **C** A detailed view of the homology model of human SERT. The protein consists of 12 transmembrane domains (TMDs). TMD1 (dark red), TMD3 (light red), TMD6 (yellow) and TMD8 (orange) are directly interacting with the substrates during the transportation cycle. TMD10 (cyan) is participating in the closure of the transporter. **D** The docking study for serotonin, performed with software MOE 2007.09, shows the position of serotonin in the central substrate binding site. The obtained poses were ranked applying a geometrical and statistical approach; furthermore, mutagenesis data were considered. **E** The docking study for cocaine was performed with the same protocol as for serotonin. The favoured binding site for cocaine is similar to the neurotransmitter binding site, as found in the study for carboxyfluorotropane, a cocaine analogue of Beuming et al. (2008). The main interacting amino acids are shown: D98 (salt bridge), I172 (hydrophobic interaction), S438 (hydrogen bonding). The residues Y95, Y176, F335, and F336 delimit the binding site. Two sodium ions (magenta) are present close to the binding site. Panels A, C, D and E were created using software VMD 1.8.7

transporters remains entirely enigmatic. Structural biology will play an important role in guiding our efforts to finally elucidate the molecular basis of the psychopharmacological action of amphetamines and cocaine and their respective congeners. Especially, molecular dynamics simulations will allow us to dynamically picture the possible conformational changes within this class of transport proteins; also, homology modelling and docking of ligands will be of relevance not only for drug discovery and design but also as the basis for the molecular dynamics simulations. There are always concerns about discrepancies between homology models and the conformations of proteins in reality; however, attempts like the one used by Beuming and colleagues (Beuming et al. 2008) are very well suited to abide such concerns and to understand *structure–function relationship*s between substrates and ligands acting at clinically important neurotransmitter:sodium symporter transporters. There is a large medical need for treatment of cocaine addiction. Antibodies against cocaine have been tested in a clinical trial without huge success (Martell et al. 2009) and evidence supports that *small chemical compound*s, refined by rational approaches and guided by homology modelling and docking studies may finally succeed (Desai et al. 2005; Loland et al. 2008). Hence, we anticipate that the effects of recreational drug use like disturbances as well as enhancement of sensory perception will finally be understood on a molecular basis.

Abbreviations

DAT – dopamine transporter
NET – norepinephrine transporter
SERT – serotonin transporter
GAT – GABA transporter
MPP^+ – Methyl-4-phenylpyridinium
MAO – monoaminooxidase
BAPTA – 1,2-bis(o-aminophenoxy)ethane-N,N,N′,N′-tetraacetic acid
PKC – protein kinase C
ADHD – attention deficit hyperactivity syndrom

References

Amara SG, Kuhar MJ (1993) Neurotransmitter transporters: recent progress. Annu Rev Neurosci 16: 73–93

Amara SG, Sonders MS (1998) Neurotransmitter transporters as molecular targets for addictive drugs. Drug Alcohol Depend 51: 87–96

Axelrod J, Whitby LG, Hertting G (1961) Effect of psychotropic drugs on the uptake of ^{3}H-Norepinephrine by tissues. Science 133: 383–384

Beuming T, Kniazeff J, Bergmann ML, Shi L, Gracia L, Raniszewska K, Newman AH, Javitch JA, Weinstein H, Gether U, Loland CJ (2008) The binding sites for cocaine and dopamine in the dopamine transporter overlap. Nat Neurosci 11: 780–789

Beuming T, Shi L, Javitch JA, Weinstein H (2006) A comprehensive structure-based alignment of prokaryotic and eukaryotic neurotransmitter/Na+ symporters (NSS) aids in the use of the LeuT structure to probe NSS structure and function. Mol Pharmacol 70: 1630–1642

Biel JH, Bopp BA (1978) Amphetamines: structure–activity relationships. In: Iversen LL, Iversen SD, Snyder SH (eds) Handbook of psychopharmacology: psychostimulants. pp 1–40, Plenum, New York

Blakely RD, Berson HE, Fremeau-RTJ, Caron MG, Peek MM, Prince HK, Bradley CC (1991) Cloning and expression of a functional serotonin transporter from rat brain. Nature 354: 66–70

Bönisch H (1984) The transport of (+)-amphetamine by the neuronal noradrenaline carrier. Naunyn Schmiedebergs Arch Pharmacol 327: 267–272

Bönisch H (1986) The role of co-transported sodium in the effect of indirectly acting sympathomimetic amines. Naunyn Schmiedebergs Arch Pharmacol 332: 135–141

Bönisch H, Trendelenburg U (1989) The mechanism of action of indirectly acting sympathomimetic amines. In: Trendelenburg U, Weiner N (eds) Handbook of Experimental Pharmacology: Catecholamines. pp 247–277, Springer, Berlin, Hamburg, New York

Bruss M, Hammermann R, Brimijoin S, Bonisch H (1995) Antipeptide antibodies confirm the topology of the human norepinephrine transporter. J Biol Chem 270: 9197–9201

Burn JH, Rand MJ (1958) The action of sympathomimetic amines in animals treated with reserpine. J Physiol 144: 314–336

Carvelli L, McDonald PW, Blakely RD, DeFelice LJ (2004) Dopamine transporters depolarize neurons by a channel mechanism. Proc Natl Acad Sci U S A 101: 16 046–16 051

Chen N, Reith ME (2000) Structure and function of the dopamine transporter. Eur J Pharmacol 405: 329–339

Chen R, Furman CA, Zhang M, Kim MN, Gereau RW, Leitges M, Gnegy ME (2009) Protein kinase Cbeta is a critical regulator of dopamine transporter trafficking and regulates the behavioral response to amphetamine in mice. J Pharmacol Exp Ther 328: 912–920

Cinquanta M, Ratovitski T, Crespi D, Gobbi M, Mennini T, Simantov R (1997) Carrier-mediated serotonin release induced by d-fenfluramine: studies with human neuroblastoma cells transfected with a rat serotonin transporter. Neuropharmacology 36: 803–809

Davis ME, Patrick RL (1990) Diacylglycerol-induced stimulation of neurotransmitter release from rat brain striatal synaptosomes. J Neurochem 54: 662–668

Desai RI, Kopajtic TA, Koffarnus M, Newman AH, Katz JL (2005) Identification of a dopamine transporter ligand that blocks the stimulant effects of cocaine. J Neurosci 25: 1889–1893

Egana LA, Cuevas RA, Baust TB, Parra LA, Leak RK, Hochendoner S, Pena K, Quiroz M, Hong WC, Dorostkar MM, Janz R, Sitte HH, Torres GE (2009) Physical and functional interaction between the dopamine transporter and the synaptic vesicle protein synaptogyrin-3. J Neurosci 29: 4592–4604

Erreger K, Grewer C, Javitch JA, Galli A (2008) Currents in response to rapid concentration jumps of amphetamine uncover novel aspects of human dopamine transporter function. J Neurosci 28: 976–989

Eshleman AJ, Henningsen RA, Neve KA, Janowsky A (1994) Release of dopamine via the human transporter. Mol Pharmacol 45: 312–316

Ewing AG, Stein TS, Lau YY (1992) Analytical chemistry in microenvironments: single nerve cells. Accts Chem Res440–447

Farhan H, Freissmuth M, Sitte HH (2006) Oligomerization of neurotransmitter transporters: a ticket from the endoplasmic reticulum to the plasma membrane. Handb Exp Pharmacol 233–249

Farhan H, Korkhov VM, Paulitschke V, Dorostkar MM, Scholze P, Kudlacek O, Freissmuth M, Sitte HH (2004) two discontinuous segments in the carboxy terminus are required for membrane targeting of the rat GABA transporter-1 (GAT1). J Biol Chem 279: 28 553–28 563

Farhan H, Reiterer V, Korkhov VM, Schmid JA, Freissmuth M, Sitte HH (2007) concentrative export from the endoplasmic reticulum of the gamma-aminobutyric acid transporter 1 requires binding to SEC24D. J Biol Chem 282: 7679–7689

Farhan H, Reiterer V, Kriz A, Hauri HP, Pavelka M, Sitte HH, Freissmuth M (2008) Signal-dependent export of GABA transporter 1 from the ER-golgi intermediate compartment is specified by a C-terminal motif. J Cell Sci 121: 753–761

Fischer JF, Cho AK (1979) Chemical release of dopamine from striatal homogenates: evidence for an exchange diffusion model. J Pharmacol Exp Ther 208: 203–209

Fishkes H, Rudnick G (1982) Bioenergetics of serotonin transport by membrane vesicles derived from platelet dense granules. J Biol Chem 257: 5671–5677

Fjorback AW, Pla P, Muller HK, Wiborg O, Saudou F, Nyengaard JR (2009) Serotonin transporter oligomerization documented in rn46a cells and neurons by sensitized acceptor emission FRET and fluorescence lifetime imaging microscopy. Biochem Biophys Res Commun 380: 724–728

Fleckenstein A, Burn JH (1953) The effect of denervation on the action of sympathomimetic amines on the nictitating membrane. Br J Pharmacol Chemother 8: 69–78

Fog JU, Khoshbouei H, Holy M, Owens WA, Vaegter CB, Sen N, Nikandrova Y, Bowton E, McMahon DG, Colbran R J, Daws L C, Sitte HH, Javitch JA, Galli A, Gether U (2006) Calmodulin kinase II interacts with the dopamine transporter C terminus to regulate amphetamine-induced reverse transport. Neuron 51: 417–429

Forrest LR, Zhang YW, Jacobs MT, Gesmonde J, Xie L, Honig BH, Rudnick G (2008) Mechanism for alternating access in neurotransmitter transporters. Proc Natl Acad Sci USA 105: 10 338–10 343

Foster JD, Pananusorn B, Vaughan RA (2002) Dopamine transporters are phosphorylated on N-terminal serines in rat striatum. J Biol Chem 277: 25 178–25 186

Furchgott RF, Kirpekar SM, Rieker M, Schwab A (1963) Actions and interactions of norepinephrine, tyramine and cocaine on aortic strips of rabbit and left atria of guinea pig and cat. J Pharmacol Exp Ther 142: 39–58

Giambalvo CT (1992a) Protein kinase C and dopamine transport–1. effects of amphetamine in vivo. Neuropharmacology 31: 1201–1210

Giambalvo CT (1992b) Protein kinase C and dopamine transport–2. effects of amphetamine in vitro. Neuropharmacology 31: 1211–1222

Giambalvo CT (2003) Differential effects of amphetamine transport vs. dopamine reverse transport on particulate pkc activity in striatal synaptoneurosomes. Synapse 49: 125–133

Giros B, el Mestikawy S, Bertrand L, Caron MG (1991) Cloning and functional characterization of a cocaine-sensitive dopamine transporter. FEBS Lett 295: 149–154

Giros B, Jaber M, Jones SR, Wightman RM, Caron MG (1996) Hyperlocomotion and indifference to cocaine and amphetamine in mice lacking the dopamine transporter. Nature 379: 606–612

Glowinski J, Axelrod J (1965) Effect of drugs on the uptake, release, and metabolism of H3-norepinephrine in the rat brain. J Pharmacol Exp Ther 149: 43–49

Gnegy ME (2003) the effect of phosphorylation on amphetamine-mediated outward transport. Eur J Pharmacol 479: 83–91

Gnegy ME, Khoshbouei H, Berg K A, Javitch JA, Clarke WP, Zhang M, Galli A (2004) Intracellular Ca2+ regulates amphetamine-induced dopamine efflux and currents mediated by the human dopamine transporter. Mol Pharmacol 66: 137–143

Gobbi M, Funicello M, Gerstbrein K, Holy M, Moya PR, Sotomayor R, Forray MI, Gysling K, Paluzzi S, Bonanno G, Reyes-Parada M, Sitte HH, Mennini T (2008) N,N-dimethyl-thioamphetamine and methyl-thioamphetamine, two non-neurotoxic substrates of 5-HT transporters, have scant in vitro efficacy for the induction of transporter-mediated 5-HT release and currents. J Neurochem 105: 1770–1780

Gobbi M, Mennini T, Garattini S (1997) Mechanism of neurotransmitter release induced by amphetamine derivatives: pharmacological and toxicological aspects. Current Topics in Pharmacology 3: 217–227

Gobbi M, Moia M, Pirona L, Ceglia I, Reyes-Parada M, Scorza C, Mennini T (2002) p-methylthioamphetamine and 1-(m-chlorophenyl)piperazine, two non-neurotoxic 5-HT releasers in vivo, differ from neurotoxic amphetamine derivatives in their mode of action at 5-HT nerve endings in vitro. J Neurochem 82: 1435–1443

Gorentla BK, Moritz AE, Foster JD, Vaughan RA (2009) Proline-directed phosphorylation of the dopamine transporter N-terminal domain. Biochemistry 48: 1067–1076

Granas C, Ferrer J, Loland CJ, Javitch JA, Gether U (2003) N-terminal truncation of the dopamine transporter abolishes phorbol ester- and substance P receptor-stimulated phosphorylation without impairing transporter internalization. J Biol Chem 278: 4990–5000

Guastella J, Nelson N, Nelson H, Czyzyk L, Keynan S, Miedel MC, Davidson N, Lester HA, Kanner BI (1990) Cloning and expression of a rat brain GABA Transporter. Science 249: 1303–1306

Hastrup H, Karlin A, Javitch JA (2001) Symmetrical dimer of the human dopamine transporter revealed by cross- linking Cys-306 at the extracellular end of the sixth transmembrane segment. Proc Natl Acad Sci U S A 98: 10 055– 10 060

Hilgemann DW, Lu CC (1999) GAT1 (GABA:Na+:Cl-) cotransport function. database reconstruction with an alternating access model. J Gen Physiol 114: 459–475

Huang X, Zhan CG (2007) How dopamine transporter interacts with dopamine: insights from molecular modeling and simulation. Biophys J 93: 3627–3639

Humphreys CJ, Wall SC, Rudnick G (1994) Ligand binding to the serotonin transporter: equilibria, kinetics, and ion dependence. Biochemistry 33: 9118–9125

Indarte M, Madura JD, Surratt CK (2008) Dopamine transporter comparative molecular modeling and binding site prediction using the LeuT(aa) leucine transporter as a template. Proteins 70: 1033–1046

Iversen L (2000) Neurotransmitter transporters: fruitful targets for cns drug discovery. Mol Psychiatry 5: 357–362

Iversen LL (1971) Role of transmitter uptake mechanisms in synaptic neurotransmission. Br J Pharmacol 41: 571–591

Jardetzky O (1966) Simple allosteric model for membrane pumps. Nature 211: 969–970

Javitch JA, Blaustein RO, Snyder SH (1984) [3H]Mazindol binding associated with neuronal dopamine and norepinephrine uptake sites. Mol Pharmacol 26: 35–44

Javitch JA, D'Amato RJ, Strittmatter SM, Snyder SH (1985) Parkinsonism-inducing neurotoxin, N-methyl-4-phenyl-1,2,3,6 – tetrahydropyridine: uptake of the metabolite N-methyl-4-phenylpyridine by dopamine neurons explains selective toxicity. Proc Natl Acad Sci USA 82: 2173–2177

Jess U, Betz H, Schloss P (1996) The membrane-bound rat serotonin transporter, SERT1, is an oligomeric protein. FEBS Lett 394: 44–46

Jones SR, Gainetdinov RR, Wightman RM, Caron MG (1998) Mechanisms of amphetamine action revealed in mice lacking the dopamine transporter. J Neurosci 18: 1979–1986

Jones SR, Joseph JD, Barak LS, Caron MG, Wightman RM (1999) Dopamine neuronal transport kinetics and effects of amphetamine. J Neurochem 73: 2406–2414

Jorgensen AM, Tagmose L, Jorgensen AM, Topiol S, Sabio M, Gundertofte K, Bogeso KP, Peters GH (2007) Homology modeling of the serotonin transporter: insights into the primary escitalopram-binding site. ChemMedChem 2: 815–826

Just H, Sitte HH, Schmid JA, Freissmuth M, Kudlacek O (2004) Identification of an additional interaction domain in transmembrane domains 11 and 12 that supports oligomer formation in the human serotonin transporter. J Biol Chem 279: 6650–6657

Kahlig KM, Binda F, Khoshbouei H, Blakely RD, McMahon DG, Javitch JA, Galli A (2005) Amphetamine induces dopamine efflux through a dopamine transporter channel. Proc Natl Acad Sci U S A 102: 3495–3500

Kalivas PW, Volkow ND (2005) The neural basis of addiction: a pathology of motivation and choice. Am J Psychiatry 162: 1403–1413

Kantor L, Gnegy ME (1998) Protein kinase c inhibitors block amphetamine-mediated dopamine release in rat striatal slices. J Pharmacol Exp Ther 284: 592–598

Kantor L, Hewlett GH, Gnegy ME (1999) Enhanced amphetamine- and k+-mediated dopamine release in rat striatum after repeated amphetamine: differential requirements for Ca2+- and calmodulin-dependent phosphorylation and synaptic vesicles. J Neurosci 19: 3801–3808

Kantor L, Hewlett G H, Park YH, Richardson-Burns SM, Mellon MJ, Gnegy ME (2001) Protein kinase C and intracellular calcium are required for amphetamine-mediated dopamine release via the norepinephrine transporter in undifferentiated PC12 cells. J Pharmacol Exp Ther 297: 1016–1024

Kazanietz MG, Caloca MJ, Aizman O, Nowicki S (2001) Phosphorylation of the catalytic subunit of rat renal Na+, K+-ATPase by classical PKC isoforms. Arch Biochem Biophys 388: 74–80

Khoshbouei H, Sen N, Guptaroy B, Johnson L, Lund D, Gnegy ME, Galli A, Javitch JA (2004) N-terminal phosphorylation of the dopamine transporter is required for amphetamine-induced efflux. PLoS Biol 2:E78

Khoshbouei H, Wang H, Lechleiter JD, Javitch JA, Galli A (2003) Amphetamine-induced dopamine efflux. a voltage-sensitive and intracellular Na+-dependent mechanism. J Biol Chem 278: 12 070–12 077

Kilic F, Rudnick G (2000) Oligomerization of serotonin tranporter and its functional consequences. Proc Natl Acad Sci USA 97: 106–3111

Kocabas AM, Rudnick G, Kilic F (2003) Functional consequences of homo- but not hetero-oligomerization between transporters for the biogenic amine neurotransmitters. J Neurochem 85: 1513–1520

Korkhov VM, Holy M, Freissmuth M, Sitte HH (2006) The conserved glutamate (Glu136) in transmembrane domain 2 of the serotonin transporter is required for the conformational switch in the transport cycle. J Biol Chem 281: 13 439–13 448

Kuhar MJ, Ritz M C, Boja JW (1991) The dopamine hypothesis of the reinforcing properties of cocaine. Trends Neurosci 14: 299–302

L'hirondel M, Cheramy A, Godeheu G, Glowinski J (1995) Effects of arachidonic acid on dopamine synthesis, spontaneous release, and uptake in striatal synaptosomes from the rat. J Neurochem 64: 1406–1409

Langston JW, Irwin I, Langston EB, Forno LS (1984) 1-methyl-4-phenylpyridinium ion (MPP+): identification of a metabolite of MPTP, a toxin selective to the substantia nigra. Neurosci Lett 48: 87–92

Loland CJ, Desai R I, Zou MF, Cao J, Grundt P, Gerstbrein K, Sitte HH, Newman AH, Katz JL, Gether U (2008) Relationship between conformational changes in the dopamine transporter and cocaine-like subjective effects of uptake inhibitors. Mol Pharmacol 73: 813–823

Lu CC, Hilgemann DW (1999 a) GAT1 (GABA:Na+:Cl-) cotransport function. kinetic studies in giant xenopus oocyte membrane patches. J Gen Physiol 114: 445–457

Lu CC, Hilgemann DW (1999 b) GAT1 (GABA:Na+:Cl-) cotransport function. steady state studies in giant xenopus oocyte membrane patches. J Gen Physiol 114: 429–444

Martell BA, Orson F M, Poling J, Mitchell E, Rossen RD, Gardner T, Kosten TR (2009) Cocaine vaccine for the treatment of cocaine dependence in methadone-maintained patients: a randomized, double-blind, placebo-controlled efficacy trial. Arch Gen Psychiatry 66: 1116–1123

Masson J, Sagne C, Hamon M, el Mestikawy S (1999) Neurotransmitter transporters in the central nervous system. Pharmacol Rev 51: 439–464

Mazei-Robinson MS, Blakely RD (2006) ADHD and the dopamine transporter: are there reasons to pay attention? Handb Exp Pharmacol 373–415

Mazei-Robison MS, Bowton E, Holy M, Schmudermaier M, Freissmuth M, Sitte HH, Galli A, Blakely RD (2008) Anomalous dopamine release associated with a human dopamine transporter coding variant. J Neurosci 28: 7040–7046

Meinild AK, Sitte HH, Gether U (2004) Zinc Potentiates an Uncoupled Anion Conductance Associated With the Dopamine Transporter. J Biol Chem 279: 49 671–49 679.

Moron JA, Zakharova I, Ferrer J V, Merrill GA, Hope B, Lafer EM, Lin ZC, Wang J B, Javitch JA, Galli A, Shippenberg TS (2003) Mitogen-activated protein kinase regulates dopamine transporter surface expression and dopamine transport apacity. J Neurosci 23: 8480–8488

Nelson N (1998) The family of Na+/Cl- neurotransmitter transporters. J Neurochem 71: 1785– 1803

Nestler EJ (2005) Is there a common molecular pathway for addiction? Nat Neurosci 8: 1445–1449

Nirenberg MJ, Chan J, Vaughan RA, Uhl GR, Kuhar MJ, Pickel VM (1997) Immunogold localization of the dopamine transporter: an ultrastructural study of the rat ventral tegmental area. J Neurosci 17: 5255–5262

Pacholczyk T, Blakely RD, Amara SG (1991) Expression cloning of a cocaine- and antidepressant-sensitive human noradrenaline transporter. Nature 350: 350–354

Pifl C, Agneter E, Drobny H, Reither H, Singer EA (1997) Induction by low Na+ or Cl- of cocaine sensitive carrier-mediated efflux of amines from cells transfected with the cloned human catecholamine transporters. Br J Pharmacol 121: 205–212

Pifl C, Agneter E, Drobny H, Sitte HH, Singer EA (1999) Amphetamine reverses or blocks the operation of the human noradrenaline transporter depending on its concentration: superfusion studies on transfected cells. Neuropharmacology 38: 157–165

Pifl C, Drobny H, Reither H, Hornykiewicz O, Singer EA (1995) Mechanism of the dopamine-releasing actions of amphetamine and cocaine: plasmalemmal dopamine transporter versus vesicular monoamine transporter. Mol Pharmacol 47: 368–373

Pifl C, Singer EA (1999) Ion dependence of carrier-mediated release in dopamine or norepinephrine transporter-transfected cells questions the hypothesis of facilitated exchange diffusion. Mol Pharmacol 56: 1047–1054

Pozzan T, Gatti G, Dozio N, Vicentini LM, Meldolesi J (1984) Ca^{2+}-dependent and -independent release of neurotransmitters from PC12 cells: a role for protein kinase C activation? J Cell Biol 99: 628–638

Ramamoorthy S, Blakely RD (1999) Phosphorylation and sequestration of serotonin transporters differentially modulated by psychostimulants. Science 285: 763–766

Ramamoorthy S, Samuvel DJ, Buck ER, Rudnick G, Jayanthi LD (2007) Phosphorylation of threonine residue 276 is required for acute regulation of serotonin transporter by cyclic GMP. J Biol Chem 282: 11 639–11 647

Robertson SD, Matthies HJ, Galli A (2009) A closer look at amphetamine-induced reverse transport and trafficking of the dopamine and norepinephrine transporters. Mol Neurobiol 39: 73–80

Ross SB, Renyi AL (1966) Uptake of some tritiated sympathomimetic amines by mouse brain cortex slices in vitro. Acta Pharmacol Toxicol (Copenh) 24: 297–309

Roux MJ, Supplisson S (2000) Neuronal and glial glycine transporters have different stoichiometries. Neuron 25: 373–383

Rudnick G (2006) Structure/function relationship in serotonin transporter. In: Sitte HH, Freissmuth M (eds) Neurotransmitter transporters pp 59–73, Springer-Verlag, Berlin Heidelberg

Rudnick G, Clark J (1993) From synapse to vesicle: the reuptake and storage of biogenic amine neurotransmitters. Biochim Biophys Acta 1144: 249–263

Rudnick G, Wall SC (1992) p-chloroamphetamine induces serotonin release through serotonin transporters. Biochemistry 31: 6710–6718

Saier MH Jr., Tran CV, Barabote RD (2006) TCDB: the transporter classification database for membrane transport protein analyses and information. Nucleic Acids Res 34:D181-D186

Saunders C, Ferrer JV, Shi L, Chen J, Merrill G, Lamb ME, Leeb-Lundberg LM, Carvelli L, Javitch JA, Galli A (2000) Amphetamine-induced loss of human dopamine transporter activity: an internalization-dependent and cocaine-sensitive mechanism. Proc Natl Acad Sci USA 97: 6850–6855

Schmid JA, Scholze P, Kudlacek O, Freissmuth M, Singer EA, Sitte HH (2001) Oligomerization of the human serotonin transporter and of the rat GABA transporter 1 visualized by fluorescence resonance energy transfer microscopy in living cells. J Biol Chem 276: 3805–3810

Scholze P, Freissmuth M, Sitte HH (2002a) Mutations within an intramembrane leucine heptad repeat disrupt oligomer formation of the rat GABA transporter 1. J Biol Chem 277: 43 682–43 690

Scholze P, Norregaard L, Singer EA, Freissmuth M, Gether U, Sitte HH (2002b) The role of zinc ions in reverse transport mediated by monoamine Transporters. J Biol Chem 277: 21 505– 21 513

Scholze P, Zwach J, Kattinger A, Pifl C, Singer EA, Sitte HH (2000) Transporter-Mediated Release: A superfusion study on human embryonic kidney cells stably expressing the human serotonin transporter. J Pharmacol Exp Ther 293: 870–878

Seidel S, Singer EA, Just H, Farhan H, Scholze P, Kudlacek O, Holy M, Koppatz K, Krivanek P, Freissmuth M, Sitte HH (2005) Amphetamines take two to tango: an oligomer-based counter-transport model of neurotransmitter transport explores the amphetamine action. Mol Pharmacol 67: 140–151

Seiden LS, Sabol KE, Ricaurte GA (1993) Amphetamine: effects on catecholamine systems and behavior. Annu Rev Pharmacol Toxicol 33: 639–677

Sen N, Shi L, Beuming T, Weinstein H, Javitch JA (2005) A pincer-like configuration of TM2 in the human dopamine transporter is responsible for indirect effects on cocaine binding. Neuropharmacology 49: 780–790

Shi L, Quick M, Zhao Y, Weinstein H, Javitch JA (2008) The mechanism of a neurotransmitter: sodium symporter–inward release of Na+ and substrate is triggered by substrate in a second binding site. Mol Cell 30: 667–677

Singh SK (2008) LeuT: a prokaryotic stepping stone on the way to a eukaryotic neurotransmitter transporter structure. Channels (Austin) 2: 380–389

Singh SK, Piscitelli CL, Yamashita A, Gouaux E (2008) A competitive inhibitor traps leut in an open-to-out conformation. Science 322: 1655–1661

Sitte HH, Farhan H, Javitch JA (2004) Sodium-dependent neurotransmitter transporters: oligomerization as a determinant of transporter function and trafficking. Mol Interv 4: 38–47

Sitte HH, Freissmuth M (2003) Oligomer formation by Na+-Cl–coupled neurotransmitter transporters. Eur J Pharmacol 479: 229–236

Sitte HH, Hiptmair B, Zwach J, Pifl C, Singer EA, Scholze P (2001) Quantitative analysis of inward and outward transport rates in cells stably expressing the cloned human serotonin transporter: inconsistencies with the hypothesis of facilitated exchange diffusion. Mol Pharmacol 59: 1129–1137

Sitte HH, Huck S, Reither H, Boehm S, Singer EA, Pifl C (1998) Carrier-mediated release, transport rates, and charge transfer induced by amphetamine, tyramine, and dopamine in mammalian cells transfected with the human dopamine transporter. J Neurochem 71: 1289–1297

Sitte HH, Scholze P, Schloss P, Pifl C, Singer EA (2000) Characterization of carrier-mediated release in human embryonic kidney 293 cells stably expressing the rat serotonin transporter: a superfusion study. J Neurochem 74: 1317–1324

Sitte HH, Singer EA, Scholze P (2002) Bi-directional transport of gaba in human embryonic kidney (HEK-293) cells stably expressing the rat GABA Transporter GAT-1. Br J Pharmacol 135: 93–102

Sorkina T, Doolen S, Galperin E, Zahniser NR, Sorkin A (2003) Oligomerization of dopamine transporters visualized in living cells by FRET microscopy. J Biol Chem 278: 28 274–28 283

Sulzer D, Sonders M S, Poulsen NW, Galli A (2005) Mechanisms of neurotransmitter release by amphetamines: a review. Prog Neurobiol 75: 406–433

Torres GE, Carneiro A, Seamans K, Fiorentini C, Sweeney A, Yao WD, Caron MG (2003a) Oligomerization and trafficking of the human dopamine transporter. mutational analysis identifies critical domains important for the functional expression of the transporter. J Biol Chem 278: 2731–2739

Torres GE, Gainetdinov RR, Caron MG (2003b) Plasma membrane monoamine transporters: structure, regulation and function. Nat Rev Neurosci 4: 13–25

Trendelenburg U, Langeloh A, Bönisch H (1987) Mechanism of action of indirectly acting sympathomimetic amines. Blood Vessels 24: 261–270

Wall SC, Gu H, Rudnick G (1995) Biogenic amine flux mediated by cloned transporters stably expressed in cultured cell lines: amphetamine specificity for inhibition and efflux. Mol Pharmacol 47: 544–550

Wang D, Deken S L, Whitworth TL, Quick MW (2003) Syntaxin 1A inhibits GABA flux, efflux, and exchange mediated by the rat brain GABA transporter GAT1. Mol Pharmacol 64: 905–913

Wolfel R, Graefe KH (1992) Evidence for various tryptamines and related compounds acting as substrates of the platelet 5-hydroxytryptamine transporter. Naunyn Schmiedebergs Arch Pharmacol 345: 129–136

Yamashita A, Singh SK, Kawate T, Jin Y, Gouaux E (2005) Crystal structure of a bacterial homologue of Na+/Cl–dependent neurotransmitter transporters. Nature 437: 215–223

Zaczek R, Culp S, De SE (1991) Interactions of [3H] amphetamine with rat brain synaptosomes. II. active transport. J Pharmacol Exp Ther 257: 830–835

Zhu CB, Carneiro A M, Dostmann WR, Hewlett WA, Blakely RD (2005) P38 MAPK activation elevates serotonin transport activity via a trafficking-independent, protein phosphatase 2A-dependent process. J Biol Chem 280: 15 649–15 658

A biological basis for musical tonality

12

Daniel Bowling and Dale Purves

Contents

Abstract

Like other sensory qualities, the human ability to perceive tonal sound stimuli has presumably evolved because of its utility. Although a variety of tonal sounds are present in the human auditory environment, the vocalizations of other humans are the most biologically relevant and the most frequently experienced. It is thus reasonable to assume that our appreciation of tonal sounds has arisen primarily for the benefits that accrue from this conspecific information. It follows that the structure and function of the tonal sounds produced by the human vocal apparatus may provide the key to understanding how and why we perceive tonality in music the way that we do. Here we consider recent evidence that bears on this idea.

Daniel Bowling
Duke-NUS Graduate Medical School
Neurobiology
8 College Road, Level 05
Singapore 169857, Singapore
e-mail: dan.bowling@duke.edu

Dale Purves
Duke-NUS Graduate Medical School
Neurobiology
8 College Road, Level 05,
Singapore 169857, Singapore
e-mail: dale.purves@duke-nus.edu.sq

1. Introduction

Tones are periodic sound stimuli perceived as having a pitch. In nature, such stimuli are produced by objects that resonate at frequencies humans can hear when acted on by a force, most commonly by animal species that generate such sounds for social communication (e. g., the sounds of stridulating insects, the vibrations produced by the songbird syrinx, and the vocalizations of many mammals). Although many such stimuli are present in the human auditory environment, the vocalizations of other humans are the most biologically relevant and frequently experienced (Terhardt 1984; Schwartz et al. 2003; Gill and Purves 2009). The prevalence and importance of vocal stimuli suggests that an appreciation of tonal sounds in music is linked to human vocalizations.

The merits of understanding musical tonality based on vocalization depend on explaining why we hear musical tones the way we do. To date, this framework has been successfully applied to a variety of phenomena including perception of the missing fundamental (Terhardt 1974; Schwartz and Purves 2004), the fundamental root of musical chords (Terhardt 1984), pitch shift of the residue, spectral dominance and pitch strength (Schwartz and Purves 2004), musical interval preferences (Schwartz et al. 2003; Ross et al. 2007), and scale preferences (Gill and Purves 2009). The work described here focuses on this framework as a way to explain the characteristic affective impact of major and minor music (Bowling et al. 2010).

2. The emotional effects of major and minor music

The affective impact of music depends on many factors including, but not limited to, intensity, tempo, rhythm and the tonal intervals used. For most of these factors the way emotion is conveyed seems intuitively clear. If, for instance, a composer wants to imbue a composition with excitement, the intensity tends to be forte, the tempo fast and the rhythm syncopated. Conversely, if a more subdued effect is desired, the intensity is typically piano, the tempo slower, and the rhythm more balanced (Cohen 1971; Bernstein 1976; Juslin and Laukka 2003). These effects on the listener presumably occur because in each case the characteristics of the music accord with the ways that the corresponding emotional state is expressed in human behavior. The reason for the different emotional impact of the tones used in music, however, is not clear.

Much music worldwide employs subsets of the chromatic scale, which divides each octave into 12 intervals defined by the frequency ratios shown in Fig.1A. In Western music over the last few centuries the most commonly used subsets are the diatonic scales in Fig.1B, the Ionian and the Aeolian scales in particular (Pierce 1962; Bernstein 1976; Randel 1986; Burns 1999; Burkholder et al. 2006). These two scales are usually referred to simply as the major and the minor scale, respectively (Fig.1C) (Aldwell and Schachter 2003). Other things being equal, music using the intervals of the major scale tends to be perceived as relatively excited, happy, bright or martial, whereas music using minor scale intervals tends to be perceived as more subdued, sad, dark or wistful (Zarlino 1558; Hevner 1935; Cooke 1959; Crowder 1984; Krumhansl 1990; Gregory et al. 1996; Peretz et al. 1998; Burkholder et al. 2006). There has been no agreement, however, about how and why these scales and the intervals that differentiate them elicit distinct emotional effects (Heinlein 1928; Hevner 1935; Crowder 1984).

3. Major vs. minor music compared to speech

Given the roots of the human tonal sense (see Introduction), an attractive hypothesis is that musical tones take on their affective qualities by mimicry of emotions expressed vocally. This general idea is supported by several observations. First, voiced speech sounds are harmonic, and has often been

A Chromatic Scale

Interval Name	Number of Semitones	Frequency Ratio
Unison (Uni)	0	1 : 1
Minor Second (m2)	1	16 : 15
Major Second (M2)	2	9 : 8
Minor Third (m3)	3	6 : 5
Major Third (M3)	4	5 : 4
Perfect Fourth (P4)	5	4 : 3
Tritone (tt)	6	7 : 5
Perfect Fifth (P5)	7	3 : 2
Minor Sixth (m6)	8	8 : 5
Major Sixth (M6)	9	5 : 3
Minor Seventh (m7)	10	9 : 5
Major Seventh (M7)	11	15 : 8
Octave (Oct)	12	2 : 1

B Diatonic Scales

Ionian	Dorian	Phrygian	Lydian	Mixolydian	Aeolian	Locrian
M2	M2	m2	M2	M2	M2	m2
M3	m3	m3	M3	M3	M3	m3
P4	P4	P4	tt	P4	P4	P4
P5	P5	P5	P5	P5	P5	tt
M6	M6	m6	M6	M6	M6	m6
M7	m7	m7	M7	m7	M7	m7
Oct	Oct	Oct	Oct	Oct	Oct	Oct
"MAJOR"					"MINOR"	

C Keyboard examples

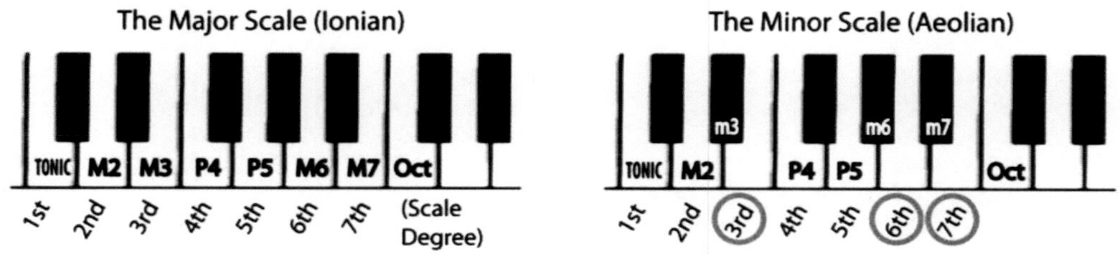

Fig. 1 Musical scales. **A** The 12 intervals of the chromatic scale showing the abbreviations used, the corresponding number of semitones, and the ratio of the fundamental frequency of the upper tone to the fundamental frequency of the lower tone in just intonation tuning. **B** The seven diatonic scales. As a result of their relative popularity, the Ionian and the Aeolian are referred to today as the major and minor scale, respectively. **C** Examples of a major and a minor scale on a piano keyboard; the circled intervals indicate the differences between the two

pointed out ratios between overtones in harmonic series correspond to the ratios that define musical intervals (Helmholtz 1885; Bernstein 1976; Rossing 1990; Crystal 1997; Schwartz et al. 2003; Ross et al. 2007; Gill and Purves 2009). Second, the physiological differences between affective states alter the spectral content of voiced speech (Spencer 1857; Juslin and Laukka 2003; Scherer 2003). Third, humans routinely extract important information about the emotional state of a speaker from vocal qualities (Johnstone and Scherer 2000; Scherer et al. 2001; Juslin and Laukka 2003). And fourth, as already mentioned, other aspects of music appear to convey emotion through mimicry of human behaviors that signify emotional state. It is therefore logical to ask whether spectral differences that specifically distinguish major and minor melodies parallel spectral differences that distinguish speech uttered in different affective states.

3.1 Empirical differences between major and minor music

In music theory, major and minor scales (see Fig.1C) differ at the third, sixth, and seventh scale degrees. In each case, the major scale is characterized by a major interval (a whole tone) whereas the minor scale is characterized a minor interval (a semitone). To determine the way these intervals actually distinguish major and minor music, we analyzed more than 4000 major melodies and 3000 minor melodies drawn from databases of monophonic Western classical and folk music (Barlow and Morgenstern 1974; Eerola and Tovianien 2004). The prevalence of intervals in the melodies was assessed with respect to the annotated tonic of the melody (tonic intervals), as well as with respect to immediately following melody notes (melodic intervals). The results of these analyses are shown in Table I.

As expected, the most salient empirical distinction between major and minor music is

Table 1

(A) Tonic Intervals

Major Melodies			Minor Melodies		
Intervals	Classical (%)	Folk (%)	Intervals	Classical (%)	Folk (%)
Unison/Octave	19.9	20.4	Unison/Octave	19.4	19.1
Minor Second	0.4	0.1	Minor Second	0.6	0.2
Major Second	12.8	15.9	Major Second	13	19.6
Minor Third	0.8	0.0	Minor Third	15.8	15.6
Major Third	18.2	16.8	Major Third	0.7	0.2
Perfect Fourth	10.6	9.5	Perfect Fourth	10.5	10.1
Tritone	1.1	0.4	Tritone	1.7	0.2
Perfect Fifth	19.1	19.6	Perfect Fifth	20.3	19.9
Minor Sixth	0.4	0.0	Minor Sixth	7.9	1.6
Major Sixth	8.4	8.9	Major Sixth	1.3	2.9
Minor Seventh	0.6	0.2	Minor Seventh	3.4	7.5
Major Seventh	7.7	8.1	Major Seventh	5.4	3.1

(B) Melodic Intervals

Major Melodies			Minor Melodies		
Intervals	Classical (%)	Folk (%)	Intervals	Classical (%)	Folk (%)
Unison	10.7	24.9	Unison	11.4	24
Minor Second	20.8	13.3	Minor Second	28.2	19.1
Major Second	36.1	29.9	Major Second	27.6	27.6
Minor Third	9.4	11.6	Minor Third	10.4	12.1
Major Third	6.9	8.6	Major Third	5.4	6.2
Perfect Fourth	7.6	7.4	Perfect Fourth	7.2	7.1
Tritone	0.4	0.2	Tritone	1.1	0
Perfect Fifth	2.7	2	Perfect Fifth	3.2	2.5
Minor Sixth	1.1	0.6	Minor Sixth	1.8	0.8
Major Sixth	1.3	0.9	Major Sixth	1.2	0.1
Minor Seventh	0.4	0.3	Minor Seventh	0.4	0.1
Major Seventh	0.1	0	Major Seventh	0.2	0
Octave	1.3	0.3	Octave	1.2	0.2
Larger	1	0	Larger	0.9	0

Frequency of occurrence of chromatic intervals in major and minor Western classical and folk music. A Tonic intervals, defined as the number of semitones between a melody note and its tonic.
B Melodic intervals, defined as the number of semitones between adjacent melody notes. The intervals that strongly distinguish major and minor music are highlighted in grey

the third scale degree. In both the classical and folk music, major thirds made up 16–18% of the intervals in major melodies and less than 1% of the intervals in minor melodies; this pattern was reversed for minor thirds, which comprised less than 1% of the intervals in major melodies and about 15% of the intervals in minor melodies. Tonic intervals of the sixth and seventh scale degrees also distinguish major and minor music, but less robustly. These intervals are only about half as prevalent in music as thirds, and their distribution in major versus minor music is less differentiated. With respect to melodic intervals, the only salient distinction between major and minor melodies is the prevalence of major versus minor seconds, major music being characterized by an increased prevalence of major seconds and minor music by an increased prevalence of minor seconds.

In short, the prevalence of major vs. minor thirds is the primary empirical distinction between major and minor music.

3.2 Comparison of music and speech spectra

To address the question of whether these empirical differences between major and minor music parallel the characteristics of vocalizations in corresponding emotional states, we compared the spectra of major and minor thirds, sixths, and sevenths with the spectra of speech uttered in a subdued or excited manner. Spectral comparisons were based on fundamental frequency and frequency ratios, two critical acoustic features in the perception of both voiced speech sounds and musical intervals. In speech, fundamental frequency carries information about the sex, age, and emotional state of a speaker (Hollien 1960; Crystal 1997; Protopapas and Lieberman 1996; Banse and Scherer 1996; Harrington et al. 2007). Frequency ratios between the first and second formants (F1, F2) differentiate particular vowel sounds, allowing them to be understood across speakers with anatomically different vocal tracts (Delattre 1952; Pickett et al. 1957; Peterson and Barney 1962; Crystal 1997; Hillenbrand et al. 1995). In music, the fundamental frequencies of tones carry the melody. The frequency ratios between notes define the intervals and provide the context that determines whether the composition is in a major or minor key.

Figure 2 shows the distributions of fundamental frequencies for ten speakers uttering speech in an excited (red) versus a subdued (blue) manner. As expected from previous studies on vocal emotion (Banse and Scherer 1996; Juslin and Laukka 2003; Scherer 2003; Hammerschmidt and Jurgens 2007), the fundamental frequencies of excited speech are significantly higher than those of subdued speech.

Musical intervals are defined by the fundamental frequency ratio between two notes. Thus a single fundamental cannot be identified in the same way as for a voiced speech sound; instead, the two notes must be considered together. The combined harmonics of any two notes comprise a subset of a single harmonic series with a fundamental given by their greatest common divisor (the implied fundamental). Figure 3 illustrates the distributions of these implied fundamental frequencies in major versus minor compositions. Comparison of Figs. 2 and 3 shows that the implied fundamentals of tonic thirds and sixths parallel the differences in the fundamental frequencies of excited and subdued speech.

The ratio of the peak harmonic frequencies in the first two formants (F1 and F2) in excited and subdued speech with the frequency ratios of the intervals that specifically distinguish major and minor music can also be compared. This analysis focused on F1 and F2 because they are the most powerful resonances of the vocal tract, and because they are necessary and sufficient for the dis-

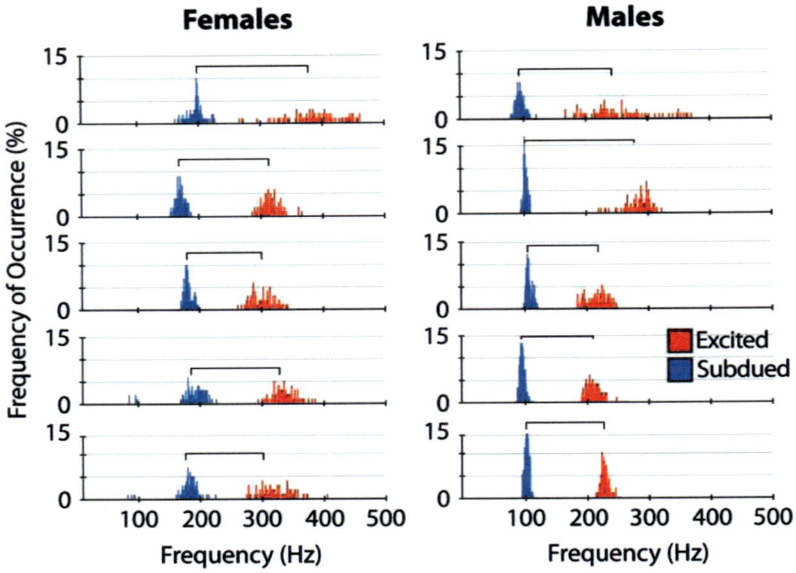

Fig. 2 The fundamental frequency distributions of excited (red) and subdued (blue) speech for 5 female and 5 male speakers derived from single word utterances; brackets indicate means. Participants were instructed to utter the words as if they were excited and happy, or conversely as if they were subdued and sad. The difference between the mean fundamentals of the excited and subdued distributions across speakers is highly significant ($p < 0.0001$ in dependent t-tests for paired samples)

crimination of vowel sounds (Delattre 1952; Pickett et al. 1957; Rosner and Pickering 1994). The distributions of F2/F1 ratios in excited and subdued speech spectra are shown in Fig. 4. In excited speech, ~36% of formant ratios correspond to major thirds, sixths and sevenths, whereas ratios corresponding to minor thirds, sixths and sevenths were entirely absent. In contrast, in subdued speech only ~20% of the formant ratios corresponded to major thirds, sixths and sevenths, whereas ~10% of the ratios corresponded to minor thirds, sixths and sevenths. These parallel differences between the occurrence of formant ratios in excited and subdued speech and the ratios of the musical intervals that distinguish major and minor melodies provide a further basis for associating the spectra of speech in different emotional states with the spectra of intervals that distinguish major and minor music.

4. Discussion

The primary tonal distinction between excited and subdued speech is the fundamental frequency of a speaker's voice; when a speaker is excited the generally increased tension of the vocal folds raises the fundamental frequency; conversely when a speaker is subdued, decreased tension lowers the fundamental frequency (see Fig. 2). In music two factors determine the frequency of a musical interval's implied fundamental: 1) the ratio that defines the interval; and 2) the pitch height at which the interval is played. The defining ratios of minor thirds and sixths (6:5 and 8:5 respectively) yield smaller greatest common divisors than the defining ratios of major thirds and sixths (5:4 and 5:3); thus minor thirds and sixths played at

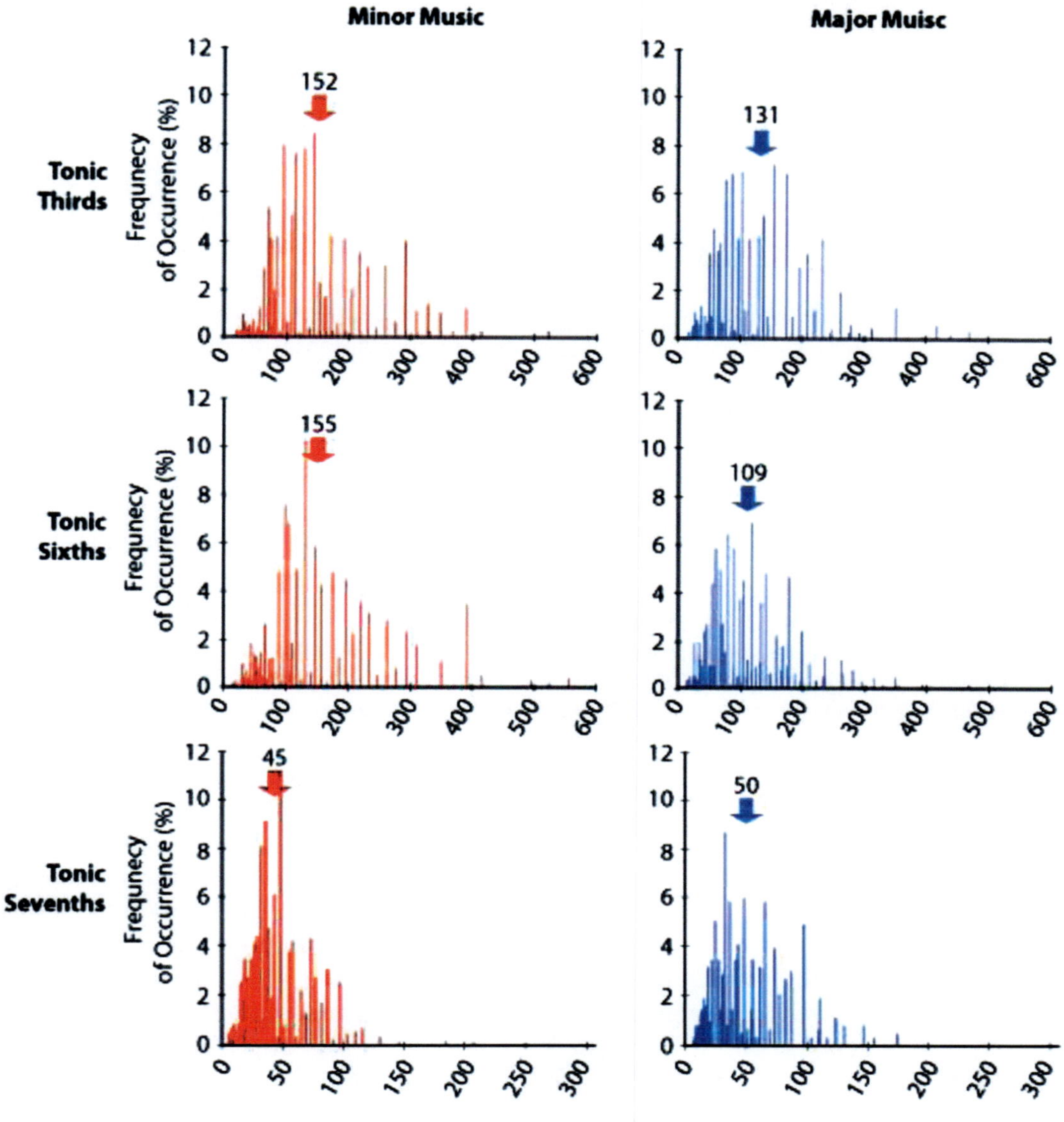

Fig. 3 The implied fundamental frequencies of tonic thirds, sixths, sevenths in major and minor melodies from Western classical music (folk music shows the same trends). Arrows indicate the mean implied fundamental frequency values for each distribution. Differences between the distributions of implied fundamental frequencies for major and minor themes are highly significant for thirds and sixths ($p<0.0075$ or less in independent two-sample t-tests)

the same pitch height as their major counterparts will always have lower implied fundamentals. Although the average pitch height of minor melodies is slightly higher than that of major melodies (see Fig. 2), the mean implied fundamentals of tonic thirds and sixths in major music are higher than those in minor music.

In contrast to the defining ratios thirds and sixths, the defining ratio of the minor seventh (9:5) yields a larger greatest common divisor than the defining ratio of the

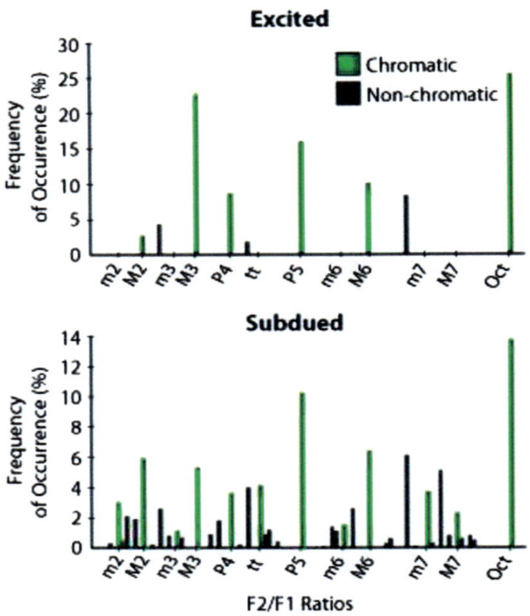

Fig. 4 Comparison of the ratios of the first two formants in excited and subdued speech derived from single word utterances. The F2/F1 ratios have been collapsed into a single octave such that they range from 1 to 2. The distribution of formant ratios in excited and subdued speech; green bars indicate ratios within 1% of chromatic interval ratios (see Table IA); gray bars indicate ratios that did not meet this criterion

major seventh (16:9), making its implied fundamental at a given pitch height higher than its major counterpart. This fact indicates why the pattern observed for tonic thirds and sixths is not apparent for tonic sevenths. Furthermore, nearly all of the implied fundamentals of tonic sevenths (~99%) fall below the range of the fundamentals in speech. These results accord with music theory. Unlike tonic thirds and sixths, tonic sevenths are not taken to play a significant role in distinguishing major and minor music (Aldwell and Schachter 2003). Rather these intervals are generally described as serving other purposes, such as creating a sense of tension that calls for resolution to the tonic (op cit.). As a result, major sevenths are commonplace in both major and minor music (see Table I).

The difference between the fundamental frequencies of excited and subdued speech also affects the prevalence of specific formant ratios. Given the same voiced speech sound, the positions of the first and second formants are relatively similar in excited and subdued speech, as they must be to allow vowel phonemes to be heard correctly. The higher fundamental frequencies of excited speech, however, increase the frequency distance between harmonics, causing lower harmonics to underlie the first and second formants. As a result the F2/F1 ratios in excited speech tend to comprise smaller numbers and thus more often represent musical intervals defined by smaller number ratios. Conversely, the lower fundamental frequencies in subdued speech decrease the distance between harmonics, causing higher harmonics to underlie the formants. Thus the F2/F1 ratios in subdued speech tend to comprise larger numbers, which more often represent musical intervals defined by larger number ratios. Intervals whose defining ratios contain only the numbers one through five (octaves, perfect fifths, perfect fourths, major thirds, and major sixths) are more prevalent in the F2/F1 ratios of excited speech, whereas intervals with defining ratios containing larger numbers (all other chromatic intervals) are more prevalent in the F2/F1 ratios of subdued speech (see Table IA and Fig. 4). In sum, the differences in the spectra of excited and subdued speech parallel differences in the spectra of major and minor music.

A final question is why the musical and emotional distinction between major and minor melodies depends primarily on tonic thirds, and how this fact aligns with the hypothesis that associations made between the spectral characteristics of music and speech are the basis for the affective impact of major versus minor music. One possibility is that among the intervals that differentiate major and minor tone collections thirds entail the lowest frequency and thus the most powerful harmonics. Accordingly, thirds are the most salient distinguishing intervals in the spectra of both voiced speech sounds and musical tones.

Conclusion

The results summarized here suggest that routine associations made between the spectra of speech uttered in different emotional states and the spectra of thirds and sixths in major and minor music are the basis for the different emotional effects of these different tone collections in music. These results support to the more general idea that tonality can be understood in terms of human vocalization. The implication of this and related work (see Introduction) is that musical aesthetics has a biological basis.

Glossary

Musical tone. A periodically repeating sound perceived as having a pitch.

Musical scale. A collection of tones that divide octaves (a doubling of frequency) into specific frequency intervals used to create music.

Scale degree. The number that specifies the position of a tone within a scale; the lowest note is the first scale degree, the second lowest tone is the second scale degree, etc.

Musical interval. The frequency relationship between two tones; specific musical intervals are defined by the ratios of their fundamental frequencies (see Fig.1A).

Major/Minor intervals. Musical intervals that differentiate major and minor scales, (see Fig.1). Major intervals are a semitone greater than the corresponding minor intervals.

Melody. An organized sequence of musical tones that expresses a musical theme or phrase.

Fundamental frequency. The lowest frequency of vibration in a harmonic series (e.g., the vibration of the full length of a plucked string).

Implied fundamental frequency. The frequency equal to the greatest common divisor of the fundamental frequencies of two or more musical tones.

References

Aldwell E, Schachter C (2003) Harmony & voice leading, 3rd edn. Wadsworth Group/Thomson Learning, Belmont, CA

Banse R, Scherer KR (1996) Acoustic profiles in vocal emotion expression. J Pers Soc Psychol 70: 614–636

Barlow H, Morgenstern S (1974) A dictionary of musical themes. Crown, New York, NY

Bernstein L (1976) The unanswered question: six talks at Harvard. Harvard University Press, Cambridge, MA

Bowling DL, Gill K, Choi JD, Prinz J, Purves D (2010) Major and minor music compared to excited and subdued speech. J Acoust Soc Am 127: in press

Burkholder JP, Grout D, Palisca C (2006) A history of Western music, 7th edn. Norton, New York, NY

Burns EM (1999) Intervals, scales and tuning. In: Deutsch D (ed) The psychology of music, 2nd edn. Academic Press, New York, NY

Cohen D (1971) Palestrina counterpoint: A musical expression of unexcited speech. J Music Theory 15: 85–111

Cooke D (1959) The language of music. Oxford University Press, Oxford, UK

Crowder RG (1984) Perception of the major/minor distinction: hedonic, musical, and affective discriminations. B Psychonomic Soc 23: 314–316

Crystal D (1997) The Cambridge encyclopedia of language, 2nd edn. Cambridge University Press, New York, NY

Delattre P, Liberman AM, Cooper FS, Gerstman LJ (1952) An experimental study of the acoustic determinants of vowel color: observation of one- and two-formant vowels synthesized from spectrographic patterns. Word 8: 195–210

Eerola T, Tovianien P (2004) Suomen Kasan eSävelmät (Finnish folk song database). Available from http://www. jyu.fi/musica/sks/ in November, 2008

Gill K, Purves D (2009) A biological rationale for musical scales. PLoS ONE 4, e8144.doi:101371/journal.pone.0008144

Gregory AH, Varney N (1996) Cross-cultural comparisons in the affective response to music. Psychol Music 24: 47–52

Hammerschmidt K, Jürgens U (2007) Acoustical correlates of affective prosody. J Voice 21: 531–540

Harrington J, Palethorpe S, Watson CI (2007) Age-related changes in fundamental frequency and formants: a longitudinal study of four speakers. In: Proceedings of Interspeech 2007, Antwerp
Heinlein CP (1928) The affective characters of the major and minor modes in music. J Comp Psychol 8: 101–142
Helmholtz H (1885) Lehre von den Tonempfindungen, 4th Edn., translated by Ellis AJ (1954) as: On the sensations of tone, 2nd edn. Dover, New York, NY
Hevner K (1935) The affective character of the major and minor modes in music. Am J Psychol 47: 103–118
Hillenbrand J, Getty LA, Clark MJ, Wheeler K (1995) Acoustic characteristics of American English vowels. J Acoust Soc Am 97: 3099–3111
Hollien H (1960) Some laryngeal correlates of vocal pitch. J Speech Hear Res 3: 52–58
Johnstone T, Scherer KR (2000) Vocal communication of emotion. In: Lewis M, and Haviland-Jones J M (eds) Handbook of Emotions, 2nd edn. Guilford, New York, NY
Juslin PN, Laukka P (2003) Communication of emotions in vocal expression and music performance: Different channels, same code? Psychol Bull 129: 770–814
Krumhansl CL (1979) The psychological representation of musical pitch in tonal context. Cognitive Psychol 11: 346–374
Krumhansl CL (1990) Cognitive foundations for musical pitch. Oxford University Press, New York, NY
Patel AD (2008) Music, language, and the brain. Oxford University Press, New York, NY
Peretz I, Gagnon L, Bouchard B (1998) Music and emotion: perceptual determinants, immediacy, and isolation after brain damage. Cognition 68: 111–141
Petersen GE, Barney HL (1962) Control methods used in a study of the vowels. J Acoust Soc Am 24: 175–184
Pickett JM (1957) Perception of vowels heard in noises of various spectra. J Acoust Soc Am 29: 613–620
Pierce JR (1962) The science of musical sound, revised edn. Freeman, New York, NY
Protopapas A, Lieberman P (1996) Fundamental frequency of phonation and perceived emotional stress. J Acoust Soc Am 101: 2267–2277
Randel DM (ed) (1986) The new Harvard dictionary of music, revised 2nd edn. Belknap Press, Cambridge, MA
Rosner BS, Pickering JB (1994) Vowel perception and production. Oxford University Press, New York, NY
Ross D, Choi J, Purves D (2007) Musical intervals in speech. Proc Natl Acad Sci 104(23): 9852–9857
Rossing TD (1990) The science of sound, 2nd edn. Addison-Wesley, New York, NY
Scherer KR (2003) Vocal communication of emotion: A review of research paradigms. Speech Commun 40: 227–256
Scherer KR, Banse R, Wallbott HG (2001) Emotional inferences from vocal expression correlate across languages and cultures. J Cross Cult Psychol 32: 76–92
Schwartz DA, Purves D (2004) Pitch is determined by naturally occurring periodic sounds. Hearing Res 194: 31–46
Schwartz DA, Howe CQ, Purves D (2003) The statistical structure of human speech sounds predicts musical universals. J Neurosci 23: 7160–7168
Spencer H (1857) The origin and function of music. Fraser's Magazine 56: 396–408
Terhardt E (1974) Pitch, consonance, and harmony. J Acoust Soc Am 55: 1061–1069
Terhardt E (1984) The concept of musical consonance: A link between music and psychoacoustics. Music Percept 1: 276–295
Zarlino G (1558) Bk. 3 of Le institutioni harmoniche, translated by Marco G, Palisca C (1968) as The Art of Counterpoint. Yale University Press, New Haven, CT

Synaesthesia and synergy in art. Gustav Mahler's "Symphony No. 2 in C minor" as an example of interactive music visualization

13

Johannes Deutsch*

Contents

Abstract

Synaesthesia occurs when stimulation of one sensory modality induces an experience in another sensory modality. This phenomenon can involuntarily emerge or be intentionally induced by artists through the synthesis of image or color and music. Consequently in the present chapter the author differentiates between the emergence of synaesthesia, understood as a neurological condition characterized by exceptional sensory perception and investigated in medicine, psychology, and recently in the neurosciences, and a long-standing tradition of synaesthetic œuvres in the history and theory of art and culture.

Johannes Deutsch
Ölzeltgasse 1/3
1030 Vienna, Austria
e-mail: office@johannes-deutsch.at

In science, synaesthesia is conceived as an involuntary experience which is perceptual in quality and not simply a form of imagery. The author follows this pathway of investigation from Gustav Theodor Fechner (1876, 1898), Sir Francis Galton (1880a, b, c; 1883), Alfred Binet (1892) to recent research by Richard Cytowic (2002; Cytowic and Eagleman 2009), Vilayanur S. Ramachandran (Ramachandran and Hubbard 2001a, b), Simon Baron-Cohen (1994; Baron-Cohen, Wyke, and Binnie 1987; Baron-Cohen and Harrison 1997), Hinderk Emrik (Emrik, Schneider, and Zedler 2002), and Julia Simner (2009). Although in synaesthesia connections between perceptions may also involve other senses like olfaction and taste, the most commonly investigated link is between auditory and visual perception. Recently, Baron-Cohen and colleagues have developed valid and reliable measures for the diagnosis of visual synaesthesia (Asher et al. 2006), as well as Eagleman for synaesthesia in general (Eagleman et al. 2007).

* I have a special debt of gratitude to my son "Chris" Léon Deutsch for his accurate scrutinizing and translating the widespread historical French literature on the relationship between vision and audition, from the early synergy between music and colors to the case reports as well as the literary elaborations of the "audition colorée".

Although some artists refer to their involuntary perceptual experience of synaesthesia, in the history and theory of art and culture synaesthesia is largely conceived of as the intentional effort to evoke sensory experiences in the audience, mainly through the synergy of music and colors (unidirectional – like in the neurological condition of synaesthesia – from the music to the colors). The author explores this long-standing history in aesthetics which connects colors to music from the Pythagoreans, through Leonardo da Vinci, the French "audition colorée", Alexander Skriabin, Wassily Kandinsky, the "sonchromatoschope" of Alexander László, to the most recent efforts in contemporary art and design.

Up to now, a major problem of realization for both the theoreticians of culture and art as well as the artists, was how to precisely recreate the simultaneity of synaesthetic perceptions so characteristic of synaesthesia as a neurological condition.

What follows shows how the author reached this goal in both avant-garde projects and in operas like Wagner's "Rheingold" and Schumann's "Manfred". Subsequently the ways in which Mahler's "Symphony No. 2 in C minor", also known as "Resurrection," was visualized are explained from both the aesthetic and the technical point of view. Regarding the technical realization, a complex computer cluster was adopted to generate real-time visualization. The cluster simultaneously processed and implemented any modulations picked up from the orchestra as nuances of the live interpretation of the music before translating them directly into visual variations.

The present chapter concludes with major theoretical questions concerning the frequency of occurrence of processes such as cross sensory coupling, the prevalence of fixed paired patterns of association as well as their relevance for art.

1. Synaesthesia as a neurological condition

In science, synaesthesia is understood as an involuntary experience which is perceptual in quality – and not simply a form of imagery – correlating to a neurological condition and seemingly having a genetic basis (Asher et al. 2009). However, imagery may additionally occur (Barnett and Newell 2008). During perceptual or cognitive activities (listening to music, reading, etc.) experiences of colors or taste automatically emerge. Synaesthetes see colors when hearing sounds ("music-color synaesthesia", Ward et al. 2006), or perceive tastes in their mouth when reading or hearing words ("lexical-gustatory synaesthesia", Simner and Ward 2006), or perceive spatial forms when reading time units (hours of the day, days of the week, months of the year, etc.) or numbers and letters ("visuo-spatial synaesthesia" and "grapheme synaesthesia", Simner 2009). A triggering stimulus termed "inducer" causes (as an effect) a secondary sensory experience termed "concurrent" (Grossenbacher and Lovelace 2001). Although in synaesthesia connections between perceptions may also involve other senses such as olfaction, or touch, the most commonly investigated relationship is the link between auditory and visual perception, especially "colored hearing" ("audition colorée") and "musical" color hearing (Cytowic 2002).

The historical pathway of investigations in synaesthesia can be traced from Gustav Theodor Fechner, Sir Francis Galton, and Alfred Binet to the contemporary research of Richard Cytowic, Vilayanur S. Ramachandran, Simon Baron-Cohen, Hinderk Emrik, and Julia Simner. Synaesthesia has been studied in medicine and psychology as a disturbance of sensory perception for about three centuries but the number of investigations increased notably between 1860 and 1930. The involuntary association of verbal sounds or phonemes, in particular vowel sounds, with colors was reported in psychology in the 19 th Century. Synaethesia research began with case studies reported by Gustav Theodor Fechner (1876), Sir Francis Galton

(1880a, b, 1883), and Alfred Binet (1892; Binet and Philippe 1892; Beaunis and Binet 1892).

Gustav Theodor Fechner investigated color synaesthesia in his "Vorschule der Ästhetik" (1876). In the later editions (e. g., 2nd edition 1898) of his treatise he also included a number of clinical case studies drawing from clinical reports of Eugen Bleuler and Karl Lehman (Bleuler and Lehman 1881). In 1913, Bleuler published a book entitled „Zur Theorie der Sekundärempfindungen" devoted to synaesthetic perception (Bleuler 1913).

Having been concerned with the question of visualized numerals (Galton 1880a, b), a "visuo-spatial synaesthesia" or "grapheme synaesthesia", the psychologist Sir Francis Galton widely discussed color-hearing in his "Inquiries into human faculty and its development" (1883). Galton published a diagram of a colored alphabet, according to which A is yellow, E is green, O is red, etc. He had to cope with a non-existent consensus regarding the color attributed to each vowel and consonant, but having already investigated the relation of synaesthesia with mental imagery (Galton 1880c), Galton reported, not without perplexity, that "when the account of one seer is submitted to another seer, who is sure to see the colors in a different way, the latter is scandalized and almost angry at the heresy of the former." (Galton 1883, p. 111) The large individual differences, which puzzled Galton and other scientists of his time, have currently become the object of systematic inquiry in their own right (Dixon et al. 2004; Dixon and Smilek 2005; Day 2007).

The renowned French psychologist Alfred Binet published a number of case studies (1892; Binet and Philippe 1892; Beaunis and Binet 1892) also remarking that even if science considers synaesthesia as a mere disturbance of sensory perception, poets value it as a new form of art (Binet 1892). Indeed, la "Revue Philosophique" published not only Binets's case studies, but a veritable explosion of case reports (Lauret and Duchaussoy 1887; Henri 1893; Philippe 1893; Claparède 1900; Daubresse 1900; Sokolov 1901; Ulrich 1903, among others). Further, medical and psychological journals in French speaking countries devoted studies to the phenomenon (Cornaz 1851; Chabalier 1864; Flournoy 1892; Clavière 1898; Lemaitre 1904, among others) and books appeared on the topic as well (e. g., Baratoux 1888; Suarez de Mendoza 1890; Flournoy 1893).

A few years later, the Viennese psychoanalyst Hermine von Hug-Hellmuth as well as the Swiss psychoanalyst Oskar Pfister also seemed attracted to investigations on audition evocating colors (von Hug-Hellmuth 1912; Pfister 1912).

As early as 1893, psychologists such as the French Edouard Gruber began trying to objectify synaesthesia by means of questionnaires (e. g., Gruber 1893) or by discriminating between authentic synaesthesia and intermodal analogy (Bos 1929). But in the course of the 20th Century, behaviorist psychology increasingly focused its efforts on observable behavior, whereas subjective experiences no longer seemed suitable for scientific research. With regards to the time prior to the behavioristic turn in psychology, Georg Anschütz (Anschütz 1925, 1926, 1927, 1931, 1936) as well as Albert Wellek (Wellek 1927, 1931a, b) are worth mentioning for their final remarkable scientific efforts in synaesthesia research of this period. For a long time thereafter, the scientific community neglected the phenomenon synaesthesia, until it was rediscovered in the 1980s. Two special issues of the journal "Cortex" attested the increasing presence of synaesthesia in cognition research: in 2006 "Cognitive neuroscience perspectives on synaesthesia", and in 2009 "Synaesthesia and visuo-spatial forms" ("Cortex" 2006, 2009).

Recently, Baron-Cohen and colleagues have developed valid and reliable measures for the diagnosis of visual synaesthesia: "Methods of diagnosing synaesthesia have undergone tremendous changes since the time of the original case report. Emphasis is shifting to a deeper analysis of synaesthetic percepts, the genetic of synaesthesia and the neural mechanisms underlying the condition." (Asher et al. 2006, p. 146) Furthermore, the neuroscientist David Eagleman created the "Synaesthesia Battery" as a diagnostic instrument and the "Texsyn Toolbox" to be used with any form of synaesthesia (Eagleman et al. 2007).

Twenty years ago synaesthesia appeared to be quite rare: the neurologist Richard Cytowic estimated the ratio between synaesthetes and non-synaesthetes to be 1:25 000 (Cytowic 1989); around 1996 Baron-Cohen and colleagues gave a ratio of 1:2000 (Baron-Cohen et al. 1996); in 2010 Julia Simner of the University of Edinburgh suggested the currently accepted frequency of synaesthesia to be 1:23 (Metzger 2010). In such a case synaesthesia can no longer be considered an exceptional sensory phenomenon.

Synaesthesia of a particular person does not change over time but is highly consistent within each individual, independent from context, automatic, and involuntary (Baron-Cohen et al. 1993). The subject is aware that the inducer is – for example – an acoustic perception (e. g., music) and that the concurrent secondary perception is a visual experience (e. g., color). Could this indicate that synaesthesia does not imply a simple merging of the sensory functions, but more complex processes? Synaethetes are usually also able to distinguish their ordinary from their synaesthetic percepts. In the same person the link between inducer (e. g., A) and concurrent (e. g., red) is an experience constant over time, whereas there are very different forms of synaesthesia (about 150 types are known) in different individuals. From Day's (2007) demographic enquiry the types of synaesthesia shown in Table 1 are the most common.

Table 1 The most common types of synaesthesia according to Day's (2007) demographic enquiry

INDUCER	CONCURRENT	PERCENTAGE
grapheme	color	64,9
time unit	color	23,1
music	color	19,5
noise	color	14,9
phoneme	color	9,2
musical note	color	9
odor	color	6,8
taste	color	6,3
noise	taste	6,1
pain	color	5,5
people	color	5,4
touch	color	4,0

Visual synaesthetes are termed "associators" if their synaesthetic experience (e. g., color) is experienced in their "mind's eye"; for other visual synaesthetes, called "projectors", the synaesthetic perception (e. g., color) is projected outside (e. g., the concurrent color overlays the inducer, e. g., letter or numeral) (Dixon et al. 2004).

Especially in grapheme-color synaesthesia, the question arises whether the concurrent color experience is primarily determined by the form of the inducer grapheme (letter or numeral) or by their meaning. Ramachandran and Hubbard (2001 b) suggested that the meaning of the inducer is not really relevant for triggering the concurrent experience, but a recent study showed that it may play a key role. According to Dixon and colleagues, for any inducer grapheme presented the concurrent color emerges independent from any variations in size or font (the meaning of the grapheme triggers the color, rather than its form) and ambiguous graphemes induce different concurrent colors, depending on whether they are read and understood as numeral or letters (Dixon et al. 2006). The debate on the relative role of meaning and form in grapheme-color synaesthesia led to two concurrent approaches to synaesthesia in science: (i) the cross-linkage, cross-activation, or "cross-talk" model, and (ii) the feedback model which will be discussed in what follows.

(i) Synaesthesia may be caused by a cross-wiring of areas in the brain, which would not normally be connected or only weakly linked. Consequently, the cross-activation model assumes that the areas of the brain that process graphemic forms such as numerals or letters (fusiform gyrus) are cross-linked to the areas of the brain that process color (fusiform areas) (Ramachandran and Hubbard 2001a, b).

(ii) The alternative model additionally postulates a direct feedback from areas of the brain that process meaning to areas that process color (Grossenbacher and Lovelace 2001; Dixon et al. 2004).

Looking for the actual neural roots of synaesthesia, Frith argued: "Brain imaging studies have shown that, in comparison to normal controls, people with colour-word synaesthesia do indeed show activity in 'visual' areas of the brain. By delineating the function of these visual areas more precisely we shall gain an understanding of the nature of this purely experiential phenomenon in physiological terms." (Frith 2001, pp. 156–159; see also Cytowic and Eagleman 2009)

Research on color experience triggered by spoken words using magnetic resonance imaging (fMRI) provided evidence that synaesthetic perceptions activate the same regions that support veridical perception. In the reported experiment, synaesthetic experiences of color activated the color-selective region V4 (Nunn et al. 2002). Hubbard and Ramachandran (2005) later validated these results. However, in March 2010 at the UK-Synaesthesia Congress in Brighton, the neuroscientist Jean-Michel Hupé (CerCo, CNRS, University of Toulouse) presented a study questioning the results of Nunn and Hubbard and claiming that he and his co-workers could not replicate the results (Hupé 2010; cf. Metzger 2010; see also Hupé et al. 2010). Further research may be required.

The neurological condition of synaesthesia is significantly represented in children. In 2006, Julia Simner asked: "What form do linguistic synaesthesias take before language acquisition? For example, is there a progression from sound-color, to phoneme-color, to grapheme-color synaesthesia during language and literacy acquisition?" (Simner 2006, p. 28) A recent study (Simner et al. 2009) provided evidence that the prevalent form among children is grapheme-color synaesthesia, which emerges along a developmental path following environmental exposure to learned linguistic units. The involvement of learned units showed that the grapheme-color synaesthesia emerged only after the acquisition of the inducer, and not from the moment of the birth.

Do results of synaesthesia-research allow drawing conclusions about the general average population? A major question remains whether synaesthetes and average population share the same neuronal processing. A number of studies provide evidence "that synaesthetes occupy an extreme end of a shared continuum on which non-synaesthetes also lie. [. . .] independent evidence shows that synaesthetic experiences reflect mechanisms found in all people, albeit to a different level of awareness." (Simner 2009, p. 1142) Therefore the associations of synaesthetes seem to rely on processes that are also present in the average population, with the difference that for synaesthetes such associations are available to consciousness. In grapheme-colors synaesthesia both synaesthetes, who experienced colors from inducers such as letters and/or numerals, and non-synaesthetes prefer certain pairings between letters and colors: "A" to red, "S" to yellow, "X" to black (Simner et al. 2005). Regarding the music-color synaesthesia Ward and colleagues (2006) indicate the tendency of synaesthetes to see lighter colors from high frequency sounds, and the same pairing between luminance and pitch is apparent in average population when making intuitive music-color associations.

2. Synaesthesia as the artist's effort to evoke sensory experiences through the synergy of music and colors

Although a few artists may rely on their own involuntary perceptual experience as synaesthetes, in the history and theory of art and culture synaesthesia refers mainly to intentional efforts of the artist to evoke sensory experiences in the audience, mostly through the synergy of music and colors. Sound indeed refers to multisensory attributes not only in synaesthetic experience, but even under average perceptual conditions. Therefore, studies on the multisensory content of sound provide a deep insight into the complexity of auditory perception (Haverkamp 2007, 2009b, 2010).

Since the early 20th century, the successful pairing music and color (and even spatial forms) began due to increasing advances of technology. But the relationship between color and sound relies on a long-standing tradition of theoretical reflections on this topic as well as on various practical attempts to realize it. Exploring this history in aesthetics linking colors to music one can start with the Pythagoreans and continue with Leonardo da Vinci, the French "audition colorée", and composers such as Alexander Scriabin. Ingenious inventors of color instruments such as the "sonchromatoscope" of Alexander László, painters like Wassily Kandinsky, and the most recent efforts in contemporary art and design are part of this history as well.

It is possible to construct correlations between perceptions and physical properties. Thus the frequency (wavelength) of light can be paired with the frequency (pitch) of a given pure tone by means of mathematical calculations. Over the centuries, philosophers, musicians, painters, as well as theoreticians have looked for the perfect match of pitch and color. As early as the 6th Century BC the Pythagoreans introduced the theory of "harmony" supervening to the universe. In the spirit of mathematical harmony, Pythagoreans created the diatonic scale ("Pythagorean") where the ratio between the highest and the lowest pitch (frequency) is 2:1, which produces the interval of an octave. Musical harmony correlated to the cosmic order (the seven planets known at the time) and had an equivalent in the seven colors (Moritz 1987, 2006). In the 4th Century BC, Aristotle's colors scale in "On sense and sensible objects" included seven colors: white, yellow, red, purple, green, blue, and black (Aristotle 1936). His scale corresponded to the tone intervals of the "Pythagorean" musical scale providing the basis for the relationship between color and sound in the ancient world (Jewanski 1999).

At the beginning of the 18th Century Isaac Newton provided experimental evidence that white light is composed of the seven colors of the spectrum (Newton 1704). According to Newton the mathematical-physical correspondence between the frequency of the tones and the much higher frequencies of the light waves could be calculated by introducing a coefficient. The ancient philosophical relation between color and sound therefore seemed to be validated by applying the laws of physics (Jewanski 1999; Topper 1999).

In his "Chromatography", Field (1835) suggested that painters should follow musicians with regard to harmony, and he provided a diagram linking the scale of color and the diatonic musical scale. In the same vein, Hay (1836) presented a scale pairing music to colors in his "The laws of harmonious coloring adapted to interior decorations, manufactures and other useful purposes".

Poets were interested in the link between sound and colors as well. In his poem "Voyelles" ["Vowels"] (1871), Arthur Rimbaud related the vowels of the alphabet to colors, proposing an intentional constructed relationship – termed "correspondance" ["correspondence"] – between sensory perceptions instead of an involuntary neuropsychological mechanism. A forerunner of the French program of the "correspondence" can be found in William Shakespeare's "A midsummer night's dream". The dream of Nick Bottom, the weaver, magnificently linked all senses: "The eye of man hath not heard, the ear of man hath not seen, man's hand is not able to taste, his tongue to conceive, nor his heart to report, what my dream was." (Shakespeare 1926 [1594–96], p. 58) Similarly, Charles Baudelaire's sonnet "Correspondances" in "Les fleurs du mal" ["The flowers of evil"] (1857) announced his program of the "correspondences", thereby paving the way for Rimbaud's proposal:

> "Comme des longs échos qui de loin se confondent
> Dans une ténébreuse et profonde unité,
> Vaste comme la nuit et comme la clarté
> Les parfums, les couleurs et les sons se répondent. "

> "As the long echoes, shadowy, profound,
> Heard from afar, blend in a unity,
> Vast as the night, as sunlight's clarity,
> So perfumes, colours, sounds may correspond."
> (Baudelaire 1993 [1857], p. 19)

Inspired by Baudelaire's program, Rimbaud created his sonnet "Voyelles" ["Vowels"], later collected in his volume "Poésies":

> "A noir, E blanc, I rouge, U vert, O bleu: voyelles [. . .]"
> "A black, E white, I red, U green, 0 blue: vowels [. . .]"
> (Rimbaud 2001 [1971], p. 135)

Indeed, in the course of the sonnet the vowel "I" becomes purple and the vowel "O" violet, whereas in synaesthesia as a neurological condition, the pairing between vowel and color is not only automatic, involuntary, and independent from context, but also remains a highly consistent fixed pattern within each individual for his or her entire life. In "Une saison en enfer" Rimbaud explicitly strengthened the intentionality of his project in the second paragraph "Alchimie du verbe" ["Alchemy of the word"] of its chapter "Délires II" ["Second delirium"]:

> "J'inventai la couleur des voyelles! – A noir, E blanc, I rouge, O bleu, U vert. Je réglai la forme et le mouvement de chaque consonne, et, avec les rythmes instinctifs, je me flattai d'inventer un verbe poétique accessible, un jour ou l'autre, à tous les sens."

> "I invented the colour of vowels!– A black, E white, I red, 0 blue, U green.–I organized the shape and movement of every consonant, and by means of instinctive rhythms, flattered myself that I was the inventor of a poetic language, accessible sooner or later to all the senses."
> (Rimbaud 2001 [1973], p. 235)

The poet aimed to extend the "correspondences" between sound and color to the other senses like touch and smell as well.

Let us now move to efforts in the past to translate theories linking color and music into practice. As early as 1500 Leonardo da Vinci projected colored light in the rhythm of music (Moritz 1987, 2006). At the court of the emperor Rudolf II, Giuseppe Arcimboldo built a sort of "color-piano" in order to support his assumption that an intimate relationship existed between heard musical sounds and perceived colors reaching from black to white (Moritz 1987). In the 18th Century, the Jesuit Louis Bertrand Castel followed Newton in suggesting that each tone (and not each interval) corresponds to a specific color. He developed an "ocular harpsichord", which was a color piano emanating colored light with every key pressed (Castel 1725, 1726a, b, 1751; Moritz 1987). It consisted "of a 6-foot square frame above a normal harpsichord; the frame contains 60 small windows each with a different colored-glass pane and a small curtain attached by pullies to one specific key, so that each time that key would be struck, that curtain would lift briefly to show a flash of corresponding color." (Moritz 1997, p. 1) From Castel's work grew a rich tradition of "color pianos" and "color organs". One example is the well known instrument of the English painter Alexander Wallace Rimington (Rimington 1911; Moritz 1987, 1997; Peacock 1988; Scheel 2006) who also provided his own "chromatic assignment". Rimington (1911) correlated a scale with twelve colors to one octave divided into 12 intervals. The most low-pitched tone (C) was paired with red, the color perceived when stimulated with the lowest frequency (longest wavelength) of visible light ("C = red").

The Russian composer Alexander Scriabin, who was said to be a "genuine" synaesthete, linked musical tones to colors in his symphony "Promethée. Le poème du feu. Op. 60" ["Prometheus. A poem of fire. Op. 60"] (1910) where he also included a part to be played by a color instrument, "clavier à lumières" ["light piano"] or "luce" ["light"], designed specifically for the performance of his tone poem. His light piano was played like a conventional piano, but it was mute and projected colored light onto a screen. It was a "Chromola", exclusively built in New York for the performance at the Carnegie Hall under the personal supervision of Preston S. Miller, at the time president of the "Illuminating Engineering Society" (and not – as sometimes reported – a color instrument of Rimington). William Moritz vividly describes the "moving lights that accompanied the 1915 New York premiere of Scriabin's synaesthetic symphony 'Prometheus. A poem of fire'. Scriabin wanted everyone in the audience to wear white clothes so that the projected colors would be reflected on their bodies and thus possess the whole room." A particularly bright light radiating at the acme of the symphony deeply impressed the audience and was even perceived as painful (Scriabin 1995 [1911]; Baker 1986; Moritz 1997).

Arnold Schönberg's "The fortunate hand. Drama with music" (1910–1914) merged colors, gestures, movements, light and music. It was first performed in Vienna in 1924. According to Schönberg, gestures, colors, and light had to be handled by the composer as if they were tones; figures and shapes were formed from individual light values and shades of color, resembling the forms, figures and motives with which musicians make music (Weibel 1987; Bidaine 2004).

A complementary effort was that of the painter Wassily Kandinsky in his tone drama "The yellow sound" (1909). It was first published in 1912 in "The Blue Rider Almanac", yellow being for Kandinsky the colour of middle-C on a piano (Kandinsky 1994). According to Kandinsky, colors correspond to the keyboard, eyes to the harmonies, and the soul to the piano with many strings, and he conceived of the painter as the hands that plays, touching the keys, in order to evoke vibrations in the soul. Kandinsky played with the harmonic connection of "musical tone and its movement, bodily-mental sound and its movement, and color-tone and its movement" (Weibel 1987). In "Concerning the spiritual in the art" (1911), he presented his own theory on color and synaesthesia (Kandinsky 2001 [1911]).

But in painting it was not only Kandinsky who was deeply concerned with synaesthesia, as several other painters interested in links between sensory perceptions can be cited. In 1913, a painter of Futurism, Enrico Prampolini published "Chromophony. The color of sounds", a chromophonic Manifesto according to which colors are the equivalent of tones and paintings corresponded to musical œuvres (Prampolini 1913). His celebrated painting "Simultaneous landscape" (1922) is also a magnificent attempt to capture sensory correspondences. Like Baudelaire and Rimbaud had earlier in literature, Futurism – this time stemming from painting – showed again the tendency to gradually extend the link between seeing and hearing to further sensory modalities. Thus Filippo Tommaso Marinetti and Luigi Enrico Colombo Fillia presented the "Manifesto of the futurist cuisine" (1930), a veritable program of cross sensory coupling (taste, touch, smell, vision – especially color-vision – as well as music-hearing). Consequently, Marinetti published his own "The futurist cookbook" (1989 [1932]). Both books deeply influenced Daniel Spoerri's contemporary "Eat Art" (Lemke 2006).

Regarding the particular synergy between hearing and seeing, a Centre Pompidou's exhibition "Sounds and light. A history of sound in the arts of the 20th century" (2004–2005) showed the synaesthetic program to be an undertaking of wide scope. It was clear that major painters such as Frantisek Kupka, Francis Picabia, Paul Klee, Piet Mondrian, Giacomo Balla, as well as many other renowned other artists, adhered to the synaesthetic program of the time (Bidaine 2004).

The long-standing tradition of "color-pianos" culminated in Alexander László's „sonchromatoscope" (1925) which enabled the accompaniment of piano music by projections. It consisted of a switchboard connected to four large and four small color projectors and was fitted with keys and levers like a sort of harmonium. This gave control over the projectors as well as the mixing of colors and images. László used slides, abstract forms, and colored light (Moritz 1997; Jewanski 1997; Scheel 2006), and correlated his color-harmonics with musical tonic intervals, therefore constructing an eight-step scale (László 1925).

Among the composers dealing with synaesthesia a few more can be briefly mentioned. Nicolai Rimski-Korsakov maintained the link between music and colors; Olivier Messiaen considered sounds deeply paired with colors moving with the music (Messiaen 1948; Johnson 1982; Nichols 1986; Samuel 1994); and finally György Ligeti, perhaps one of the few confirmed synaesthetes, linked his compositions of noises to shapes and colors, explicitly and accurately testifying to the phenomenology of his own perceptions in interviews (Ligeti 1958, 1983).

In the early 1930s, the steady ascent of film obscured the previous interest of audience and critics in color instruments, and a history of the relationship between music and film began. Between 1921 and 1967, Oskar Fischinger created about thirty visual music films, first without sound, later with sound and colors (Moritz 1997). After 1945 synaesthetic art movements arose such as "Happening" and "Fluxus" – both deeply inspired by Fischinger –, as well as the already mentioned "Eat Art" of Daniel Spoerri, and not least the "Theatre of Orgies and Mysteries" of Hermann Nitsch, one of the "Vienna Actionists". Nitsch further conceived a synaesthetic design for sets and costumes staging Jules Massenet's opera "Hérodiade" (Nitsch 1995).

This tendency to link hearing and seeing did not remain confined to avant-garde, but reached the popular contemporary club scene: from DJ to VJ. VJing – as a DJ does with music pieces – applies technology advances to mixing electronic music and moving images, which are generated separately. However, this genre soon transcended its popular field of origin by merging with design, architecture, and art (Behne 1987; Scheel 2006; Fischer 2008).

A further effort at going beyond the link between audition and vision in order to involve all sensory modalities is apparent in the field of design. In the 1920s, already the Bauhaus attempted synaesthetic correlations between design and architecture (Düchting 1996). The heritage of the intentional synaesthetic program of art and culture has carried on through to synaesthetic design (Riccò 1999, 2008).

Michael Haverkamp has been studying synaesthesia also as a neurological condition in order to apply its principles to engineering automobiles (Haverkamp 2009a) or other objects of every day life. He explained how the perceptual system aims to identify physical objects by generating perceptual objects. If that physical object is already known and represented in memory, its cross sensory attributes can be recalled by stimulation of only one modality: "learning to handle physical objects of daily life requires testing of all sensory properties like vision, audition, smell, taste, surface structure, hardness, and many others." (Haverkamp 2009b, p. 3)

Interactive real-time processes of contemporary media art not merely represent a substantial new development, but a significant shift in synaesthetic design feasibility. Through real-time interactivity, one art form can simultaneously react with another art form. The visual design enables us to create vision evoking analogies to each musical passage. In addition, for the first time advanced software makes real-time interaction possible so that the live performance and the unique subjective interpretation of the music associated with it can shape and modify this visual design. This synergy of art forms allows a highly concentrated and direct access to multisensory art perception.

3. Visualizing Mahler's "Symphony No. 2 in C minor"

Some of the author's own projects are conceived as synaesthetic œuvres. They comprise of avant-garde projects such as "Gesichtsraum" ["FaceSpace"] (Deutsch 2002; Maresch 2010; Zuckriegl 2010), as well as the visualization and staging of operas like Wagner's "Rheingold" (Binder 2004; Deutsch 2004, 2006a; Scheel 2006; Trenks 2005; Waltz 2008; Winkler 2005, 2006) or Schumann's "Manfred" (von Leliwa 2010). Another example of a synaesthetic œuvre is the landscape art project "Der unsichtbare Garten" ["The invis-

Fig. 1 Das Rheingold (Virtual Realm of the Gods), Donner and Wotan, the tempest in Scene 4, Concert performance at Brucknerhaus Linz, 2004

Fig. 2 Gustav Mahler "Symphony No. 2 in C minor", II. Movement, Concert performance at Philharmonic Concert Hall of Cologne, 2006

ible garden"] (Deutsch 2010, 2011). This is, however, not only a garden for visually impaired people, but rather one in which visually impaired people can bring new sensory perceptions to the sighted and where visually impaired and sighted people can exchange experiences. To enable this to happen, every visitor must put on a blindfold before entering the garden and then be led through it under the expert guidance of a visually impaired person. The garden provides a synaesthaetic link between touching and smelling the leaves on the trees, as well as the pairing of these perceptions with the hearing of the leaves' movement: "Within it the non-visual senses (touch, smell etc.) are exposed – along a determined route – to the various young trees from the region which have been chosen for their tactile, olfactory and acoustic potential." (Deutsch 2010, p. 21; Deutsch 2011)

The idea behind the presentation of a symphonic work like Mahler's "Symphony No. 2 in C minor", also known as "Resurrection", as a live and visually interactive performance was to create a new way to present classical music in media (Deutsch 2006c, 2008; Pelzmann 2008).[1] The media concept created for this live concert, for broadcast on television, and additionally for two films (Deutsch 2006b, 2007), sought to answer the questions such as how interactive media are able to create artistic content alongside classical music and how an artistic experi-

[1] Commissioned by the WDR, West German Broadcast Corporation to mark its 50th anniversary, for both a gala concert given at the Kölner Philharmonie (Philharmonic Concert Hall of Cologne) and the live broadcast on television on the 1st of January 2006.

Fig. 3 Gustav Mahler "Symphony No. 2 in C minor", V. Movement, Concert performance at Philharmonic Concert Hall of Cologne, 2006

ence involving several senses simultaneously can be evoked in the audience.

In the following the author's path to the visualization is described in some detail from both the aesthetic and the technical point of view.

First, the Symphony had to be analyzed and interpreted. Inspired by Mahler's music, a total of eighteen three dimensional objects were developed to provide the basis of the 3-D visualization. The design of the visual images was based on the statements made in the music and the emotions released by it. Two interpretations had to be blended: one expressed musically and the other visually. In the given case the cultural-historical ambience of Gustav Mahler and his own sources of inspiration were not the subject of the visualization.

In order to understand how the authors's visualizations have been able to reach his audience, it is helpful to mention Haverkamp's commentary on the visualization of the author. Haverkamp (2010, p. 2) argues that indeed "an incalculable number of visualizations of music can be generated on the basis of computerized analysis of sound. Algorithms generate visual elements, whose features are – partly or as a whole – correlated to selected parameters of the music." But they simply fail to become an art work. Consequently, Haverkamp raises the question how the artist has been able to create not merely an "entertaining" but an "œuvre d'art". Apart from the already outlined steps which Haverkamp (2010, p. 2) termed the intentional "conscious concepts" of the artist, he also described at least three main steps of "intuitive coupling" which together enable the artist to reach the audience. All three steps are influenced by the context of perception (thus differing from genuine

Fig. 4 Gustav Mahler "Symphony No. 2 in C minor", V. Movement, Concert performance at Philharmonic Concert Hall of Cologne, 2006

synaesthesia): (i) Intermodal or cross sensory analogies, including correlations of single attributes (e. g., generic attributes such as intensity, brightness, sharpness etc.; motion; synchronicity). The concept of "intermodal analogy" was introduced by Robert Melara, an experimental psychologist, who investigated the dimensional interactions between color and pitch experimentally (Melara 1989 a, b). (ii) Iconic coupling is based on a concrete association referring to known physical object, meaning the identification of sources of stimuli. Iconic coupling refers to objects in memory and is based on learning and experience (e. g., the sound of crackling refers to the image of fire). (iii) Symbolic connections implying semantic correlations (meanings of the auditory and visual symbols as well as meanings of their connections) (Haverkamp 2010).

On an abstract level eighteen 3-D objects represent the world of the Second Symphony as a kind of virtual domain. Like the music, the objects of the visualization undergo a transformation that Mahler brings to us as a musical experience across five movements. The specific thematic junctures and stages of the five movements of the symphony – suffering, romance, irony, love, doubt, and hope – are associated with the transformation of these virtual objects. As Haverkamp (2010, p. 5) remarks commenting on the visualization of the author: "The animation refers to the composition by means of a color concept which supports narrative aspects of the music. Without losing their abstract appearance, the visual elements correlate to the thematic stages of this Symphony [. . .]". For example, the transformations occur in the first movement when the objects rise up and then

Fig. 5 "The invisible garden", location of the garden, draft 2006

break down, or in the second movement when they dance, radiate, and shine. In the third movement, irony influences the shapes and causes them to be distorted, while love causes them to glow in the fourth movement, whereas in the fifth movement follow tremors, earthquake, and destruction, with the objects bursting and triggering an exodus in the virtual world. Redemption does not come until the very end when everything shimmers, sails, and soars.

Regarding the spatial shapes created by the author, Haverkamp remarked: "Eighteen elementary forms carry the colors and are subject of various form transformations and groupings in space. Those forms were constructed physically and have been scanned into the virtual environment. The whole scenery, however, remains abstract. The only exception is given during the 4 th movement ("Urlicht" ["Primary or early light"]) where an iconic ambience of romanticism reminds to how the moon appears above a nightly landscape, roughly comparable to surreal sceneries unfolded by the painter Max Ernst." (Haverkamp 2010, pp. 5–6)

What is different in the author's visualisation of Mahler's "Symphony No. 2 in C minor" with respect to other current animations? According to Haverkamp, in "contrast to being a predetermined sequence of animation, the presentation is interactive in nature. Sounds of musical instruments directly influence the scenery by initiating synchronous light effects, object pulsation, and movement. For that purpose, sensors were attached to a variety of instruments throughout the orchestra." (Haverkamp 2010, p. 6)

As the objects evolve through the stages of the symphony, a pulsing becomes visible on their surfaces as if their hearts were beating or their lungs breathing. The degree to which the objects pulsate, as well as how and even whether their radiance visibly manifests itself, depends upon the music and its interpretation during the actual performance. The vision is real-time and interactively linked to the music, enlivening the virtual objects like a pace-maker. In this way, the visualization reflects the extreme contrasts of the dramatic and lyrical movements, ranging from playful to religious; from powerful expressions of release and joy through to profound emotional turbulence.

The objects' movements are a key element of the author's visualization, because they enhance the movements of the music by visualizing them. In order to understand why movement is so crucial for communicating music in performance and thus for really reaching the audience let me refer again to Haverkamp. In a previous paper by Haverkamp (2007) the analogy between music and motion is shown acting according to the rule that "motion is the primal element of music," a statement expressed by the recently rediscovered German musicologist Alexander Truslit in his "Gestaltung und Bewegung in der Musik" ["Shaping and motion in music"] (Truslit 1938). Truslit provided a pioneering theory of the basis of common music experience of the audience, a theory of "musical" listening of the audience – as distinct from the "musicological" listening of expert musicologists (Cook 1990) –, studying subjects' responses to music as well as their inner physiological motion (Repp

Fig. 6 "The invisible garden", leafs, summer 2010. The garden has been conceived on the model of interaction with and immersion in a virtual world and so only is apparent after specific navigation. This special garden borders on the herb garden. It is visually and hermetically sealed from the outside by a thicket-like hedge of pines. Within it, the non-visual senses (touching, smelling, hearing etc.) are exposed – on a predetermined path – to various young trees from the region which have been chosen and planted on haptic, olfactory, and acoustic criteria. Our hands (as also our hearing and our smelling) become "data gloves" and navigators through this immersion in the natural and yet unusual world of perception.
Our visually impaired museum and art guides are more experienced in navigating this difficult thicket. By exchanging experiences they help us to come to grips with this world. They also arouse our attention in an extension of our normal awareness for the specific plastic, tactile, olfactory, and acoustic qualities of the museum garden (Deutsch 2010, 2011)

1992). Truslit's assumption of the transmission of motion information from the musicians to the audience can be compared to the theory of Liberman and Mattingly for speech perception (Liberman and Mattingly 1985; cf. Repp 1992). According to Truslit, music is a biologically conditioned form of motion shaped by its inner movement. Indeed, just by means of its inner movement music enables the listener to experience what music brings to sound. Truslit assumed a close relationship between the hearing organ, the vestibular organ, and the body muscles: the auditory sensations are paired with an inner motion experience and corresponding muscular reactions (Haverkamp 2007). Truslit's proposal that the vestibulum is the organ of music motion is supported by contemporary researchers in acoustics such as Todd (1992). In general, Truslit's biological hypothesis can be compared to the neuropsychological approach of Clynes (1986; Clynes and Nettheim 1982).

The ability of the 3-D objects to change makes it possible to capture the distinctive symphonic quality of the Second Symphony. While Gustav Mahler's tonal settings are generally well known for their ideal synthesis of word and sound, the aspect that fascinates the media artist most about the Symphony is the direct translation of meaning into music. The sequential order of the music allows Mahler's statements to be understood, in a sense, word for word, and the series of emotions generated by the music thus evoke a kind of mental filmstrip. Based on this inherent potential, the 3-D space in the Philharmonic Hall was used as a visually immersive counterpoint encompassing all the senses, while radiance and darkness, majesty and turbulence pass over and through all the objects.

3.1 The technology

A complex computer cluster, designed together with the Ars Electronica Futurelab of Linz, was used to generate the real-time visualization of Mahler's Symphony No. 2. The cluster simultaneously processed and implemented any modulations picked up from the orchestra. Its 56 musical instruments were fitted with microphones and individual soundtracks and linked by sensors directly to the computer cluster providing all the nuances of the live interpretation of the music before translating these directly into variable visual elements.

The computer systems generated a virtual world in real time resulting from the synthesis of music and vision, which meant that the performers (under the direction of conductor Semyon Bychkov) interactively shaped the visual dramaturgy with their own interpretation. They could indeed influence the appearance of the objects, their movement, and their intensity of movement. The audience inside the Philharmonic Concert Hall – equipped with 3-D glasses – entered an all-embracing virtual world projected onto a curved panoramic screen. It thus had the opportunity to perceive the musical and the visual simultaneously and to see how the musician's own interpretation was visualized.

Additionally, the computing system simultaneously generated a variant of the virtual world for the live television feed using dramaturgically-driven virtual cameras. These cameras, also developed at Futurelab, opened up the digitally-designed virtual world to the television viewer by moving independently through the artificial environment according to choreographed criteria. In addition to the real cameras located in the Philharmonic Concert Hall, this gave the television production an innovative and cinematic dimension that in terms of content was closely linked to the music.

Information systems computed when, where, and from what perspective the interactive fusion of music and vision could best be achieved. In this way, a TV variation of the visualization could be broadcast live.

The computer also supplied clips of the projection adapted to the 16:9 TV format, giving them an exceptional depth. With the help of 3-D glasses this effect was further enhanced.

The special achievement of the WDR production was to assemble all three visual levels – the real visual level in the Philharmonic Concert Hall in Cologne, the virtual camera tracking shots, and the clip level. This meant that the television audience had the opportunity to visualize the Symphony from several different angles.[2]

2 At the director's console, the four direct TV feeds of the visualization were mixed live with images captured by TV cameras set up in the Philharmonic Concert Hall to provide an exceptional multi-layered visual experience. This particular visual experience has been published as the first DVD "Vision Mahler. Artist's edition – Johannes Deutsch 1" (2006 b).

Conclusion

Has cross sensory coupling to be considered as a common type of neuronal processing in both synaesthetes and average population? Are intermodal associations immediately available to consciousness for synaesthetes only? The answer is: possibly yes. Another question is: Do the results of scientific research provide evidence that non-synaesthetes show fixed patterns of coupled association as well? Some data seems to corroborate this assumption. Simner and colleagues showed that in grapheme-colors synaesthesia both synaesthetes and non-synaesthetes prefer certain pairings of letters and colors: "A" to red, "S" to yellow, "X" to black (Simner et al. 2005). According to recent studies synaesthetes tend to see lighter colors when exposed to increasingly higher frequency sounds and the same pairing between luminance and pitch was found in average people making intuitive music-color associations (Ward et al. 2006).

In art, cross sensory coupling seems not only to be a common form of processing sensory information, but something that artists intentionally employ as a tool. A key-question then obviously refers to the involuntary, automatic coupling. Do artists show fixed patterns of paired associations? Rimski-Korsakov employed F frequently for his pastoral music, because it recalls green leaves and grass, maintaining that his own associations rooted in his biography. For Kandinsky yellow was the colour of middle-C on a piano (Kandinsky 1994). On the other side, Baudelaire pleaded for the intentionality as well as the variability of the correspondences in poetry and art. Rimbaud explicitly strengthened this program in his literary work. In music, Rimington constructed mathematical-physical correlations coupling C, the lowest-pitched tone, with red, the color resulting from the lowest frequency (longest wavelength) light (Rimington 1911). Finally, Hermann Nitsch also strongly assumed intentional construction in visual art. Does the complex aesthetic and technical conception of contemporary art, especially of interactive media art, not provide evidence enough that art has little to do with involuntariness?

Knowing that genuine and spontaneous synaesthesia is far more common than previously assumed does not help to simply link the involuntary neurological condition to the production of art. However, on the other side, it could allow the formulation of a tentative hypothesis about the experience of the audience.

References

Anschütz G (1925) Untersuchungen zur Analyse musikalischer Photismen. Arch Gesamte Psychol 51: 155–218

Anschütz G (1926) Untersuchungen über komplexe musikalische Synopsie. Arch Gesamte Psychol 54: 129–273

Anschütz G (ed) (1927) Farbe-Ton-Forschungen. vol 1. Akad Verlagsges, Leipzig

Anschütz G (ed) (1931) Farbe-Ton-Forschungen. vol 3. Bericht über den 2. Kongress für Farbe-Ton-Forschung. Verlag der Psychologisch-ästhetischen Forschungsges, Hamburg

Anschütz G (ed) (1936) Farbe-Ton-Forschungen. vol 2. Verlag der Psychologisch-ästhetischen Forschungsges, Hamburg [published after vol 3]

Asher J E, Aitken M R F, Farooqi N, Kurmani S, Baron-Cohen S (2006) Diagnosing and phenotyping visual synaesthesia: a preliminary evaluation of the revised Test of Genuineness (TOG-R). Cortex 42: 137–146

Asher J E, Lamb J A, Brocklebank D, Cazier J B, Maestrini E, Addis I, Mallika S, Baron-Cohen S, Monaco A P (2009) A whole–genome scan and fine-mapping linkage study of auditory-visual synaesthesia reveals evidence of linkage to chromosomes 2q24, 5q33, 6p12 and 12p12. Am J Human Genet 84: 279–285

Aristotle (1936) On sense and sensible objects. In: Goold GP (ed) Aristotle in twenty-three volumes, vol 8. Harvard Univ Press, Cambridge MA, pp. 205–283

Baker J M (1986) The music of Alexander Scriabin. Yale Univ Press, New Haven, London

Baratoux J (1888) De l'audition colorée. Chaix, Paris

Barnett K J, Newell F N (2008) Synaesthesia is associated with enhanced, self-rated visual imagery. Conscious Cogn 17: 1032–1039

Baron-Cohen S (1994) Synaesthesia: an account of colored hearing. Leonardo 27: 343–346

Baron-Cohen S, Burt L, Smith-Laittan F, Harrison J, Bolton P (1996) Synaesthesia: prevalence and familiarity. Perception 25: 1073–1079

Baron-Cohen S, Harrison J (eds) (1997) Synaesthesia: classic and contemporary readings. Blackwell Publishers, Cambridge MA

Baron-Cohen S, Harrison J, Goldstein L H, Wyke M A (1993) Coloured speech-perception: is synaesthesia what happens when modularity breaks down? Perception 22: 419–426

Baron-Cohen S, Wyke M A, Binnie C (1987) Hearing words and seeing colors: an experimental investigation of a case of synaesthesia. Perception 16: 761–767

Baudelaire C (1993 [1857]) The flowers of evil. Translated by McGowan J and with an introd by Culler J. Oxford Univ Press, Oxford, New York

Beaunis H, Binet A (1892) Sur deux cas d'audition colorée. Rev Philos 33: 448–461

Behne K – E (ed) (1987) Film-Musik-Video oder Die Konkurrenz von Auge und Ohr. Gustav Bosse Verlag, Regensburg

Bidaine P (ed) (2004) Sons & Lumières. Une histoire du son dans l´art du XXe siècle. Éditions du Centre Pompidou, Paris

Binder T (2004) Auf Entdeckungsreise: „Das Rheingold" als virtuelle, dreidimensionale Götterwelt. Interview mit Johannes Deutsch. In: Das Rheingold. Oper visualisiert. Programmheft Brucknerfest Linz 2004. Publisher Brücknerhaus, Linz, pp 4–8

Binet A (1892) Le problème de l'audition colorée. Rev Deux Mondes 63: 586–607

Binet A, Philippe J (1892) Étude sur un nouveau cas d'audition colorée. Rev Philos 33: 461–464

Bleuler E (1913) Zur Theorie der Sekundärempfindungen. Z Psychol 65: 321–401

Bleuler E, Lehman K (1881) Zwangsmäßige Lichtempfindungen durch Schall. Fues's Verlag, Leipzig

Bos M C (1929) Über echte und unechte audition colorée. Z Psychol 111: 321–401

Castel L B (1725) Clavecin pour les yeux. Mercure de France. November: 2552–2577

Castel L B (1726a) Démonstration géométrique du clavecin pour les yeux et pour tous les sens. Mercure de France. February: 277–292

Castel L B (1726b) Difficultés sur le clavecin oculaire, avec leurs réponses. Mercure de France. March: 455–465

Castel L B (1751) Du clavecin oculaire. Mercure de France. Dec: 7–21

Chabalier C (1864) De la pseudochromesthésie. J Méd Lyon 2: 92–102

Claparède E (1900) Sur l'audition colorée. Rev Philos 49: 515–521

Clavière J (1898) L'audition colorée. Année Psychol 5: 161–178

Clynes M (1986) Music beyond the score. Communication and Cognition 19: 169–194

Clynes M, Nettheim N (1982) The living quality of music: neurobiological patterns of communicating feeling. In: Clynes M (ed) Music, mind, and brain. Plenum Press, New York, pp 47–82

Cook N (1990) Music, imagination, and culture. Clarendon Press, Oxford

Cornaz E (1851) De l'hyperchromatopsie. Ann Ocul (Paris) 25: 3–9

Cortex (2006) Cognitive neuroscience perspectives on synaesthesia (special issue) 42/2

Cortex (2009) Synaesthesia and visuo-spatial forms (special section) 45/10: 1190–1277

Cytowic R E (1989) Synaesthesia: a union of the senses. 1st edn. Springer, New York

Cytowic R E (2002) Synaesthesia: a union of the senses. 2nd edn. MIT Press, Cambridge, MA

Cytowic R E, Eagleman D M (2009) Wednesday is indigo blue: discovering the brain of synaesthesia. MIT Press, Cambridge, MA

Daubresse M (1900) L'audition colorée. Rev Philos 49: 300–306

Day S A (2007) Demographic data http://home comecast.net/-sean.day

Deutsch J (2002) Gesichtsraum – CAVE. In: Unplugged – Art as the scene of global conflicts. Ars Electronica 2002. Hatje Cantz Verlag, Ostfildern-Ruit, p 347

Deutsch J (2004) Das Rheingold – virtual realm of the gods. In: Timeshift – the world in twenty-five years. Ars Electronica 2004. Hatje Cantz Verlag, Ostfildern-Ruit, pp 219–221

Deutsch J (2006a) Das Rheingold – virtuelle Götterwelt. In: Louis E, Stoos T (eds) Kunst auf der Bühne – Art on stage. Les grands spectacles II. Publisher Museum der Moderne Salzburg, Salzburg, pp 144–146

Deutsch J (2006b) Vision Mahler. Artist's edition – Johannes Deutsch 1. DVD. Art Haus Musik, Leipzig, Berlin

Deutsch J (2006c) Vision Mahler. In: Simplicity – the art of complexity. Ars Electronica 2006. Hatje Cantz Verlag, Ostfildern-Ruit, pp 296–299

Deutsch J (2007) Vision Mahler. Artist's edition – Johannes Deutsch 2. DVD. Art Haus Musik, Leipzig, Berlin

Deutsch J (2008) Vision Mahler, interactive visualization. In: Migrating reality. Publisher KHM Kunsthochschule für Medien, Köln, Berlin, Vilnius, pp 54–60

Deutsch J (2010) Der unsichtbare Garten 2007. In: Braunsteiner M, Peters C (eds) Play Admont. Stift Admont. Bibliothek der Provinz, Weitra, pp 22–23

Deutsch J (2011) www. johannes-deutsch. at
Dixon M J, Smilek D, Merikle P M (2004) Not all synaesthetes are created equal: projector versus associator synaesthetes. Cogn Affect Behav Neurosci 4: 335–343
Dixon M J, Smilek D (2005) The importance of individual difference in grapheme-color synaesthesia. Neuron 45: 821–823
Dixon M J, Smilek D, Duffy P L, Zanna M P, Merikle P M (2006) The role of meaning in in grapheme-color synaesthesia. Cortex 42: 243–252
Düchting H (1996). Farbe am Bauhaus. Synthese und Synästhesie. Neue Bauhausbücher. Gebrüder Mann, Berlin
Eagleman D M, Kagan A D, Nelson S S, Sagaram D, Sarma A K (2007) A standardized test battery for the study of synaesthesia. J Neurosci Methods 159: 139–145
Emrich H M, Schneider U, Zedler M (eds) (2002) Welche Farbe hat der Montag? Synästhesie: das Leben mit verknüpften Sinnen. S. Hirzel Verlag, Stuttgart, Leipzig
Fechner G T (1898) Vorschule der Ästhetik. 2 vol. (2nd edn) Breitkopf and Härtel, Leipzig
Field G (1835) Chromatography; or, a treatise on colours and pigments and of their powers in painting. Charles Tilt, London
Fischer E (ed) (2008) Sound: Frame, Festival zur Visualisierung elektronischer Musik. Czernin Verlag, Wien
Flournoy T (1892) Enquête sur l'audition colorée. Extrait des archives des sciences physiques et naturelles, troisième période, Genève 28
Flournoy T (1893) Des phénomènes de synopsie. Alcan, Paris and Charles Eggimann & Cie, Genève
Frith C D (2001) Brain imaging psychological disorders. In: Underwood G (ed) The Oxford guide to the mind. Oxford Univ Press, Oxford, pp 154–159
Galton F (1880a) Visualised numerals. Nature 21: 494–495
Galton F (1880b) Visualised numerals. J R Anthropol Instit 10: 85–102
Galton F (1880c) Statistics on mental imagery. Mind 5: 301–318
Galton F (1883) Inquiries into human faculty and its development. E P Dutton, New York
Grossenbacher P G, Lovelace C T (2001) Mechanisms of synaesthesia: cognitive and physiological constraints. Trends Cogn Sci 5: 36–41
Gruber E (1893) Questionnaire psychologique sur l'audition colorée, figurée et illuminée. Rev Philos 35: 499–502
Hay D R (1836) The laws of harmonious colouring, adapted to interior decorations, manufactures, and other useful purposes. 3rd edn. W & R Chambers, Edinburgh
Haverkamp M (2007) Bewegung als Ursprung der Musik. Alexander Truslits frühe Studien zur Musikwahrnehmung. In: Forschritte der Akustik. Plenarvorträge und Fachbeiträge der 33. Deutschen Jahrestagung für Akustik DEGA. Publisher Dtsch Ges f Akustik DEGA, pp 557–558
Haverkamp M (2009a) Synästhetisches Design, kreative Produktentwicklung für alle Sinne. Hanser Verlag, München
Haverkamp M (2009b) Look at that sound! Visual aspects of auditory perception. Paper at the 3rd Congreso intyernacional de sinestesia, cienncia y arte, Grenada
Haverkamp M (2010) Visualizing auditory perception: correlations, concepts, synaesthesia. Paper at the Galayev readings 2.–5. 10. 2010, Prometheus-Institute, Kazan, Russia
Henri V (1893) Notes sur un cas d'audition colorée. Rev Philos 35: 554–558
Hupé J-M (2010) Psychophysics and fMRI of grapheme-color synaesthesia. Paper at the UKSA [United Kingdom Synaesthesia Association] Congress, March 2010, University of Sussex, Brighton
Hupé J-M, Bordier C, Dojat M (2010) Colors in the brain and synaesthesia. Abstract. Paper at the 33th ECVP [European Conference on Visual Perception] August 2010, Lausanne, Switzerland. Perception 39 (Supplement): 159
Hubbard E M, Ramachandran V S (2005) Neurocognitive mechanisms of synaesthesia. Neuron 48: 509–520
Jewanski J (1997) Die Farblichtmusik Alexander Lászlós. Z Kunstgesch 1: 12–43
Jewanski J (1999) Ist C = Rot? Eine Kultur- und Wissenschaftsgeschichte zum Problem der wechselseitigen Beziehung zwischen Ton und Farbe: von Aristoteles bis Goethe. Berliner Musik Studien 17. Studio Schewe Verlag, Sinzig
Johnson B D (1975) Messiaen. Univ California Press, Berkeley, Los Angeles
Kandinsky W (2001 [1911]) Concerning the spiritual in art. In: Glew A (ed) MFA Publications, Tate Publishing, London New York
Kandinsky W (1994) Complete writings on art. In: Lindsay K C, Vergo P (eds) Da Capo Press, Cambridge, MA

László A (1925) Die Farbenlichtmusik. Breitkopf & Härtel, Leipzig

Lauret J, Duchaussoy L (1887) Un cas héréditaire d'audition colorée. Rev Philos 23: 222–224

Lemaitre A (1904) Un cas d'audition colorée hallucinatoire. Arch Psychol (Genève) 3: 164–177

Lemke H (2006) Das Manifest der futuristischen Koch-Kunst. Annäherungen an die Eat Art. In: recenseo, Texte zu Kunst und Philosophie, Saarbrücken http://www. recenseo. de/index. php?id =91&kategorie=artikel&nav=Inhalt

Liberman A M, Mattingly I G (1985) The motor theory of speech revisited. Cognition 21: 1–36

Ligeti G (1958) Artikulation. Electronic composition. Hörpartitur by Wehinger R (1970). Schott's Söhne, Mainz

Ligeti G (1983) Ligeti in conversation. Eulenburg Books, London

Maresch P (2010) Conversation with Johannes Deutsch on 16 th January 2003. In: Hattinger G, Bogner P (eds) Space inventions. Der künstliche Raum. Schlebrügge, Wien, pp 128–131

Marinetti F T (1989 [1932]) The futurist cookbook. Trans by Brill S and with an intr by Chamberlein L. Bedford Art Publisher, San Francisco

Marinetti F T, Fillia L E C (1930) Manifesto della cucina futurista [Manifesto of the futurist cuisine]. Gazzetta del Popolo, Torino, 28. 12. 1930

Melara R D (1989a) Dimensional interaction between color and pitch. J Exp Psychol Hum Percept Perform 15/1: 69–79

Melara R D (1989b) Dimensional interaction between color and pitch. J Exp Psychol Hum Percept Perform 15/2: 212–231

Messiaen O (1948) L'ascension: quatre méditations symphoniques pour orchestre. Alphonse Leduc, Paris

Metzger J (2010) Synästhesie: Ich sehe etwas, was du nicht siehst. Psychologie heute 37: 44–49

Moritz W (1987) Der Traum von der Farbmusik. In: Bódy V, Weibel P (eds) Clip, Klapp, Bum. Von der visuellen Musik zum Musikvideo. DuMont Verlag, Köln, pp 17–51

Moritz W (1997) The dream of color music, and machines that made it possible. Animation World Magazine 2/1 April 1997 [Internet Journal]

Moritz W (2006) Der Traum, Musik sichtbar zu machen. In: Vision Mahler. Programmheft zum Festkonzert zum 50-Jahr Jubiläum des Westdeutschen Rundfunk WDR, Köln, pp 28–29

Newton I (1704) Opticks or a treatise of the reflections, refractions, inflections and colours of light. Smith and Walford Printers to the Royal Society, London

Nichols R (1896) Olivier Messiaen. 2nd edn. Oxford University Press, Oxford, New York

Nitsch H (1995) Vom Orgien Mysterien Theater zur Operninszenierung. In: Massenet – Hérodiade. Programmheft der Wiener Staatsoper, Wien

Nunn J A, Gregory L J, Brammer M, Williams S C R, Parslow D M, Morgan M J, Morris R G, Bullmore E T, Baron-Cohen S, Gray J A (2002) Functional magnetic resonance imaging of synaesthesia: activation of V4/V8 by spoken words. Nat Neurosci 5: 371–375

Peacock K (1988) Instruments to perform color-music: two centuries of technological experimentation. Leonardo 21: 397–406

Pelzmann B (2008) Vision Mahler – Bilder einer Sinfonie. EIKON Internat Z Photographie & Medienkunst 61: 71–73

Pfister O (1912) Die Ursache der Farbenbegleitung bei akustischer Wahrnehmungen und das Wesen anderer Synästhesien. Imago 1: 265–275

Philippe J (1893) Résumée d'une observation d'audition colorée. Rev Philos 36: 330–334

Prampolini E (1913) Gazzetta Ferrarese, 26th August 1913, Ferrara, Italy

Ramachandran V S, Hubbard E M (2001a) Psychophysical investigations into the neural basis of synaesthesia. Proc R Soc Lond B Biol Sci 268: 979–983

Ramachandran V S, Hubbard E M (2001b) Synaesthesia: a window into perception, thought and language. JCS 8: 3–34

Repp B H (1992) Music as motion: a synopsis of Alexander Truslit's (1938) „Gestaltung und Bewegung in der Musik". Haskins Laboratories Status Report on Speech Research 111/112: 265–278

Riccò D (1999) Sinestesie per il design. Le interazioni sensoriali nell'epoca dei multimedia. Etas, Milano

Riccò D (2008) Sentire il design. Carrocci, Roma

Rimbaud A (2001 [1871]) Vowels. In: Sorrell M (ed) Arthur Rimbaud. Collected Poems. Transl, with an introd and notes by Sorrell M. Oxford University Press: Oxford, New York, pp 134–135

Rimbaud A (2001 [1873]) Alchemy of the word. In Sorrell M (ed) Arthur Rimbaud. Collected Poems. Translated, with an introd and notes by Sorrell M. Oxford Univ Press, Oxford, New York, pp 234–237

Rimington, A W (1911) Color-music. The art of mobile color. Hutchison & Co, London

Samuel C (1994) Olivier Messiaen: music and color. Conversations with Claude Samuel. 2nd edn. Amadeus Press, Portland, Oregon

Scheel S (2006) Music visualization – the interplay of color and sound. In: Simplicity – the art of complexity. Ars Electronica 2006. Hatje Cantz Verlag, Ostfildern-Ruit, pp 273–289

Shakespeare W (1926 [1594–96]) A midsummer night's dream. In: Quiller-Couch A, Wilson J D (eds) The works of Shakespeare. Cambridge Univ Press, Cambridge

Scriabin A (1995 [1911]) Poem of ecstasy. Prometheus: poem of fire. Dover, New York

Scriabin A (1911) Prometheus. Poem of fire. At the Tonhalle Düsseldorf: http://www. youtube. com/ watch?v=9CruEBhdxJE

Simner J (2006) Beyond perception: synaesthesia as a psycholinguistic phenomenon. Trends Cogn Sci 11: 23–29

Simner J (2009) Synaesthetic visuo-spatial forms: viewing sequence in space. Cortex 45: 1138–1147

Simner J, Ward J, Lanz M, Jansari A, Noonan K, Glover L, Oakley D A (2005) Non-random associations of graphemes to colors in synaesthetic and non-synaesthetic populations. Cogn Neuropsychol 22: 1069–1085

Simner J, Ward J (2006) The taste of words on the tip of the tongue. Nature 444: 438

Simner J, Harold J, Creed H, Monro L, Foulkes L (2009) Early detection of markers for synaesthesia in childhood populations. Brain 132: 57–64

Sokolov P (1901) L'individuation colorée. Rev Philos 51: 36–46

Suarez de Mendoza F (1890) L'audition colorée. Octave Doin, Paris

Todd N P (1992) The dynamic of dynamics: a model of musical expression. J Acoustic Soc Am 91: 3540–3550

Topper D (1999) Newton and the number of colors in the spectrum. Stud Hist Philos Sci 21: 269–279

Trenks B (2005) 4.5 Johannes Deutsch. Exemplarische Analyse der Lichtgestaltung eines Bühnenwerkes (Richard Wagners „Das Rheingold"). In: Inszeniertes Licht im Bühnenraum. Diplomarbeit. Akademie der Bildenden Künste, Wien, pp 78–86

Truslit A (1938) Gestaltung und Bewegung in der Musik. C F Vieweg Verlag, Munich

Ulrich A (1903) Phénomènes de synesthésies chez un épileptique. Rev Philos 56: 181–187

von Hug-Hellmuth H (1912) Über Farbenhören. Imago 1: 228–264

von Leliwa E (2010) Theater im Kopf. Interview mit Regisseur Johannes Deutsch. In: Robert Schumann. Manfred. Programmheft der Tonhalle Düsseldorf, Düsseldorf, pp 4–17

Walz S B (2008) 4. Das Rheingold. In: Mediale Analogien von Musik und Bild. Theatrale Abstraktionen in Musikinszenierungen. Diplomarbeit. Ludwig-Maximilians-Univ, München, pp 60–78

Ward J, Huckstep B, Tsakanikos E (2006) Sound-color synaesthesia: to what extent does it use cross-modal mechanisms common to us all? Cortex 42: 264- 280

Weibel P (1987) Von der visuellen Musik zum Musikvideo. In: Bódy V, Weibel P (eds) Clip, Klapp, Bum. Von der visuellen Musik zum Musikvideo. DuMont, Köln, pp 53–163

Wellek A (1927) Die Farbe-Ton-Forschung und ihr erster Kongress. Z Mw 9: 576–584

Wellek A (1931 a) Zur Geschichte und Kritik der Synästhesie-Forschung. Arch Gesamte Psychol 79: 325–384

Wellek A (1931 b) Zur Entwicklung unserer Notenschrift aus der Synopsie. In: Anschütz G (ed) Farbe-Ton-Forschungen. vol 3. Bericht über den 2. Kongress für Farbe-Ton-Forschung. Verlag der Psychologisch-ästhetischen Forschungsgesellschaft, Hamburg, pp 143–153

Winkler W (2005) Visualized music – an old theme, a new theme. In: Hybrid – living in paradox. Ars Electronica 2005. Hatje Cantz Verlag, Ostfildern-Ruit, pp 325–328

Winkler W (2006) The visualization of music. In: Simplicity – the art of complexity. Ars Electronica 2006. Hatje Cantz Verlag, Ostfildern-Ruit, pp 300–301

Zuckriegl M (2010) An image of space. In: Hattinger G, Bogner P (eds) Space inventions. Der künstliche Raum. Schlebrügge, Wien, pp 132–139

Perception and memory: conscious and unconscious processes

IV

Introductory remarks

Section IV is devoted to perception and memory as both conscious and unconscious processes. The following chapters outline the controversial early investigations in this field and focus on current interdisciplinary research. Since the end of the 18th Century (1775) clinical investigations in dynamic psychiatry, psychosomatics, and finally psychoanalysis have explored unconscious processes (Ellenberger 1970), whereas through the centuries philosophers only sporadically adventured to gain deeper insight into non-conscious states of mind (White 1960). For most of the last century, scientists tended to avoid this topic as well. Only recently disciplinary barriers and mutual distrust have begun to decrease.

After years of neglect by neurobiology and cognitive psychology, a shared commitment to study consciousness and the unconscious emerges. As George Mandler (2003, p. 15) remarks: "Freud, in particular, has contributed much to the finer distinctions among shadings of the unconscious. However, if we are to make a fresh start within the experimental investigation of consciousness, we shall probably also have to rediscover these distinctions within the new realm of discourse." Furthermore, analytic philosophy, which had mainly been committed to linguistic analysis before, has made a "mentalistic turn" and evolved into philosophy of mind. Philosophy of mind strives to integrate data on the mind derived from experimental cognition research (cognitive science and neurobiology) into its study of consciousness. The relevance of unconscious processes for an understanding of consciousness has been seen as well. In Chapter VI, 19 Gerhard Roth outlines these recent developments in philosophy.

The neurobiologist Eric Kandel calls upon psychoanalysis to cooperate with biology, especially with cognitive neuroscience. He offers multiple examples providing evidence why a closer relationship seems promising. In its 100-year history, a key tenet of psychoanalysis has been the idea that we are unaware of much of our mental life, including perceptions, thoughts, and memories. According to Kandel (2005, pp. 70–73) biology should join the psychoanalytic efforts to understand unconscious psychic processes as well as consciousness. The chapters of Section IV of the present volume show the relevance of such a request.

1. Consciousness and the unconscious

In her contribution *"Perception, consciuos and unconscious processes"* Patrizia GIAMPIERI-DEUTSCH introduces Freud's conception of the mind focusing on his understanding of consciousness and relating it to current debates on this topic in philosophy and cognition research.

Like in Freud's original work, perception and consciousness have often been conflated in philosophy as well. However, current philosophers of mind learn from neuropsychological studies of brain-damaged patients and now question the assumption that perception and consciousness of perception are always inseparable. Some neuropsychological syndromes involve dissociations between perception (mostly vision) and awareness of perception. Examples are blindsight, covert recognition of faces in prosopagnosia, unconscious perception of neglected stimuli, and implicit reading in alexia[1]. All these brain-damaged patients possess a "covert knowledge", which means access to knowledge of something without being aware of it. However, their non-conscious vision is merely latently rather than dynamically unconscious. The latent unconscious activity functions more or less as conscious activities do, however it lacks awareness. The dynamic unconscious, on the other hand, is psychological, it is active, and it can be different in character from conscious psychological processes (cf. also below). For this reason subliminality research raises additional questions for cognitive psychology with regard to the understanding of dynamic unconscious processes. Whereas Freud tended to conflate perception and consciousness, psychoanalysis later abandoned his point of view acknowledging unconscious perception. It now investigates subliminal perception by incorporating the methodologies and concepts of neighbouring disciplines (cognitive psychology as well as neurobiology). Howard Shevrin has been testing psychoanalytic assumptions about the unconscious by way of subliminally presented stimuli (Shevrin 2004, 2005). Eric Kandel refers to "the important and long-standing tradition of work by Howard Shevrin, [that] correlate[s] the perception of subliminal and supraliminal stimuli with event-related potentials in the brain in an attempt to analyze aspects of unconscious mental processes". Kandel holds that investigations on the mind, following the seminal work of Shevrin, "need to develop creative ways of studying subjective phenomena" (Kandel 2005, p. 96). Shevrin's subliminality experiments show that the dynamic unconscious clinically described by psychodynamic psychiatry and psychoanalysis is in fact experimentally observable through controlled investigations into unconscious conflict, unconscious affect, and unconscious anxiety. Howard Shevrin provides a definition of the dynamic unconscious: "It is *psychological*, it is *active*, and it can be *different* in character from conscious psychological processes." (Shevrin and Dickman 2003, p. 542) (i) The dynamic

---

1 "Blindsight" patients have blind areas in their visual fields, but they demonstrate certain visual abilities when required to guess in indirect testing. Despite their residual abilities the patients say that they do not see anything. "Prosopagnosics" report to be unable to recognize even the faces of their own relatives. However, indirect tests show evidence of "covert knowledge" of their facial expressions. Patients with "neglect" cannot report on the stimuli occurring on the contralesional side of space: even though their behavior suggests that they do not perceive them, they have non-conscious perception of neglected stimuli. Finally, "alexia" patients can no longer read words at a glance, but experiments provide evidence that they can read implicitly.

unconscious is *psychological*, meaning that the descriptive terminology used for conscious processes can be used for unconscious processes: perception, thought, affect, motivation, etc. Not only conscious processes, but also unconscious processes are correlated with brain processes. (ii) The dynamic unconscious is *active*, meaning that unconscious processes have a bearing on behavior and experience, even though the subject may be unaware of this. (iii) The dynamic unconscious can be *different* to conscious processes and follows different principles of organization than those which characterize processes occurring during the state of consciousness.

2. Consciousness

Consciousness has become a major focus for research in science. According to Howard SHEVRIN, the recent rediscovery of consciousness leads to both problems as well as new opportunities for bringing together cognitive science, neurophysiology, and psychoanalysis. In his contribution *"Consciousness, states of consciousness, unconscious psychological processes, and psychological states"*, *Shevrin* addresses four questions: i) Is there an ambiguity in the current use of the term consciousness? ii) How are mental contents related to mental processes? iii) How are consciousness, states of consciousness, and unconscious processes related to each other? iv) How are consciousness, mind, brain, and person related to each other?

i) In order to overcome current confusion in the use of the term "consciousness", Shevrin proposes to unambiguously identify consciousness as experience, distinguishing it from consciousness as system. *Shevrin's* suggestion converges on similar proposals of philosophers of mind such as Ned Block (1995), or David Chalmers (1995, 2007).

ii) Consciousness should not be considered to be made up of mental contents but of mental processes. When a person perceives an object, he or she is experiencing a process which has as its external referent an actual stable object. The referent (external object) is not to be confused with its internal representation which is usually subject to far more variability and instability than the referred-to-object. According to *Shevrin*, once mental contents are redefined as experienced processes, or internal representations of external referred-to-objects, much greater conceptual flexibility is made possible so that a much broader range of phenomena can be explained.

iii) Conscious processes are one important subset of processes constituting psychological states, which also include unconscious psychological processes. The distinction between consciousness as experience and consciousness as system allows conceptualizing how consciousness, states of consciousness and unconscious processes are related to each other. *Shevrin* points out that there is always a preceding unconscious phase and some hypothetical conditions that determine the particular emergence into conscious experience.

iv) If the term "consciousness" were restricted to its meaning as experience, it would make for a clearer discourse and research so that "psychological state" would then become the superordinate hypothetical structure. The notion of "person" within the proposed frame of reference, is a higher order construct referring to the necessarily unique organization of psychological states characterizing each individual with its enacting and adaptive capacities primarily in mind.

According to *Shevrin*, his proposals allow an interdisciplinary cooperation between cognitive science, neurophysiology, and psychoanalysis on the investigation of consciousness.

3. Explicit and implicit memory

The explicit-implicit distinction in memory involves not one but a complex of phenomena. From a more general point of view, implicit memory shows the non-conscious influence of a past event on further performance: "Implicit memory occurs when subjects are exposed to a target information and are later influenced by that information even when they cannot recollect it consciously." (Schacter and Curran 2000, p. 829) In his chapter "*Explicit and implicit memory*" Matthew Hugh ERDELYI, a psychologist studying cognition and deeply committed to research on subliminality, focuses on the distinction between explicit and implicit memory. He remarks that the contraposition explicit-implicit has become ubiquitous in psychology, such as conscious-unconscious, intentional-unintentional, direct-indirect. According to the author such distinctions are not always accurate enough. Therefore their justification is an issue. The polarities addressed nevertheless point to classic phenomena investigated in psychology and neuroscience. Sigmund Freud's neuroscience monograph "On aphasia" (1891) prefigures crucial issues in cognitive psychology: The consciously inaccessible may be available unconsciously; unconscious contents may be reflected indirectly (through symptoms, associations, behavior) and may be recovered into consciousness; twilight states (dreams, free-associations, art) may be indirect indicators of unconscious content. Experiments back to Otto Pötzl and Charles Fisher show that dreams and fantasy can yield reliable and valid measures of inaccessible content and, also, that Freud's manifest-latent distinction is scientifically viable. Recent psychophysical measurements of sensitivity (e. g., "d'" in *Erdelyi's* chapter: the chance-limen or objective threshold) have indeed emerged from studies of latent-content sensitivity. This research reveals individual differences among subjects in sensitivity to latent contents. Experimental work also shows that memory waxes and wanes over time, suggesting that observed dissociations between indicators of memory may also increase or decrease, or even reverse, over time.

According to *Erdelyi*, memory is both local and provisional. It is local because any particular measure of memory, even if reliable internally, may not be in agreement with other measures of memory; a direct, explicit test may yield an index of memory drastically different from that shown by an implicit or indirect test. Memory is also provisional in that it varies over time. Additionally, memory reflects an interaction between reminiscence, an up-tendency, and oblivescence, a down-tendency. One does have some psychological control over these functions, i. e. an ability to modulate explicit memory over time without an experimenter's intervention. One may engage some of these controls for defensive or for cognitive reasons. *Erdelyi* finally remarks that accessibility is not unconscious or conscious, implicit or explicit, subliminal or supraliminal, but varying up and down over time. **P G-D**

References

Block N (1995) On a confusion about a function of consciousness. Behav Brain Sci 18: 227–247 [reprint in: Block N, Flanagan O, Güzeldere G (eds) The nature of consciousness. Philosophical debates. MIT Press, Cambridge, MA 1997, pp 375–415]

Chalmers D (1996) The conscious mind. Oxford University Press, Oxford

Chalmers D (2007) The hard problem of consciousness. In: Velmans M, Schneider S (eds) The Blackwell companion to consciousness. Blackwell Publishing, Malden MA Oxford, pp 225–235

Ellenberger H F (1970) The discovery of the unconscious. The history and evolution of dynamic psychiatry. Basic Books, New York

Freud S (1891b) On aphasia. In: SE vol 14, pp 206–215 [excerpt]

Kandel E R (2005) Biology and the future of psychoanalysis. In: Psychiatry, psychoanalysis and the new biology of the mind. American Psychiatric Publishing, Washington, DC, London, pp 63–106

Mandler G (2003) Consciousness: respectable, useful, and probably necessary. In: Baars BJ, Banks W P, Newman J B (eds) Essential sources in the scientific study of consciousness. MIT Press, Cambridge, MA, pp 15–33

Schachter D L, Curran T (2000) Memory without remembering and remembering without memory: implicit and false memories. In: Gazzaniga M S (ed) The new cognitive neurosciences. 2nd edn. MIT Press, Cambridge, MA, pp 829–840

Shevrin H (2004) Die experimentelle Untersuchung von unbewusstem Konflikt, unbewusstem Affekt und unbewusster Signalangst. In: Giampieri-Deutsch P (ed) Psychoanalyse im Dialog der Wissenschaften. Anglo-amerikanische Perspektiven, vol 2. Kohlhammer, Stuttgart, pp 114–142

Shevrin H (2005) Toward a theory on consciousness based on recent developments in subliminal research. In: Giampieri-Deutsch P (ed) Psychoanalysis as an empirical, interdisciplinary science. Verlag Österr Akad Wiss, Wien, pp 57–74

Shevrin H, Dickman S (2003) The psychological unconscious: a necessary assumption for all psychological theory? In: Baars BJ, Banks WP, Newman JB (eds) Essential sources in the scientific study of consciousness. MIT Press, Cambridge, MA pp 541–538

White L L (1960) The unconscious before Freud. Basic Books, New York

Perception, conscious and unconscious processes

14

Patrizia Giampieri-Deutsch

Contents

Abstract

This chapter outlines Freud's conceptions of the mind and focuses on his understanding of consciousness and perception and of the unconscious mind. It also considers and discusses current debates on this topic in philosophy and cognitive science.

In philosophy perception and consciousness have often been conflated. In the wake of neuropsychological studies of brain-damaged patients and subliminality studies, the assumption that perception and consciousness of perception are always inseparable had to be revised. Several neuropsychological syndromes clearly involve dissociations between perception (mostly vision) and awareness of perception, examples of which are blindsight, covert recognition of faces in prosopagnosia, unconscious perception of neglected stimuli, and implicit reading in alexia. In the case of these examples, all of the respective brain-damaged patients are processing visual information without the conscious experience of vision. Accordingly, after having successfully completed an experimental task by indirect testing they each claim to be merely "guessing". However, their non-conscious vision is latently rather than dynamically unconscious. Although the latent unconscious activity functions more or less like conscious activities do, it lacks awareness (for example, a latent unconscious perception is a non-conscious or a weak form of conscious perception; the dynamic unconscious is psychological, active, and can be different in character from conscious psychological process-

Patrizia Giampieri-Deutsch
University of Vienna
Institute of Philosophy
Universitätsstr. 7, room C0215
1010 Vienna, Austria
e-mail: patrizia.giampieri-deutsch@univie.ac.at

es). The impact of the dynamic unconscious on human mind and behavior is one of the main tenets of Freud's theory of the unconscious. That is why subliminality studies raises additional challenging questions for cognitive neurobiology and cognitive psychology with regard to the understanding of unconscious dynamic processes. Experiments on subliminality suggest that the unconscious clinically described by dynamic psychiatry and psychoanalysis is indeed experimentally observable, and that it is in fact possible to investigate unconscious conflict, unconscious affect, and unconscious anxiety in a well controlled manner.

1. Introduction

Current psychoanalysis is characterized by its efforts to enter a dialogue with its neighboring disciplines such as cognitive psychology and cognitive neurobiology (as outlined by Giampieri-Deutsch 2002, 2004, 2005b, 2009, and 2010). In turn, neuroscientists like Kandel (2005), Panksepp (1999), Damasio (1999), Gallese (2006), and Roth (2004) among others, acknowledge basic psychoanalytic insights.

To counter their neglect of research on the mind, cognitive neurobiology and cognitive psychology recently started to investigate conscious as well as non-conscious processes. Their key interest no longer focuses on questions of the existence of unconscious processes alone, but is directed toward investigating experimentally their contribution to the mind, including the function of perception.

2. Freud's models of the mind: the three phases

In order to clarify the relationship between perception and consciousness and to move toward a distinction between the two, it is worth surveying Freud's conceptions of perception and consciousness over time, as elaborated in his models of the mind. Freud developed his conception of the mind, which he also termed "mental apparatus", progressively. Rapaport (1959) originally identified three phases and his point of view was further adopted by Sandler, Dare, and Holder: During a first pre-analytic phase (1886 to 1897) Freud elaborated his "affect-trauma" model; in a second phase (1897 to 1923) the topographical model emerged; and finally the third phase (after 1923) generated the structural model (Sandler et al. 1972).[1] In all his conceptions of the mind, Freud assumed the "mental apparatus" to be a tool for adaptation to demands from both the internal and external world.

2.1 The "affect-trauma" model

During the first pre-analytic phase between 1886 to 1897, Freud's conception of the mind was not explicitly formulated, but can be taken from his works and was first termed "affect-trauma" model by Sandler et al. (1972). At that time Freud emphasized the adaptation to the traumas deriving from external reality. He claimed that the occurrence of real external traumatic events

[1] Joseph Sandler, Alex Holder and Christopher Dare published eleven articles in the *British Journal of Medical Psychology* between 1972 and 1982 framing Freud's conceptualization of the mind; Holder then provided an outline in German (Holder 1982); finally the authors also published a synopsis in English (1997). Part 2 of this chapter is based on the results of their conceptual research. I am, furthermore, drawing from two articles (Giampieri-Deutsch 2008a and 2008b) as the starting point for part 2 and part 3 of this chapter.

arouses affects which the mental apparatus is not able to absorb or discharge and assumed a conscious perception of the real trauma.

In case they are incompatible with the consciousness of the subject, these conscious percepts, memories, representations as well as their related affects have to be dissociated by the ego and therefore become unconscious. The unconscious "constricted" affects along with the memories and associated representations could then again be released by bringing them back into consciousness. At that early stage Freud termed his treatment of patients the "cathartic method", a therapy through abreaction and assimilation of the unconscious mental contents actively dissociated by the conscious part of the mind.

In his "Project for a scientific psychology", Freud (1950c [1895], p. 308) first outlined how consciousness is linked to perception: "Consciousness gives us what are called *qualities*-sensations which are *different* in a great multiplicity of ways and whose *difference* is distinguished according to its relations with the external world. Within this difference there are series, similarities and so on, but there are in fact no quantities in it. It may be asked *how* qualities originate and *where* qualities originate." Freud assumed the existence of a system of perceptual neurones ("omega" neurons) "whose states of excitation give rise to the various qualities – are, that is to say, *conscious sensations*." (Freud 1950c, p. 309)

2.2 The topographical model

Although during the first phase, Freud had already distinguished between the conscious and unconscious mind, this distinction was only fully developed in his topographical model of the mental apparatus in the second phase from 1897 to 1923. The model is composed of three systems: the System Unconscious, the System Preconscious, and the System Perception-Consciousness.

(i) The "System Unconscious" contains instinctual drives and wishes. These would represent a threat if they were allowed to emerge into consciousness, because they constantly strive toward discharge, gratification, and relief of painful tension according to what Freud termed the "pleasure principle". Freud ascribed a number of characteristics to the System Unconscious, such as timelessness, the disregard of external reality, the overestimation of "psychic reality", meaning that imagined experiences are – unconsciously – not distinguished from real events, the absence of contradiction, the absence of negation, and the equation of words with "things" (like a rebus).

Freud´s own contribution to the inquiry into the mind is the discovery of the dynamic unconscious. This marks his difference from Leibniz, Kant, and Helmholtz, and in "A note on the unconscious in psycho-analysis" Freud (Freud 1912g, p. 262) pointed out the novelty of his own concept of the unconscious: "The term *unconscious*, which was used in the purely descriptive sense before, now comes to imply something more. It designates not only latent ideas in general, but especially ideas with a certain dynamic character, ideas keeping apart from consciousness in spite of their intensity and activity."

Regarding the role of aggression and sexuality in unconscious mental life, the legacy of the Darwinian revolution strongly impacted on Freud. In the words of Sulloway (1979, p. 276): "it was Darwin who handed Freud the most powerful instrument – namely, evolutionary theory´s stress upon the dynamic, the instinctual, and, above all, the *nonrational* in human behaviour." Since neuroscientists like LeDoux (1996), Damasio (1999), and Edelman (1992) praise both Darwin and Freud as congenial partners for securing a privileged place for emotion in science, contemporary evolutionary biology too as well as current psychoanalysis could benefit from rediscovering Freud.

(ii) The "System Preconscious" contains unconscious thoughts and memories that were not defended against. Some functions and capacities of the System Preconscious are: the preconscious scrutinizing of representations and feelings, the censoring of instinctual wishes and their derivatives, the storing of organized memory systems, the testing of reality, the binding of psychic energy, the control of access to consciousness and action, the control over the development of affects, the use of defense mechanisms, as well as the elaboration of creative productions and symptom formation.

(iii) The "System Perception-Consciousness" receives perceptual stimuli from both the external world and one's own body (kinesthetic, proprioreceptive, visceral – amongst others) and exercises the function of attention. External perceptual stimuli have to overcome a certain threshold or stimulus barrier before becoming conscious.

2.3 The structural model

The structural model, Freud's conception of the mind during the third phase (after 1923), is composed of the three structures: id, ego and superego. It only partially replaced the topographical model, which continued to be applied where convenient.

In "The ego and the id" (Freud 1923), inaugurating this phase, the question of perception was at the core: "All perceptions which are received from without (sense-perceptions) and from within – what we call sensations and feelings – are *Cs.* [P. G.-D.: conscious] from the start. But what about those internal processes which we may – roughly and inexactly – sum up under the name of thought-processes?" (Freud 1923 b, p. 19)

Therefore Freud asked how does something become conscious and answered that the first step is to ask how something becomes preconscious. Freud's tentative reply to this second question was: "Through becoming connected with the word-presentations corresponding to it." He continued further: "These word-presentations are residues of memories; they were at one time perceptions, and like all mnemic residues they can become conscious again. Verbal residues are derived primarily from auditory perceptions, so that the system *Pcs.* [P. G.-D.: the System Preconscious] has, as it were, a special sensory source." (Freud 1923 b, pp. 20–21)

(i) The "id" is a structure already available at birth. It is the reservoir of the instinctual drives and wishes, as well as of repressed contents. Id contents are entirely unconscious and can be considered similar to those of the System Unconscious of the topographical model.

(ii) The "superego" has been differentiated from the id and from the ego, but a large part of it is still deeply unconscious. It functions as the agency of ideals and values, and as consciousness-prescribing rules transmitted by the parental figures.

(iii) The "ego" is the structure which developed from the id interacting with the external world in the interests of self preservation. "The ego is that part of the id which has been modified by the direct influence of the external world through the medium of the *Pcpt.-Cs.* [P. G.-D.: the System Perception-Consciousness]; in a sense it is an extension of the surface-differentiation." Here Freud underscored once more the topic of perception: "Moreover, the ego seeks to bring the influence of the external world to bear upon the id and its tendencies, and endeavours to substitute the reality principle for the pleasure principle which reigns unrestrictedly in the id. For the ego, perception plays the part which in the id falls to instinct." (Freud 1923 b, p. 25) The ego copes with the conflicting demands of the id, the superego and the external world. A great part of the ego is also unconscious, meaning the ego can no longer be equated with consciousness, even if the id and the superego only have access to consciousness through the ego.

Freud discussed the relationship between perception and the body: "A person's own body, and above all its surface, is a place from which both external and internal perceptions may spring. It is *seen* like any other object, but to the *touch* it yields two kinds of sensations, one of which may be equivalent to an internal perception. Psychophysiology has fully discussed the manner in which a person's own body attains its special position among other objects in the world of perception. [...] The ego is first and foremost a bodily ego; it is not merely a surface entity, but is itself the projection of a surface. If we wish to find an anatomical analogy for it we can best identify it with the 'cortical homunculus' of the anatomists, which stands on its head in the cortex, sticks up its heels, faces backwards and, as we know, has its speech-area on the left-hand side." (Freud 1923b, pp. 25–26) In a later footnote of 1927 Freud further commented upon the relationship between the ego and the body: "I. e. the ego is ultimately derived from bodily sensations, chiefly from those springing from the surface of the body. It may thus be regarded as a mental projection of the surface of the body, besides, as we have seen above, representing the superficies of the mental apparatus." (Freud 1923b, p. 26)

In the structural model, consciousness is an ego function, which is connected with perception, and includes the metaphor of consciousness as a sense-organ of the ego.

Let us now address current interpretations. For a technically sound and impartial evaluation of Freud´s complex understanding of the unconscious we turn to a comment by the neurobiologist Eric Kandel (2005, pp. 70–71). He distinguished three different uses of the concept of the unconscious by Freud: "First, [Freud] used the term in a strict or structural way to refer to the *repressed* or *dynamic unconscious*. This unconscious is what the classical psychoanalytic literature refers to as the *unconscious*. It included not only the id but also that part of the ego which contains unconscious impulses, defenses, and conflicts and therefore is similar to the dynamic unconscious of the id. In this dynamic unconscious, information about conflict and drive is prevented from reaching consciousness by powerful defensive mechanisms such as repression."

Kandel then summarizes the second use of this term by Freud as being a second part of the ego that is unconscious, although not repressed, and therefore not involved with unconscious drives or conflicts. Unlike the "preconscious unconscious" – Freud´s third use of the term unconscious, a merely descriptive one – this second unconscious part of the ego cannot become conscious, even though it is not repressed or conflicted. Kandel refers to this part as the "procedural unconscious", because it is concerned with habits as well as perceptual and motor skills. Kandel suggests that this second unconscious part of the Freudian ego appears to map onto what cognitive neurobiologists call "procedural memory".

A fundamental neurobiological insight is the distinction between procedural (implicit) and declarative (explicit) memory. Studying her amnesic patient H. M., Brenda Milner discovered in 1954 that the medial temporal lobe and the hippocampus mediate the declarative (explicit) memory storage (Scoville and Milner 1957). Declarative memory is the conscious recall of people, objects, and places: "It depends on the integrity of the medial temporal lobe and affords the capacity for conscious recollections about facts and events. Declarative memory is propositional – it can be true or false" (Milner et al. 1998, p. 450). In 1962 Milner discovered that even in spite of the patient H. M.'s loss of his declarative memory, his procedural memory was preserved. H. M. was thus able to memorize implicitly and learn further perceptual and motor skills. His procedural (implicit) memories were unconscious: if their conscious recall failed, they were nonetheless evident in performance (Milner et al. 1998). Kandel connects the procedural processes of memory to different brain systems: "priming or recognition of recently encountered stimuli, is a function of sensory cortices; the acquisition of various cued feeling states involves the amygdale; formation of new motor (and perhaps cognitive) habits requires the neostriatum; learning new motor behavior or coordinates activities depends on the cerebellum" (Kandel 2005, pp. 70–73; see also Squire and Kandel 1999 as a general introduction). Commonly the two memory systems are used together: constant repetitions transform declarative conscious recollection into a procedural memory (e.g. learning to play the piano involves conscious memory at first, but playing becomes a non-conscious motor activity). According to Kandel, procedural memory provides a biological example of one component of unconscious mental processes.

Kandel also refers to Robert Clyman, who investigated memory in the context of emotion (Clyman 1991), and seeks to integrate the procedural unconscious into Freud's conceptions of the mind. After 1923, Freud's structural model of the mind distinguished between id, ego and superego. According to Kandel the procedural unconscious should be considered a part of the unconscious ego: "this unconscious part of the ego is never accessible to consciousness even though it is not repressed. Since this unconscious is concerned with habits and perceptual and motor skills, it maps onto procedural memory." (Kandel 2005, p. 71)

Similarly, Fonagy (1999, p. 218) linked procedural memory to the unconscious in order to explain the efficacy of psychoanalytic treatment. Contrary to a still widespread assumption, Fonagy did not consider the therapeutic action of psychoanalysis as dependent upon the recovery of repressed memories of childhood, which are autobiographical declarative memories. In turn, he was also struck by the role of procedural memory in treatment, saying that "[p]sychic change is a function of a shift of emphasis between different mental models of object relationships. Change occurs in implicit memory leading to a change of the procedures the person uses in living with himself and with others."

3. Consciousness in the different phases of Freud's conceptions of the mind

In "The Oxford companion to the mind", Dennett (1987, p. 162) acknowledged that it was Freud who introduced the idea of "repressed unconscious activity [. . .] driven out of 'sight' of consciousness". At the same time, Dennett erroneously added that it was not until more recently that one began to understand that unconscious mental processes do not originate from active repression, and that on the contrary, the mind is unconscious from the beginning. Dennett equally mistakenly suggested that it is contemporary philosophy of mind which first asked the question: "Why does mental activity become conscious?"

From the beginning to his late work, Freud maintained two assumptions: first, that mental processes are in themselves unconscious, and second, that consciousness is available. Likewise he did not assume that only some mental activity is unconscious, for example a repressed representation, but that mental processes are unconscious in themselves.[3] Indeed, it was already implied in a remark at the beginning of his second paper on "The neuro-psychoses of defence" (Freud 1896b). In 1900, in "The interpretation of dreams," Freud suggested very emphatically that *"the most complicated achievements of thought* [...] *are possible without the assistance of consciousness* – a fact which we could not fail to learn in any case from every psycho-analysis of a patient suffering from hysteria or from obsessional ideas." (Freud 1900a [1899], p. 593, Freud's italics) Similarly, Freud suggested, also in "The unconscious" (Freud 1915e, p. 171) and in "A difficulty in the path of psychoanalysis" (Freud 1917a, p. 143), that "mental processes are in themselves unconscious". Further, in 1925 Freud defended his view in "The resistances to psychoanalysis", pointing out that "what is mental is in itself unconscious and that being conscious is only a quality, which may or may not accrue to a particular mental act." (Freud 1925e, p. 216)

Returning to Dennett's misconception, Freud spelled out very clearly in "The ego and the id" that the unconscious does not overlap with the repressed: "We recognize that the *Ucs.* [P. G.-D.: the System Unconscious] does not coincide with the repressed; it is still true that all that is repressed is *Ucs.*. [P. G.-D.: unconscious], but not all that is *Ucs.*.[P. G.-D.: unconscious] is repressed. A part of the ego, too–and Heaven knows how important a part–may be *Ucs.*[P. G.-D.: unconscious], undoubtedly is *Ucs.* [P. G.-D.: unconscious]." (Freud 1923b, p. 18)

In turn, against the widely spread assumption that Freud scarcely paid any attention to consciousness, his few statements on the topic are worth mentioning here. According to Freud even if mental processes are unconscious in themselves "if anyone speaks of consciousness we know immediately and from our most personal experience what [it] is meant to be." (Freud 1940a [1938], p. 157) Further, when talking about the behaviorism of John B. Watson, who eliminated the term consciousness from his theoretical frame of reference, Freud stated: "One extreme line of thought, exemplified by the American doctrine of behaviorism, thinks it possible to construct a psychology, which disregarded this fundamental fact!" (Freud 1940a [1938], p. 157)

Regarding the relationship between consciousness and perception, it is worth noting that Freud described consciousness as emerging during the function of perception: "This system (*Pcpt.-Cs.*[P. G.-D.: System Perception-Consciousness]) is turned towards the external world, it is the medium for the perceptions arising thence, and during its functioning the phenomenon of consciousness arises in it." (Freud 1933a [1932] p. 75) However in 1915 in "The unconscious" Freud had already conceded that "becoming conscious is no mere act of perception, but is probably also a *hypercathexis* [P. G.-D.: meaning a large amount of psychic energy], a further advance in the psychical organization." (Freud 1915e, p. 194)

According to Freud's "Outline of psychoanalysis," consciousness is "a fact without parallel, which defies all explanation or description." (Freud 1940a [1938], p. 157) "There is no need to characterize what we call 'conscious': it is the same as the consciousness of philosophers and of everyday opinion." (Freud 1940a [1938], p. 159) Freud here is addressing "our most personal experience" – as Chalmers would put it – the "hard problem of consciousness", a phenomenon which according to Chalmers cannot

[2] In the same vein see also Mark Solms (Solms 1997); Solms' main topic in his article was indeed a Kantian understanding of Freud's theory of consciousness that I do not discuss here.

be explained in terms of computational or neural mechanisms according to the standard methods of cognitive science or cognitive neurobiology (Chalmers 1996, 2007). To illustrate this issue further Chalmers adds: "Sometimes, terms such as 'phenomenal consciousness' or 'qualia' are also used here, but I – writes Chalmers – find it more natural to speak of 'conscious experience' or simply 'experience'." (Chalmers 2007, p. 226).

And yet, consciousness still remains one of the most neglected topics in psychoanalysis. Freud investigated the subject in one of the twelve writings that were to be collected in a comprehensive "Metapsychology". On the 12th of August 1915 Freud wrote to Ferenczi (letter F 560, Freud and Ferenczi 1996, p. 75): "The twelve papers are – so to speak – finished." However, Freud only published six chapters, and the article on consciousness disappeared. Freud wrote to Ferenczi on the 24th of March 1916: "I am thinking of abandoning the paper about the cs. [P. G.-D.: consciousness] and replacing it with a more appropriate one, for example, 'The Three Viewpoints of MΨ'" [P. G.-D.: Metapsychology] (letter F 603, Freud and Ferenczi 1996, p. 121)

3.1 Consciousness as a quality of mental events

By avoiding misleading accounts of Freud's work on consciousness, Shevrin (1997) provides a reliable classification of Freud's conceptions of consciousness: consciousness as a quality of mental events; consciousness as a system in the topographical model; and finally consciousness as an ego function in the structural model, which is connected with perception and includes the metaphor of consciousness as a sense organ.

The first conception (consciousness being a quality of mental events) is close to the clinical experience and cannot be ascribed to one of the three phases of Freud's models of the mind. Although this model is available from the early pre-analytic writings to "The ego and the id" (Freud 1923b), it is best described in one of Freud's last writings, the posthumously published "An outline of psycho-analysis." In the chapter devoted to "Psychical qualities" he says: "Thus we have attributed three qualities to psychical processes: they are either conscious, preconscious or unconscious. The division between the three classes of material which possess these qualities is neither absolute nor permanent. What is preconscious becomes conscious, as we have seen, without any assistance from us; what is unconscious can, through our efforts, be made conscious, and in the process we may have a feeling that we are often overcoming very strong resistances." (Freud 1940a [1938], p. 160)

Shevrin points out how, according to this model, a representation can be "in three states vis-à-vis consciousness: it could be conscious, preconscious (easily acquiring the quality of consciousness), or unconscious (acquiring the quality of consciousness with difficulty). On this model one can say that the relationship of an idea to consciousness was operationally defined." (Shevrin 1997, p. 748)

Consciousness is a condition of subjective awareness. Both a threshold (or barrier) and intensity are part of this model. The threshold for a preconscious representation to become conscious is low. For a preconscious representation to become conscious is a function of its intensity. Our everyday perceptions develop from an initial preconscious to a conscious quality. The threshold for an unconscious representation (such as in repression) to become conscious, is much higher and may counteract even the most intense representation.

Fisher (1957) and Shevrin (1992, 1997) pointed to the compatibility of this model of consciousness with the findings of subliminal perception (with which Freud's two other models – the System Perception-Consciousness and the sense organ of the ego – have trouble).

The model of consciousness as a quality of mental events is also clinically valuable: for example, if a patient cannot remember core events from the preceding session, the analyst assumes repression. Within this model, consciousness is a property of mental processes emerging under certain conditions. As the research of Shevrin (Shevrin et al. 1996) has shown, the questions of how consciousness emerges as a quality of a representation, and what psychological and neurophysiological conditions should be present for that to happen, can be studied in the consulting room as well as in the laboratory.

3.2 Consciousness as a system in the topographical model

In the topographical model, Freud elaborated the System Perception-Consciousness from the time of his "Project for a scientific psychology" (Freud 1950a [1895]) which equated perception with consciousness. Shevrin suggests that in this model the normal waking state is given primacy and provided with specific functions close to waking perception, such as reality testing[4], attention and a capacity for self-reflection. Shevrin also remarks that something was lost in this model, Edelman's "primary consciousness" (1989) or Damasio's "core consciousness" (1999), that is to say, immediate subjective awareness without any accompanying reflectiveness, that which James (1890) described as the "stream of consciousness."

When conflating consciousness and perception, the latter always had to be conscious first, and any interaction between the external world and preconscious as well as unconscious processes became impossible. This did not correspond to our everyday psychoanalytic clinical experience. In addition, the results of neuropsychological investigations of brain-damaged patients (see further below) also ask for a revision of the assumption that perception and consciousness of perception are always inseparable. Also the whole tradition of psychoanalytic research into subliminal perception from Pötzl (1917) through Fisher (1954, 1956, 1957, 1959; Shevrin and Fisher 1967) to Howard Shevrin, shows how seriously researchers took the topic of unconscious percepts. Identifying perception with consciousness did not fit with the impressive results of subliminal research, in which it was repeatedly shown that external stimuli can be perceived subliminally and interact with unconscious processes, thereby producing dreams, images, and associations without any waking-state awareness of their source. Indeed, as early as 1919, in his re-edition of his "Interpretation of dreams", Freud himself made a positive comment on Otto Pötzl's findings on subliminality: "In a series of experiments Pötzl required the subjects to make a drawing of what they had consciously noted of a picture exposed to their view in a tachistoscope [P. G.-D.: an instrument for exposing an object to view for an extremely short time]. He then turned his attention to the dreams dreamt by the subjects during the following night and required them once more to make drawings of appropriate portions of these dreams. It was shown unmistakably that those details of the exposed picture which had not been noted by the subject

3 According to Freud the "testing of reality" is the capacity by which a distinction can be made between what is "unreal" (such as fantasies, memories, and dreams) and what are perceptual representations of something in the external world (Sandler et al. 1997, p. 92).

provided material for the construction of the dream, whereas those details which had been consciously perceived and recorded in the drawing made after the exposure did not recur in the manifest content of the dream." (Freud 1900a, p. 181–182)

3.3 Consciousness as an ego function, the relation to perception and the metaphor of consciousness as a sense organ

In psychoanalysis, mental phenomena are explained as the outgrowth of conflict between (i) an instinctual (aggressive or sexual) wish and a defense against the wish, (ii) different intrapsychic structures (id, ego, superego), or (iii) an impulse in opposition to internalized demands of external reality. Therefore wishes are warded off by defensive techniques like repression, isolation, undoing, reaction formation[5], among others. This has subsequently been known as the dynamic aspect of psychoanalysis and has also been adopted by current psychodynamic psychiatry (Hartmann 1959; Gabbard 2005). Freud developed the structural model of id, ego and superego because he had realized that the defenses against aggressive or sexual wishes belonging to the System Perception-Consciousness worked unconsciously. Following this clinical observation, Freud noticed the paradox of an unconscious process referring to a system controlled by consciousness. Consequently, by establishing the ego as a superordinate structure, Freud not only assumed conscious, but also unconscious processes such as ego defenses within the ego.

Nonetheless consciousness belongs to the ego too, as Freud stated in his book "The ego and the id": "It is to this ego that consciousness is attached [...]" (Freud 1923b, p. 17). Although the notion of consciousness as a specific ego function taking on the role of a sense organ was also available in Freud's earlier writings, it was in the structural model that the "sense organ" metaphor was fully developed.

Shevrin criticizes Freud's metaphor of consciousness as a sense organ, comparing it to similar analogies proposed by Francis Crick and other neuroscientists, who describe consciousness as a sort of light illuminating the dark regions of the world or our mind. According to Shevrin, the problem lies in the implicit assumption of a dualistic solution of the problem of consciousness. As Shevrin maintains, metaphors like a sense organ or light, lead to a new edition of the "homunculus" and cannot be supported by experimental evidence. Shevrin is one of few psychoanalytic scholars who resisted projecting dualism into Freud, who in fact regarded the physical and the mental as a causally connected "continuum".[5]

---

4 For example, the defense "repression" operates by segregating unacceptable wishes, feelings, memories, or fantasies from consciousness; "isolation" of affect detaches affect from representation (e.g., a traumatic memory may be recalled but will be disconnected by any accompanying intense feelings); regarding "undoing", a symbolic thought or action have to be performed to cancel out an unacceptable thought or action; "reaction formation" is characterized by defending against an unacceptable wish or impulse (e.g., aggressive wish) by adopting a character trait (e.g., compassion) completely opposed to it (Gabbard 2005).

5 We have evidence that Freud was fully aware of the philosophical discussion about the problems of the dualistic positions to the mind-body question. For example in "The unconscious," in 1915, when he referred to "the insoluble difficulties of psycho-physical parallelism." (Freud, 1915e, p. 168)

4. Perception is not only a conscious process

As early as 1910 Freud investigated "hysterical blindness" and for the first time clearly assumed the existence of a sort of unconscious perception: "Appropriate experiments have shown that people who are hysterically blind do nevertheless see in some sense, though not in the full sense. [. . .] Thus hysterically blind people are only blind as far as consciousness is concerned; in their unconscious they see. It is precisely observations such as this that compel us to distinguish between conscious and unconscious mental processes." (Freud 1910, p. 212) Freud objected to French researchers such as Charcot, Janet, and Binet, who claimed that the cause of hysterical blindness was nothing but autosuggestion: "The hysterical patient is blind, not as the result of an autosuggestive idea that he cannot see, but as the result of a dissociation between unconscious and conscious processes in the act of seeing; his idea that he does not see is the well-founded *expression* of the psychical state of affairs and not its *cause.*" (Freud 1910, p. 212)

Hysterical blindness is not due to a lesion (as in the case of blindsight), the patients are physiologically entirely healthy. Freud assumes psychic conflict between desire and defence resulting in a temporary psychodynamic disturbance to be entirely solved by means of psychoanalytic treatment. This case study is relevant for Freud for the following reasons: not only does it differentiate between conscious and unconscious processing, but in particular it finally leads him to the assumption of unconscious perception.

Even though a number of neuroscientists like Edelson, Damasio, LeDoux, Panksepp, and others, are referring to Freud or psychoanalysis, cognitive neurobiologists as well as cognitive psychologists in general only rarely appeal to psychoanalytic case studies when discussing conscious and unconscious processes (excepting Kandel and Roth). Recently, neuro-psychoanalysts upgraded the value of clinical research in psychoanalysis through the investigation of changes in personality, motivation, and emotion of brain-damaged patients in psychoanalysis or psychoanalytic therapy (Kaplan-Solms and Solms 2000; Solms 2004; Levin 2004, 2009).

In what follows we discuss far more conventional neuropsychological studies of brain-damaged patients, as their findings strongly impact on contemporary philosophy of mind and question the assumption that perception and consciousness of perception are always inseparable.

4.1 Non-conscious perception and neuropsychological syndromes

According to Weiskrantz "Blindsight has made us aware that there is more to vision than seeing, and more to seeing than vision." (Weiskrantz 2007, p. 179)

The interest in "blindsight" was elicited by an article on clinical cases of four war veterans with cortical damages and residual visual function (Poppel et al. 1973). All these patients suffered from "cortical blindness", i. e. they were blind because that part of the cortex providing visual consciousness did not work anymore. Poppel and colleagues described how blind patients were nevertheless reacting to visual stimuli. Weiskrantz, Warrington, Sanders, and Marshall termed this phenomenon "blindsight" (1974). Weiskrantz further investigated it. As he showed, in blindsight damaged primary visual cortex (striate cortex or area V1) causes blindness in parts of the affected visual fields. Even though blindsight patients have blind areas in their visual fields, they demonstrate certain visual abilities when they are asked to make a "forced choice" and required to guess in indirect testing. Despite their residual abilities the patients say that they do not see anything and so are obviously unaware of their perception (Weiskrantz 1986, 1990, 1997, 1998, 2007).

In the case of "blindsight" as well as in the case of some other neuropsychological syndromes the function of perception (mostly vision) and the conscious experience of perception fall apart, visual processing and visual consciousness diverge. Since its discovery this phenomenon has continuously attracted the attention of a number of philosophers of mind such as van Gulick (1994), Block (1995), Chalmers (1996), Dretske (1995), Flanagan (1992; Güzeldere et al. 2000), and Searle (1979, 1992) (cf. Giampieri-Deutsch 2005 a, pp. 17–18). Over the last 30 years, starting with Searle (1979) on blindsight, philosophers have discussed a number of other cases of brain-damaged patients with specific perceptual and cognitive deficits, but with access to stimuli that are not perceived consciously. "Prosopagnosics" report to be unable to recognize even the faces of their own relatives. However, indirect tests show evidence of "covert knowledge" of their facial expressions. Patients with "neglect" cannot report on the stimuli occurring on the contralesional side of space: even though their behavior suggests that they do not perceive them, they have non-conscious perception of neglected stimuli. Finally, "alexia" patients can no longer read words at a glance, but experiments provide evidence that they can read implicitly. All these patients are processing visual information, i. e. they have non-conscious visual perception without conscious experience of vision, so that when they succeed in completing an experimental task by indirect testing they claim to be "guessing" (cf. Farah 1994, 2001 for details on indirect testing).

Are these residual capacities to be related to a damaged but partly functioning primary visual pathway, or are they indicating alternative pathways for the visual system? The neurophysiological research on blindsight is still ongoing. The validity of blindsight has been questioned and some researchers explained residual functions by artifacts (Campion et al. 1983), or by spared cortical tissue (incomplete lesion) (Fendrich et al. 1992; Gazzaniga et al. 1994).

Undoubtedly, these phenomena are of great interest to philosophers of mind, because neuropsychological findings lead to a revision of the long-standing philosophical assumption that perceiving something is being aware of it. According to current philosophers of mind, phenomena such as blindsight allow an understanding of the nature and function of consciousness. Commenting on blindsight, Güzeldere, Flanagan and Hardcastle ask: "Does consciousness *necessarily* play a role in mental life, or could we do just as well without it?" (Güzeldere et al. 2000, p. 1279) Blindsight patients are missing consciousness of the stimuli in the blind part of the visual field and so fail to engage in some expected behavior. Therefore philosophers propose that consciousness has the capacity to cause that behavior. Consequently the philosophers answer their initial question, suggesting that "the lack of conscious awareness of the blind field partly explains the inability to bring the knowledge the system possesses into normal, high-quality play in inference, reporting, and action." (Güzeldere et al. 2000, p. 1283)

The non-conscious vision of brain-damaged patients is unconscious in a merely descriptive sense; it is latently unconscious rather than dynamically unconscious. The latent unconscious activity functions more or less as conscious activities do, however it lacks awareness (e. g., a latent unconscious perception is a non-conscious or a weak form of conscious perception). The dynamic unconscious is psychological, it is active, and it can be different in character from conscious psychological processes (cf. also below "4.2. Unconscious perception and subliminality studies"). But unconscious perception is not simply a weak form of conscious perception. That is why neuropsychological investigations as well as their philosophical inferences reach their limit here.

4.2 Unconscious perception and subliminality studies

Howard Shevrin (see also Chapter IV, 6 by H. Shevrin) provides a definition of the dynamic unconscious: "It is *psychological*, it is *active*, and it can be *different* in character from conscious psychological processes." (Shevrin and Dickman 2003, p. 542). (i) The dynamic unconscious is *psychological*, meaning that the descriptive terminology used for conscious processes can be used for unconscious processes: perception, thought, affect, motivation, etc. Not only conscious processes are correlated with brain processes, but unconscious processes also. (ii) The dynamic unconscious is *active*, meaning that unconscious processes have a bearing on behavior and experience, even though the subject may be unaware of this. (iii) The dynamic unconscious can be *different* to conscious processes and follows different principles of organization than those which characterize processes occurring during the state of consciousness.

Shevrin reminds us that not just psychoanalysis, but also psychodynamic psychiatry as well as most other psychotherapy schools (and not merely the classical Jungian, Adlerian, and Sullivanian, but the more recent forms such as transactional analysis, gestalt therapy, and many others) share such a definition of the dynamic unconscious.

Regarding psychology at large, Shevrin asks himself about the possibility of formulating two definitions of the unconscious – a "weak" and a "strong" definition. His "weak" definition of the unconscious recognizes that "psychological" unconscious processes exist and "actively" affect conscious processes. Shevrin's "strong" definition of the unconscious points out, in addition, that "psychological" unconscious processes follow different principles of organization. In both his definitions of the unconscious Shevrin intentionally leaves out the role of motivational factors (e. g., drives) which psychodynamically oriented clinicians usually assume to be relevant.

According to Shevrin cognitive psychology seems to imply *psychological* unconscious processes in experiments on selective attention, subliminal perception as well as in visual phenomena involving perceptual processing (retinal image stabilization, binocular rivalry, and backward masking), thus providing indirect evidence in support of the *psychological* unconscious. Cognitive experiments on selective attention confirmed that cognitive activity occurs outside awareness, and that unconscious processes affect ongoing and subsequent conscious processes, and are therefore consistent with the *weak* definition of the unconscious. In cognitive experiments on subliminal perception it was additionally shown that unconscious complex activity is characterized by properties differing from those of conscious cognitive activity, thus supporting both the *weak* and the *strong* definition of the unconscious. Visual phenomena involving perceptual processing such as retinal image stabilization, binocular rivalry, and backward masking can be considered to indirectly support the *weak* definition of unconscious. From his detailed review of these above mentioned experiments Shevrin drew three conclusions: first, the initial cognitive stage for all stimuli occurs outside of consciousness; second, the initial non-conscious cognitive stage is psychological, active on consciousness and can follow principles of organization differing from conscious cognition; and third, consciousness of stimulus is a later and optional stage in cognition. Consequently Shevrin poses the crucial question: what factors may determine the emergence of a stimulus into consciousness or inhibit it, respectively. Shevrin's answer is: first, stimulus factors such as loudness, brightness, figural coherence, etc.; second, factors concerning the state of the receiver, such as level of arousal, sleep stage, fatigue, distractibility, etc.; and third motivational factors, such as avoidance of anxiety, guilt, conflict, etc. (Shevrin and Dickman 2003).

Indeed, there are plenty of research studies on unconscious processes (Marcel 1983; Velmans 1991; Merikle and Daneman 2000; Cheesman and Merikle 2003) in cognitive psychology, but little consensus regarding unconscious motivational factors. Although psychodynamic psychiatry and psychoanalysis have provided a wide range of clinical evidence since the New Look's debate in psychology (cf. Erdelyi 1974) to the present day, topics such as "dynamic unconscious" or "unconscious motivational factors" are still treated with caution.

Apart from clinical investigations in psychoanalysis, a less known ongoing experimental tradition of psychoanalytic research is worth mentioning. Carried out by Pötzl (1917), Benussi (1923, 1925, 1927), Fisher (1954, 1956, 1957, 1959), Luborsky (1988; Shevrin and Luborsky 1958), and Holt (1968, 1989), its efforts culminated in the longstanding research program of Howard Shevrin (1992, 2000, 2004, 2005; Wong et al. 1994; 1997; Snodgrass and Shevrin 2006), and aimed at proving the activity of the dynamic unconscious by experimental findings. Therefore, having moved from Freud to contemporary psychoanalysis and to the most recent experiments within the long tradition of subliminal perception in psychoanalysis, the work of Howard Shevrin has been well known and valued for many years, and quoted not only by cognitive neurobiologists such as Kandel, Panksepp and Damasio, but also by cognitive psychologists such as Merikle and Erdelyi (see also Chapter IV, 16 by MH Erdelyi). A few years ago, the cognitive psychologist Erdelyi published a key-article on subliminal perception followed by a discussion providing a particularly striking appreciation and substantiation of Shevrin's approach (Erdelyi 2004a, 2004b; Kihlstrom 2004; Bachmann 2004; Snodgrass 2004; Reingold 2004).

Over the last decades, the main clinical research approach or "case study method" in psychoanalysis has grown fast by incorporating the methodologies and concepts of its neighboring disciplines. Neuropsychoanalysis is now bridging the neurosciences and psychoanalysis, enhancing the image of Freud as "biologist of the mind" (Sulloway 1979). Within cognitive neurobiology and cognitive psychology, research of unconscious processes has become a field of increasing importance. It concerns central topics, such as implicit learning, procedural memory, and subliminal perception. Until now, methodological and theoretical difficulties have complicated the seminal transfer of knowledge between disciplines. Shevrin and his colleagues made a substantial contribution to the convergence of cognitive neurobiology, cognitive psychology, and psychoanalysis (cf. Shevrin et al. 1996; Shevrin 2004).

Assuming a continuum between physical and mental phenomena, Freud had to forgo a description of neurobiological processes underlying the mind because the neurobiology of his day was not yet sufficiently developed. To find neurobiological correlations for mental states seems possible now and some insights of contemporary neurobiology converge creating novel paths for research in psychoanalysis. In an interdisciplinary dialogue, three areas of psychoanalysis are backed up by consolidated findings of neurobiology: (i) The importance of early object relations and early experiences and the correlating vulnerability for both mental disorders and physical diseases (Fonagy 2001; Hofer 2001; Kandel 2005, cf. pp. 79–86; Trevarthen 2001; Trevarthen and Reddy 2007). (ii) Psychoanalytic therapy has been shown to be a treatment of the mind bringing about not only mental but also neurobiological modification. Kandel suggested that these structural changes in the brain could be visualized by studies using brain imaging techniques (Kandel 2005, cf. pp. 91–93, p. 386–387). (iii) The dynamic unconscious (still controversially discussed) as part of the mind can be made observable (e. g. by studies on subliminal perception) and does not have to be merely inferred.

Shevrin's remarkable research endeavors using the subliminality approach has been praised by the neurobiologist Eric Kandel, who said that "the important and longstanding tradition of work by Howard Shevrin, [that] correlate[s] the perception of subliminal and supraliminal stimuli with event-related potentials in the brain in an attempt to analyze aspects of unconscious mental processes." Kandel (2005, p. 96) suggests following the seminal work of Shevrin because we "need to develop creative ways of studying subjective phenomena".

In Shevrin's research design, first-person information (patient accounts of their understanding of their own symptoms) is combined with third-person cognitive psychometrics as well as neurophysiological measures. In this way evidence is provided for dynamic motivational unconscious processes, showing that they are active and controlled rather than merely automatic (Shevrin 2000, cf. Velmans 2000, p. 5; Shevrin 2004). Shevrin could also show that in general, all unconscious processes are individual and deeply influenced by the history of the subject (Shevrin 2004, 2005).

Even though the problem of unconscious affect is discussed in cognitive neurobiology, in cognitive psychology, and in psychoanalytic theory, a unified theory of affects is still a desideratum. The findings of Shevrin and colleagues provide evidence for the assumption of actual unconscious affects and against the concept of affects existing as mere "dispositions". This result was gained by an interdisciplinary pluralism of methods meeting the strictest scientific standards. It yields important possibilities for the integration of theoretical aspects coming from different research traditions toward a more coherent theory of mind.

Among Shevrin's countless subtle experiments dealing with unconscious affects, at least two samples are worth mentioning here. In a first experiment, subjects were exposed subliminally and supraliminally to twenty words having either positive or negative emotional meaning. These emotional valences were determined by having subjects rate the words on the evaluative scales developed by Osgood et al. (1975). The event-related potential (ERP) of each subject, derived from the electroencephalographic (EEG) record, was measured. During the supraliminal presentation, the ERPs following the presentation of emotionally negative words were more positive in voltage than those to the emotionally positive words. Interestingly, the difference of subliminal ERPs between emotionally negative and positive words is greater on the left than on the right side of the brain; there is a brain lateralization effect. This is not the case for the supraliminal presentation. On the basis of these results the experiment shows that the meaning of each affect is registered unconsciously. Shevrin concludes that unconscious processes differentiate between positive and negative meanings or valences without consciousness (Bernat et al. 2001; Shevrin 2000, 2004).

Since affects have expressive and physiological dimensions, in another experiment

Shevrin and colleagues have shown that a schematic face aversively conditioned subliminally causes greater facial tension in particular in the corrugator muscle above the eyes which indicates a negative emotional expression as in a frown (Bunce et al. 1999; Shevrin 2000, 2004).

Finally, in his paper "Toward a theory on consciousness based on recent developments in subliminal research", Shevrin holds that the relationship between conscious and unconscious processes is a key question in current cognitive psychology and neurobiology. Outlining a number of his own studies, Shevrin provides experimental evidence for the following four propositions: To begin with, Shevrin thinks that consciousness cannot be fully understood without a concomitant understanding of unconscious processes. Further, unconscious processes are dynamic that is complex, representational, and highly interactive with conscious processes, and not merely dispositional or latent. Then, the interaction between conscious and unconscious processes can best be understood when both person and context are taken into account. Finally, there are quantitative and qualitative differences between conscious and unconscious processes that can best be seen as a function of these individual differences (Shevrin 2005).

Outlook

In his "Biology and the future of psychoanalysis" of the year 1999, Kandel not only praised psychoanalysis as still being "the most coherent and intellectually satisfying view of the mind" (Kandel 2005, p. 64), but also complained about the "failure of psychoanalysis to provide objective evidence that it is effective as a therapy", including the evidence that it works better than non-analytically oriented therapy (Kandel 2005, p. 97). However, the researchers inside the psychoanalytic community were quite well acquainted with the seminal publication of Peter Fonagy, "An open door of outcome studies in psychoanalysis," which summarizes these efforts, but was hardly noticed from the outside (Fonagy 2002).

In the meantime the situation has changed. A number of independent key-articles published in medical, psychiatric, and psychological – not psychoanalytical – journals provide empirical evidence of the efficacy of long-term psychodynamic therapy. Additionally, therapeutic gains were shown to be lasting and patients continue to improve after their treatment (Leichsenring and Rabung 2008; 2011; Leichsenring et al. 2009; Shedler 2010).

Even short psychodynamic therapies yield significant, large, and stable improvements, just as cognitive-behavioral therapies do (Leichsenring et al. 2009). Furthermore, non-psychodynamic therapies may be successful in part because the more effective therapists utilize techniques that have long been central to psychoanalytic theory and practice (Shedler 2010).

In a recent article, the cognitive psychologist Jonathan Shedler concludes: "Presentations that equate psychoanalysis with dated concepts that last held currency in the psychoanalytic community in the early 20th century are similarly misleading: they are at best uninformed and at worst disingenuous." (Shedler 2010, pp. 106–107)

References

The chronological order of Freud's writings is based on the bibliography:

Meyer-Palmedo I, Fichtner G (eds) (1989). Freud-Bibliographie mit Werkkonkordanz [Freud-bibliography with concordance of his publications]. Fischer, Frankfurt am Main, pp 15–90

Bachmann T (2004) Inaptitude of the signal detection theory, useful vexation from the microgenetic view, and inevitability of neurobiological signatures in understanding perceptual (un) awareness. Conscious Cogn 13: 101–106

Benussi V (1923) La suggestione e l'ipnosi come mezzo di analisi psichica reale. In: Atti del IV Congresso Nazionale di Psicologia. Bandettini, Firenze 1925, pp 35–65

Benussi V (1925) La suggestione e l'ipnosi come mezzo di analisi psichica reale. Zanichelli, Bologna

Benussi V (1927) Zur experimentellen Grundlegung hypnosuggestiver Methoden psychischer Analyse. Psychol Forsch 9: 197–274

Bernat E, Bunce SC, Shevrin H (2001) Event-related brain potentials differentiate positive and negative mood adjectives during both supraliminal and subliminal processing. Int J Psychophysiol 42: 11–34

Block N (1995) On a confusion about a function of consciousness. Behav Brain Sci 18: 227–247 [reprint in: Block N, Flanagan O, Güzeldere G (eds) The nature of consciousness. Philosophical debates. MIT Press, Cambridge, MA 1997, pp 375–415]

Bunce SC, Bernat E, Wong PS, Shevrin H (1999) Further evidence for unconscious learning: preliminary support for the conditioning of facial EMG to subliminal stimuli. J Psychiatr Res 33: 341–347

Campion J, Latto R, Smith Y (1983) Is blindsight an effect of scattered light, spared cortex, and near-threshold vision? Behav Brain Sci 6: 423–86

Chalmers D (1996) The conscious mind. Oxford University Press, Oxford

Chalmers D (2007) The hard problem of consciousness. In: Velmans M, Schneider S (eds) The Blackwell companion to consciousness. Blackwell Publishing, Malden MA Oxford, pp 225–235

Cheesman J, Merikle PM (2003) Distinguishing conscious from unconscious perceptual processes. In: Baars BJ, Banks WP, Newman JB (eds) Essential sources in the scientific study of consciousness. MIT Press, Cambridge, MA, pp 519–540 [original in: Canadian J of Psychology 40: 343–367]

Clyman R (1991) The procedural organization of emotion: a contribution from cognitive science to the psychoanalytic theory of therapeutic action. J Am Psychoanal Assoc 39: 349–381

Damasio A (1999) The feeling of what happens. Body and emotion in the making of consciousness. William Heinemann, London

Dennett DC (1987) Consciousness. In: Gregory RL (ed) The Oxford companion to the mind. Oxford University Press, Oxford

Dretske F (1995) Naturalizing the mind. MIT Press, Cambridge, MA

Edelman GM (1989) The remembered present. A biological theory of consciousness. Basic Books, New York

Edelman GM (1992) Bright air, brilliant fire. On the matter of the mind. Basic Books, New York

Erdelyi MH (1974) A new look at New Look: Perceptual defense and vigilance. Psychol Rev 81: 186–190

Erdelyi MH (2004a) Comments on commentaries: Kihlstrom, Bachmann, Reingold, and Snodgrass. Conscious Cogn 13: 430–433

Erdelyi MH (2004b) Subliminal perception and its cognates: theory, indeterminacy, and time. Conscious Cogn 13: 73–91

Farah MJ (1994) Visual perception and visual awareness after brain damage: a tutorial overview. In: Umiltà C, Moscovitch M (eds) Attention and performance. MIT Press, Cambridge, MA, pp 37–75 [reprint in: Block N, Flanagan O, Güzeldere G (eds) The nature of consciousness. Philosophical debates. MIT Press, Cambridge, MA 1997, pp 203–236]

Farah MJ (2001) Consciousness. In: Rapp B (ed) The handbook of cognitive neuropsychology: what deficits reveal about the human mind. Psychology Press/Taylor & Francis, London

Fendrich R, Wessinger CM, Gazzaniga MS (1992) Residual vision in a scotoma: implications for blindsight. Science 258: 1489–1491

Fisher C (1954) Dreams and perception: The role of preconscious primary modes of perception in dream formation. J Am Psychoanal Assoc 2: 389–445

Fisher C (1956) Dreams, images, and perception: a study of unconscious-preconscious relations. J Am Psychoanal Assoc 4: 5–48

Fisher C (1957) A study of the preliminary stages of the construction of dreams and images. J Am Psychoanal Assoc 5: 60–87

Fisher C (1959) Further observations on the Pötzl phenomenon: The effects of subliminal visual stimulation on dreams, images, and hallucinations. Psychoanalysis and Contemporary Thought 11 (1984): 3–56 [original in: Evol Psychiatr 4 (1959): 551–566]

Flanagan O (1992) Consciousness reconsidered. MIT Press, Cambridge, MA

Fonagy P (1999) Memory and therapeutic action. Int J Psychoanal 80: 215–224

Fonagy P (2001) Attachment theory and psychoanalysis. Other Press, New York

Fonagy P (ed) (2002) An open door review of outcome studies in psychoanalysis. 2nd edn. International Psychoanalytical Association, London

Freud S (1896b) Further remarks on the neuro-psychoses of defence. In: The Standard Edition of the complete psychological works of Sigmund Freud (SE) vol 3. London, Hogarth Press, pp 162–185

Freud S (1900a) The interpretation of dreams. In SE vol 4–5

Freud, S. (1910i) The psycho-analytic view of psychogenic disturbance of vision. In: SE vol 11, pp 211–218
Freud, S. (1912g) A note on the unconscious in psycho-analysis. In: SE vol 12, pp 260–266
Freud S (1915e) The unconscious. In: SE vol 14, pp 166–204
Freud S (1917a) A difficulty in the path of psychoanalysis. In: SE vol 17, pp 137–144
Freud S (1923b) The ego and the id. In: SE vol 19, pp 12–59
Freud S (1925e) The resistances to psychoanalysis. In: SE vol 19, pp 213–224
Freud S (1933a [1932]) New introductory lectures on psycho-analysis. In: SE vol 22, pp 1–182
Freud S (1940a [1938]) An outline of psycho-analysis. In: SE vol 23, pp 144–207
Freud S (1950c [1895]) Project for a scientific psychology. In: SE vol 1, pp 295–397
Freud S, Ferenczi S (1996) The correspondence of Sigmund Freud and Sándor Ferenczi 1914–1919, vol 2, Brabant E, Falzeder E, Giampieri-Deutsch P (eds) Harvard University Press, Cambridge, MA
Gabbard G O (2005) Psychodynamic psychiatry in clinical practice. 4th edn American Psychiatric Publishing Inc, Washington, London
Gallese V (2006) Mirror neurons and intentional attunement: Commentary on Olds. J Am Psychoanal Assoc 54/1: 47–57
Gazzaniga MS, Fendrich R, Wessinger CM (1994) Blindsight reconsidered. Curr Dir Psychol Sci 3: 93–96
Giampieri-Deutsch P (ed) (2002) Psychoanalyse im Dialog der Wissenschaften. Europäische Perspektiven, vol 1. Kohlhammer, Stuttgart
Giampieri-Deutsch P (ed) (2004) Psychoanalyse im Dialog der Wissenschaften. Anglo-amerikanische Perspektiven, vol 2. Kohlhammer, Stuttgart
Giampieri-Deutsch P (2005a) Approaching contemporary psychoanalytic research. In: Giampieri-Deutsch P (ed) Psychoanalysis as an empirical, interdisciplinary science. Verlag Österr Akad Wiss, Wien, pp 15–53
Giampieri-Deutsch P (ed) (2005b) Psychoanalysis as an empirical, interdisciplinary science. Collected papers on contemporary psychoanalytic research. Verlag Österr Akad Wiss, Wien
Giampieri-Deutsch P (2008a) Die Erkenntnis des Geistes, Mentalen in der Psychoanalyse. In: Appel K, Weber HP, Langthaler R, Müller S (eds) Naturalisierung des Geistes? Beiträge zur gegenwärtigen Debatte um den Geist. Königshausen & Neumann, Würzburg, pp 135–150
Giampieri-Deutsch P (2008b) Zur Erkenntnis des Bewusstseins: Perspektiven aus der Psychoanalyse und aus der analytischen Philosophie des Geistes unter Berücksichtigung der experimentellen Wissenschaften. In: Gloy K (ed) Kollektiv- und Individualbewusstsein. Königshausen & Neumann, Würzburg, pp 87–107
Giampieri-Deutsch P (ed) (2009) Geist, Gehirn, Verhalten: Sigmund Freud und die modernen Wissenschaften. Königshausen & Neumann, Würzburg
Giampieri-Deutsch P (2010) Some remarks on psychoanalytic research and universities. Int Forum Psychoan 19/4: 210–217
Güzeldere G, Flanagan O, Hardcastle VG (2000) The nature and function of consciousness: lessons from blindsight. In: Gazzaniga M (ed) The new cognitive neurosciences. MIT Press, Cambridge, MA, pp 1277–1284
Hartmann H (1959) Psychoanalysis as a scientific theory. In: Hook S (ed) Psychoanalysis, scientific method, and philosophy. New York University Press, New York, pp 3–37
Hofer M (2001) Origins of attachment and regulators of development within early social interactions: from animal to human. In: Kalverboer A F, Gramsbergen A (eds) Handbook of brain and behaviour in human development. Kluwer, Dordrecht, Boston, London, pp 821–840
Holder A (1982) Freuds Theorie des psychischen Apparats. In: Eicke D (ed) Tiefenpsychologie vol 1. Beltz Verlag, Weinheim, Basel
Holt RH (1968) Manual for the scoring of primary-process manifestations in Rorschach responses. 10th edn Research Center for Mental Health, New York University, New York
Holt RH (1989) Freud reappraised. A fresh look at psychoanalytic theory. The Guilford Press, New York, London
James W (1890) The principles of psychology. Harvard University Press, Cambridge, MA 1981
Kandel E (2005) Psychiatry, psychoanalysis and the new biology of the mind. American Psychiatric Publishing, Washington, DC, London
Kaplan-Solms K, Solms M (2000) Clinical studies in neuro-psychoanalysis. Introduction to a depth neuropsychology. Karnac Books, London, New York
Kihlstrom JF (2004) Availability, accessibility, and subliminal perception. Conscious Cogn 13: 92–100

LeDoux J (1996) The emotional brain. The mysterious underpinnings of emotional life. Simon and Schuster, New York

Leichsenring F, Rabung S (2008). Effectiveness of long-term psychodynamic psychotherapy: a meta-analysis. J Am Medical Assoc 300 (13): 1551–1565

Leichsenring F, Rabung S (2011) Long-term psychodynamic psychotherapy in complex mental disorders: update of a meta-analysis. Br J Psychiatry 199: 15–22

Leichsenring F, Salzer S, Jaeger U, Kächele H, Kreische R, Leweke F, Rueger U, Winkelbach C, Leibing E (2009). Short-term psychodynamic psychotherapy and cognitive-behavioral therapy in generalized anxiety disorder: a randomized, controlled trial. Am J Psychiatry 166(8): 875–81

Levin F (2004) Das exekutive Aufmerksamkeitsnetzwerk (ECN), selektive Aufmerksamkeit und kognitives Defizit: Mögliche Mechanismen der Lernstörung (LS) und der Aufmerksamkeitsdefizitstörung (ADS). In: Giampieri-Deutsch P (ed) Psychoanalyse im Dialog der Wissenschaften. Anglo-amerikanische Perspektiven, vol 2. Kohlhammer, Stuttgart, pp 192–212

Levin F (2009) Emotion and the psychodynamics of the cerebellum (CB): a neuro-psychoanalytic analysis and synthesis. In: Giampieri-Deutsch P (ed) Geist, Gehirn, Verhalten: Sigmund Freud und die modernen Wissenschaften. Königshausen & Neumann, Würzburg, pp 67–83

Luborsky L (1988) Recurrent momentary forgetting: its contents and its context. In: Horowitz M (ed) Psychodynamics and cognition. University of Chicago Press, Chicago, pp 223–251

Marcel AJ (1983) Conscious and unconscious perception: an approach to the relations between phenomenal experience and perceptual processes. Cogn Psychol 15: 238–300

Merikle PM, Daneman M (2000) Conscious vs. unconscious perception. In: Gazzaniga M (ed) The new cognitive neurosciences. MIT Press, Cambridge, MA, pp 1295–1303

Milner B, Squire L R, Kandel E R (1998) Cognitive neuroscience and the study of memory. Neuron 20: 445–468

Musatti C (1957) Coscienza e inconscio nelle ricerche sperimentali di Vittorio Benussi. Riv Psicol 51: 3–23

Osgood CE, May WH, Myron MS (1975) Cross-cultural universals of affective meaning. University of Illinois Press, Urbana

Panksepp J (1999) Emotions as viewed by psychoanalysis and neuroscience: an exercise in consilience. Neuro-psychoanal 1: 15–38

Poppel E, Held R, Frost D (1973) Residual function after brain wounds involving the central visual pathways in man. Nature 243: 295–96

Pötzl O (1917) Experimentell erregte Traumbilder in ihrer Beziehungen zum indirekten Sehen. Z Gesamte Neurol Psychiatr 37: 278–349

Rapaport D (1959) A historical survey of psychoanalytic ego psychology. Psychol Issues, vol 1, Monograph 1. International Universities Press, New York

Reingold EM (2004) Unconscious perception: assumptions and interpretive difficulties. Conscious Cogn 13: 117–122

Roth G (2004) Wie das Gehirn die Seele macht. In: Giampieri-Deutsch P (ed) Psychoanalyse im Dialog der Wissenschaften. Anglo-amerikanische Perspektiven, vol 2. Kohlhammer, Stuttgart, pp 171–191

Sandler J, Dare C, Holder A (1972) Frames of reference in psychoanalytic psychology. II. The historical context and phases in the development of psychoanalysis. Br J Med Psychol 45: 133–142

Sandler J, Holder A, Dare C (1997) Freud's models of the mind. Karnac books, London

Scoville W B, Milner B (1957) Loss of recent memory after bilateral hippocampal lesions. J Neurol Neurosurg Psychiatry 20: 11–21

Searle J (1979) Expression and meaning. Studies in the theory of speech acts. Cambridge University Press, Cambridge, MA

Searle J (1992) The rediscovery of the mind. MIT Press, Cambridge, MA

Shedler J (2010) The efficacy of psychodynamic psychotherapy. Am Psychol 65: 98–109

Shevrin H (1992) Subliminal perception, memory, and consciousness: Cognitive and dynamic perspectives. In: Bornstein RF, Pittman TS (eds) Perception without awareness: cognitive, clinical, and dynamic perspectives. Guilford Press, New York, pp 123–142

Shevrin H (1997) Commentaries. J Am Psychoanal Assoc 45: 746–753

Shevrin H (2000) The experimental investigation of unconscious conflict, unconscious affect, and unconscious signal anxiety. In: Velmans M (ed) Investigating phenomenal consciousness. John Benjamins Publ Comp, Amsterdam, Philadelphia, pp 33–65

Shevrin H (2004) Die experimentelle Untersuchung von unbewusstem Konflikt, unbewusstem Affekt und unbewusster Signalangst. In: Giampieri-Deutsch P (ed) Psychoanalyse im Dialog der Wissenschaften. Anglo-amerikanische Perspektiven, vol 2. Kohlhammer, Stuttgart, pp 114–142

Shevrin H (2005) Toward a theory on consciousness based on recent developments in subliminal research. In: Giampieri-Deutsch P (ed) Psychoanalysis as an empirical, interdisciplinary science. Verlag Österr Akad Wiss, Wien, pp 57–74

Shevrin H, Bond J, Brakel L, Hertel R, Williams W (1996) Conscious and unconscious processes. Psychodynamic, cognitive, and neurophysiological convergences. Guilford Press, New York, London

Shevrin H, Dickman S (2003) The psychological unconscious: a necessary assumption for all psychological theory? In: Baars BJ, Banks WP, Newman JB (eds) Essential sources in the scientific study of consciousness. MIT Press, Cambridge, MA pp 541–538

Shevrin H, Fisher C (1967) Changes in the effects of a waking subliminal stimulus as a function of dreaming and nondreaming sleep. J Abnorm Psychol 72(4): 362–368

Shevrin H, Luborsky L (1958) The measurement of preconscious perception in dreams and images: An investigation of the Poetzl phenomenon. J Abnorm Soc Psychol 56: 285–294

Snodgrass M (2004) The dissociation paradigm and its discontents: How can unconscious perception or memory be inferred? Conscious Cogn 13: 107–116

Snodgrass M, Shevrin H (2006) Unconscious inhibition and facilitation at the objective detection threshold: Replicable and qualitatively different unconscious perceptual effects. Cognition 101(1): 43–79

Solms M (1997) What is consciousness? J Am Psychoanal Assoc 45: 681–703 and 765–778

Solms M (2004) Ein Beispiel aus der neuropsychoanalytischen Forschung: Das Korsakow-Syndrom. In: Giampieri-Deutsch P (ed) Psychoanalyse im Dialog der Wissenschaften. Anglo-amerikanische Perspektiven, vol 2. Kohlhammer, Stuttgart, pp 213–228

Squire L R, Kandel E R (1999) Memory: from mind to molecules, Scientific American Library, New York

Sulloway F J (1979) Freud, biologist of the mind. Burnett, London

Trevarthen C (2001) The neurobiology of early communication: Intersubjective regulations in human brain development. In: Kalverboer A F, Gramsbergen A (eds) Handbook of brain and behaviour in human development. Kluwer, Dordrecht, Boston, London, pp 841–881

Trevarthen C, Reddy V (2007) Consciousness in infants. In: Velmans M, Schneider S (eds) The Blackwell companion to consciousness. Blackwell Oxford, pp 41–57

van Gulick R (1994) Deficit studies and the function of phenomenal consciousness. In: Graham G, Stephens GL (eds) Philosophical psychopathology. MIT Press, Cambridge, MA

Velmans M (1991) Is human information processing conscious? Behav Brain Sci 14: 651–726

Velmans M (2000) An introduction to investigating phenomenal consciousness. In: Velmans M (ed) Investigating phenomenal consciousness. John Benjamins Publ Comp, Amsterdam, Philadelphia, pp 1–15

Weiskrantz L (1986) Blindsight: a case study and implications. Oxford University Press, Oxford

Weiskrantz L (1990) Outlooks for blindsight: Explicit methodologies for implicit processes. Proc R Soc Lond B Biol Sci 239: 247–78

Weiskrantz L (1997) Consciousness lost and found. Oxford University Press, Oxford

Weiskrantz L (1998) Blindsight: a case study and implications. 2nd edn. Oxford University Press, Oxford

Weiskrantz L (2007) The case of blindsight. In: Velmans M, Schneider S (eds) The Blackwell companion to consciousness. Blackwell Publishing, Malden MA Oxford, pp 175–180

Weiskrantz L, Warrington EK, Sanders MD, Marshall J (1974) Visual capacity in the hemianopic field following a restricted occipital ablation. Brain 97: 709–728

Wong PS, Bernat E, Bunce SC, Shevrin H (1997) Brain indices of nonconscious associative learning. Conscious Cogn 6, 519–544

Wong PS, Shevrin H, Williams WJ (1994) Conscious and nonconscious processes: An ERP index of an anticipatory response in a conditioning paradigm using visually masked stimuli. Psychophysiol 35, 87–101

Consciousness, states of consciousness, unconscious psychological processes, and psychological states

15

Howard Shevrin

Contents

1. Introduction

As psychology under the auspices of cognitive science once again welcomes consciousness to sit at its table, it becomes important to know what kind of guest we have invited back. Is he the same unmanageable critter he was in the 19th Century? Will he eat us out of scientific house and home while our behaviorist in-laws tell us, "I told you so"? And what about some of his old friends – mental contents, states of consciousness, and unconscious processes – will they break up the party and force a return to the Spartan fare of psychology's preconscious era?

In my view these dangers exist as well as new opportunities for bringing together such disparate enterprises as cognitive science, neurophysiology, and psychoanalysis. There are at least four questions which need to be addressed: i) Is there an ambiguity in the current use of the term consciousness? ii) How are mental contents related to men-

Howard Shevrin
University of Michigan
Department of Psychology
1012 East Hall
530 Church Street
Ann Arbor, MI 48109-1043, USA
e-mail: shevrin@umich.edu

tal processes? iii) How are consciousness, states of consciousness, and unconscious processes related to each other? iv) How are consciousness, mind, brain, and person related to each other? Clearly, there can be no definitive answers to these questions, in particular, as offered by me at this time, but we can hope for empirically useful, conceptually consistent clarifications. In what follows I will draw on several different sources including cognitive research, neurophysiology, and psychoanalytic clinical practice and theory.

2. Four questions

(i) Is there an ambiguity in the current use of the term consciousness?

There is indeed a significant ambiguity in the use of the term which needs to be clarified before any useful discussion can take place. I refer to the dual and often confounded uses of the term consciousness to refer to *experience*, or awareness, on the one hand, and to *psychological system* or *structure*, on the other hand. Thus, Sperry at one point speculates that there are "... central processes specifically organized for conscious awareness." (Sperry 1969, p. 535) Consciousness is defined as a "holistic systemic property and an active dynamic part of high-order brain functioning." (Sperry 1977, p. 117) Elsewhere, however, he stresses that the "... laying down, storage, cataloging, and retrieval of memories seem to proceed very largely on the basis of their holistic properties (i. e., consciousness) rather than those of the neuronal infra-mechanisms." (Sperry 1977, p. 122) In other words, "holistic properties" not only are associated with consciousness as awareness, but also with those systematic properties having to do with memory formation and retrieval. Problems develop with this dual usage when we employ the term consciousness without any clear indication as to which meaning we have in mind. When James spoke of a "split-off" consciousness and when neo-dissociationists speak of "dissociated states of consciousness," is it intended to mean that the individual is *experiencing* that "split-off" state, or is it rather that the "split-off" state exists in some significant systematic sense albeit without the attribute of being experienced at the same time? Moreover, if it is experienced at the time, in what sense is it then "split-off"? Certainly we can have *alternating* experiences, but then once again we can ask in what sense is one state "split-off" while the other is being experienced? (I will return to this question in the section below dealing with consciousness, states of consciousness and unconscious processes.)

One can also apply this distinction between consciousness as experience and consciousness as system to split-brain patients, who have remarkably different experiences linked to the two hemispheres. In extreme cases when a female split-brain patient was shown a nude woman to the left hemisphere the patient described it as a nude woman without any apparent emotional reaction; when shown the same picture to the right hemisphere she became visibly embarrassed but could not describe the picture. The evidence shows that these patients, to a considerable extent (and only to an extent, not totally or absolutely), have two different *experiences* of stimuli appearing in the left and right· half fields. But when Sperry goes on to argue that split-brain patients therefore have two separate *minds*, he is then saying that split-brain patients not only suffer from a "split" in conscious *experience* but also suffer from a division in psychological structure beyond experience as such. Moreover, some theoreticians have in fact gone beyond this and argue that if there are two minds, then there are also two persons and we are all one variation or another of multiple personalities, some more dramatically multiple than others. But as Marks has argued, in my view persuasively, disunified consciousness (that is, disunified *experience*) does not necessarily imply a disunified mind or a mitotic personality (Marks 1981). Without at this point going into the details of Marks's argument, which will concern us later, I wish simply to underscore the importance conceptually of distinguishing consciousness as experience from consciousness as psychological system or structure.

Psychoanalysis has not escaped this ambiguous usage of the term consciousness. In Freud's early formulation of the relationship of unconscious to conscious processes (the topographic model), he talked simply about an idea *becoming* conscious: Freud, however, had also developed a *systems* model of conscious and unconscious processes so that instincts, for example, were located in the system Unconscious and all of what we ordinarily consider conscious experience was located in the *system* Consciousness. Moreover, different principles of mental organization prevailed in the *system* Unconscious from the *system* Consciousness – the so called primary and secondary processes. But then what was Freud to do with defenses such as repression which were *unconscious* in their operation, not experienced, but did not belong with the instincts in the system Unconscious because they were in fact directed at controlling those very instincts. He tried to resolve this dilemma by creating a *third* model which is the one that is most widely known – the so-called structural model composed of Id, Ego, and Super-Ego. In this model one could have *systematically* unconscious processes in the Ego as well as in the Super-Ego along with conscious experience.

The distinction between consciousness as experience and consciousness as system is also useful in understanding much recent work on attention. There is, for example, a strong tendency among attention theorists to talk about consciousness as having certain structural characteristics which in effect implies that it is more than simply experience. Thus, to become conscious of a stimulus also involves a *single channel* (what Freud called the "defile of consciousness"). Now, it may be true that in the process of becoming aware of a stimulus only one stimulus can become conscious at a given time, but these are separable issues and in fact it is likely that this restriction may not be generally true. We are also conscious in the experiential sense during dreaming, states of intoxication, and psychosis. The fascinating characteristic of these states is the bewildering multiplicity of stimuli which present themselves simultaneously, so much so that words often fail us in describing what we are in fact conscious of. Once we distinguish between consciousness as experience and consciousness as system, the process of becoming conscious and what the process of *being* conscious can be clearly distinguished.

Similarly, in subliminal research, the effects of stimuli can be detected while the person remains unaware of what is affecting him and will provide rationalizations of his perceptions much as hypnotic subjects will after executing a post-hypnotic suggestion. Nisbett and Wilson (1977) describe how subjects may choose a particular association and provide a reason for this choice which is at variance with what the investigators know to be the real reason for the choice based on previous associative priming. Again it is necessary to distinguish between consciousness as experience and the process of becoming conscious. (I will leave aside the question whether the real cause is subject to introspective recapture.)

In short, distinguishing consciousness as experience from consciousness as system increases the flexibility and precision of our thinking with respect to a number of interesting questions.

(ii) How are mental contents related to mental processes?

It would be a mistake to equate experience, or awareness, with mental content and to equate the systems meaning of consciousness with mental process. Some such distinction is relied on when it is asserted that we can only become aware of contents and not of processes. As James (1890) observed, "a permanently existing 'idea' which makes its appearance before the footlights of consciousness at periodic intervals is as mythological as an entity as the Jack of Spades." Rather, the *experience* of objects changes as a function of affect, motive, arousal, need, age, etc. Experience is therefore as much a process as any other except that it is a process associated with awareness. We must also be careful not to equate process or system with what is neurophysiological and experience with what is psychological. There are psychological processes which are not to be identified on a one to one basis with neurophysiological processes. Sperry, in particular, has insisted on the importance of this distinction. Moreover, experienced processes must also be associated in some way with neurophysiological events so that there can be no true symmetry present when we equate experience with the psychological and processes with the neurophysiological. Nevertheless there are constants in experience. These constants are the *referred-to-objects* not the experiences themselves which are highly variable psychological events in which a variety of factors play a role. When a person *perceives* an object, he is experiencing a process which has as its external referent an actual stable object. We should not confuse the referent with its internal representation which is usually subject to far more variability and instability than the referred-to-object. The importance of this distinction is borne home in the clinical realm in the study of phobic and delusional patients for whom objects may be associated with internal representations at great variance with reality. Whether a phobia be understood in terms of conditioning or unconscious conflict the internal representation is greatly influenced by powerful affective forces so that the apparent internal representation of the object has as its true referent not the object "out there" but some other internal state, whether it is the memory of a previously feared object or a symbolic representation of a feared, conflict-laden impulse. If, on the other hand, we link rigidly the experienced object, or internal representation, to the referred-to-object and omit experience as a variable process, then it becomes difficult to account for these clinical observations; one is left solely with an explanation based on *misperception* as a function of some *external* set of factors – poor lighting, structural ambiguity of the stimulus, etc. Once mental contents are redefined as experienced process, or internal representations of external referred-to-objects, much greater conceptual flexibility is made possible so that a much broader range of phenomena can be explained. Clearly, this position leaves unexplained how we know the real object from its distorted representations. My only defense is to cite the fact that psychologists, spanning about a hundred years of psychological thought, from James (1890) to Fodor (1981) have suggested that this problem is beyond psychology. Fodor, in fact, makes the solution of this problem contingent on the complete development of all other sciences first! – a position that has its difficulties. Perhaps a provisional solution is to rely on a consensus judgment on the nature of the external object.

(iii) How are consciousness, states of consciousness, and unconscious processes related to each other?

With the distinction between consciousness as experience and consciousness as system discussed above in hand, it becomes an easier task to conceptualize how consciousness, states of consciousness, and unconscious processes are related to each other. The superordinate term would need to be states of consciousness in which the term consciousness was used in its systematic and not its experiential sense. However, since most states of consciousness also include experienced consciousness, terminological confusion can be avoided by referring instead to psychological states and by restricting the term consciousness to its experiential sense. I will be following this convention throughout the remainder of the paper. A psychological state is a system or structure. There can be interesting sequences of such psychological states as occurs, for example, during the sleep-dream cycle. Of special interest is the fact that there appear to be neurophysiological markers for these shifts in psychological states. A desynchronized EEG accompanied by rapid eye movements marks a psychological state in which dream consciousness is highly likely to occur, along with those psychological processes giving rise to the dream experience. Psychoanalysts have offered a variety of hypotheses as to what these psychological processes are like. Interestingly, others have hypothesized (McCarley and Hobson 1977) that there are no meaningful psychological processes involved at all in dream experience but that the dream results from a random firing of cortical neurons which would account for the bewildering and incoherent nature of dream experience. Interestingly, James came up with the same hypothesis, likening what happens during dreams to short circuits in the customary neural pathways. On the basis of this explanation we would no longer be dealing with a psychological state but with the dream as an epiphenomenal event not in need of a psychological explanation. Reductionism of this stripe has its serious drawbacks, not the least of which is that it dismisses out of hand an interesting question: What do dreams mean? or, perhaps more precisely put: "Of what psychological event is the dream an internal representation?", much as we may ask of what the phobic experience is an internal representation.

Of any psychological state we can ask: How do certain representations achieve consciousness, or enter awareness? Much research has been devoted to the normal, waking, alert psychological state in which representations of objects appear to become conscious automatically and are usually quite veridical. However, as cognitive research has explored the seeming automatic consciousness of external objects, it has learned that a great deal has to happen first psychologically before that internal representation is experienced as such. Elsewhere (Shevrin and Dickman 1980) I have reviewed several bodies of research (attention, subliminal perception, stabilized retinal image, and binocular rivalry) bearing on the relationship of unconscious processes to consciousness and arrived at three propositions which in my view summarize the main conclusions to be drawn from these investigations:

1. The initial cognitive stage for all stimuli occurs outside of consciousness.
2. This initial cognitive stage outside of consciousness is psychological in nature, active in its effects on consciousness, and can be different from conscious cognition in its principles of operation.
3. Consciousness of a stimulus is a later and optional stage in cognition.

I also suggested that three sets of factors determine the conditions under which a shift from unconscious processes to experience will occur:

1. Stimulus factors (e.g., loudness, brightness, figural coherence, etc.).
2. State factors (e.g., level of arousal, sleep stage, fatigue, distractability, etc.).
3. Motivational factors (e.g., avoidance of anxiety, guilt, conflict, etc.).

It will be noted that the psychodynamic hypothesis concerning motivation appears as one of a set of conditions not out of keeping with the view of some attention theorists that biologically relevant stimuli (e.g., those related to drives) have privileged access to consciousness through the mechanism of permanently lowered thresholds. Note also that state variables are distinguished from motivational factors insofar as state is taken to refer to *dispositions*. A *disposition*, which is itself an intrinsic part of a given psychological state, can combine with a motive as when a person decides not to pursue a task because he recognizes that he is too tired.

Once psychological state is established as the keystone hypothetical structure, it becomes easier to locate systematically a number of important concepts and to visualize their relationships. I refer to all those processes, unconscious and conscious, that are integrated in some unique pattern so that a particular psychological state emerges. One can begin to sketch a taxonomy of such states: The normal waking state, the rapid eye movement dreaming state, the Stage II sleep state, the Stage IV sleep state, psychotic states, states of intoxication, etc. Clearly once we go beyond these markedly different states the boundaries become less clear and not so easily demarcated. I believe, however, that it would be a useful empirical enterprise to develop such criteria, starting with those states which are markedly different as a basis for developing differentiating criteria. A taxonomy of this kind would be of great help in psychiatric diagnosis. Lastly, I believe we would need to assume that at any given time only one psychological state as such prevails, otherwise the concept of state fragments into any number of ad hoc conditions and the important principle of integration as related to psychological phenomena would be violated. (I will deal with this principle in the next and concluding section.) It follows from this assumption of integration that there is no such psychological state as a truly dissociated condition, or "split-off consciousness," or different minds based in the right and left hemisphere, or multiple personalities in the sense of several truly different persons dwelling in the same body. Rather, it would follow from this principle of integration that in conditions of dissociation, split-brain disorders, and multiple personalities, that there is in fact a functional interaction present, much as in repression the psychoanalyst assumes that the disturbing impulse or fantasy is *kept from* consciousness; it is not simply "split-off ", latent and inactive. In fact, in clinical practice the psychoanalyst relies on implicit influences of the repressed on consciousness to detect the underlying presence of repression. A patient of mine in the midst of a difficult erotic transference observed that she would hate to become pregnant because of the distortions her body would need to go through. She then commented that this was an important issue for her and she would need to pursue it further. The next session, less than 24 hours later, she was failing to get back to what was so important from the previous session until she said that somehow her thoughts wandered to the Pearl Buck novel, "The good earth", and the scene in which the heroine gives birth in a field. Despite the obvious link to her previous observations on pregnancy, the association failed to bring back the memory and with it the opportunity to pursue further the important issue of her attitude toward pregnancy. In cognitive terms the "good earth" association was an *implicit* memory related to the previous session's explicit memory. In psychoanalytic terms it was a *derivative* memory caused by the influence of the repressed memory about pregnancy from the previous session. Whether or not the psychoanalyst chose to point this out or not would depend on other factors bearing on the status of the psychoanalysis. What needs to be stressed in terms of our present considerations, is that the repressed memory does not remain latent and inactive but influence the flow of thought occurring in another psychological state. In this important respect the unity or integration of psychological states is to an extent compromised, and this may be one of the hallmarks of psychopathology.

Lastly, if it is indeed true that the initial stage for all stimuli occurs unconsciously, then this is equally the case for each "split-off" personality or state. Each face of Eve has its unconscious aspect. Thus to speak of multiple states of consciousness or multiple selves, does not rule out the need to deal with the role of unconscious processes. For if consciousness as experience is always optional this principle would apply across the full range of experience, from a single idea to an alternate self. But no matter how major or minor the vicissitudes of conscious experience there is always a preceding unconscious phase and some hypothetical conditions that determine the particular emergence into conscious experience.

(iv) How are consciousness, mind, brain, and person related to each other?

If psychological states are central to our understanding of the role of consciousness as experience and consciousness as system how are psychological states related to our understanding of mind, brain, and person? In his critique of Sperry's view of the psychology of the split-brain person, Marks takes the position that although such patients suffer from a *disunified* consciousness they possess one mind. Following Fodor, Grice, and Perry, Marks defines mind as "those entities the states of which explain our 'propositional attitudes'" (1981, p. 34); by propositional attitudes Marks refers broadly to memory, belief, desire, intention, etc. He concludes that in the case of the split-brain patient that it is the entire brain, rather than the divided hemispheres, which is the "best candidate for having the states the true human psychology countenances." (1981, p. 35) He then offers a case of multiple personality (Mary Reynolds) as a striking opposite instance in which he feels it could be argued that two minds do indeed exist in one body because there is a total breakdown in the unity of propositional attitudes so that one personality has no interaction or knowledge of the other, apparently sharing no memories, beliefs, desires, intentions, which makes it possible to talk about two minds in one body. I would like to extend further Marks's definition of mind while taking exception to his characterization of multiple personalities as possessing two independent minds.

Elsewhere Shevrin (1992) has proposed that what one ordinarily thinks of as *psychological* in nature, which we can for present purposes equate with Marks's definition of mind, be defined as the consequence of brain integration. Although this has some similarity to Sperry's definition, it differs from his in one important respect: It is not consciousness as such that is identified as the consequence of brain integration but *all things psychological*; this is done in order to avoid the ambiguity in the use of the term consciousness already discussed. It follows from this definition of the psychological that the integrated activity of the brain can best be known through psychological means. It also follows that neurophysiological processes are always to be considered as contributory *parts* to this psychological integration. Thus, there can only be partial correlations between neurophysiological and psychological measures. By the same token these partial correlations can be theoretically quite significant. Take for example the desynchronized electroencephalogram (EEG) and rapid eye movements present during dreaming sleep, or Libet's findings relating a certain duration of cortical activation and awareness of a stimulus (1973), or my own research on evoked potential correlates of unconscious processes (Shevrin 2001). In psychiatry a psychopharmacological agent may have an unequivocal biochemical affect in the brain, but how the person actually responds to that effect will be a function of the total integration of brain activity, or of that person's psychology. Thus, a manic patient given lithium may begin to experience the moderating influence of the medication on his manic state but he may paradoxically stop taking the lithium because its beneficial effects disrupt a previous integration serving important adaptive psychological purposes. In the clinical instance I have in mind, the patient, a young sexually active woman, gave up lithium despite is beneficial effects on her manic states, because the lithium also dampened her libido, a not unusual side effect. The flagrant failure of upwards of 50% of patients to take effective medications can in large measure be attributed to this one factor, how the medication disrupts current psychological integrations.

One other consequence follows from defining the psychological as a function of brain integration: No psychological process can in and of itself be localized in a particular part of the brain; thus, this must be true for unconscious psychological processes which are also functions of brain integration and in this sense they are as psychological as conscious processes. It follows from this position that the psychological cannot be equated with consciousness as experience and the unconscious with neurophysiological processes. Both conscious and unconscious processes are fully psychological because they are both a function of brain integration and are significant, interacting processes constituting psychological states.

Lastly, it is inconsistent with this view of the mind and the relationship of the psychological to the neurophysiological, to suppose as Marks does, that multiple personalities have plural minds. Rather, it must follow from this position that multiple personalities are characterized by a remarkable, exotic, and recondite form of integration. I believe this to be a logical necessity given my assumptions. However, I also believe that there is evidence to support this view as well as a clinical theoretical approach that is consistent with the evidence. Often in the close examination of multiple personalities it is found that there is a hierarchical relationship of acquaintanceship among the various personalities which in itself suggests that there is an underlying integration at work. Usually this hierarchical ordering of acquaintanceship is unidirectional, with one personality knowing more about the second than the second knows about the first. To understand this fact I would invoke psychoanalytic clinical theory which would hypothesize that this kind of "splitting" must serve defensive needs for the *individual* understood in this instance to be one *superordinate* person possessing one mind albeit fractionated into several self-identified "personalities." It would make better sense to consider each "personality" as a certain special subset of psychological states, rather than as a separate mind or person in the true sense.

The notion of "person" within the proposed frame of reference, is a higher order construct referring to the necessarily unique organization of psychological states characterizing each individual with its enacting and adaptive capacities primarily in mind. By this definition "person" could not refer to one part of the brain, as in split-brain patients, or to one set of psychological states as in multiple personalities and dissociated conditions, but to one overarching organization.

Conclusion

To return to my opening remarks, I would welcome consciousness back to psychology with the following provisos which should be clearly spelled out in the invitation: i) Consciousness should be identified unambiguously as experience or awareness; ii) It should not be considered to be made up of mental contents but of mental processes; iii) Conscious processes are one important subset of processes constituting psychological states, which also include unconscious psychological processes; iv) That it would make for clearer discourse and research if the term consciousness were restricted to its meaning as experience or awareness, so that *psychological state* would then become the superordinate hypothetical structure. If these provisos were accepted I would be confident that the festivities would go well and that at the very least cognitive science, neurophysiology, and the clinical and conceptual contributions of psychoanalysis, would find some good work to do together.

References

Fodor J A (1981) Representations. Cambridge, MA, Bradford Books, MIT Press

James W (1890) The principles of psychology. Boston, MA, Henry Holt & Co

Libet B (1973) Electrical stimulation of cortex in humans and conscious sensory aspects. In: Iggo A (ed) Handbook of sensory physiology, vol. 2. New York, Springer, pp 744–790

Marks C E (1981) Commissurotomy, consciousness and unity of mind. Cambridge, MA, MIT Press

McCarley R, Hobson J (1977) The neurobiological origins of psychoanalytic dream theory. Am J Psychiatry 134(11): 1211–1221

Nisbett R E, Wilson T D (1977) Telling more than we can know: verbal reports on mental processes. Psychol Rev 84: 231–259

Shevrin H (1992) Subliminal perception, memory and consciousness: Cognitive and dynamic perspectives. In: Bornstein R F, Pittman T S (eds) Perception without awareness. New York, NY, Guilford Publications, pp 123–142

Shevrin H (2001) Event-related markers of unconscious processes. Internat J Psychophysiol 42: 209–218

Shevrin H, Dickman S (1980) The psychological unconscious: A necessary assumption for all psychological theory? Am Psychol 35: 421–434

Sperry R W (1969) A modified concept of consciousness. Psychol Rev 76(6): 532–536

Sperry R W (1977) Forebrain commissurotomy and conscious awareness. J Medicine and Philosophy 2(2): 101–146

Explicit and implicit memory 16

Matthew Hugh Erdelyi*

Contents

Abstract

The recent explicit-implicit distinction has become ubiquitous in psychology, being aligned with other contrapositions such as conscious-unconscious, intentional-unintentional, direct-indirect. Since such distinctions are not always coterminous – or clear-cut – construct-validity is an issue. Nevertheless, these polarities point to classic phenomena of psychology and neuroscience. Freud's (1891) neuroscience monograph "On aphasia" prefigures crucial themes in cognitive psychology, including psychoanalysis: The inaccessible may be available unconsciously; unconscious materials may be reflected indirectly (in symptoms, associations, behavior) and may be recovered into consciousness; twilight states (dreams, free-associations, art) may be indirect indicators of unconscious content. Laboratory work extending back to Pötzl (1917) and Fisher (1956; 1960a; 1988) shows that dreams and fantasy can yield reliable and valid measures of inaccessible content and, also, that Freud's manifest-latent distinction is scientifically viable, with recent psychophysical measurements of sensitivity (e. g., d') having been made of latent-content sensitivity. This nascent "psychophysics of the third ear" reveals individual differences in sensitivity to latent contents. Laboratory work also shows that memory waxes and wanes over time, suggesting that observed dissociations between indicators of memory may also increase or decrease, or even reverse, over time.

Matthew Hugh Erdelyi
The University of New York
Department of Psychology, Brooklyn College,
2900 Bedford Avenue, Brooklyn, NY 11 210, USA
e-mails: iyledre@comcast.net;
Erdelyi@brooklyn.cuny.edu

* For helpful comments on earlier drafts of this paper, I thank Ran Hassin, Arthur Reber, Jim Uleman, Ralph Erich Schmidt, David Shanks, and Diane Zizak.

1. Introduction

Even before Wilhelm Wundt's official founding of scientific psychology in 1879 in Leipzig (Wundt 1879), there was appreciation in scientific circles of the existence of complex cognitive processes that were not accessible to consciousness but which could be inferred from their effects. The great 19 th century physicist-neurologist-psychologist, Hermann von Helmholtz–an intellectual grandfather of sorts of Sigmund Freud, since Helmholtz was a teacher of Ernst Brücke, who in turn was a teacher of Freud–introduced the notion of *unconscious inference* to explain the remarkable constancies observed in the perception of, for example, size and shape, despite the continuously varying images that real-world stimuli subtend on the retina. Helmholtz suggested that from optic rules not consciously accessible to naive perceivers, a series of corrective transformations (also inaccessible to consciousness) on the raw percepts imparted psychological constancy to the perception of the wildly unstable visual field.

Despite such illustrious foreshadowing, scientific psychology was highly suspicious of the notion of the unconscious for various reasons, including its association with questionable traditions like "psychic phenomena" and *Mesmerism* (see Ellenberger 1970; Erdelyi 1985) and concerns over scientific rigor. Although he was not always consistent on the subject, William James famously excoriated the notion of unconscious mentation as "the sovreign means for believing what one likes in psychology, and of turning what might become a science into a tumbling-ground for whimsies" (1890, p. 163). Similarly, Titchner, who transplanted Wundt's psychology to America (at Cornell University), warned that "when we invent an unconscious mind to give coherence and continuity to the conscious . . . we voluntarily leave the sphere of fact for the sphere of fiction" (1917, p. 40).

Actually, the unconscious was "dead-on-arrival" in Wundt's system since psychology was defined by him as *the science of mind* and the reigning philosophical tradition of those days identified mind with consciousness. Thus, non-conscious activity was, *by definition* (as Freud not infrequently noted) outside the purview of the mental or psychological. When such a concept was called for–i. e., when a term was needed for structured processes that were not accessible to the subject's consciousness, terms such as "latent" or "state-of-tendency" were typically employed in this era. A question that will be examined in the next section is whether more modern terms like "implicit" are being used in a corresponding fashion to dodge still-troublesome (to some) associations attached to the "unconscious."

The behaviorists, as is well known, carried psychology's rejectionism even further, dismissing not only the "unconscious" but "consciousness" as well. Psychology was no longer the science of mind/consciousness but the science of *behavior* (Watson 1924) and psychoanalytic concepts, including the unconscious, were dismissed out of hand as intellectual aberrations on the plane with phrenology whose concepts were throwbacks to demonology (p. 297). Echoes of this rejectionist stance are to be found among more modern figures who would count themselves on this side of the "cognitive revolution." Bandura and Walters (1963), for example, echoing Watson's stance (and tone), characterize basic psychoanalytic concepts, including the notion of unconscious psychodynamic forces, as forms of "demonology" and "mystical thinking" that harken back to the "dark ages" (p. 30).

The "unconscious" is highly discordant with behaviorist dogma not only because of behaviorism's rejection of the conscious-unconscious dimension but also because the implicit processes giving rise to it are not directly observable but must be inferred. Here, actually, would be an advantage of "implicit" over "unconscious": The unconscious may or may not exist, but structured behavior (perception of constancy, perception and production of grammatical speech) does unquestionably exist that *implies* (and hence is implicit of) underlying structured processing, even if such processing is not accessible to consciousness.

A question raised in the next section (on "Meaning and theory") is whether we are only engaging in a tiresome cavil over terms or whether actual conceptual differences are involved. For this reason, as developed below, a formalization of concepts like "implicit", "unconscious", "subconscious", "automatic" is essential to push the scientific argument forward, beyond the predilections of parochial schools of psychology in which the labels function as subtle battle flags.

It will be argued that *dissociation* is the fundamental, undergirding paradigm (e. g., Erdelyi 1985) and that the various terms deployed over the years are not substantive but short-hands for observed psychological dissociations that are unquestionably real, observable, and reliable.

Consider constancy scaling or the perception of grammatical rules (Chomsky 1957; Reber 1967; 1993). The perceiver claims not to be aware of the rules and, yet, the ruleful behavior implies that knowledge of the rules, at some level, is available to the perceiver. It could be said that these rules are unconscious; or, alternately, that they are implicit (since the ruleful behaviours / percepts imply ruleful substrates). Similarly, the widely cited publication of Claparède (1911) involving a Korsakoff patient, which is often used in the modern literature as an example of the explicit-implicit memory distinction, hinges on a dramatic dissociation: The patient, suffering from dense anterograde amnesia, cannot remember Claparède from one introduction to another. Claparède plays a nasty trick on her, hiding a pin between his fingers and pricking her on shaking hands. On the next meeting, the patient still fails to recognize Claparède but does, in some sense, recognize him by refusing to shake hands with him. Implicitly she remembers him or, at least, she remembers the painful hand-shake of the previous encounter. In alternate terms, she has an unconscious memory of Claparède, which has an impact on her behavior, even if not on her conscious memory. Illustrating the same theme, the case of H. M. (e. g,., Milner 1965; Gabrieli et al. 1993; Corkin 2002), perhaps the most influential case from clinical neurology in modern times, shows the anterograde amnesic, H. M., who could not recall factual and experiential events subsequent to his operation decades earlier (involving the bilateral removal of parts of his medial temporal lobes, including the hippocampus) nevertheless being able to learn, and therefore remember, new motor skills. For example, H. M. got 10 trials in each of 7 sessions, spread over a full year, in tracing a star while watching his hand in a mirror. H. M.'s performance improved dramatically over the year, even though H. M. could not remember on succeeding sessions ever having done mirror tracing before (Gabrieli et al. 1993). Similar dissociations will be noted in the discussion below in clinical and experimental studies of aphasia, subliminal perception, dreams, jokes, and the normal ups-and-downs of memory.

First, however, we turn to a focused discussion of the explicit versus implicit memory distinction, which was my assignment to expound on in this chapter.

2. Meaning and theory

The explicit-implicit memory distinction was introduced relatively recently (Graf and Schacter 1985; Schacter 1987) but has become ubiquitous in cognitive psychology and neuroscience. Hundreds of publications involving the distinction have appeared in the last generation, and if we include cognate domains – implicit learning, implicit attitudes, implicit perception, implicit impression formation – the number extends out to the horizon. Obviously, the explicit-implicit distinction is popular and is treated in many textbooks as fundamental. But, what does it mean?

Here is how one of its originators, Schacter (1987), explains it:

> Implicit memory is revealed when previous experiences facilitate performance on a task that does not require conscious or intentional recollection of those experiences; explicit memory is revealed when performance on a task requires conscious recollection of previous experiences (Schacter 1987, p. 501).

In addition to conscious-unconscious or intentional-unintentional, other contrapositions have been aligned with the explicit-implicit distinction, including, direct vs. indirect, controlled vs. automatic, analytic vs. nonanalytic.

Why this profligacy of terms? If, for example, the explicit-implicit distinction is chiefly the distinction between conscious and unconscious, why not just stick to the traditional nomenclature? Reber, who introduced earlier-on the notion of "implicit learning" of grammars (Reber 1967) – we learn, remember, and produce grammatical speech with little consciousness for the complex rules that guide such grammatical behavior – opted for "implicit" over "unconscious" because, as he suggests with his characteristic candor, he sought to avoid entanglements with Freudian notions which were at that time, and continue still, to be unpopular in mainstream psychology (Reber 1993).

Schacter's explanation for shifting to new terminology was that traditional terms such as unconscious, are "saturated with multiple and possibly misleading meanings" (1987, p. 502). And this is true. In my 1985 book, "Psychoanalysis: Freud's cognitive psychology," I was obliged to attempt an exegesis of the 7 ± 2 significations of "unconscious" in psychoanalysis. This untrammelled polysemy produced confusion not only in psychology but in psychoanalysis also. Freud realized that this was an unacceptable situation and attempted in "The Ego and the Id" (1923) – probably too late – to simplify. The capitalized unconscious (The "Unconscious" or "Ucs.") was reconceptualized as a structure or system of the mind, henceforth to be designated the Id (das Es), and contraposed to the capitalized "Conscious" (or "Preconscious") – Cs./Pcs. – as the Ego (das Ich). When Freud speaks of the laws of the "unconscious" differing from the laws of the "conscious," he is speaking – confusingly – of the capitalized "Unconscious" and "Conscious," i. e., the Id versus the Ego which, by definition, differ qualitatively from one another in basic goal (respectively, fun vs. reality) and in cognitive style (primary process vs. secondary-process functioning).

Unfortunately, the new terminologies of cognitive psychology, including implicit-explicit, have in short-order generated their own "multiple and possibly misleading meanings." It is by no means clear, for example, whether the various polarities used to define the implicit-explicit polarity are themselves theoretically coherent or whether they always covary. Does, for example, conscious-unconscious – explicit-implicit – necessarily align conceptually or empirically with intentional-unintentional or controlled-automatic? Almost surely not. As

John Bargh (1994) has shown, the popular notion of "automaticity" has at least four meanings. Consequently, it may be necessary to go "beyond the false dichotomy of automatic versus controlled processes" (Uleman et al. 2008, pp. 337–338). In a recent book edited by Hassin, Uleman, and Bargh (2005), "The new unconscious," several proposals seem further to undercut the polarities we have used to define implicit-explicit. For example, Glaser and Kihlstrom (2005) argue that volition can be unconscious (e. g., as in the experimentally demonstrated "automatic correction for unintended bias," p. 180) and Hassin (2005) similarly argues that "control" has multiple significations and can include "nonconscious control" (pp. 215ff.). Have we, perhaps, actually managed to compound our difficulties by trying to define one problematical polarity with a tethered set of other problematical polarities?

There are conceptual problems as well. One widely debated issue in the field (see, for example, Roediger and Craik's edited volume of 1989; Roediger 1990) is whether explicit-implicit are to be viewed as distinct memory systems (like Freud's capitalized structures), or whether they are merely short-hand labels for the different results obtained with a variety of different interrogations of memory – involving a variety of different stimuli, instructions, adopted strategies, types of participants, and so on. There is discernible rebellion building against such semantic and theoretical ambiguities, which may stifle or derail scientific progress (e. g., Richardson-Klavehn and Bjork 1988; Merikle and Reingold 1991; Whittlesea and Price 2001; Erdelyi 2004a,b;2006a,b; Berry et al. 2006). For example:

> The implicit/explicit dichotomy divides performance into that which is based on a nonintentional, nonconscious influence of memory and that which is based on an intentional, conscious use of memory. We believe that it is a distinction that does more harm than good…. We see a real danger of reifying [the explicit-implicit dichotomy]. In this way of describing the issue, we believe that two quite separate issues are becoming mixed up, memory as store and memory as use of that store (Whittlesea and Price 2001, pp. 243–245).

In sum, the problem with the implicit-explicit distinction is that it tries simultaneously to capture a set of sometimes co-occurring phenomena and therefore necessarily loses definition, both semantically and conceptually. The explicit-implicit distinction involves not one but a complex of phenomena (as Schacter 1987 acknowledges, p. 511) and we must decide, for the sake of clarity, what it is that we are trying to study.

I believe – and this seems to be the consensus view – that the core issue in the "implicit-explicit" distinction is consciousness – or the lack of consciousness. Implicit memory, shorn of nonessential meanings, comprises, as in the case of Claparède's amnesic patient or H. M., "remembering without awareness," or "the dissociation of memory and awareness (Jacoby and Witherspoon 1982, pp. 300, 302). In short, implicit memory is unconscious memory.

And so, we seem to be back where we started, with the problematical construct, "unconscious". We obviously cannot, as scientists, proceed by endlessly pasting a new label on every new problem (or on every unresolved old problem) and must eventually formalize our concepts. I faced this issue in my work on subliminal perception and in my book on psychoanalysis (Erdelyi 1985). It was clear at this earlier period that, as now, terms were being bandied about ambiguously – and sometimes questionably. There were more than a dozen ways in which the "unconscious" was applied to diverse and far-flung phenomena and I sought to discover the unifying conceptual structure or paradigm that defined the construct. It was obvious from the start that the unconscious was not a "box" of some sort from which a read-out would provide us a measure of what was unconscious. Rather, the concept is inextricably linked to consciousness (for an incisive recent discussion, see Boag 2008). Everything in memory – perception, cognition – is unconscious but for the sliver subset that is conscious. Hence, a definition of unconscious (implicit) could only be realized when it could be demonstrated that more was available in memory than accessible to consciousness. The concept was formalized symbolically by the inequality,

$\varepsilon > \alpha$,

where ε is an indicator of availability and α an indicator of conscious accessibility, and the formalization was designated the "dissociation paradigm of the unconscious" (Erdelyi 1985; 1986; 2004a,b). In plain terms, if more information can be demonstrated in a system than can be accessed to consciousness by the system, the unconscious information (memory, perception, cognition) will have been demonstrated.

In real-life situations, in contrast to the laboratory, the dissociation paradigm typically involves complex indicators (Erdelyi 1985). For example, if we wish to ascertain whether a person is angry, we usually are not satisfied with a single indicator (e.g., ε_a = he is raising his voice) but additionally evaluate other indicators (ε_b = he is turning red; ε_c = his teeth are clenched; ε_d = he is breaking furniture; and so on). Now, if the complex indicator, i.e., the set of all the evaluated indicators, $\{\varepsilon\}$, suggests anger, which, however, the person denies experiencing (consciously), i.e., $\alpha \approx 0$, then we have instantiated the dissociation paradigm with a complex indicator, i.e., $\{\varepsilon\} > \alpha$. Complex indicators, which are the rule in the real-world if not the laboratory, play a critical role, as I will suggest later on, in the distinction between manifest (surface) and latent (deep) semantic contents in normal, everyday human communication – as well as in poetry, jokes, dreams.

My dissociation paradigm of the unconscious was, actually, an elaboration of the treatment of unconscious or subliminal phenomena by the great critic of subliminal perception, Charles Eriksen (1958), which he sometimes referred to as the partial-correlation model:

> I strongly suspect that a considerable amount of the evidence that the clinician finds for unconscious processes is in the nature of discrepancies between concurrent responses. The clinician may note that the participant's autonomic responses and evidences of emotion do not correspond, i.e., do not correlate very well, with his verbal statements. This lack of correlation is the basis for his inference of unconscious processes or defense mechanisms at work. (Eriksen 1958, p. 202)

What Eriksen treated as a criticism, I turned into a formalization of the unconscious. The unconscious is, indeed, an inference derived from the partial correlation (association) – and therefore partial dissociation – between two concurrent indicators (or sets of indicators), one linked to consciousness and another to additional information in the system. This formalization has the advantage of not being amenable to reification – or to false dichotomization, with r ranging continuously from -1 to +1. This is an important nuance because many of the well-known formulations of unconscious/subliminal/implicit processes imply – or insist on – null consciousness (i.e., $\alpha = 0$).

Consensus is emerging, however, that this "special case of the dissociation paradigm" (Erdelyi 1986; 2004a,b),

$\varepsilon > \alpha \mid \alpha = 0$

(i.e., $\varepsilon > \alpha$ *when* $\alpha = 0$) is unattainable or indeterminate (Reingold and Merikle 1988; Erdelyi 1986; Macmillan 1986; Merikle and Reingold 1991; Erdelyi 2004a,b; Haase and Fisk 2004; Holender and Duscherer 2004; Reingold 2004; Reder et al. 2009; Erdelyi, 2010; for a dissenting view, however, see Snodgrass et al. 2004a,b).

The reasons for abandoning the special case of the dissociation paradigm, in which $\alpha = 0$, are interesting. One important consideration is that memory waxes and wanes over time and, also, varies with testing condition. Thus, any measure of "accessibility" is necessarily provisional (it changes, up and down, over time) and parochial (it is specific to a particular technique of measurement or retrieval strategy) (Erdelyi 2004a,b; for an even more relativistic conception of "memory," see Roediger 2008). Further, as suggested by a remarkable experimental finding of Snodgrass et al. (1993), very strange things happen at near-zero levels of consciousness, including sub-chance perception: As perceptual sensitivity begins to approach zero, sensitivity d′ can actually dip below chance levels (e. g., $d' < 0$) (for an overview, see Erdelyi, 2004a,b).

Van Selst and Merikle (1993) attempted to replicate the sub-chance effect and – to their surprise – succeeded. Other replications or variations on the sub-chance effect have appeared (e. g., Greenwald et al. 1995; Snodgrass and Shevrin 2006; Snodgrass et al. 2004a,b; see Erdelyi 2004a, for an overview), and the evidence suggests that individual differences in retrieval strategy, different experimental instructions, different levels of motivation, and other factors, play a role in whether measures of sensitivity, such as d′, will be slightly above or slightly below zero (and not because of chance variations around zero). Thus, even when $\alpha = 0$ as measured by d′ or a homologue, we have not ruled out some consciousness since $\alpha = 0$ could be an averaging artifact of d′ values that are slightly above zero or slightly below zero.

These developments might suggest a hopeless situation for a science of subliminal/unconscious/implicit processes. Instead, they point to a breakthrough. We might have been, for over half a century, barking up the wrong theoretical tree and can now begin to redirect our efforts in profitable directions. It may not be possible to dichotomize perception or memory into two neat categories or states – conscious-unconscious, explicit-implicit, subliminal-supraliminal. The real action lies on a grainy continuum of conscious accessibility. We may continue to speak loosely of conscious vs. unconscious or explicit vs. implicit but what we have, really, is a range or continuum of conscious accessibility. We might do better to think in terms of liminal phenomena occurring within a range of greater or lesser obscuration of consciousness and explore the range itself – from high-liminal, to mid-liminal, to low-liminal, to very low-liminal in order to extract lawful patterns that might not emerge from a mere two-fold approach. We might retain the term "subliminal" for some arbitrarily defined very low level of liminality. A recent publication by Ruys and Staple 2008 on "subliminal priming" by face-stimuli moves in this direction, with two posited levels of subliminality ("super-quick subliminal" [40 ms] and merely "quick subliminal" [120 ms]) shown to produce different response patterns, one more affective and the other more conceptual.

Thus, there exists no "clean boundary between conscious and unconscious processes or a sharp division between implicit and explicit epistemic systems . . . " (Reber 1993, p. 23). The special case of the dissociation paradigm, as first suggested in my commentary on Holender's (1986) critique of perception without awareness, is unrealizable. Nevertheless, the simple dissociation paradigm ($\varepsilon > \alpha$), which depends on the demonstration of more information being available in the system than accessible to consciousness, is logically sufficient for demonstrating unconscious mentation. Weiskrantz (1989), approaching the question from a neuroscience perspective, reached essentially the same conclusion: "the dissociations are just as real when they are relative as when they are absolute" (1989, p. 106). And, from psychophysical considerations similar to my own, Reingold and Merikle (1988; Merikle and Reingold 1991) and Reingold (2004) have been advocating a relative sensitivity approach in which it is not necessary to stipulate null consciousness in the explicit (direct) indicator: it is enough to demonstrate that "indirect sensitivity > direct sensitivity ≥ zero" (Reingold 2004, p. 883).

3. Aphasia

Much of the modern interest in the explicit-implicit distinction arose with the cross-fertilization in the late 20 th century between psychology and clinical neurology. Interestingly, a very similar cross-fertilization occurred during the late 19 th century, which, unfortunately, has been largely lost to mainstream psychology because of the ice age that descended on our field – I am referring to the era of behaviorism – which cleaved these two temporal continents apart. It is difficult to believe – I was there so I believe it – that in the 1950's-1960's there were serious arguments about whether neurology belonged in a graduate psychology curriculum.

It turns out, as I have been intimating, that the earlier research program on the conscious-unconscious distinction yielded substantially similar conclusions as today's efforts and these convergences may be seen as a type of replication of one epoch by the other. The replication (not only of findings but of problems) should not, perhaps, be surprising since, after all, the brain is the brain, and brain traumas – from strokes, to tumors, to gun-shot wounds – should produce the same effects in the 19 th as in the 20 th and 21st centuries. Of interest, however, are the different directions which the psychological research of these two periods took and my own definite scientific predilection is not to dissociate the "new unconscious" from the old (Hassin et al. 2005), but to integrate them, particularly since they substantially complement one another. For example, the notion of "remembering without awareness" ubiquitously encountered in modern cognitive psychology and neuroscience (e. g., Jacoby and Witherspoon 1982; Roediger and Craik 1989; Gabrieli et al 1993) or in psychoanalytic contexts (e. g., Breuer and Freud 1895; Fisher 1956; 1988) are equivalent and derive from the same class of phenomena. Similarly, the distortions observed in narrative recall (e. g., Bartlett 1932; Erdelyi 1996 Chapter 8) are the same whether the distortions arise from cognitive schematization (in line with logic, cultural habits) or defensive reconstructions (in line with wishes, defensive needs); even the terms used by Bartlett and Freud to refer to the distortions observed by them in, respectively, the laboratory and therapy settings, are virtually the same (see Erdelyi 2006 b, especially Table 1, pp. 509–511, 540–542). There is no discernible scientific rationale for fragmenting a coherent psychological picture that has taken over a century to come into focus.

Some of the exciting late 19 th – early 20 th century cognitive neuroscience developments unfolded in Vienna. A Privatdozent at the University of Vienna, Sigmund Freud, published an integrative neuroscience monograph, "On aphasia," in 1891. This was only a few years before Freud was obliged to leave neuroscience research for clinical work as a physician and, soon, as a psychologist (clinical psychologist or psychiatrist, in today's parlance), and developed his psychological system, psychoanalysis.

"On aphasia" contains many of the basic elements of psychoanalysis but also some of the contemporary cognitive psychology themes in memory we have been discussing. For example, there is a frequent emphasis in Freud's neuroscience monograph on the "dissociations" in the brain that must be posited to explain some dozen aphasias discussed in the monograph. Brain localization approaches would identify specific aphasias with damage to specific brain loci. However, the large variety of aphasias suggested (to Freud and others) that many of the aphasias resulted not from problems in a particular brain center but from disconnections of "pathways" "spread widely over the brain" (Freud 1891, pp. 1, 17). My own reluctance to espouse categorical dichotomies of memory and to be skeptical about the realizability of the special case of the dissociation paradigm of the unconscious (where $\alpha = 0$) is in line with Freud's mostly anti-localization position and his observation that disturbances of speech and memory functions are "distributed in various degrees, instead of one being completely lost and another having remained intact" (p. 9).

Freud discusses some cases in which patients recover percepts or memories – sometimes by unexpected techniques, like closing of the eyes – which dovetail with the increasing modern emphasis on the processing strategies the participant uses in memory retrieval as a result of individual differences or experimental instructions (e. g., Graf et al. 1984; Merikle and Reingold 1991; Snodgrass et al. 1993; Berry et al. 2006). Pierre Charcot (as Freud noted toward the end of his aphasia monograph, pp. 98–100) placed great stress on individual differences in processing strategies.

The phenomenon of *paraphasia* – "a speech disorder in which the appropriate word is replaced by a less appropriate one, which, however, still retains a certain relationship to the correct word" (p. 21), clearly prefigures the later Freud seeing meaning in "slips of the tongue" and other errors. I will cite some examples of paraphasias from a 1956 article of Heinz Werner, "Microgenesis and Aphasia". Heinz Werner was, at the time, the most prominent developmental psychologist after Jean Piaget, and had applied some of his notions to the micro-development or "microgenesis" of perception of briefly flashed stimuli. Werner (1956) wondered whether some of his laboratory findings could find counterparts in the clinical neurology literature. Since Werner had, in 1956, a wider network of clinical examples to draw on than the Freud of 1891, Werner's examples are sometimes richer than Freud's.

Many of the neurological cases sampled by Werner show partial, if often distorted, recoveries of target materials. Specific concepts may be inaccessible but their "spheres of meaning" (Werner 1956) are often accessible. A patient, for example, trying to read the word "cigar" produced, instead, "smoke" (p. 348). Another patient (from Conrad 1948), trying to read "butter" made motions of spreading butter but then responded with "Brot" (bread). In this case, after some stumbling effort, the stimulus was recovered by the patient: "Butter... Butterbrot" (bread and butter) (p. 349). The recovery of inaccessible memories over time has been a major experimental and clinical issue in psychology for over a century – it is the central memory issue in Freud's first psychological book, "Studies on hysteria," co-authored with another famous son of Vienna, Josef Breuer (Breuer and Freud 1895). I have devoted a lifetime of laboratory work to this problem (Erdelyi 1996; 2010). The recovery of memories by aphasics, hysterics, or normal college participants reflects a temporal version of the dissociation paradigm, sometimes called the recovery paradigm of the unconscious (Erdelyi 1985), in which a later test of accessibility, α_2, yields more information than an earlier test of accessibility, α_1: thus, $\alpha_2 > \alpha_1$.

Another patient (cited by Werner 1956, from Weisenburg and McBride 1935) could sometimes access "scissors" after he had made finger motions resembling cutting. Behavioral recollections – which may lead to explicit verbal ones – tend to be treated in the contemporary literature as *procedural memory*, which in turn is usually treated as a type of implicit memory. Even when the patient cannot correctly produce the stimulus (words or phrases) his incorrect responses imply that he has grasped its sphere of meaning. One of Conrad's (1948) patients, for example, responded to the stimulus, Zunge (tongue), with: "Something on the body, where and what I do not know." The same patient, trying to read the expression, "Eine flinke Katze kriegt die Mäuse" (A quick cat gets the mice) produced: "Fox and rabbit, or something; a proverb it was, I think; something to do with hunting" (Werner 1956, p. 351). Thus, although the correct percept or memory cannot be quite accessed to consciousness, the patient implicitly shows contact with the inaccessible material, or at least some aspect of it, and claims often to have a "feel" for the inaccessible material. A "feeling" for the unconscious stimulus has also been noted by Weiskrantz (2002a) in blindsight, in which patients claiming no awareness of perceptual stimuli nevertheless show, to their surprise, residual capacities, e.g., correctly reaching for a stimulus they cannot see. Normal participants also show kindred effects, as in "feeling of knowing" (Metcalfe and Wielbe 1987), "visual sensing without seeing" (Galpin et al. 2008; Resnik 2004), and "tip-of-the-tongue" phenomena (Ballard 1913; Brown and McNeill 1966; Brown 1991).

The aphasic patient – indeed, all of us – draws on multiple substrates (pathways, circuits, modules, processing stages) to produce memory responses. These substrates include verbal substrates, images, behaviors (e.g., spreading butter), "feelings," "intuitions," and so on. In some cases – and this is an important nuance – we must subtract-out defective or undesirable subsystems for retrieval success. For example, Lube's patient (cited by Freud 1891, p. 43), could perform better at reading a word when the stimulus was removed from sight (thereby, apparently, preventing him from trying the self-defeating tactic of spelling out the word). Another patient, mentioned by Freud (p. 80), could correctly name an object if he was allowed to hold the object and to close his eyes. More recently, Weiskrantz found, to his surprise, that when one of his blindsight patients, on closing his eyes after the stimulus went off, reported a conscious after-image of the stimulus for which "he had reported absolutely no awareness... " (2002b, p. 573). Late-developing consciousness for initially subliminal stimuli in, for example, after-images or hypnagogic imagery, is actually a classic phenomenon, going back to the works of Helmholtz, Purkinje, Johannes Müller, Urbantschitsch, among others, with the recovered materials undergoing "transformations, e.g., condensations, fragmentations, displacements, rotations, etc., similar to those found in dreams" (Fisher 1960b, p. 3).

4. Subliminal perception

I would like to touch on the work of another important Viennese figure, the clinical neurologist Otto Pötzl (1917), who for many years was a member of the Vienna Psychoanalytic Society. Pötzl had undertaken a project that was the reverse of Werner's, seeking to determine whether he could duplicate the symptoms of his gun-shot wound occipital patients in laboratory tachistoscopic studies on normal participants. (A "tachistoscope" is any instrument that can flash stimuli for very short durations, e. g., 1, 10, 100 ms.) One of these neurological patients, for example, could not perceive central stimuli but would sometimes recover, albeit in distorted form, the previously unseen central stimuli in ongoing peripheral vision.

Pötzl succeeded in his project – as had Werner in his – demonstrating that tachistoscopic stimuli, which the participants could not consciously perceive ($\alpha \approx 0$), nevertheless turned up, in various distorted forms, in the content of their dreams ($\varepsilon > \alpha$). In lovely studies later on, during the New Look era, the psychoanalyst, Charles Fisher, replicated Pötzl's findings on dreams and extended them to other implicit domains such as fantasy and free-associations (Fisher 1956; 1960a; 1988) (for a critical overview, see Erdelyi 1996). In his 1956 paper, Fisher was able to produce visual versions of the types of paraphasias Werner reports. Participants were tachistoscopically presented for 10 ms the picture of a parakeet in-between two Siamese cats. None of the participants reported seeing the parakeet, but their drawings and associations provided implicit evidence for the information having been registered and stored. One participant, after the 10 ms flash, claimed to have seen little, maybe "two black and white animals which resembled dogs or pigs" (p. 23) but produced a drawing with bird-like characteristics in the two animals. In a subsequent free-associative response, the participant drew an image of what she claimed to be "a watchdog standing in front of a house" (p. 25) but produced a "dog" that was even more bird-like (luxuriant feathers for tail, a beak for its mouth). This time she herself noticed the bird-like features of her drawing and, actually, expressed bafflement about her performance, stating that she was quite capable of drawing a dog, and had done so many times before.

A general theme emerges from both the older and the newer studies: Effects observed in neurological patients can be (usually) demonstrated in the laboratory with normal participants (e. g., Erdelyi 1996; Roediger and Craik 1989; Weiskrantz 2002a).

5. Symptoms, dreams, jokes, memory

In turning from aphasic distortions and those observed with tachistoscopic stimuli, which often seem to be the same, I wish to note another striking convergence in the data: The types of distortions observed in aphasia and in subliminal perception are also often the same as those posited by Freud for symptoms, dreams, jokes, and defense mechanisms.

In transforming himself from neurologist to psychologist in the 1890's, Freud parlayed his insights on aphasic-amnesic neurological cases to his pseudo-neurological patients, i. e., "neurotics" or "hysterics." His 1895 book with Breuer (Breuer and Freud 1895), comprising five case-histories, focused on the meaning of patients' superficially puzzling utterances or behaviors. His formula, "Hysterics suffer mainly from reminiscences" (p. 7) is both classic – reflecting the past neurological literature on aphasia and amnesia – and modern – reflecting the notion of "remembering without awareness" (Jacoby and Whiterspoon 1982), i. e., implicit memory. Freud's "Interpretation of dreams" (1900) can also be seen as a vast extrapolation of his experiences with aphasia-amnesia, including the paraphasic distortions of stimuli. Dreams are, in effect, implicit articulations – albeit, often extremely noisy articulations – of materials for which the patient may have little explicit knowledge. The "interpretation" part of Freud's theory is the other side of the "implicit" coin, since, by definition, the "implicit" needs to be interpreted to get at the inaccessible content.

If inaccessible percepts or memories, as Freud, Pötzl, Fisher, and others claimed, emerge in dreams, why has this question not been engaged by modern experimental psychologists of memory? Dreams should be a fundamental "implicit medium" for cognitive psychology since dreams (along with daydreams and fantasy) form an integral part of our daily – and nightly – psychological lives, and reflect memories and emotions often inaccessible to explicit probes. Perhaps critiques of Freud's cognitive dream theory, especially those of Hobson and his associates (e. g., Hobson and McCarley 1977; Hobson 1988), have led to the premature closing of a fabulous goldmine.

In a work in preparation, "The Interpretation of Dreams, and of Jokes", I try to demonstrate that the Hobson-McCarley critique of cognitive dream theory is not logically tenable. One sticking point of cognitive dream theory for Hobson and his colleagues, is that it involves a depth-psychology distinction not usually met with in standard laboratory research, namely the distinction between the manifest and latent content of dreams. Actually, contemporary work on subliminal priming effects on the disambiguation of polysemous stimuli (see, Marcel 1983a,b; Roediger and Craik 1989; Hassin et al. 2005) begins to converge on the larger issue of the semantic multi-layeredness of stimuli in human communication which lies at the heart of Freud's cognitive psychology.

Surprisingly, jokes make what seems to be problematical in dreams quite obvious (Erdelyi 1985; 1999; 2006a,b; 2010; 2011 in preparation). Freud himself, actually, provided this insight in his 1905 book on jokes which both experimental psychologists and psychoanalysts have consigned to a preternatural silence. Yet, as I have suggested elsewhere (Erdelyi 1999), jokes are Goldilocks stimuli – they are not too simple like lists of nonsense syllables or words, nor too complex, like case histories or dreams; they are just right. For example, jokes demonstrate unmistakably the difference between manifest (surface) and latent (deep) semantic content and, moreover, prove the necessity and the feasibility of interpreting latent meanings (Erdelyi 1985; 1999; 2001; Bergstein and Erdelyi 2008).

I have provided a formalization of the manifest-latent content distinction (Erdelyi 1985; 2004a,b; 2006a,b) which renders the distinction obvious and, also, scientifically tractable. The manifest-latent content distinction involves a "vertical" dissociation which, incidentally, needs not be a dissociation between conscious and unconscious. It is the dissociation between the surface semantic content of an event, and the selfsame event in context. As in depth perception, where context yields depth, the event-in-context yields a deeper semantic content than the event-in-isolation. Context can be conceptualized as a complex indicator, i. e., a set of other events, $\{\varepsilon\}$, both external and internal, and the interaction of the target event, ε, with the context, $\{\varepsilon\}$, yields the latent content (Erdelyi 1985; 2010; 2011 in preparation). Thus, the manifest vs. latent distinction is formalized by the inequality,

$$\varepsilon \neq \varepsilon \mid \{\varepsilon\}$$

which, in plain language, means that an event in isolation, ε, may not convey the same meaning as the selfsame event in context, $\varepsilon \mid \{\varepsilon\}$. This is such an obvious point that it is almost embarrassing to have to make it. Yet, this is the simple point of Freud's "depth psychology," which scientific psychology has perennially contested or avoided. (For recent, surprisingly variegated views on the subject, see Macmillan 2001 in "Psychological Inquiry.")

Interpretation still has a bad reputation in scientific psychology – strange since interpretation is the necessary other side of "implicit"– but interpretation, certainly with jokes, can be highly reliable. Jackie Bergstein and I (2008) independently generated the latent content (the meaning) of some 200 cartoons and we come up with essentially the same latent content in some 90% of the cases. Further, we applied standard Signal Detection Theory techniques to measuring sensitivity to latent contents (Swets 1964; Macmillan and Creelman 2005): In a forced-choice procedure, participants had to choose between correct latent contents and distractor latent contents for a series of cartoons, from which hit and false-alarm rates were calculated, and the sensitivity index, d′, computed for each participant. With this technique we were able to demonstrate recognition hypermnesia (enhancement of recognition memory) for latent contents over time with repeated testing. In a small side-bar to the project we were able to document significant individual differences in sensitivity to latent contents (split-half reliabilities were $r = 0.61$). Recently, Kozbelt and Nishioka (2010) were also able to show individual differences in humor comprehension and production. Thus, well-established psychophysical tools can be applied to non-physical latent contents (e.g., the underlying meaning of jokes) and the results point to the possibility of a "psychophysics of the third ear," a psychophysics of insight and not merely of sight, which would have both theoretical and applied implications.

6. The ups and downs of memory

We now draw to the last section of my chapter, which deals with the "provisional" nature of any measure of memory and, therefore, casts further doubt on the possibility of any crisp dichotomy between explicit and implicit – or conscious-unconscious, or accessible-available.

In one of the most underappreciated experimental publication on memory, "Oblivescence and Reminiscence," Ballard (1913) demonstrated that if participants attempt to recall some stimulus not once but at least twice (with usually one or two days' interval between the tests), it is invariably the case for groups of participants that on Test 2 (T2), some materials will be forgotten that had been remembered on Test 1 (T1) – this is what Ballard called "oblivescence" – but also, some materials will be recovered on T2 that had not been recalled on T1 – this is what Ballard called "reminiscence." Thus, memory is perverse over time; it both decreases and increases. Ballard also showed that different stimuli tended to produce different balances between oblivescence and reminiscence. When the balance is positive (as with poetry, pictures), reminiscence outpaces oblivescence over time, and we get overall "improvement," or, as we call it today, hypermnesia (Erdelyi 1996). When the balance is negative (as it tends to be with nonsense syllables), oblivescence outweighs reminiscence, and we get amnesia. "Amnesia," as I use it, has no special link to brain syndromes or pathology. Amnesia, simply, is the opposite of hypermnesia; it is a balance phenomenon we get when oblivescence (forgetting) exceeds reminiscence (recovery) and hypermnesia is a balance phenomenon obtained when reminiscence outpaces oblivescence.

Although the balance phenomena, hypermnesia and amnesia, tend to be linked to stimulus types and, therefore, are found in some cases but not in others, the underlying down-tendency, oblivescence, and the up-tendency, reminiscence, are both bedrock phenomena that, regardless of stimulus class, are invariably found, if to different degrees. Moreover, oblivescence and reminiscence vary over time, and not necessarily in tandem (Erdelyi 1996; 2004a; 2010), and so the balance between them may also change over time, even yielding, sometimes, reversals in their balance over different time-frames, with amnesic memory in one-time frame becoming hypermnesic memory over a different time-frame and then again amnesic memory over a still different time-frame. Thus, conscious ("explicit") memory varies up and down over time and any crisp demarcation between "explicit" and "implicit" memory necessarily breaks down over time.

Conclusion

Memory is both parochial and provisional. It is parochial because any particular measure of memory, even if highly reliable internally, may not be concordant with other measures of memory; a direct, explicit test may yield a drastically different index of memory than an implicit or indirect test. Memory also is provisional in that it varies – both up and down – over time. Memory is fundamentally dialectical over time, reflecting a dance between an up-tendency (reminiscence) and a down-tendency (oblivescence). To the extent that we have some psychological control over the up-function, reminiscence, and the down-function, oblivescence – and we do have some control – we can, without an experimenter's intervention, modulate explicit memory over time (Erdelyi 1996; 2006a,b; Erdelyi and Zizak 2004). We may engage some of these controls for defensive reasons (e.g., we might prefer not to have too much access to some nasty materials) or for cognitive reasons (different levels of accessibility may be more suitable for different tasks). The critical point is that conscious accessibility is not all-or-none – unconscious-conscious, implicit-explicit, subliminal-supraliminal – but more-or-less, varying up and down over time.

References

Ballard PB (1913) Oblivescence and reminiscence. Br J Psychol (Monogr Suppl) 2: 1–82

Bandura A, Walters RH (1963) Social learning and personality development. Holt, Rinehart, & Winston, New York

Bargh JA (1994) The four horsemen of automaticity: Awareness, intention, efficiency, and control in social cognition. In: Wyer RS, Srull TK (eds) Handbook of social cognition (Vol. 1): Basic Processes (2nd Ed.). Erlbaum, Hillsdale NJ, pp 1–40

Bartlett FC (1932) Remembering. Cambridge University Press, Cambridge, England

Bergstein J, Erdelyi MH (2008) Recognition hypermnesia: How to get it. Memory 16: 689–702

Berry CS, Shanks DR, Henson RNA (2006) On the status of unconscious memory: Merikle and Reingold (1991) revisited. J Exp Psychol Learn Mem Cogn 32: 925–934

Boag S (2008) Making sense of subliminal perception. In: Columbus AM (ed) Advance in psychological research. Nova, New York, pp 117–139

Breuer J, Freud S (1895) Studies on hysteria. In: Strachey J (ed and trans) The standard edition of the complete psychological works of Sigmund Freud (Vol. 2). Hogarth Press, London, 1955

Brown AS (1991) Spatial and verbal components of the act of recall. Can J Psychology 22: 349–368

Brown R, McNeill D (1966) The "tip of the tongue phenomenon." J Verb Learn Verb Be 5: 325–337

Chomsky N (1957) Syntactic structures. Mouton, The Hague

Claparède E (1911) Reconnaissance et moiité. Arch Psychologie 11: 79–90

Conrad K (1948) Über Gestalt-und Funktionswandel bei der sensorischen Aphasie. Arch Psychiat 181: 53–92

Corkin S (2002) What's new with the amnesic patient, H. M.? Nature Rev Neurosci 3: 153–160

Ellenberger HF (1970) The discovery of the unconscious. Basic Books, New York

Erdelyi MH (1985) Psychoanalysis: Freud's cognitive psychology. Freeman, New York

Erdelyi MH (1986) Experimental indeterminacies in the dissociation paradigm of subliminal perception. Behav Brain Sci 9: 30–31

Erdelyi MH (1996) The recovery of unconscious memories: Hypermnesia and reminiscence. The University of Chicago Press, Chicago

Erdelyi MH (1999) The unconscious, art, and psychoanalysis. Psychoanal Contemp Thought 22: 609–626

Erdelyi MH (2001) Studies in historicism: Archaeological digs will not resolve the scientific questions of validity and reliability in free-association and interpretation. Psychol Inq 12: 133–135

Erdelyi MH (2004a) Subliminal perception and its cognates: Theory, indeterminacy, and time. Conscious Cogn 13: 73–91

Erdelyi MH (2004b) Comments on commentaries: Kihlstrom, Bachmann, Reingold, Snodgrass. Conscious Cogn 13: 430–433

Erdelyi MH (2006a) The unified theory of repression. Behav Brain Sci 29: 499–511

Erdelyi MH (2006b) The return of the repressed. (Response to commentaries.). Behav Brain Sci 29: 535–551

Erdelyi MH (2010) The ups and downs of memory. Amer Psychol 65: 623–633. doi: 1037/a0020440

Erdelyi MH (2011, in preparation) The interpretation of dreams, and of jokes. Unpublished manuscript

Erdelyi MH, Zizak DM (2004) Beyond gizmo subliminality. In: Shrum LJ (ed) Blurring the lines: The psychology of entertainment media. Erlbaum, Mahwah, pp 13–43

Eriksen CW (1958) Unconscious processes. In: M. R. Jones (ed) Nebr Symp Motiv (Vol. 6) University of Nebraska Press, Lincoln, pp. 169–27

Fisher C (1956) Dreams, images, and perception: A study of unconscious-preconscious relationships. J Am Psychoanaly Ass 4: 380–445

Fisher C (1960a) Subliminal and supraliminal influences on dreams. Am J Psychiat 116: 1009–1017

Fisher C (1960b) Introduction. In: Wolff J, Rapaport D, Annin SH (eds and trans.) Preconscious stimulation in dreams, associations, and images: Classical studies by Otto Pötzl, Rudolf Allers, and Jakob Teler. Psychol Issues, 2 (No. 3), Monograph 7. International University Press, New York, 1960, pp 1–40

Fisher C (1988) Further observations on the Pötzl phenomenon: The effects of subliminal visual stimulation on dreams, images, and hallucinations. Psychoanal Contemp Thought 11: 3–56

Freud S (1891) On aphasia. (Stengel E, trans.). International University Press, New York, 1953

Freud S (1900) The interpretation of dreams. In: Strachey J (ed and trans) The standard edition of the complete psychological works of Sigmund Freud (Vols 4 & 5, pp i-338 and pp iii-627). Hogarth Press, London, 1953

Freud S (1905) Jokes and their relation to the unconscious. In: Strachey J (ed and trans) The standard edition of the complete psychological works of Sigmund Freud (Vol 8, 1–258). Hogarth Press, London, 1958

Freud S (1917) A general introduction to psychoanalysis. (Riviere J, trans) Liveright, New York, 1963. Also: Introductory lectures on psychoanalysis. In: Strachey J (ed and trans) The standard edition of the complete psychological works of Sigmund Freud (Vol 15 and 16). Hogarth Press, London, 1961 and 1963

Freud S (1923) The ego and the id. (Riviere J, trans) In: Strachey J (ed) The standard edition of the complete psychological works of Sigmund Freud (Vol 19, pp 1–66). Hogarth Press, London, 1961

Gabrieli JDE, Corkin S, Mickel SF, Growdon JH (1993) Intact acquisition and long-term retention of mirror tracing-skill in Alzheimer's disease and in global amnesia. Behav Neurosci 107: 899–910

Galpin A, Underwood G, Chapman P (2008) Sensing without seeing in comparative visual search. Conscious Cogn 17: 672–687

Glaser J, Kihlstrom JF (2005) Compensatory automaticity: Unconscious volition is not an oxymoron. In: Hassin RR, Uleman JS, Bargh JA (eds) The new unconscious. Oxford University Press, Oxford, pp 171–195

Graf P, Squire LR, Mandler G (1984) The information that amnesic patients do not forget. J Exp Psychol Learn Mem Cogn 10: 164–178

Graf P, Schacter DL (1985) Implicit and explicit memory for new associations in normal and amnesic participants. J Exp Psychol Learn Mem Cogn 11: 501–518

Greenwald A, Klinger M, Schuh E (1995) Activation by marginally perceptible ("subliminal") stimuli: Dissociation of unconscious from conscious cognition. J Exp Psychol Gen 124: 22–42

Haase SJ, Fisk GD (2004) Valid distinctions between conscious and unconscious perception? Percept Psychophys 66: 868–871

Hassin RR (2005) Nonconscious control and implicit working memory. In: Hassin RR, Uleman JS, Bargh JA (eds) The new unconscious. Oxford University Press, Oxford, pp 196–222

Hassin RR, Uleman JS, Bargh JA (2005) The new unconscious. Oxford University Press, Oxford

Helmholtz H von (1867) Physiological optics. In: Southall JPC (ed and trans) Helmholtz's treatise on physiological optics. Dover, New York

Holender D (1986) Semantic activation without conscious identification in dichotic listening, parafoveal vision, and visual masking: A survey and appraisal. Behav Brain Sci 9: 1–66

Holender D, Duscherer K (2004) Unconscious perception: The need for a paradigm shift. Percept Psychophys 66: 872–881

Hobson JA (1988) The dreaming brain. Basic Books, New York

Hobson JA, McCarley RW (1977) The brain as a dreams-state generator: An activation-synthesis hypothesis of the dream process. Am J Psychiat 134: 1135–1168

James W (1890) The principles of psychology. Holt: New York

Kozbelt A, Nishioka K (2010). Humor comprehension, humor production, and insight: An exploratory study. Int J Humor Reg 23: 375–401

Jacoby LL, Witherspoon D (1982) Remembering without awareness. Can J Psychol 32: 300– 324

Macmillan M (2001) Limitations to free-associations and interpretation. Psychol Inq 12: 113–175

Macmillan NA (1986) The psychophysics of subliminal perception. Behav Brain Sci 9: 38–39

Macmillan N, Creelman CD (2005) Detection theory: A user's guide (2nd Ed.). Lawrence Erlbaum Associates, Mahwah, NJ

Marcel AJ (1983a) Conscious and unconscious perception: Experiments on visual masking and word recognition. Cogn Psychol 15: 197–237

Marcel AJ (1983b) Conscious and unconscious perception: An approach to the relation between phenomenal experience and cognitive processes. Cogn Psychol 15: 197–237

Merikle PM, Reingold EM (1991) Comparing direct (explicit) and indirect (implicit) measures to study unconscious memory. J Exp Psychol Learn Mem Cogn 17: 224–233

Metcalfe J, Wielbe D (1987) Intuition and insight and non-insight problem solving. Mem Cogn 15: 238–246

Milner B (1965) Memory disturbance after bilateral hippocampal lesions. In: Milner P, Glickman, S (eds) Cognitive processes and the brain. Van Nostrand, Princeton

Pötzl O (1917) The relationship between experimentally induced dream images and indirect vision. In: Wolff J, Rapaport D, Annin SH (eds and trans.) Preconscious stimulation in dreams, associations, and images: Classical Studies by Otto Pötzl, Rudolf Allers, and Jakob Teler. Psychol Issues, 2 (No. 3), Monograph 7. International University Press, New York, 1960, pp 41–120

Reber AS (1967) Implicit learning and artificial grammars. J Verb Learn Verb Be 6: 317–327

Reber AS (1993) Implicit learning and tacit knowledge: An essay on the cognitive unconscious. Oxford University Press, New York

Reder LM, Park H, Kieffaber PD (2009) Memory systems do not divide on consciousness: Reinterpreting memory in terms of activation and binding. Psychol Bull 135: 23–49

Reingold EM (2004) Unconscious perception and the classic dissociation paradigm: A new angle? Percept Psychophys 66: 882–887

Reingold EM, Merikle PM (1988) Using direct and indirect measures to study perception without awareness. Perecept Psychophys 44: 563–575

Rensink RA (2004) Visual sensing without seeing. Psychol Sci 15 (1): 27–32

Richardson-Klavehn A, Bjork RA (1988) Measures of memory. Annual Rev Psychol 39: 475–543

Roediger HR (1990) Implicit memory: Retention without remembering. Amer Psychol 45: 1043–1056

Roediger HL (2008) Relativity of remembering: Why the laws of memory vanished. Annual Rev Psychol 59: 225–254

Roediger HL, Craik FIM (eds) (1989) Varieties of memory and consciousness: Essays in honor of Endel Tulving. Erlbaum, Hillsdale, NJ

Ruys IK, Stapel DA (2008) Emotion elicitor or emotion messenger? Subliminal priming reveals two faces of facial expressions. Psychol Sci 19: 593–600

Schacter DL (1987) Implicit memory: History and current status. J Exp Psychol Learn Mem Cogn 13: 501–518

Snodgrass M, Shevrin H (2006) Unconscious inhibition and facilitation at the objective detection threshold: Replicable and qualitatively different unconscious perceptual effects. Cognition 101: 43–79

Snodgrass M, Shevrin, H, Kopka, M (1993) The mediation of intentional judgments by unconscious perceptions: The influences of task strategy, task preference, word meaning, and motivation. Conscious Cogn 2: 169–193

Snodgrass M, Bernat E, Shevrin H (2004a) Unconscious perception: A model-based approach to method and evidence. Percept Psychophys 66: 846–867

Snodgrass M, Bernat E, Shevrin H (2004b) Unconscious perception at the objective detection exists. Percept Psychophys 66: 888–895
Swets JA (ed) (1964) Signal detection and recognition in human observers. Wiley, New York
Titchner EB (1917) A textbook of psychology. Macmillan, New York
Uleman JS, Saribay SA, Gonzalez CM (2008) Spontaneous inferences, implicit impressions, and implicit theories. Annual Rev Psychol 59: 329–360
Van Selst M, Merikle P (1993) Perception below the objective threshold? Conscious Cogn 2: 194–203
Watson JB (1924) Behaviorism. Norton, New York
Weisenburg T, McBride KE (1935) Aphasia. Commonwealth Fund, New York
Weiskrantz L (1989) Remembering dissociations. In: Roediger HL, Craik FIM (eds) Varieties of memory and consciousness. Essays in honour of Endel Tulving. Erlbaum, Hillsdale, NJ, pp 101–120
Weiskrantz L (2002a) Blindsight: A case study and implications. Oxford University Press, Oxford
Weiskrantz L (2002b) Prime-sight and blindsight. Conscious Cogn:11: 568–581
Werner H (1956) Microgenesis and aphasia. J Abnorm Soc Psychol 52: 347–353
Whittlesea BWA, Price JR (2001) Implicit/explicit memory versus analytic/nonanalytic processing: Rethinking the mere exposure effect. Mem Cognit 29: 234–246
Wundt W (1902) Outlines of psychology (4th German ed) (CH Judd trans). Engelmann, Leipzig

The history of sensory worlds V

Introductory remarks

Most people are not really aware of the fact that senses and perception also have an important historic aspect. One time scale along which senses are changing is individual development and age. Another one is much larger and marked by the epochs of human history. In addition there are significant differences between people from different societies and cultural background.

1. Color systems

With his chapter on *"Colour ordered and disordered"* John GAGE adds to and enriches the contribution of Christoph von Campenhausen and Jürgen Schramm entitled "Understanding color vision with comments on mind and matter" (see Chapter III, 9)

As shown by a recent study of colour-order systems there is only little close agreement among the nearly one hundred examples from the twentieth century alone as to what the salient features of a colour system should be. What is it about colour that has made it so resistant to a coherent treatment? One reason for the difficulties lies in the fact that colour is first and foremost a psychological phenomenon, which is subject to all the social and individual variations that psychology implies. Language is one of the keys to these psychological variations. In his contribution *John Gage* focuses on a few of the many anomalies in the languages of Aboriginal Australia, where the traditional restricted colour palette has survived in painting and ritual until the present day, but has also seen an efflorescence of highly polychromatic art in recent years.

Gage's essay suggests that the notorious, but puzzling, vagueness in colour-categorization follows from the essential purposiveness of colour-language: we see what we are interested in seeing. This has long been recognized, for example, among peoples engaged in animal husbandry, where the number of colour-terms for hides far exceeds that used in other contexts. In the seventeenth-century European optical science shifted its focus of attention to the hue-characteristics of colour. Before that, colour names, and hence colour perceptions, were invariably anchored in the names of substances offering those colours, as they still are in non-industrialized societies. In many Oceanic and Australian languages abstract colour-terms are generated from the duplication of terms used for particular substances. For example, in Manam, the language of the Northern New Guinea island, daradara, red, is formed from dara, blood (as, indeed, it is in the Indo-Germanic languages of Europe), and in several Central Desert languages in Australia, karntawarra-karntawarra, yellow, derives from a particular precious yellow ochre, karntawarra. It is striking that an acrylic painting by the Pintubi artist Yanjatjarri Tjakamarra, representing the main karntawarra mine near Papunya in the Central Desert, makes only limited use of yellow and instead gives more space to red. The image of the mine itself, in the centre of the painting, is an orange-red, which is where in colour-space karntawarra is located in Warlpiri, a language of a community neighbouring on the Pintubi. The substance from which the colored paint is derived is the primary source of the word used for it, and the terms actually used cannot be considered to be 'basic' in the sense of Berlin and Kay (1991).

Modern attempts to standardize colour for commercial purposes have been bolstered by a rhetoric of 'objectivity'. However, we only have to compare the coloured images of the same programme running on the screens of competing television companies in our local electronics store to see that 'uniformity' and 'fidelity' are not regarded as problems by the consumer who chooses one of the sets and takes it home. Much of the research into standardization and perception of colour is based on exercises of matching, but outside the world of industrial colour production matching obviously does not play a vital role in the way we respond to colour. We need to return to some version of the relativism against which Berlin and Kay argued (Berlin and Kay 1991).

2. History of smell

Robert JÜTTE's contribution *"The sense of smell in historical perspective"* is about the history of a specific sense which may deserve particular attention because of its closeness to our emotions. The diachronic implementation confronts the historian with huge, though not necessarily insurmountable, problems. There may be conceptual difficulties, but a particular additional problem is the lack of written records. The history of smell can only draw from a small base of evidence, from which the modes of use of this sense and its frames of reference may be deduced. The available sources consist of normative texts, such as conduct books and health manuals, moral-theological and philosophical treatises and, not least, so-called "self-documentations" that give us glimpses of cultural practices and bodily experiences. Despite the obvious difficulty of assembling a persuasively comprehensive corpus of source material for a period extending all the way from antiquity to the present, the author opted for a largely diachronic account of culturally influenced sensory functions and practices connected with them. According to *Robert Jütte* the process of deodorization, which the French cultural historian Alain Corbin has described as a product of the medical enlightenment and increasing hygienic awareness of the late eighteenth and early nineteenth centuries, is now giving way to a process of "re-odorization". This transformation is marked by the suppression of stenches of any kind by the modern technology. So regardless of where the nose pokes itself in, the only smells it encounters are fragrant. Viewing the long-term character of this development, Constance Classen, an American ethnologist and the author of a remarkable cultural sociology of smelling, poses the obvious question: 'As odours, like roses, have long-standing associations with both spirituality and sensuality, one wonders if this postmodern interest in smell is evidence of a quest for spiritual and/or sensual fulfilment.' The use of pheromones in the production of perfume may perhaps be regarded as part of this quest. **H-D K**

References

Berlin B, Kay P (1991) Basic color terms: their universality and evolution. (2nd ed) Univ Calif Press, Berkeley

Colour ordered and disordered 17

John Gage †

Contents

1. Introduction

A recent study of colour-order systems has shown that among the nearly one hundred examples from the twentieth century alone there has been little close agreement as to what the salient details of a colour system should be. An exhibition this year in New York has, likewise, shown that contemporary painters stimulated by the idea of scientific colour-systems, have nevertheless wanted to maintain their individuality. What is it about colour that has made it so resistant to a coherent treatment? One answer must lie in the fact that colour is first and foremost a psychological phenomenon, subject to all the social and individual variations that psychology induces. Language is one key to these psychological variations, and this essay will look particularly at some of the many anomalies in the languages of Aboriginal Australia, where the traditional restricted colour-palette has survived in bark- painting and ritual until the present, but which has also seen in recent years an efflorescence of highly polychromatic art.

John Gage
† deceased February 10, 2012
University of Cambridge
Department of History of Art
e-mail: jsg1000@cam.ac.uk

2. Colour ordered and disordered

In "Color ordered: a survey of color order systems from antiquity to the present" (2008, p. 333), Rolf Kuehni and Andreas Schwarz present a remarkable analysis of more than 170 colour-systems, nearly 100 of which were devised in the twentieth century alone; and in March of that year Ann Temkin mounted "Color chart: reinventing color, 1950 to today" at the Museum of Modern Art in New York, an exhibition featuring the work of artists anxious to use industrial materials in quasi-systematic ways in order to exclude hierarchy and individuality from their colour compositions. In the substantial book of this exhibition, the art-historian Briony Fer mused:

> If color is supposedly so direct and immediate, then it is striking how difficult it is to discuss, and if color is so systematic, then why is it so hard to contain within a given system?
> (Temkin 2008, p. 28)

Kuehni and Schwarz show that colour has, in fact, never been "so systematic"; but they also show that the desire to organize it into systems has been felt in many cultures and many periods in Europe, Asia and the Americas. One systematiser rather surprisingly absent from their survey is the English colour-maker George Field, one of the most influential colour-theorists for artists in the first half of the nineteenth century, whose most important book, "Chromatography" (Field 1835), when it was translated into German at Weimar in 1836, was naturally linked to Goethe's "Farbenlehre" (Goethe 1810) (Fig. 1).

But Field operated with two mutually incompatible systems. His linear scale, between white and black formed a showcase for the pigments he himself manufactured, and is essentially an Aristotelian arrangement, charting progressive intermixtures, from white, then the three subtractive primaries in light and dark versions, and through the secondary and tertiary mixtures up to black. It is a scale which is likely to have been of most interest to painters; and it is also a scale which reflects Field's anti-Newtonianism, for Newton's theory of the colours inherent in white light was, for Field, contrary to nature, reason, and "the authority of God" (Gage 2001 a, p. 215). Field believed that the three primary colours, red, yellow and blue, were evidence of the Holy Trinity operating throughout the natural world, and they are the key to the structure of the colour-circle placed immediately above his linear scale. This circular arrangement was, paradoxically, derived ultimately from Newton, although it had become standard, refined, and made symmetrical by the close of the eighteenth century, as Kuehni and Schwarz abundantly show. Field's version emphasizes contrasts, those subjective colour-effects which had been increasingly investigated since Newton. His yellow-purple axis identifies advancing and retreating colours, perhaps for the first time in diagrammatic form; and his blue-orange axis signals colour-temperature, a phenomenon which was probably first explored by painters early in the eighteenth century (Gage 2008; for some more recent instances of the idea of warmth and coolness in colour: Gage 2001 b, pp. 34–35). Thus Field's system is, in the fullest sense, over-determined and hierarchical, and it is notable that, like Goethe, whose "Farbenlehre", in the English translation by the painter Charles Lock Eastlake (Goethe 1810), became familiar to Field only after the publication of his own theory, he placed red at the top of his circle, and it was a madder-red on which his reputation as a colour-manufacturer primarily rested. His system has none of the apparently random distribution of samples in commercial paint-charts, so attractive to the German painter Gerhard

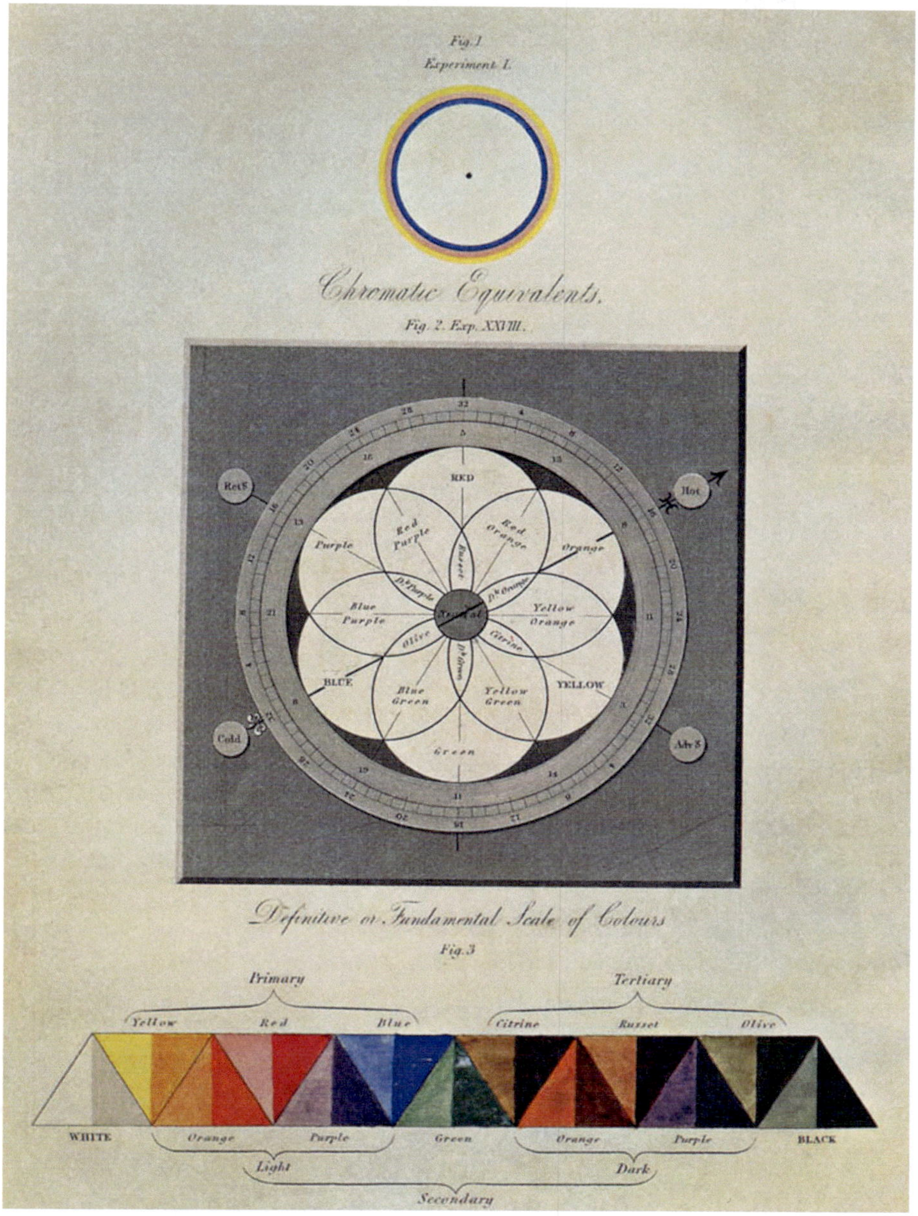

Fig. 1 George Field (1835), frontispiece to *Chromatography; or a Treatise on Colours and Pigments, and of their Powers in Painting.* Cambridge University Library

Richter (Fig.2) and other artists in the New York show.

One important recent development in the study of colour-perception which was ignored by Field was the investigation of after-images from strong colour-stimuli, contrasts which came to be regarded as complementary (Roque 1994). After-images had, of course, been noticed by many observers since Classical Antiquity (Berry 1922, p. 309; Buckley-Green 1990, chapter 5), and notably in the early eighteenth century by the French

Fig. 2 Gerhard Richter (1973), *1024 Colors* (CR 350–3), lacquer on canvas. Paris, Centre Georges Pompidou

naturalist Buffon, who called them "couleurs accidentelles". But they were more systematically studied in the 1780s by Charles Darwin's father, Robert Waring Darwin, whose account was taken up by Goethe, and made the lynchpin of his theory of harmony, since, he felt, the after-image of a given primary stimulus was the colour of the sum of the two remaining primaries, and established the eye's yearning for chromatic wholeness.

Goethe, like Field, edited the visual phenomena in the interest of symmetry, though other researchers, including Newton, had seen that the opposite of, e. g., red is not green, but blue-green (Gage 1999, p. 22). That Field, who, like Goethe again, believed that a harmonious picture required a balance of all three primary colours (Gage 2001a, p. 221), did not allude to Darwin's "ocular spectra" (Fig. 3), suggests once more that his theory was specifically tailored to the needs of artists. Field believed that his system was objective, indeed divinely ordained; and in their vast survey, Kuehni and Schwarz have found far more systems depending on analyses of the stimuli, than on analyses of perceptions, depending, as these do, not only on a variety of environmental factors, such as context and lighting, but also on the individual observer (Kuehni and Schwarz 2008, p. 333). This predominance is, of course, largely due to the fact that the stimuli lend themselves far more readily than perceptions to quantification.

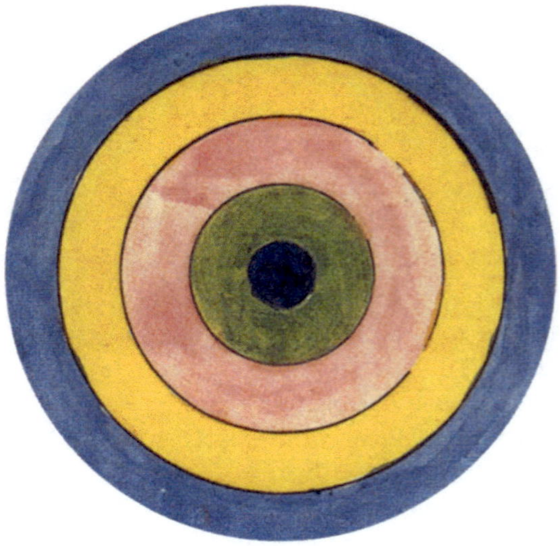

Fig. 3 Robert Waring Darwin (1786) *Ocular Spectra*. Phil Trans R Soc LXXVI

Several other phenomena related to the subjectivity of colour were investigated in the eighteenth century, for example, simultaneous contrast, and it's contrary, colour-constancy. Others, such as the therapeutic effect of colours, had to wait for the more developed lighting technologies of the nineteenth century. But one very significant area was not substantially broached until the second half of that century, largely as a consequence of Darwinism and the rise of experimental psychology: I mean the light that colour-language may throw on matters of function and purpose.

Looking is, of course, intrinsically purposive, and although most humans have all the visual tools for colour-discrimination, not all have an interest in using these tools to the full. In my view, artists know most about colour, because they are most directly involved in using it, but as a specialized minority in society, they have rarely been consulted as subjects for psychological experiment. The German psychologist G. J. von Allesch did indeed include artists in his enormous sample of subjects to investigate colour-preferences in the early years of the twentieth century, and although he reached no firm conclusions, the material he published in "Die ästhetische Erscheinungsweise der Farben" in 1925 – a title which reminds us of David Katz's classic study of 1911, "Die Erscheinungsweise der Farben, und ihre Beeinflussung durch die individuelle Erfahrung" (Katz 1911) – is still a rich source for the study of colour-meanings (von Allesch 1925).

In 1981 the Swiss scientists A. von Wattenwyl and H. Zollinger published a revealing study of colour-recognition by art and science students, which, contrary to what might have been expected, found that the art students responded far more hesitantly than the sample of science students, because their own speciality obliged them to identify specific nuances far more precisely (von Wattenwyl and Zollinger 1981).

Katz's 1911 study, amplified in 1930, looked at the idiosyncratic perceptions of colours by individuals; but more important, perhaps, are the very various ways in which whole cultures have classified colours, revealed primarily through the study of language. The realization that different cultures classify colours in different ways goes back, again, at least to Classical Antiquity, when, in the second century A. D., the Roman writer Aulus Gellius described a sort of language-game between a philosopher and a politician about the relative colour-vocabularies of Greek and Latin. It was already recognized that colour-perceptions vastly outnumber colour-terms (current estimates vary between seven and ten million perceptions to fewer than two hundred terms in any language), and the philosopher Favorinus adduced four Greek terms to nuance the single Latin term, "rufus" (red). The politician Fronto retorted by listing seven terms for red in Latin, some of them, notably "flavus", which seem to us to be rather doubtful (Gage 2001a, p. 80). There the matter seems to have rested; but in more modern times the possibility that different societies see colours in different ways was grounded more empirically. In one of his didactic letters to a German princess, the eighteenth-century Swiss mathematician Leonhard Euler related variations in perception to the eternally problematic analysis of the colours in the rainbow.

Euler argued that the number of colours perceived in the bow might depend on the number of words to name them. Other nations, he suggested, might be able to distinguish more than the canonical six, and yet others fewer than these; he noted that the purple which is seen by some observers at the extremity of the red band, to others seemed simply to be a nuance of red itself. Purple was in fact essential to Euler's own chromatics, since he argued that the vibratory character of the various colours of light was closely analogous to musical pitch, and to effectively compare the spectrum with the diatonic musical scale, he needed, as had Newton, a seventh division (Euler 1823; for Newton and eighteenth-century thought on the relationship between colour and music, see Jewanski 1999; for the difficulties in seeing the rainbow, see Gage 2001a, chapter 6).

Although the Spanish conquest of the Americas in the sixteenth century highlighted some asymmetries between the colour-vocabularies of the invaders and those of the indigenous peoples (Gage 1999, chapter 7), these anomalies were not, it seems, made the subject of analysis. It was not until the intensive European colonization in Africa, Asia and Oceania which began in Euler's day that exposure to indigenous languages made the nature of human colour-perception seem problematic. This was also the period when defective colour-vision was first identified and investigated, not least by Goethe. Possibly the earliest record of a very limited colour-vocabulary in a hunter-gatherer society was by a settler in the Port Jackson region of South-East Australia (now Sydney) in 1790, two years after the British invasion. John Hunter reported that "the only colours we have as yet discovered [the Aborigines] have any knowledge of are red . . . white . . . black . . . [and] green" (Hunter 1793, p. 234). A few years later another British settler recorded the same list, with slightly different transcriptions of the indigenous terms (Collins 1798, p. 508). If this set was indeed characteristic of the language of the Port Jackson Aborigines, it corresponds precisely to the Stage III in the influential modern study of colour-language, "Basic color terms: their universality and evolution" by Brent Berlin and Paul Kay (1969; see also the bibliographical appendix by L. Maffi in the 2nd edition, 1991; and for a critical appraisal of these ideas, Tornay 1978), where green or yellow are said to be introduced after red has been added to the most fundamental terms, black and white.

Hunter's brief reference is vitiated by its complete lack of context: we learn nothing of how these few colour-words were used. The same goes for Berlin and Kay's study; nor do the three concepts in their title, "basic", "evolution" and "universality" carry much conviction, although their book has stimulated a large and important body of ethno-linguistic research (Berlin and Kay, 2nd ed. 1991, Appendix). Australia offers abundant materials for addressing the very complex question of the colour-terminologies of indigenous peoples, not simply because of its many indigenous languages, but also because of the modern renaissance of Aboriginal painting, which is now well-known throughout the world. The ethnolinguistics of colour has been compromised by its lack of attention to the use of colour in artifacts, since the pioneering work of the Austrian anthropologist R. Hochegger in his study, "Die geschichtliche Entwicklung des Farbsinnes" (1884). Three of the four colours reported by Hunter at Port Jackson, black, white and red, were, and are, those most highly charged with meaning in ceremonial contexts, for body or ground painting; and if green and yellow are parts of the same category, as they often are in indigenous, and even European thought (see, for example, Murrinh-Patha, a language from Port Keats at the top of the Northern Territory: "wudanil", yellow, orange, pink, green, sandy, ginger; Hargrave 1982, p. 209; Arrernte, a language from the Central Desert: "tierga" or "turga", yellow, green, blue: Spencer and Gillen 1927, p. 552; Kriol, a widely used English-based creole, spoken in the Roper River region of Western Australia: "yela" for "grin": Hargrave 1982, p. 216). M. Bornstein has produced some remarkable figures for the assimilation of green and blue. Languages of two hundred societies tested show that a majority have the same term for green and blue, or blue and black. 75% of American cultures have the same term for green and blue; 28% of Asian cultures, 18% of Australian cultures and 39% of African cultures do not distinguish blue and black linguistically, and 25% have a single term for green, blue and black. Assimilations of red and yellow are also common (Bornstein 1978; see also Gage 2001a, p. 61), the "green" term may even have been used to refer to the yellow paint, which was the fourth important colour for ritual purposes.

At Port Jackson, whose tribes and their languages were effectively extinguished in the nineteenth century, a three- or four-colour palette was almost certainly used exclusively for body-painting, but in the tropical North-East and far West it was – and still is – employed in bark designs (Fig. 4).

Natural ochres, with all their sacred significance, are still used everywhere for body- and ground- painting, but outside the North-East they have largely been supplanted in portable or large canvas paintings by modern acrylics.

Yet in the Central and Western Deserts, which have seen the most dynamic of the recent painting movements, these modern paints are still referred to by their traditional ochre names, which in many languages are the only words for these colours, and would thus not qualify as 'basic' in Berlin and Kay's scheme, since they are the colours of particular substances.

It is estimated that at the time of the 1788 invasion there were about 250 languages and some 700 dialects among the circa 600 indigenous tribes of Australia. About one hundred languages have survived in some form, and about sixty are strong, with many speakers. Aborigines traditionally had many opportunities to learn several languages, whether at the many- tribal gatherings, sometimes called "corroborees", or on ochre-expeditions of often several hundreds of kilometers and crossing many other tribal territories. The State and Federal Governments' paternalistic and assimilationist policies up until the 1960s, as well as the operations of the Christian Missions, also brought most Outback Aborigines from many tribes together into large centres, and these factors may account for the vastly expanded multilingualism which is such a feature of current Aboriginal culture. On a more intimate level, language-switching is also a major feature of indigenous social interaction. To quote the ethnolinguist Barry Alpher:

Fig. 4 Irvala (Yirawala, Billy, Gunwiggu group, 20th CE) (1963) *Two Mimi men* (two reclining rock spirits resting inside a cave),Croker Island, Arnhem Land, Australia, *Eucalyptus* bark, pigments. Paris, Musée du Quai Branly

> Australian Aborigines often speak more than one Aboriginal language, more than one regional variety of the same Aboriginal language, and one or more varieties of English. If they live in a traditional community, they are also likely to have available different varieties of their own regional dialect, which they use for different purposes and with persons of different social categories. These speech varieties include forms of the language used with or about relatives with whom

> people maintain relations of respect or avoidance (for example, mothers-in-law and certain kinds of cousins); forms used with relatives with whom people maintain relations of jesting familiarity; forms used with very young children [...]; forms learned by very young men in the course of their initiation and used only in that context; and sign language, in which gestures made with the hands are used instead of spoken words. (Alpher 1993, p. 97)

As if this were not enough, one category not mentioned by Alpher is the "secret" song-vocabulary accompanying some ceremonial, which is often not understood, even by the singers themselves (Keen 1994, pp. 146, 239, 275)

One of the best-known of the modern Central Desert painters, the late Clifford Possum Tjapaltjari (Fig.5), claimed to speak six indigenous languages, dialects of three distinct language groups; and another recently deceased artist, Jimmy Robertson Jampijinpa, from a region slightly to the North-West of Possum's, ten.

When talking about his method of work (which he, like most Aboriginal artists, did very rarely), Possum sometimes used colour-terms peculiar to more than one of his languages, for example, his red-ochre term, "yulpa", common to five languages, but in one of them connoting red or yellow, and in other cases, to one only. He also characterized the prestigious red ochre "karrku" with the English words "pink" for the light grade and "black" for the dark (Johnson 1994, p. 142). But we know very little about the function of language-switching in indigenous discourse, particularly whether it is more than casual. These are not topics which indigenous people are likely to discuss.

Since there are no written records of indigenous languages before the invasion, it is almost impossible to reconstruct the histories of these languages, and to assess where the many colour-terms originated, notably the fundamental four. Colour-words such as "karrku" (red ochre) and "maru" (black) occur in many widely-scattered languages. "Karrku", a particularly glossy ochre, is found in – and widely exported from – a mine in the Central Desert, the subject of a very large painting by thirty-four artists from Yuendumu, a major Government Settlement near this mine, now in an American collection (Dussart 1999) (Fig. 6).

But in the form "karko", red ochre is recorded in the mid-nineteenth century as coming from a quite different mine, many hundreds of kilometers away in South Australia. This may be because lexicographers were active among the indigenous peoples of South Australia long before the Centre was explored by Europeans; but it is also conceivable that the Central Desert mine took its name from a term for any fine red-ochre, rather than the other way round. Or, again, that the name migrated northwards with the possible migration of peoples, or through trade. "Maru" is even more problematic as a widely-used term, since black pigment was usually made locally from charcoal, and hence available everywhere, and it was ritually far less significant than red.

Fig. 5 Clifford Possum Tjapaltjarri (1982) *Bushfire Dreaming*, acrylic on canvas. Adelaide, Art Gallery of South Australia

Fig. 6 Warlukurlangu Artists (1996) *Karrku,* acrylic on canvas. Charlottesville, University of Virginia, Kluge-Ruhe Collection

So it is not easy to interpret the ubiquity of these loan-words. Possum stated that "karrku", coming as it did from a sacred site could only be mined and used by tribes whose totems were represented at that site; he could not use it himself: his red ochre (for body-painting of course; his canvases were painted in acrylics) came from elsewhere, and had a different name. Some languages, indeed, have many words for "red ochre", and even for the abstract colour "red", although "maru" in most languages I have studied is the only term for abstract black. It looks as though "karrku" is a widely borrowed term because it refers to something highly valued, and "maru" for the opposite reason, that some tribes feel it is not worth inventing a new term for something relatively insignificant. However that may be, these two colour-terms are clearly perceived in a very particular way.

It is also clear that indigenous Australians sometimes adopted the invaders' terminology as well as their products. Blue, for example, used as a paint in the nineteenth century in the form of imported washing-blue, was also described with terms derived from English (Lee 1987, p. 93; Goddard 1987, p. 197, s. v. "puluwana"). Gidjingali, a language spoken at Maningrida, on the far North-East coast, has terms only for "light" and "dark", representing Berlin and Kay's most primitive stage, but its modern speakers use English loan-words for shopping (Jones and Meehan 1978; for the very complex linguistic situation at Maningrida, see Elwell 1982). One of Berlin and Kay's sources for Australian languages was the work of the Cambridge anthropologist W. H. R. Rivers, who studied the speech of the Torres Strait Islanders and the indigenous peoples of North Queensland around 1900. Rivers found that, on one island, some natives "showed extraordinary ingenuity in devising special names, apparently on the spur of the moment, for different shades of colour" in the anthropologist's Lovibond Tintometer. Most of these names derived from natural objects, and witness a remarkable capacity for colour abstraction. Rivers also noted that, in describing the rainbow – an especially potent symbol in indigenous mythologies – some informants included blue, green and indigo; but although double bows were frequently seen in this region, no-one had noticed the reversal of colours in the secondary bow (Rivers 1901 a, pp. 69–70). Yet, paradoxically, the anthropologist concluded that the limitations in the standard indigenous colour-vocabulary were evidence of defective colour-vision (Rivers 1901b, p. 47).

Borrowing or inventing colour-terms is surely evidence of perceptual capacity, and a restricted vocabulary only of restricted needs. As Rivers conceded, it "may be merely due to lack of interest". Like other early anthropologists in Australia, he surmised that the chief focus of thought about colour among indigenous people was the pigments used in body, ground, rock or bark painting, namely black, white, red and yellow. But the polychrome acrylic canvases pioneered in the Central Desert in the 1970s were by artists who were interested in the traditional set, but whose languages, for the most part, also include terms for blue and green.

The salience of black, white and red is by no means confined to the languages of hunter-gatherer societies; as recently as the 1980s T. J. A. Bennett gathered a rich harvest of English proverbial colour usages which show very much the same pattern. Black was by far the most common colour used metaphorically, followed by red and white, in that order. Yellow, on the other hand, came rather far down the list of frequencies, behind blue and green (Bennett 1988). This is, of course, only the roughest of guides, although it agrees substantially with a study of the use of colour-words in seventeen English-language novels, carried out in the 1940s (Evans 1948, pp. 230–231). It is remarkable that the most interesting colours in highly industrialized and highly consumer-oriented societies such as Great Britain and the United States should be the same as those in cultures which, less than a century ago, were still being described by anthropologists as "stone-age".

It is not surprising that the most recent colour-order system examined by Kuehni and Schwarz, the CIEDE 2000 of the Commission Internationale de l'Eclairage, published in 2001, should, according to them, be far from perfect, and, in particular, yield only "approximately 65% accuracy in predicting average visual data" (Kuehni and Schwarz 2008, p. 169). Nor is it unexpected that industrial paint manufacturers have wished to mask the elaborate colorimetric background to their products under attractive names and the randomly arranged paint-charts which have encouraged some modern artists to feel that, taking them as a starting-point, they are working systematically and thus impersonally, but are also free to show their individuality.

As one of the most prominent, and most individual of these artists, the late Sol le Witt, told Ann Temkin in 2007: "I never tried to arrange the color or the other forms to please the eye. In fact, I tried to use the system or randomness to avoid preconceived notions of aesthetic 'beauty' or other color statements" (Temkin 2008, p. 141) (Fig. 7).

Since these apparently incompatible aims and these imprecisions of colour-language have helped to stimulate the production of works of great beauty and sensitivity among both indigenous and western artists, we have no reason to complain.

Conclusion

What I have tried to suggest in this essay is that the notorious, but puzzling, vagueness in colour-categorization is a consequence of the essential purposiveness of colour-language: we see what it is in our interest to see. This has long been recognized, for example, among peoples engaged in animal husbandry, where colour-terms for hides far exceed those used in other contexts (Gage 2001a, p. 79). Until seventeenth-century European optical science shifted the focus of attention to the hue-characteristics of colour, colour names, and hence colour perceptions, were invariably anchored in the names of substances of some cultural significance offering those colours, as they still are among non-industrialized societies. In many Oceanic and Australian languages abstract colour-terms are generated from the re-

Fig. 7 Sol le Witt (1970) All single, double, triple and quadruple combinations of lines and color in four directions in one-, two-, three-, and four-part combinations. Hessel Museum of Art, Center for Curatorial Studies, Bard College, Annandale-on Hudson, New York

duplication of substance-terms; for example, in the Northern New Guinea island language Manam, "daradara", red, is formed from "dara", blood (as, indeed, it is in the Indo-Germanic languages of Europe), and in several Central Desert languages in Australia, "karntawarra-karntawarra", yellow, derives from a particular precious yellow ochre, "karntawarra" (Blust 2001, pp. 25–26; for the various functions of reduplication, see Fabricius 1998, p. 145). It is striking that an acrylic painting by the Pintubi artist Yanjatjarri Tjakamarra, representing the main "karntawarra" mine, near Papunya in the Central Desert (Myers 2002, pp. 88, 92), makes a very limited use of yellow, and gives more space to red.

The image of the mine itself, in the centre of the painting, is an orange-red, which is where, in colour-space, "karntawarra" is located in Warlpiri, a language of a community neighbouring on the Pintubi (Hargrave 1982, p. 209). The substance is primary, and these terms cannot be considered to be "basic" in Berlin and Kay's sense. The modern attempts at colour-standardization for commercial purposes have, as usual, been bolstered by a rhetoric of "objectivity"; but we have only to compare the varied coloured images of the same programme running on the screens of competing brands of television receivers in our local electronics store to see that "uniformity" and "fidelity" are not regarded as problems by the consumer who chooses one set and takes it home. Much of the research into standardization and perception of colour has been done with excercises in matching, but it seems to me that, outside the world of industrial colour production, matching plays no vital role in the way we respond to colour. We need to return to some version of the relativism which it was Berlin and Kay's endeavour to combat.

References

Allesch GJ von (1925) Die ästhetische Erscheinungsweise der Farben. Psychologische Forschung 6: 1–91, 215–281

Alpher B (1993) Out of the ordinary ways of using a language. In: Walsh M, Yallop C (eds) Language and culture in Aboriginal Australia. Aboriginal Studies Press, Canberra, pp 97–106

Bennett TJA (1988) Aspects of English color collocations and idioms. Carl Winter, Heidelberg, pp. 147–301

Berlin B, Kay P (1969) Basic color terms: their universality and evolution. University of California Press, Berkeley, 2nd ed. 1991
Berry W (1922) The flight of colors in the after image of a bright light. Psychol Bull 19: 307–37
Bickley-Green C (1990) Afterimage in painting (PhD Diss). University of Georgia
Blust R (2001) Reduplicated colour terms in Oceanic languages. In: Pawley A, Ross M, Tryon D (eds) The boy from Bundaberg. Studies in Melanesian linguistics in honour of Tom Dutton. Pacific Linguistics, Canberra, pp 23–49
Bornstein M (1978) Considerations sur l'organisation des tonalities chromatiques. In: Tornay S (ed) Voir et nommer les couleurs. Laboratoire d'ethnologie et de sociologie comparative, Nanterre
Collins D (1975) An account of the English colony in New South Wales, with remarks on the dispositions, customs, manners, etc, of the native inhabitants of that country, Vol. 1. Fletcher BH, Reed A. H. and A. W. (eds), in association with the Royal Australian Historical Society. Terrey Hills, New South Wales (first publ. in 1798–1802. T. Cadell & W. Davis, London)
Dussart F (1999) What an acrylic can mean: on the meta-ritualistic resonances of a central desert painting. In: Morphy H, Smith Boles M (eds) Art from the land. Dialogues with the Kluge-Ruhe collection of Australian Aboriginal art. University of Virginia, Charlottesville, pp 193–218
Elwell VMR (1982) Some social factors affecting multilingualism among Aboriginal Australians: a case study of Maningrida. Int j sociol lang 36: 83–103
Euler L (1823) Letter XXXI. In: Brewster D (ed) Letters of Euler on different subjects in natural philosophy, addressed to a German princess, 3rd ed., Vol. I. W & C Tait, Edinburgh
Evans RM (1948) Introduction to color. Wiley, New York
Fabricius AH (1998) A comparative study of reduplication in Australian languages. Lincom Europa, München
Field G (1835) Chromatography; or, a treatise on colours and pigments and of their powers in painting. Charles Tilt, London
Gage J (1999) Die Sprache der Farben: Bedeutungswandel der Farbe in der bildenden Kunst. Ravensburger Buchverlag, Ravensburg
Gage J (2001a) Kulturgeschichte der Farbe: von der Antike bis zur Gegenwart. E. A. Seemann, Leipzig
Gage J (2001b) Status or status-symbol? The future of colour (and a little of its past). In: Alliez E, Samsonov E (eds) Chroma drama: Widerstand der Farbe. Turia+Kant, Wien, pp 34–40
Gage J (2008) When warm was cool: on the history of colour-temperature. In: Busch W (ed) Verfeinertes Sehen: Optik und Farbe im 18. und frühen 19. Jahrhundert. Schriften des Historischen Kollegs, Kolloquien 67. Oldenbourg, Munich, pp 91–99
Goddard C (1987) A basic Pitjantjatjara/Yankunytjatjara to English dictionary. Institute for Aboriginal Development, Alice Springs
Goethe JW von (1810) Zur Farbenlehre. Cotta'sche Verlagsbuchhandlung, Stuttgart ([1840/1970] Theory of colours. Engl. transl. Eastlake CL. MIT Press, Cambridge Mass)
Hargrave S (1982) A report on colour-term research in five Aboriginal languages. In: Hargrave S (ed) Language and culture. Work papers of the Summer Institute of Linguistics, Australian Aboriginal Branch, Ser. B, Vol. 8. Summer Institute of Linguistics, Darwin, pp 201–226
Hunter J (1793) An historical journal of the transactions at Port Jackson and Norfolk Island. John Stockdale, London
Hochegger R (1884) Die geschichtliche Entwicklung des Farbensinnes. Wagner, Innsbruck
Jewanski J (1999) Ist C = Rot? Eine Kultur- und Wissenschaftsgeschichte zum Problem der wechselseitigen Beziehung zwischen Ton und Farbe, von Aristoteles bis Goethe. Verlag Schewe, Sinzig
Johnson V (1994) The art of Clifford Possum Tjapaltjarri. Gordon and Breach Arts International, Basel
Jones R, Meehan B (1978) Anbarra concepts of colour. In: Hiatt LR (ed) Australian aboriginal concepts. Australian Institute of Aboriginal Studies, Humanities Press, Canberra, pp 26–32
Katz D (1911) Die Erscheinungsweise der Farben und ihre Beeinflussung durch individuelle Erfahrung. Zeitschr f Psychol, Ergänzungsband 7
Keen I (1994) Knowledge and secrecy in an aboriginal religion. Clarendon Press, Oxford
Kuehni R, Schwarz A (2008) Color ordered: a survey of color order systems from antiquity to the present. Oxford University Press, New York
Lee J (1987) Tiwi today: a study of language change in a contact situation. Department of Linguistics, Research School of Pacific Studies, Australian National University, Canberra
Maffi L (1991) Bibliographical appendix. In: Berlin B, Kay P (eds) Basic color terms: their universality

and evolution. University of California Press, Berkeley, 2nd ed
Myers FR (2002) Painting culture: the making of an aboriginal high art. Duke University Press, London
Rivers WHR (1901a) Vision. In: Reports on the Cambridge anthropological expedition to the Torres Strait, Vol. II, pt. I, Physiology and psychology. Cambridge University Press, Cambridge, pp 1–140
Rivers WHR (1901b) Primitive color vision. Popular Science Monthly 59: 44–58
Roque G (1994) Les couleurs complementaires: un nouveau paradigme. Rev Hist Sci Paris 47: 405–433
Spencer B, Gillen FJ (1927) The Arunta: a study of a stone-age people, Vol 2. Macmillan, London
Temkin A (ed) (2008) Color chart: reinventing color, 1950 to today. Museum of Modern Art, New York
Tornay S (ed) (1978) Voir et nommer les couleurs. Laboratoire d'ethnologie et de sociologie comparative, Nanterre, pp 77–79
Wattenwyl A von, Zollinger H (1981) Color naming by art students and science students: A comparative study. Semiotica 35: 303–315

The sense of smell in historical perspective

18

Robert Jütte

Contents

1. Introduction

The writing of a history of a particular sense is both an original and risky undertaking. Its diachronic implementation confronts the historian with huge, though not necessarily insurmountable, problems. There may be certain conceptual difficulties, but not the least of these is the lack of written records.

Robert Jütte
Institute for the History of Medicine
Robert Bosch Foundation
Straussweg 17
70184 Stuttgart, Germany
e-mail: robert.juette@igm-bosch.de

The history of smell, in comparison, can draw from a small base of evidence from which the historicity of this sense, its modes of use, and its frames of reference, may be ascertained. These sources consist of normative texts, such as conduct books and health manuals, moral-theological and philosophical treatises and, not least, so-called 'self-documentations,' that give us glimpses of cultural practices and the experience of possessing a body. Despite the obvious difficulty of assembling a persuasively comprehensive corpus of source material for a period extending all the way from antiquity to the present, I have opted for a largely diachronic account of culturally influenced systems of sensory functions and practices connected with them.

2. The Greek tradition

In Timaeus, Plato (428/27–348/47 B.C.) deals systematically with the senses, in the order of taste, smell, hearing and vision. Unlike the other senses, the sense of touch is not attached to a specific physical organ. Sensations of pleasure and pain and other qualities perceptible to the senses, such as soft and hard, warm and cold, heavy and light or rough and smooth feature as 'disturbances that affect the whole body in a common way' (Plato, Timaeus, 65c). In the case of smell Plato allows just the one distinction, between pleasant and unpleasant, 'because a smell is always a "half-breed". None of the elemental shapes, as it happens, has the proportions required for having any odour' (Plato, Timaeus, 66d). In his view, it is only when certain substances become damp, begin to decompose, or melt and vaporize that smells arise which are then experienced as fragrant or foul.

The object of the sense of smell is an odour, although Aristotle (384–322 B.C.) admits that it is not quite as easy to devise categories for this sense as for a more differentiated perception such as taste. He believes that it is always possible to distinguish between sweet, sharp (hot), bitter, pungent and oily smells. Smelling is again rendered possible by a medium (air or water). Aristotle thought that the air breathed in through the nose, and the pneuma flowing from the heart into the vessels of the brain, were mixed together in the poroi (channel-like passages between the nose and the brain). He expounds his ideas on the seat of the organ of smell in De sensu (438b, 25f.). Here, the olfactory faculty is sited in the region of the brain. This opinion was later endorsed by Galen (129-c199/216), who managed not only to locate the sense of smell anatomically, but also to explain why the detection of smells should take place in that particular place: 'What remains is the organ of smell, which is not in the nasal passages, as the majority believe, but in the tips of the anterior ventricles of the brain, to which the nasal passages ascend; for at this point its ventricles are most vaporous' (Galen 1984, 463). These ideas must be seen in the context of the rather odd-sounding contemporary notion that cerebral fluid can flow into the nose and throat. The Greeks referred to this phlegm discharged by the brain as catarrh. Blessed with the authority of the great Galen, the idea of the nose as the 'cloaca of the brain' (la cloaca del cerebro) – in Francesco Sansovini's (1521–1587) graphic phrasing – persisted into the modern era. With its complex, though logically coherent, development of Aristotelian ideas, Galen's theory of smell did much to ensure that the physiological event of sensation as we know it today – stimulation of the sensory cells, transmission through the corresponding neural conductors (nerves) and perception in the brain – was not appreciated for centuries (Kassel 1914–1922; Seifert 1969; Benedum 1988;). It was not until around 1500, when the dissection of corpses became more common, that the fila olfactoria were identified as olfactory nerves. In 1655, the Wittenberg professor of medicine, Conrad Victor Schneider (1614–1680), finally succeeded in proving scientifically that the smell receptors were located in the nasal mucous membrane. However, the first accurate dissection of the olfactory nerve branchings from the mucous membrane to the brain was performed towards the end of the eighteenth century by the Italian anatomist Antonio Scarpa (1752–1832).

The renowned Arab physician Avicenna (980–1037) deals only briefly with the five external senses. In addition to the objects proper to its character as a particular sensory faculty, each of the five senses grasps five other qualities or modalities: shape, number, size, movement and repose. The same applies to the sense of smell, although Avicenna admits that in the case of human beings, whose sense of smell is not very highly developed, this acute power of discernment is barely or only partially present. The process of smelling is explained as follows: We inhale the air which has absorbed the smell of an odorous body. This affected air is received by the nose and comes into contact with the frontal part of the brain, where it is identified by the olfactory faculty.

The intermediate position of the sense of smell has rarely been questioned since antiquity (Jütte 2005). Aristotle had placed it between the long-range senses of seeing and hearing, and the two short-range senses of taste and touch. In his treatise on the soul, Albertus Magnus (1193/1206–1280) followed Avicenna in assuming that the object of perception was present in two forms (material and immaterial), and believed this to be corroborated by the ambivalent position of the sense of smell. He too was unwilling to contemplate a higher ranking, since traditional theory held that the faculty of smell was of minor importance to the inner senses. Albertus Magnus sees confirmation of this in the fact that people almost never dream of smells. Avicenna, who ranked the senses according to their usefulness, with the senses of touch and taste in first position, also places smell in the middle of his somewhat utilitarian hierarchy. Konrad von Megenberg (1309–74) came up with an interesting variant of this in-between position: 'The sense of smell occupies the middle position between the three, with sight above it, and hearing to one side of it' (Megenberg 1897, 9). Here, smell inhabits a point equidistant from the so-called higher senses. The hierarchical scheme employed by René Descartes (1596–1650), on the other hand, is almost classical. He also puts smell in an intermediate position on the grounds that, although less crude than touch and taste, it lacked the perceptual refinement of hearing, to say nothing of vision. The absolute primacy of smell is rarely postulated, however, and is usually encountered in literary rather than natural philosophical texts. One example is the Jewish-Spanish poet, Abraham Ibn Esra (c. 1092–1167) who, in a commentary on Isaiah 11, 3, emphasized the infallibility of smell compared to other senses (Kaufmann 1884, 155). It is also stressed repeatedly (and not only in the Jewish tradition) that smell carries the seal of the Lord (see Gen. 2,7), which may explain the great importance attached to scent in Christian worship (Albert 1990) – in the cults surrounding saints, for example, or the ritual use of incense, myrrh and scented chrism. In Judaism there are the spice boxes containing fragrant herbs, which are savoured at the close of the Sabbath as a sign of the difference between the secular weekday and the sacred day of rest. But while noting the widespread regard for the sense of smell in the religions of the world, we should not overlook the fact that, in Christianity at least, there were others, notably the missionary early Fathers of the Church, who considered it to be a source of temptation, and were only prepared to accept it – if at all – in a purely spiritual form, purged of the dross of profanity.

3. Early olfactory imagery

The symbolic vocabulary of the sense of smell can refer to both, the organ itself, the nose, and the object of its perceptions, namely smell. The range is impressive. Indeed, the palette of metaphors covers everything from the medieval scented garden to the Baroque baby's bottom (Palmer 1993, 61) (Fig. 1).

Fig. 1 Allegory of smell, engraving after N. van der Horstl

In the Bible, the sense of smell occurs largely in association with fragrance. In early modern visual representations of the five senses (for example, those by the Dutch painter Marten de Vos, 1532–1603) we find treatments of the scene in which God breathes soul into Adam (Gen. 1,7) or where Mary Magdalene anoints the feet of Jesus ('The house was filled with the odour of the ointment', John 12,3). These two passages reveal how breath and fragrance were considered symbolic of divine proximity or presence. Hrabanus Maurus (c. 780–856) makes a point of stressing that Jesus was sweet-smelling, a reading for which there is plenty of support in the New Testament: 'Now thanks be unto God, which always causes us to triumph in Christ, and maketh manifest the savour of his knowledge by us in every place. For we are unto God a sweet savour of Christ, in them that are saved, and in them that perish'(Cor. 2,14–15). This also explains why the French founder of the Salesian Order, Francis of Sales (1567–1622), uses images such as odeur, onguents or parfums in his vocabulary of devoutness (Tietz 1990, 93). Fragrance can also be a sign of paradise, particularly in medieval cults of saints and relics (Angenendt 1997). It would also be possible to cite numerous examples of the fragrance of the sacred in non-Christian cultures (Lohmeyer 1919).

In view of the figurative meanings attached to pleasant smells in the religious context, it is hardly surprising that in the Middle Ages scent was considered a sign of power and status: The medieval poet Rudolf von Ems (c 1200–1245) writes: 'He ordered the roads to be strewn with pure and pleasant-smelling things in honour of his rule' (Quoted in Wenzel 1995, 177).

Besides the sphere of power, smell was also read as symbolic of the dark realm of sexuality, the best-known example being a passage in the Old Testament: 'The mandrakes give a smell, and at our gates are all manner of pleasant fruits' (Song of Solomon 7, 13). Similar sexual imagery (flowers, fruit) appears in medieval and early-modern representations of smell, as, for instance, in the five-senses cycle comprising the Lady with the Unicorn at the Musée de Cluny in Paris, or Jan Breughel's Smell (now in the Prado Museum in Madrid) (Fig. 2).

Bad or evil smells can be a token of the approach of death (2 Cor. 2,14). According to the Viennese court chaplain Abraham a Santa Clara (1644–1709), they are also signs of imminent plague. Apart from this,

Fig. 2 Allegory of smell with sexual connotations, engraving, 17th century

images of stench were used to stigmatize strangers and outsiders – particularly Jews – until well into the modern era. The foetor judaicus, a stench allegedly peculiar to Jews and a reminder of their proximity to the Devil, notoriously persists as a stereotype into the twentieth century (Geller 1992, 1997). But supposedly 'typical' smells or stenches were not just used to expose and isolate Jews, prostitutes and witches. The association of certain – usually unpleasant – smells with such purely social categories as sex, age and nationality continues to this day (Wesselski 1934, 15).

The physical configuration of the olfactory sense itself is fraught with allegorical potential (Fig. 3). The Cabbalists believed the triangular shape of the nose represented the first letter (ש = Schin) of one of the Hebrew names of God (Schaddai). As David Kaufmann – one of the great figures of Jewish studies in the latter part of the nineteenth century – notes in his book on the senses in the Jewish Middle Ages, many Jews saw the nose as 'the

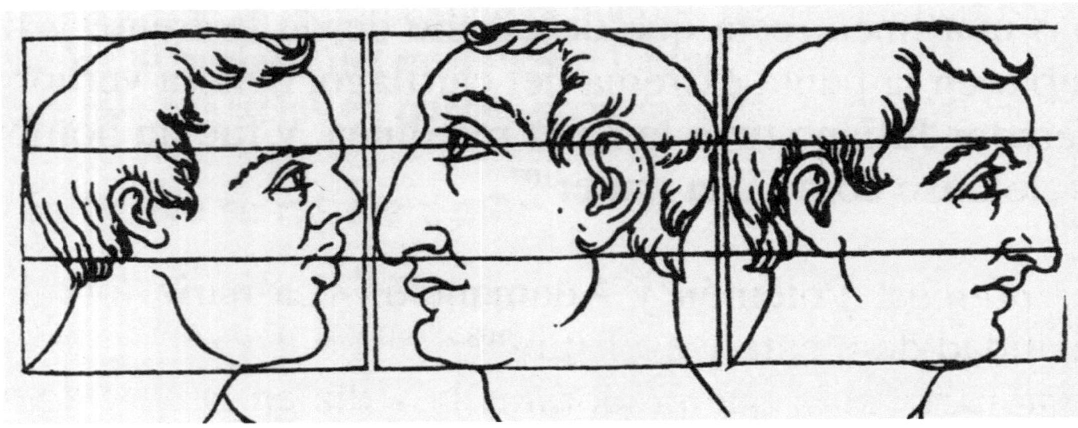

Fig. 3 Nasal shapes according to G. B. della Porta's *Physiognomia* (1586)

seal of God set on our faces' (Kaufmann 1884, 156). But the practice of attaching human characteristics and behavioral dispositions to specific nasal shapes was not confined to Judaism. The physiognomic works of the Neapolitan scholar Giovanni Baptista della Porta (1540–1610) contain a profusion of relevant examples. Porta interprets a whole series of typical 'nasal forms' by means of quotations from classical literature (Porta 1593, 227). Aristotle is his main authority for the then widespread notion that a long nose reaching down to the mouth was a sign of boldness and bravery. Porta construes the nose that is bent like the beak of a crow as a sign of greed, whereas a hooked or aquiline nose betokens courage. A big-nosed individual, on the other hand, is typically one who judges others too severely and is interested only in his own concerns. However, as a reference to Polemon (315–266/5 B. C.) suggests, the big nose can also be a sign of bravery. Polemon is again Porta's source for the idea that the owner of a small nose is politically unreliable (mutabilis consilii vir). Albertus Magnus is summoned in support of his interpretation of the pointed nose as a sign of moral laxity, while his diagnosis of the flat or ape-like nose as a symptom of voluptuousness again rests on the authority of Aristotle. This latter preconception has, of course, been with us for a long time. Late in the nineteenth century, Wilhelm Fließ (1858–1928), a colleague and correspondent of Freud (1856–1939), speculated about the connections between the size of the nose and the length of the penis (Shorter 1992, 67–68). The so-called 'nasal reflex theory', which posited an interaction between the nose and the genital apparatus, enjoyed widespread support in the medical circles of the time. People often had operations on their noses for quite non-cosmetic reasons. On a final, legal-historical, note it is perhaps worth mentioning here that a hundred years before Porta's time, the amputation of the nose by the local executioner was performed as a token of atonement for sexual transgressions (Groebner 1995, 371).

4. Loss of smell

Petrarch's (1304–74) popular book on the right 'spiritual medicine' for good times and bad, typically bemoans deafness and hardness of hearing, but has nothing to say about loss of sense of smell. Instead, there is a chapter entitled 'On sweet and agreeable smells' (I, 22). The total or partial loss of the olfactory sense was evidently not considered to be as serious a loss as deafness or blindness. Unlike Christianity, Judaism attached ritual significance to the ability to distinguish smells, notably in the Havdala ceremony that concluded the Sabbath. When Rabbi Jacob Reischer (d. 1733), one of the leading rabbinical authorities of the modern era, was asked whether a man with no sense of smell was entitled to pronounce the prescribed benediction over the sweet spices at the end of the Sabbath, he replied that it was permitted because, although he might be incapable of physically savouring the fragrance, such a man was none the less capable of spiritual delectation (Zimmels 1952, 47).

Zedler's Universal Lexicon (1735) provides an early example of the occasional interest of the medical community in this physiological problem. The headword, 'Smell, loss of' is itself an indication that the science of medicine was rather at a loss with cases of this kind: 'Those who lose their sense of smell are unlikely to recover it, particularly if they are old. The most useful treatment in these cases is marjoram, which may be applied in any way' (Zedler 1735, col 1206). The author of this relatively short entry also recommends rosemary, anime resin and amber as stimulants for weakened olfactory nerves. In his Neues Arzneibuch ['New Pharmacopoeia', 1568], Christoph Wirsung (1500–1571) was similarly sceptical of the possibility of treatment: 'This is a by no means minor defect of the human body, and the learned say it cannot be remedied if we have been born with it or have grown old. And even if it [the sense of smell, R. J.] could be restored, this could not be achieved without a great deal of time and effort' (quoted in Kassel 1914, I, 244). In those times, only the temporary loss of smell brought on by heavy colds and the like could be treated more or less quickly with various kinds of expectorants and fragrant substances (Le Guérer 1993, 63–101).

Although of inferior rank, the sense of smell was nonetheless conceded a certain importance in everyday life, and was by no means left out of account in preventative healthcare. On reaching sixty, the alderman of the City of Cologne, Hermann Weinsberg (1518–1597) confided in his journal that: 'sense of smell good, exercise it often when I am sitting or thinking, I smell lavender, rosemary, flowers and herbs' (quoted in Jütte 1988, 267)

Aromatic substances have played a part in the treatment of a host of illnesses since antiquity and are today back with a vengeance in the form of aromatherapy. One of the main proponents of this type of therapy was Caelius Aurelianus, who practised in Rome in the third century A. D. and dispensed strong-smelling substances such as vinegar, rose oil and castoreum to epileptics (Kassel 1914, I, 51). This Roman physician was also able to distinguish between loss of smell due to paralysis and the similar symptoms of ozaena, a disorder of the nose which renders the patient insensitive to a foul-smelling nasal discharge. The opposite case of an extremely sensitive nose has scarcely ever been seen as problematic by doctors. Nevertheless, this particular medical phenomenon stimulated novelist Patrick Süßkind, to write his best-selling novel *Perfume* (1985), and has also been a source of inspiration to several other writers (Rindisbacher 1992; Brüggemann 1995).

5. The refinement of smell

In literature dealing with the senses we occasionally encounter the assumption that, like the sense of taste, the sense of smell was largely spared the 'civilizing dressage of the body' (Kampe and Wulf 1984, 13; Barlösius 1987, 368). The historical facts do not support this, although it is clearly apparent that the refinement of the sense of smell was not accorded the same priority as the perfection of vision and hearing. The crown witness is Jean-Jacques Rousseau (1672– 1747). In his system of sensory education, the sense of smell is typically the lowest item on the agenda, the reason being the dwindling evolutionary importance of the sense of smell in the discovery of sources of nutrition: 'Indeed I believe that if children were trained to scent their dinner as a dog scents game, their sense of smell might be nearly as perfect; but I see no very real advantage to be derived from this sense, except by teaching the child to observe the relation between sense and smell' (Rousseau 1911, 122). Rousseau therefore saw little purpose in training the sense of smell from infancy – which would, of course, have involved restoring the child to that original state of nature, in which his disciples fancied they had discovered the 'wild child' of Aveyron and Kaspar Hauser a few decades later.

Rousseau was thus simply concerned not to allow the sense of smell to atrophy any further, and this then became the purpose of the enlightened health-care pedagogy pursued by educationalists after the end of the eighteenth century. In 1859, Dr Daniel Moritz Schreber (1808–1861) recommended exercising the sense of smell by 'not only comparing (with closed eyes) thousands of differently scented blossoms, leaves and roots, but also the smells of various kinds of wood (both living and dead) and earth etc' (Quoted in Rutschky 1977, 467) (Fig. 4).

The consensus of the times was that badly aired sick-rooms and foul-smelling lavatories were the principal sources of wear and tear on the olfactory nerves, for which the best antidote was frequent exposure to fresh air. So olfactory education on the one hand cultivated sensitivity to pleasant, less overpowering odours, and on the other hand put its faith in the avoidance of places where dreadful stenches lurked. An example from the city of Zurich shows that this dual strategy enjoyed a certain success even in the nineteenth century. As part of the city's mid-nineteenth public hygiene programme, the majority of houses acquired new and relatively odourless flushing toilets in place of the old earth closets. A few years later, the city council was complaining of the lack of tolerance shown by the users of the new sanitary installations: 'Since the conversions, [many people] have become oblivious of the frightful conditions of the past and are suddenly very quick to criticize, and become very incensed if the privies – whose windows they often keep permanently closed, while carelessly leaving the seats up and allowing the bowls to become caked in excrement – are not exactly free of smells' (Illi 1987, 103; Gleichmann 1979). According to a survey completed in 1896, only 20 per cent of households in the first district of Zurich, which was home to the wealthier middle classes, lacked water closets. On the other hand, the vast majority (90 per cent) had toilet siphons to expel smells.

An investigation conducted in 1897 by the Frankfurt Tenants' Association demonstrates clearly that, as more and more city dwellers began to feel disturbed by the permanent presence of the smell of other people's excrement and urine, the purification of the air of living areas developed into one of the concrete utopias of the late nineteenth century. A now privately owned collection of

Fig.4 Physician examining the stool, steel engraving, c. 1826

eight phials dating from the Rococo Age, each bearing the inscription 'Air des Femmes' – airtight jars containing, in other words, 'women's farts' – would have probably struck people even then as the bizarre hobby of a smell fetishist.

As we have already indicated, even fragrant smells were thought to be harmful to the sense of smell, particularly when used to excess. Rousseau had already cautioned against the bewitching scents of the boudoir with their promise of erotic adventure, and praised the 'man of dull feelings' with no nose for seductive perfumes. The extremely ambivalent character of smell, particularly in the intimate realm, is demonstrated by the books of marital advice and related publications, which appeared around 1800. In one of them, we encounter the informative observation that very few men regarded the 'anchovy-butter smell of the stinking whore' (Quoted in Wernz 1993, 119) as an aphrodisiac. In the context of the connection between smell perception and sexual behaviour, the theory of the now forgotten American surgeon, John Noland Mackenzie (1853–1925), deserves a passing mention. According to Mackenzie, there were a number of correlations between a sensitive nose and the stimulation of the male and female genitals. This was a version of the nasal reflex theory mentioned in an earlier chapter. At the turn of the twentieth century, its adherents included such distinguished psychiatrists as Wilhelm Fließ, Sigmund Freud and Joseph Breuer (1842–1925).

The French cultural historian Alain Corbin is perhaps best known for his pioneering book on smell entitled ´Le miasme et la jonquille´ ['The Foul and the Fragrant', 1994]. The central thesis of this book is that the period between the mid-eighteenth century and the end of the nineteenth century was a revolutionary era in the history of smell, during which time human attitudes to odours were radically transformed. According to Corbin, the new form of sensory perception is reflected in the numerous attempts made first to describe smells, then to analyse and classify them socially, and finally to eradicate them, and not simply repress them or douse them with perfume. Corbin describes this process as a 'deodorization' of the environment that brought about permanent changes to the social life of cities.

In this context the French cultural historian is not primarily interested in changes in house-building methods and transformations of traditional townscapes (e. g., as a result of extensive redevelopment) or the banishment of 'stinking' industrial plants to the outskirts of cities. Corbin's interest centers on the discourse of social hygiene which began long before the period of intensive industrialization. He is concerned with the changing attitude of the bourgeois elites to the lower strata of society, that is, to people who were supposedly responsible for spreading the 'stench of the poor'. Contemporary opinion saw the poor as a serious threat to hygiene, and one of the unfortunate consequences of this attitude, writes Corbin, was dissociation from the ordinary people, who smelt of death, decay and sin (Corbin 1994, 142–160).

Not only in France, but in Germany too, smell tolerance levels began to fall towards the end of the eighteenth century. In 1796, for example, the physician Carl Georg Kortum (1765–1847) spoke out against the unacceptable conditions in the town of Stolberg near Aachen, which was a centre of early industrial metal processing: 'Small wonder [. . .] that notwithstanding the spaciousness of the valley, the thick, metallic and sulphurous smoke which pours from the oven chimneys eventually pervades the whole atmosphere and can be smelled everywhere' (Quoted in Stolberg 1994, 23).

But it was not just the stench caused by the rapidly expanding metal-working and chemical industries that Kortum and his colleagues perceived as exceptionally unpleasant and even classified as unhealthy. They also turn their noses up at the putrid smell of stagnant water, swamps, sewers and cemeteries. In the course of the nineteenth century, complaints were also raised about the stinking dunghills and cesspools of the countryside, but contemporary olfactory specialists constantly stressed the relative purity of country air. At the same time, however, the lowered smell tolerance levels of city people were increasingly evident: 'Excrement is the smallest but most disgusting and dangerous component of urban waste,' wrote one observer at the beginning of the 1870s. 'The "soul of agriculture" may cause no offence in small farming towns, but it nauseates the more finely-tuned senses of city dwellers. This is simply the hygienic instinct, although odourless is not of course synonymous with healthy, and bad smells are just one of the signs of danger' (Quoted in Rodenstein 1988, 103).

Thus a smell was not just a smell. There were fine distinctions between smells as Corbin shows in the case of France. Let us also note the words of a Berlin medical officer, who in 1878 published a book entitled Gesundheitslehre für Jedermann ['Hygiene For All']: 'There is no lack of "scents" of all kinds out in the countryside, particularly in the manuring season, but they merely brush our nostrils superficially and pass away. Some enthusiasts even find them pleasant, for they represent the "smell of the countryside". In the city, on the other hand, smells seem to be condensed and concentrated in an oppressive mass, and are so intrusive that even the most patient of people let drop the euphemism 'foul smell' and complain of the stench. And their diagnosis is perfectly correct: city air stinks!' (Niemeyer 1878, 189). In the Middle Ages, the phrase 'town air makes you free' was an allusion to the privileged legal position of the towns. In the industrial age, however, the image of the city was shaped less by the social and professional opportunities it continued to offer than by the foul air of its huge conurbations.

If Corbin is basically right in assuming that olfactory perception had become manifestly more discriminating among the urban upper- and middle-classes since the end of the eighteenth century, then we need to ask what kind of 'deodorizing' measures were either contemplated or put into effect at the time? One rather curious method was reported in one of the 1864 issues of the Monatsblatt für medizinische Statistik und öffentliche Gesundheitspflege [Monthly Bulletin of Medical Statistics and Public Hygiene]. After what appears to have been a general collection of rubbish, during which all the smelly refuse had been swept to the edge of the gutters, the mayor of an unnamed town in Saxony was forced to admit that the operation had resulted in the creation of 'an enormously disgusting smell.' This would not, in itself, have moved him to take further action, had he not received the unexpected news that the King was about to visit the town. In this state of emergency, and given the need for rapid action, the 'brilliant' idea of one of the town councillors seemed to offer a solution. It was no sooner said than done: 'The prisoners and vagrants were released from the jail and the town-hall cellars and told to line up in front of the evil-smelling places and breathe in the smell, rhythmically opening and closing their nostrils as ordered' (Quoted in Göckenjahn 1985, 114). This story may be too good to be true, but it does have the merit of underlining the importance of the organ of smell in those days as a weapon in the struggle against air pollution. The sense of smell retained its time-honoured function throughout the nineteenth century. The stink informed the citizens and the city fathers, as well as distinguished medical experts like the Munich professor of hygiene Max von Pettenkofer (1818–1901), that the filth of the streets, the foul-smelling sewage, the contaminated rivers and the polluted air, harboured dangerous miasmata and the germs of disease.

When the stench of the rapidly growing towns became even worse and people literally 'had it up to their noses,' the roads were torn up everywhere, and sewage pipes were laid to carry the refuse into the rivers, with the result that although inner cities became less smelly in the nineteenth century, people in the surrounding areas began to complain of stinking rivers. By 1907, forty German cities already had discharge tunnels. Scientific reports based on water samples and chemical analysis played a large part in the expensive extension and improvement of municipal water supply and sewage systems in the second half of the nineteenth century.

6. The physiology of smell

Compared to the work completed on the senses of sight and hearing, the physiologists of the eighteenth and nineteenth centuries were only marginally concerned with smell. The schematic division of smells into seven different types (aromatic, scented, ambrosial, garlic-like, stinking, repulsive and disgusting) by the Swedish doctor and botanist, Karl von Linné (1707–81) was still being discussed seriously in the nineteenth century. The first comprehensive work on the sense of smell emerged from the pen of the Paris anatomist Hippolyte Cloquet (1784–1840) in 1821. A German translation appeared three years later under the somewhat elaborate title Ophresiologie oder Lehre von den Gerüchen, von dem Geruchsinn and den Geruchsorganen und deren Krankheiten ['Ophresiology or the theory of smells; on the sense of smell and the organs of smell and their diseases]. From an anatomical and physiological basis, Cloquet devotes more than 500 pages to the 'nature' and classification of smells, as well as the olfactory nerves and the 'conditions of smelling'. Concerning the seat of the sense of smell, for instance, he writes: 'Henceforward, we are no longer dealing with guesswork. We may state as a general principle that in humans and most of the animal vertebrates the nostrils and mucous membranes are assuredly the parts where the sense of smell has its seat, and that they perform the work necessary for the exercise of this particular sensory power' (Cloquet 1821).

Just how little was known about the physiology of the sense of smell in those days is clear from Cloquet's extremely vague remark about the olfactory nerves: 'The nerves of the mucous membrane are manifestly of two kinds; some serve the purpose of smelling, and these are the branches of the olfactory nerves, or the first pair, while others perform the task of preserving the life of the smell in the skin' (Cloquet 1821, 173). The separation of the olfactory mucosa (regio olfactoria) from the ordinary nasal mucous membrane that is touched on here was not examined experimentally until the 1850s, when several studies appeared (Seifert 1969, 309–310). Among them was an essay whose title 'Untersuchung des Retinazapfens und des Riechhautepithels bei einem Hingerichteten' ['Investigation of the retinal ganglion and olfactory epithelium of an executed male'] blandly reveals the use of experimental materials which would nowadays be regarded as ethically questionable in the highest degree.

The connection between the fibres of the nervus olfactorius and the sensory cells of the mucous epithelium remained a contentious issue for a long period of time, and was only clarified histologically at a much later date. The tiny sensory hairs of the olfactory cells – whose structure is now fully known, following examination by modern electron microscopes, were discovered and described by Max Schultze (1825–74) in 1856.

Thus, the physiology of the sense of smell made little progress until well into the second half of the nineteenth century. Nevertheless, the number of pages devoted to the subject of smell in the physiological textbooks began to increase. Whereas it had been dismissed in eleven pages in Rudolf Wagner's Handwörterbuch der Physiologie [Concise dictionary of physiology] of 1844, it managed to receive 62 pages in the physiological lexicon published by Ludimar Hermann (1838–1914) in 1880. The smell receptors in the nasal mucous membrane were first described by Max Schultze in 1863. In 1886, Emil Aronsohn (1863-?) published his experimental investigations into olfactory physiology, in which he dealt with the habituating effects of smelling. It was until Hendrik Zwaardemaker's invention of the olfactometer in 1884 that research in the field received another boost. Using Zwaardemaker's smell gauge, it was at last possible to measure keenness of smell. His new method enabled Fechner's law of the proportional increase of sensitivity thresholds to be applied to the sense of smell.

The use of rhinoscopy (Fig. 5) for the diagnostic examination of the nostrils begins with Joseph Czermak (1825–72). Czermak introduced the new procedure in an article in the prestigious Wiener Medizinische Wochenschrift ['Vienna Medical Weekly'] in 1858. The instrument consisted of a double tube which could hold the tongue flat at the

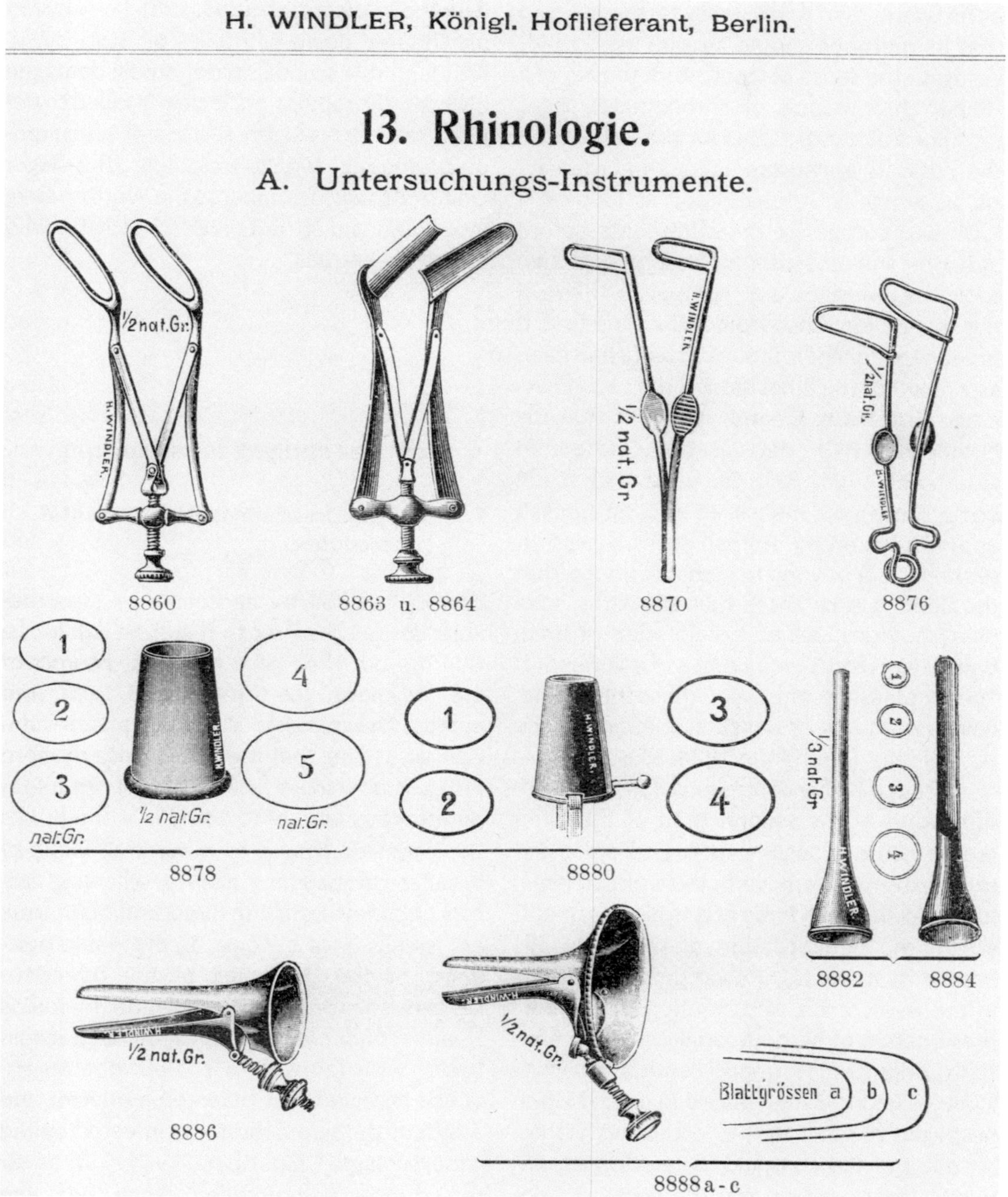

Fig. 5 Instruments for rhinoscopy, 1909

same time. After the 1860s, doctors also increasingly used funnel-shaped mirrors to examine the nasopharynx. In the second third of the nineteenth century, so-called 'mental chronometry' found a role in the discipline of physiological psychology established by Wilhelm Wundt (1832–1920). The biologist Gustav Jäger (1832–1917) – or 'Soul Sniffer,' as he was known to his detractors – believed that its methods should also be used to investigate the sense of smell. With the aid of a Hipper chronoscope, or chronometer, Jäger examined a host of scents for their effects on the nervous apparatus. Jäger's neuroanalysis, as he called it, proceeded as follows: 'I took two successive measurements before and after the inhalation of a scent, and then took the averages and compared them. If the second figure was lower than the first, it proved that my nervous excitability had risen as a result of the inhalation; if the second average value was greater, it meant that my excitability had decreased' (Quoted in Kaufmann 1984, 31). Once he had made enough of these measurements on himself and his voluntary human guinea pigs, it seemed clear beyond reasonable doubt 'that the fleeting substances that affect the nose pleasantly produce an acceleration of time, or in other words an increase of excitement. The unpleasant ones, on the other hand, have a retarding or depressing effect on the sensibilities' (Kaufmann 1984, 31).

Although highly controversial in the scientific world of the second third of the nineteenth century, Jäger's theory of smell became extremely familiar to the world at large, for it provided the basis of a hygienics oriented on the deodorization of the body. His popularity made Jäger even more suspicious in the eyes of his opponents. His scientific presentation of his neuroanalysis to the fifty-sixth Congress of German Scientists and Physicians at Baden Baden ended in uproar when he began to wind up his lecture with a description of the so-called 'defecation experiment.' Jäger's words and the heckling from the audience were recorded in the proceedings. Just as Jäger was turning to his neuroanalytical scent graph and explaining that the 'two first averages [were taken] before the act of defecation, the other three averages post actum defecationis [. . .] and the last after the exhalation of this smell into the fresh air' (Quoted in Weinreich 1993, 196), he was simply shouted down with cries of 'Stop, stop!' Although this scandal undoubtedly damaged his scientific reputation, it clearly failed to depress sales of his highly successful 'perspiration-promoting' woollen clothing (the 'Jäger Uniform'), which numbered the Württemberg industrial magnate Robert Bosch (1861–1942) among its patrons.

7. From deodorizing to reodorizing

7.1 The politics of smell: environmental protection

'Smells can still be agreeable or disagreeable,' writes the French historian Annick Le Guérer, 'but they have lost their powers of life and death' (Le Guérer 1993, 205). And indeed, the miasmic 'stench of plague', the sign of a very real menace in pre-modern times, is nowadays just a rather worn metaphor for any kind of bad smell ('it smells like the plague'). Thanks to a hundred years of popular campaigning against billowing factory chimneys, stinking rivers and open sewers, smells have become an avoidable nuisance. At the beginning of the twentieth century the German Association of Public Health conducted a survey of all German towns with more than 15,000 inhabitants, which revealed that between a quarter and a fifth of the population was infected by the 'smoke plague' (Spelsberg 1984, 75), as air pollution was then called. Since then, the improvement of the quality of the atmosphere and the reduction of smell pollution have become major issues in science and technology and local and environmental politics.

Since industry was slow to install air filters to deal with oppressive smells and dangerous but odourless emissions, and normally only acted in response to enormous pressure, the legislature was forced to intervene at an early stage. The Civil Code of 1900 expressly assured property owners that they were not obliged to put up with 'excessive' nuisance, which included smells (§ 907). All the same, an owner could only seek legal redress if the smell interfered seriously with the use of his property or exceeded 'normal local levels.' After 1915, the courts no longer took the average use of the property as their yardstick but 'what the residents of the area concerned regarded as normal' (Quoted in Spelsberg 1984, 152). This implied a new, 'sensory' interpretation of 'local norms' which remained in force until 1959, when the passage referring to 'normal local levels' was amended. Since then, nuisances of any kind do not have to be tolerated 'if they may be prevented by measures for which those who are responsible may reasonably be expected to pay' (Quoted in Spelsberg 1984, p. 211).

Translating all this into action was both crucial and difficult. The problem is highlighted by the so-called 'technical provisions' of the 1895 trading regulations, which actually remained in force until 1964. The clean air provisions of these regulations deal with noise and smell pollution in very general terms and only specify limits in a few exceptional cases (such as sulphur emissions from glassworks). However, they did not preclude the prescription of limits in individual cases by competent licensing authorities. In the 1920s, for example, the municipal council of the city of Berlin considered the enactment of 'municipal air-protection regulations' to stem a rising tide of dissatisfaction with the much-lauded 'Berlin air.' In the event, the chief constable of Berlin found himself forced to drop this impractical legal device 'because it would be impossible to implement' (Quoted in Wey 1982, 115).

The monotonous, 'international' smell of car-exhaust fumes and industrial effluvia hung over industrial urban areas until well into the post-war period. Given that a large part of the population continued to be preoccupied with everyday cares, it was some time before environmental awareness was again strong enough in Germany to compel the attention of politicians. The 1970s and 1980s were a period of industrial resistance and administrative foot-dragging, and almost two decades elapsed before a German minister of the environment was able to demonstrate the improved quality of the waters of German rivers by taking a bath in the Rhine under the carefully staged gaze of the media. The clean air legislation was also slow to take effect.

7.2 The olfactory imagination

Olfactory perception has become a political issue in the twentieth century and not only in the context of environmental protection. The German-Jewish sociologist Georg Simmel (1858–1918) rightly describes smell as a 'dissociating sense,' for it not only creates 'much more repulsion than attraction' between individuals in everyday social life, but there is also something 'radical and non-negotiable' about its emotional judgements (Simmel 1992, 736).

Turning up one's nose at someone is hardly a new phenomenon, but it is a behavioral gesture that may be encountered in almost all cultures throughout history. In the nineteenth and early twentieth centuries, for instance, the 'hereditary' political enemies France and Germany 'couldn't stand the smell of each other' – which did not prevent the middle classes of both countries from scorning the 'stinking working class.' In the United States the notion that the blacks stank, and therefore had to be segregated, persisted even after the abolition of slavery. Supporters of apartheid in South Africa followed a similar path of reasoning.

In Germany the idea that Jews were dirty and spread a repellent smell (foetor judaicus) was not invented by Hitler. 'A bad smell is clearly an extension of the attribute of dirtiness,' writes the American historian John Efron, 'and the copious testimony of popular culture allows us to conclude that the foul-smelling Jew was largely the creation of the "Christian olfactory imagination"' (Efron 1998, 76). An anti-Semitic Franconian proverb runs: 'Anständige Juden und Juden, die nicht stinken,/ kannst du wohl suchen, aber nicht finden' ['Decent Jews and Jews that don't stink/ Are harder to find than you think']. Even children's rhymes played their part in carrying the negative stereotypes deep into the twentieth century: 'Der Jude Isaak Meier / Der stinkt wie faule Eier [Isaak Meyer, the dirty Yid/ Smells just like a pile of shit'; lit: 'The Jew Isaak Meyer stinks like a rotten egg']' could be heard from the mouths of children during the Third Reich. The traditional prejudice acquired a new and lethal resonance when Hitler used it as one of the building blocks of his racial anti-Semitism. In *Mein Kampf* the future dictator claimed that the smell of an orthodox Jew would often turn his stomach during his years in Vienna. One of the consequences of this 'olfactory imagination' is now very well known: millions of Jews were murdered in gas chambers disguised as disinfection units.

Yet even though it was tackled with incredible bureaucratic energy and German precision, the 'final solution' failed to free the National Socialists from 'the smell of the Jew.' One of the most striking manifestations of the Nazi regime's obsession with the sense of smell was its attempts to conceal from the outside world the smell of burning flesh issuing from the chimneys of the ovens of Auschwitz and other death camps. Most of these efforts were futile. One Auschwitz doctor said later that his wife had been given the usual 'fictions' reserved for visitors: 'in such a large place it was inevitable that many people died, so that a crematorium was needed, and the smoke was due to the fact that it was not working' (Quoted in Lifton 1986, 319). However, it was impossible to deceive the victims' sense of smell. As one Auschwitz survivor relates: 'And then of course immediately you realized what the unbelievable smell was [. . .] that you have been smelling [. . .]. Somehow or other they [the inmates, R. J.] were [. . .] already so inured' (Lifton 1986, 164). Even the murderers themselves had needed time to adjust to it, as another anonymous Auschwitz doctor admitted in an interview with the medical historian Robert Jay Lifton: 'When you have gone into a slaughterhouse, where animals are being slaughtered, [. . .] the smell is also part of it, [. . .] not just the fact that they [the cattle] fall over [dead] and so forth. A steak will probably not taste good to us afterwards. And when you do that [stay in the situation] every day for two weeks, then your steak again tastes as good as before' (Lifton 1986, 197).

So for all their hankering after German perfection not even the Nazis managed to find a way of disposing of corpses without creating a smell. Even in the murderous twentieth century, a monstrous and perfidious project of this kind could only be imagined as science fiction. Strangely enough, however, the idea actually appears as a theme in the work of a Jewish writer. In 1911, Salomo Friedländer (1871–1946) published a grotesque entitled Von der Wollust, über Brücken zu gehen ['On the delights of walking over bridges'] under the pseudonym Mynona. The plot is reminiscent of a modern science-fiction novel. A German scientist named Dr. van der Krendelen discovers a chemical formula for ridding the earth of bad smells. But few manage to survive in the perfectly pure air. Most people die and have to be cremated. And then something quite miraculous occurs: 'Furthermore, the corpses burned in the wonderful air of the early spring without giving off the faintest smell of decomposition' (Mynona 1965, 20).

7.3 Fighting mouth- and body odour

The story of the Germans' preoccupation with bad breath – that most intimate of physical subjects – would make a gripping study of mental attitudes. It is inextricably entwined with the history of a global company, whose name is still on everyone's lips: 'Odol'. Since the turn of the twentieth century, when Odol first became synonymous with mouthwash as such, bad breath has been a subject that could be talked about openly in Germany without giving rise to embarrassment. In England and the United States, on the other hand, people were still reluctant to talk about bad breath in the 1920s, so the advertising department of the Lambert Pharmaceutical Company had to find a scientific name for it. It quickly came up with the word 'halitosis', which proved to be a great success. Following the launch of a mouthwash named 'Listerine,' the firm's turnover rose from $100,000 in 1920 to over $4 million in 1927 (Fig. 6).

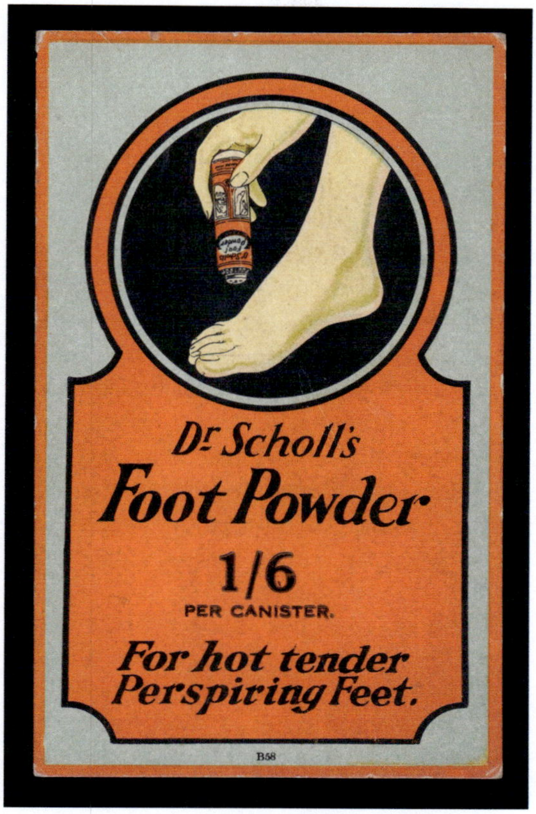

Fig. 6 Powder for foot odour, poster, circa 1930

Under normal circumstances, people are a little more sparing in their use of deodorants. According to an article in the *Frankfurter Allgemeine Zeitung* of 23 November 1998, only one in two Frenchmen uses a deodorant – in a country which is one of the leading consumers of perfume. A non-representative survey conducted by the psychologist Ingelore Ebberfeld in 1998 revealed that people are particularly fond of smelling their partners just after they have emerged from the shower (61.6 per cent). If this percentage seems rather high, we should bear in mind that in Germany and Britain no less than 70 per cent of the population now bathe or shower daily. The figure for France is said to be just 47 per cent. Rising expenditures on body care products suggest that the citizens of the Federal Republic are smelling better and attaching more importance to a clean appearance.

7.4 Going to the smellies

Although still a project of the future, the sensuous 'Scented TV' with a remote control unit containing aromatic capsules activated by a radio signal already haunts the nightmares of the critics of our media society, who fear that the five senses are gradually being monopolized by multimedia. However, the sense of smell has in fact already entered the world of cinema. As the media specialist Anne Paech has noted: 'In its pursuit of the ultimate illusion, cinema has already made several serious attempts to heighten certain visual effects for the spectator by introducing corresponding olfactory sensations' (Paech 2009).

The first film to appeal to the olfactory nerves of the audience was screened at the New York World Fair in 1940. Its title, My Dream, alluded to a perfume that was very popular at the time. Using the so-called 'Odorated Talking Pictures' process developed by the Swiss Hans E. Laube, it first became possible to reproduce smells during the performance of a film using a fully automated system. A cinema of scent was a long-standing dream, and the reader may recall the synchronized olfactory experiences created by the famous 'scent organ' in Aldous Huxley's dystopian novel *Brave New World* (1932).

There was another successful attempt to appeal to the film-goer's sense of smell in 1962. The 'Aroma-Rama' process developed by Charles Weiss involved piping smells into the auditorium through the air-conditioning system, and was used in a film documentary about the Great Wall of China. However, owing to the 'olfactory chaos' (Paech 2009) produced by the relatively primitive technology it found no immediate imitators.

In the same year, another 'smelly' made film history. This was Mike Todd Jr's thriller Scent of Mystery. The American stage- and film producer had taken Hans E. Laube's process a stage further, renaming it 'Glorious-Smell-O-Vision'. However, this unique olfactory experience was available only to audiences at specially converted cinemas in Chicago, New York and Los Angeles, where the seats were connected to a system of tubes. A signal on the film's soundtrack triggered the emission of the appropriate smell in the auditorium at a given point in the action. This was then followed immediately by the release of a neutralizing scent, prior to the next sequence of smells. 'The olfactory information,' writes Anne Paech, 'matched by and large the images on the screen, which were connected with things like garlic, gunpowder, wine, peppermint, shoe polish, lemons, fish, bananas, pipe tobacco, perfume and more than twenty other smells' (Paech 2009). Despite all the expense, this state-of-the-art version of olfactory cinema was also only a modest commercial success.

Outlook

It seems that the process of deodorization, which the French cultural historian Alain Corbin has described as a product of the medical enlightenment and increasing hygienic awareness of the late eighteenth and early nineteenth centuries, is now giving way to a process of 'reodorization.' This transformation is marked by the more or less definitive suppression of stenches of any kind by the latest modern technology. So regardless of where the nose is poked, the only smells it encounters are fragrant. Viewing the long-term character of this development, Constance Classen, an American ethnologist and the author of a remarkable cultural sociology of smelling, poses the obvious question: 'As odours, like roses, have long-standing associations with both spirituality and sensuality, one wonders if this post-modern interest in smell is evidence of a quest for spiritual and/or sensual fulfilment.' (Classen 1993, 35). The use of pheromones – mysterious, sexually enticing substances secreted by the body – in the production of perfume may perhaps be regarded as part of this quest.

References

Albert JP (1990) Odeurs de sainteté: La mythologie chrétienne des aromates. Éd. de l'École des Hautes Études en Sciences Sociales, Paris

Angenendt A (1997) Heilige und Reliquien. Die Geschichte ihres Kultes vom frühen Christentum bis zur Gegenwart. Beck, Munich

Barlösius E (1987) Riechen und Schmecken – Riechendes und Schmeckendes. Kölner Zeitschrift für Soziologie und Sozialpsychologie (Wiesbaden) 39: 367–375

Benedum J (1988) Das Riechorgan in der antiken und mittelalterlichen Hirnforschung und die Rezeption durch S. Th. Soemmering. In: Mann G,

Dumont F (eds) Gehirn – Nerven – Seele. Anatomie und Physiologie im Umfeld S. Th. Soemmerings. Fischer, Stuttgart, pp 11–54
Brüggemann R (ed) (1995) Das Schnüffelbuch. Reclam, Stuttgart
Classen C (1993) Worlds of sense: exploring the senses in history and across cultures Routledge, London and New York
Cloquet H (1821) Osphrésiologie: ou, traité des odeurs, du sens et des organs de l'olfaction; avec l'histoire détailée des maladies du nez et des fosses nasals, et des operations qui leur conviennent, 2nd edn. Méquignon-Marvis, Paris
Corbin A (1994) The foul and the fragrant: odour and the social imagination. Picador, London
Descartes R (1632, 1648) Über den Menschen (1632) sowie Beschreibung des menschlichen Körpers (1648). Nach der ersten französischen Ausgabe von 1664 übersetzt von Rotschuh K E (1960). Schneider, Heidelberg
Efron J (1998) Der reine und der schmutzige Jude. In: Gilman S, Jütte R, Kohlbauer-Fritz G (eds) 'Der schejne Jid'. Das Bild des 'jüdischen Körpers' in Mythos und Ritual. Picus, Vienna, pp 75–85
Galen (1984) On the doctrines of Hippocrates and Plato. De Lacy P (ed and trans), 2nd edn. Akademie Verlag, Berlin
Geller J (1992) (G)nos(e)ology: the cultural construction of the other. In: Eilberg-Schwartz H (ed) People of the body: Jews and judaism from an embodied perspective. Albany NY, State Univ of New York Press, pp 243–282
Geller J (1997) Aromatics of Jewish difference; or, Benjamin's allegory of aura. In: Boyarin J, Boyarin D (eds) Jews and other differences. Univ of Minnesota Press, Minneapolis et al., pp 202–56
Gleichmann P (1979) Die Verhäuslichung körperlicher Verrichtungen. In: Gleichmann P, Goudsblom J, Korte H (eds) Materialien zu Norbert Elias' Zivilisationstheorie. Suhrkamp, Frankfurt am Main
Groebner V (1995) Das Gesicht wahren: Abgeschnittene Nasen, abgeschnittene Ehre in der spätmittelalterlichen Stadt. In: Schreiner K, Schwerhoff G (eds) Verletzte Ehre: Ehrkonflikte in Gesellschaften des Mittelalters und der Frühen Neuzeit. Böhlau, Weimar Vienna, pp 361–380
Göckenjahn G (1985) Kurien und Staat machen. Gesundheit und Medizin in der bürgerlichen Welt. Suhrkamp, Frankfurt am Main
Illi M (1987) Von der Schîssgruob zur modernen Stadtentwässerung. Verlag Neue Zürcher Zeitung, Zurich
Jütte R (1988) Aging and body image in the sixteenth century: Hermann Weinsberg's (1518–1597) perception of the aging body. European History Quarterly 18, pp 259–290
Jütte R (2005) A History of the senses. From antiquity to cyberspace. Polity Press, Cambridge
Kampe D, Wulf C (eds) (1984) Das Schwinden der Sinne. Suhrkamp, Frankfurt am Main
Kassel K (1914–22) Geschichte der Nasenheilkunde, 2 vols. Kabitzoch, Würzburg
Kaufmann D (1884) Die Sinne. Beiträge zur Geschichte der Physiologie und Psychologie im Mittelalter aus hebräischen und arabischen Quellen. Brockhaus, Leipzig
Kaufmann E (1984) Gustav Jäger 1832–1917. Arzt, Zoologe und Hygieniker. Juris Verlag, Zürich
Landauer S (1876) Die Psychologie des Ibn Sina. In: Z Dtsch Morgenländ Ges 29: 335–418
Le Guérer A (1993) Scent: the mysterious and essential powers of smell. Trans. Richard Miller. Chatto and Windus, London
Lifton R (1986) The Nazi doctors: medical killing and the psychology of genocide. Basic Books, New York
Lohmeyer E (1919) Vom göttlichen Wohlgeruch. In: Sitzungsber Heidelberger Akad Wiss Phil-hist section, article 9. Springer, Heidelberg, pp 1–52
Megenberg C (1897) Das Buch der Natur. Schulz HJ (ed). Abel, Greifswald
Mynona [Salomo Friedländer] (1965) Rosa die schöne Schutzmannsfrau und andere Grotesken. Otten E (ed). Die Arche, Zurich
Niemeyer P (1878) Aerztliche Sprechstunden. Gesundheitslehre für Jedermann, vol 1. Costenoble, Jena
Palmer R (1993) In: Bad odours: smell and its significance in medicine from antiquity to the seveneeth century. In: Bynum W, Porter R (eds) Medicine and the five senses. Cambridge Univ Press, Cambridge
Paech A (2009) Das Aroma des Kinos. Filme mit der Nase gesehen: Vom Geruchsfilm und Düften und Lüften im Kino. Retrieved June 11, 2009from http://www.uni-konstanz.de/FuF/Philo/LitWiss/MedienWiss/Texte/duft.htm.
Porta J (1593) De humana physiognomia libri iii. Fischer, Hannover
Plato (1997) The complete works. John M Cooper (ed). Hackett, Indianapolis, Cambridge
Rindisbacher H (1992) The smell of books. A cultural-historical study of olfactory perception in literature. Univ of Michigan Press, Ann Arbor

Rodenstein M (1988) 'Mehr Licht, mehr Luft'. Gesundheitskonzepte im Städtebau seit 1750. Campus-Verlag, Frankfurt am Main New York

Rousseau JJ (1911) Emile or education. Foxly B (trans). Dent, London; Dutton, New York

Rutschky K (1977) Schwarze Pädagogik. Quellen zur Naturgeschichte der bürgerlichen Erziehung. Ullstein, Frankfurt am Main Berlin Vienna

Scarpa A (1800) Anatomische Untersuchungen des Gehörs und Geruchs. Aus dem Lateinischen. Raspe, Nürnberg

Schneider C V (1655) Liber de osse cribriformi et sensu ac organo odoratus et morbis ad utrumque spectantibus. Mevius & Schumacher, Würzburg

Seifert K (1969) Geschichte und Bibliographie der Erforschung des peripheren Geruchsorgans. Clio Medica,Oxford et al. 4: 303–337

Simmel G (1992) Soziologie. Untersuchungen über die Formen der Vergesellschaftung. Rammstedt O (ed). Suhrkamp, Frankfurt am Main

Shorter E (1992) From paralysis to fatigue: a history of psychosomatic illness in the modern era. Free Press, New York

Spelsberg G (1984), Rauchplage. Hundert Jahre saurer Regen. Alano, Aachen

Stolberg M (1994) Ein Recht auf saubere Luft? Umweltkonflikte am Beginn des Industriezeitalters. Fischer, Erlangen

Tietz M (1990) Nicht-verbale Überzeugungsstrategien bei François de Sales. In: Kapp V (ed) Die Sprache der Zeichen und Bilder. Hitzeroth, Marburg, pp 90–101

Weinreich H (1993) Duftstofftheorie. Gustav Jaeger (1832–1917): Vom Biologen zum 'Seelenriecher'. Wiss Verlagsges, Stuttgart

Wenzel H (1995) Hören und Sehen, Schrift und Bild. Kultur und Gedächtnis im Mittelalter. Beck, München

Wernz C (1993) Sexualität als Krankheit. Der medizinische Diskurs zur Sexualität um 1800. Enke, Stuttgart

Wesselski A (1934) Der Sinn der Sinne. Ein Kapitel der ältesten Menschheitsgeschichte. Orientální Ústav, Praha; Harrassowitz, Leipzig

Wey K-G (1982) Umweltpolitik in Deutschland. Kurze Geschichte des Umweltschutzes in Deutschland seit 1900. Westdeutscher Verlag, Opladen

Zedler J (1735) Grosses vollständiges Universal-Lexicon, vol 10 (reprinted Graz 1993–1998). Zedler, Halle Leipzig

Zimmels H (1952) Magicians, theologians and doctors. Studies in folk-medicine and folk-lore as reflected in the Rabbinical Responsa (12th–19th Centuries). Goldston, London

List of figures

Nr.	Source	ID-Nr. 9
1	© Wellcome Library	L0 019 094
2	© Wellcome Library	V007 699
3	© Institut für Geschichte der Medizin der Robert Bosch Stiftung	
4	© Wellcome Library	L0 049 796
5	© Institut für Geschichte der Medizin der Robert Bosch Stiftung	
6	© Wellcome Library	L0 040 545

Neurophilosophy and free will VI

Introductory remarks

Neurophilosophy is a programme that has been intensively studied for the last few decades. It strives towards a unified mind-brain theory in which the mind can be understood as a function of the brain.

The reader may be surprised that in this section of the book we also speak about free will. The problem of free will is related to the overall theme of the present publication because it is becoming ever more clear that sensory perception imperatively requires the reference to motor function. It is only possible to verify that the physical effect of the outside world on a self-sustaining system elicits sensory perception, if we can show that the respective system responds to physical affections with reactions that are mostly motor actions but can also be chemical or neuronal. These reactions have to be interpreted as functional in the service of self preservation of the respective system. Man experiences, at least subjectively, the world as if his/her motor skills were in many ways steered by free will. Because of the close connection of sensory and motor skills it must be assumed, that the human species-specific peculiarities of sensory perception are also affected by the requirements of the species-specific steering strategies of its motor function. Therefore, we decided to make space for the problem of free will in an anthology on sensory perception. The problems that emerge in this context may also be relevant for a number of animal groups where, at the least, indications of similar behavioural steering strategies can be ascertained, e. g. primates or ravens.

1. Neurophilosophy

In his contribution on "*Neurophilosophy*" Gerhard ROTH traces the boundaries of this rather recent discipline. Neurophilosophy includes all philosophical discussion about aims, methods, empirical findings, and consequences of neuroscience, as well as current attempts to clarify the relationship between brain states and mental states. According to *Roth*, major topics of neurophilosophy are: (i), the "mind-brain"-problem, including the nature of mental states and their relationship to brain states, particularly with respect to consciousness and mental causation; (ii), the relationship between the "outer" or "objective" and "the inner" or "phenomenal" world including the epistemological quest for certainty of perception and cognition; (iii), the problem of free will and culpability; (iv), the question of unique human as opposed to animal traits including the presence of mind and consciousness in animals; (v), the origin and function of the "self"; (vi), the origin of social and moral behaviour, and finally, (vii), ethical questions of neuroscience. In regard to the mind-brain problem, *Roth* pleads for the position of non-reductionist monism or physicalism. His examples for neurophilosophy present a current productive interaction between philosophy, psychology and neuroscience in the context of empirical mind-brain or consciousness research. However, further experiments need to be carried out to solve the mind-brain problem in a way acceptable for the representatives of all three disciplines. *Roth* leaves the question open, whether already existing physical concepts will be able to explain the "nature of mind" or whether new laws have to be discovered.

2. Buddha´s conception

In *"Sensory perception, body and mind in Indian Buddhist philosophy"* Ernst STEINKELLNER presents Buddha's conception of the nature of sentient beings. Some interpretations in subsequent Buddhist philosophical traditions appear astonishingly modern. Buddha's usage of defined abstract notions was limited to those necessary for presenting soteriological (related to salvation) and practical teachings and his approach proved to be philosophically useful. Basic concepts and theorems were soon developed into systems of ontology and psychology, which were followed by a normative epistemology that considered the kinds of cognition that can warrant valid knowledge. The relationship between mind or mental phenomena and the body was discussed with different proponents of Indian Materialism in order to prove the correctness of Buddha's analysis of the empirical person. Steinkellner surveys the concept of the senses, their nature and function according to a particular system of mainstream Buddhism, as well as of the ideas about the relationship between mental phenomena and the material body, as can be seen in the attempt of the influential Indian philosopher Dharmakirti to refute the reductionistic explanation of consciousness. *Steinkellner* leaves open the question of monism versus dualism. Instead he proposes to differentiate between the study of the physiological nature of the mind and the study of the social and epistemological nature of the mind and its functions useful to the species. Whereas the former study explains the mind's nature, the latter deals with the mind's function. Such a distinction, if considered as conventional practice, would neither devalue the progress of science in its efforts for a natural explanation of the mind, nor would it devalue ordinary human practice, which relies on the mind as one of the best tools in the strivings for survival.

3. Eye of the mind, eye of the body

In her chapter *"The 'eye of the mind' and the 'eye of the body': Descartes and Leibniz on truth, mathematica, and visuality"* Sybille KRÄMER discusses the conflict between rationalism and empiricism. Even though the rationalist is regarded as a philosopher whose source of knowledge is reason rather than sensory perception, the terminology of "sight" plays a striking role in rationalist philosophy. This paradox of the simultaneous "devaluation" and "valuation" of seeing can be explained in terms of the difference between the "eye of the mind" and the "eye of the body". The rationalist transforms sight into the activity of reasoning, whereby the "intellectual eye" sees all the more clearly the more the body's eyes remain blind. *Krämer's* contribution corrects this understanding by looking at the epistemologies of Descartes and Leibniz. Her investigation into the epistemological meaning of the mathematical innovations of both philosophers helps rehabilitate the role of bodily sight in rationalist forms of knowing. *Krämer* suggests that the calculisation in mathematics, to which Descartes' analytical geometry and Leibniz's infinitesimal calculus contributed, promotes a specific type of visuality which she calls "tactile seeing" or "seeing with the hand." It is also shown, that traces of calculisation, in the form of the core rationalist move of reducing truth to correctness, can be found in epistemology. (i), according to *Krämer*, Hindu-Arabic arithmetic with its decimal positional notation is the prototype of the technique of calculisation. The basic idea behind the practice of calculation is the operative separation of construction and interpretation. Mathematics becomes a model for rationalist epistemology by virtue of its calculising problem-solving techniques ("more calculo") and not its axiomatic-deductive techniques of reasoning ("more geometrico"). (ii), *Krämer* shows the mathematical innovations: Cartesian analytical geometry translates geometric figures into algebraic formulas, so that only what can be rendered algebraically is considered part of scientific geometry. Descartes transforms mathematics into a kind of "language"; one, however, that "speaks" graphically to the eyes. With the help of a graphic signifying system, Leibniz's infinitesimal calculus introduces algorithmic arithmetic to higher analysis. Leibniz separates the operation with infinite values from their interpretation, so that the truth of a statement comes to rest solely on its internal correctness. (iii), the origins of Descartes' method are traced: it is no longer the object that determines the method, but the method that determines the object. This idea has its origins in algebraisation. Descartes' "Regulae ad directionem ingenii" demonstrates the exemplary role of algebra in two ways: Descartes appeals to the tradition of problem-solving analysis understood as "ars inveniendi" and not to the Aristotelian proving analysis of "ars iudicandi". Further, Descartes introduces an extensional, graphic symbolism meant to work like a calculus into his "mathesis universalis". Only that which can function as an object of reference of this symbolism can be admitted as an object of universal science. Visualization and methodology become enmeshed. (iv), Leibniz's reduction of truth to correctness is discussed. Leibniz articulates the concept of calculation as an interpretation-neutral operation with graphic signs and tries to apply calculisation as a universal process onto knowing itself. This shows that the objects of knowledge are constituted by the order of the process of knowing, according to Leibniz's "generative definition." Since the world is made up of unique things, but calculising knowledge only refers to abstract constructs of the mind, the epistemological consequences are that our formal knowledge can no longer refer to the real world, but only to our models of the world. (v), *Krämer* shows the process of visualising the invisible or "seeing with the hand": The heart of science is not simply "abstraction", but the sensualisation of abstractions and the visualisation of the invisible. Modern algebraisation stands for the formation of an "operative" or "tactile" kind of visuality. The rationalists devalue "ocular seeing," since it is closely tied with the illusionary, but they value "tactile seeing," which is not a "seeing with the mind," but a "seeing with the hand."

4. Free will

In his contribution *"Free will"* Hans-Dieter KLEIN pleads for compatibilism. Free will can be defined at one extreme as libertarian free will, according to which free actions are unconstrained and uncaused by any physical process. It is therefore incompatible with the concept of the physical world as a closed causal system. However, due to the shortcomings of the common opinion that free will and physicalism are simply incompatible, the more sophisticated approach of the compatibilists rejects libertarian free will, because free will for compatibilists is not the absolute libertarian freedom. Rather, it is the unrestricted ability to act without external or internal (e.g. mental diseases) constraints. Moreover, compatibilists are aware that even in the determined world some processes are chaotic and their complex outcome cannot be predicted, although they are causally related to their starting conditions. This approach can be supported by a long philosophical tradition inaugurated by Aristotle. The conception that our free will is free in as much as the decision of the subject is not hindered by inner or outer compulsions goes back especially to Aristotle. Free of such compulsion, the freely deciding subject, in his opinion, determines himself, that is, through following the logical rules he relates the possible actions to self-produced norms, and makes his decision accordingly. *Klein* shows that free will so understood is not in opposition to a continuous physical causality. In fact, as the example of computers shows, thought processes can be physically realised. **H-D K**

Neurophilosophy

19

Gerhard Roth*

Contents

Abstract

Neurophilosophy includes all philosophical discussion about aims, methods, empirical findings, and the possible consequences of neuroscience. Also, it denominates all attempts to clarify the relationship between brain states on the one hand and mental states on the other. Major topics of neurophilosophy are: (1) The "mind-brain"-problem, i. e., the nature of mental states and their relationship to brain states, particularly with respect to consciousness and mental causation, (2) the relationship between the "outer" or "objective" and "the inner" or "phenomenal" world including the epistemological quest for certainty of perception and cognition, (3) the problem of free will and culpability, (4) the question of unique human as opposed to animal traits including the presence of mind and consciousness in animals, (5) the origin and function of the "self", (6) the origin of social and moral behavior, and (7) ethical questions of neuroscience. As to the mind-brain problem, only versions of non-reductionist monism or physicalism appear worth discussing. Empirical evidence underlines that mental states obey physical laws and can be predicted on the basis of knowledge about brain states. It has to be left to the future as to whether already existing physical concepts will suffice to explain the "nature of mind", or whether new laws, while being compatible with the existing ones, have to be discovered.

Gerhard Roth
University of Bremen
Institute for Brain Research
P.O. Box 330440
Leobener Str.
28359 Bremen, Germany
e-mail: gerhard.roth@uni-bremen.de

* I thank Anna-Lena Dicke (University of Tübingen), John Dylan Haynes (Charité Berlin and Bernstein Center for Computational Neuroscience) and Michael Pauen (Humboldt University Berlin) for helpful criticism.

1. Introduction

The term "neurophilosophy" was coined by Patricia Churchland in 1986 in her book "Neurophilosophy: Towards a unified science of the mind-brain". Here, Churchland put forward a specific philosophical concept called "eliminative materialism" (see below). Subsequently, the meaning of the term "neurophilosophy" has been substantially extended to indicate any kind of philosophical discussions of neurobiological subjects as well as a treatment of philosophical subjects on the basis of neurobiological knowledge or findings.

Such attempts to connect philosophy with biology and medicine are as old as Western philosophy starting with Plato, who in his dialogue "Timaios" speculated on the site of the mind (in the sense of intellect) as well as of emotional and affective states inside the human body. They first culminated in the antique concept of the "nature of mind" and the "ventricle theory", first developed by the antique physician Claudius Galenos (131–200 A. D.). This theory assumed that mind was located in the cerebral ventricles filled with a subtle substance called *pneuma* or *spiritus*, and it dominated Western philosophy until the beginning of the 19th century, when modern neuroscience started. Descartes developed the idea of the pineal organ as an "interface" between mind/soul and brain/body. Leibniz, in his "monadology", presented the famous "mill argument" to explain that we cannot understand how material systems may reveal mental properties. Newton believed that mind was identical with an electric substance called ether, and this mostly speculative tradition ended with Soemmering's concept of the brain as the "organ of the soul". In the 19th and early 20th century, mind-brain concepts were increasingly based on new findings in brain anatomy, and eminent scholars tried to develop a unified theory of how brain structures and functions lead to mental states. Among them were Franz Gall and Caspar Spurzheim, the founders of modern neuroanatomy as well as of obscure "phrenology", the physicist and physiologist Hermann von Helmholtz, the physiologists Ewald Hering, Sigmund Exner and Wilhelm Wundt (the founder of experimental psychology), the neuroanatomist Theodor Meynert and even young Sigmund Freud and his "Entwurf einer Psychologie" of 1895. This tradition was continued by Paul MacLean's concept of the "Triune Brain" (1990), Bernhard Rensch's "Biophilosophy" (1968) and by the neurophysiologist John Eccles in his books, partly co-authored by the philosopher Karl Popper ("The Self and its Brain", 1982). Today, the discipline called "philosophy of mind", originally derived from analytical philosophy, has developed a strong overlap with neurophilosophy.

Neurophilosophy has two albeit related meanings. First, it includes all philosophical discussion about the aims, methods, tacit and explicit assumptions, concepts, empirical findings, and possible consequences of neuroscience. This should more accurately be called "philosophy of neuroscience" as a sub-discipline of philosophy of science. Second, it denominates all attempts to clarify the relationship between brain structures and functions on the one hand and mental states and functions on the other.

Major topics of present-day neurophilosophy are: (1) The "mind-brain"-problem, i. e., the nature of mental, including psychic, states and their relationship to brain structures and functions, particularly with respect to consciousness and the problem of mental causation, (2) the relationship between the "outer" or "objective" and "the inner", "subjective" or "phenomenal" world including the epistemological quest for certainty of perception and cognition, (3) the problem of free will and culpability, (4) the question of unique human as opposed to animal traits including the presence or absence of mind and consciousness in animals, (5) the origin and function of the "self", (6) the origin of social and moral behavior, and (7) ethical questions of neuroscience.

In the following, I will concentrate on the mind-brain and the closely related consciousness problem, because it is regarded as the dominant one by representatives from the relevant disciplines, i. e. philosophers, psychologists and neuroscientists. In this article I will not so much present my own solution to that problem, but rather demonstrate an *approach* to a solution acceptable to the three disciplines.

2. The mind-brain or consciousness problem

The mind-brain and more specifically the brain-consciousness problem essentially addresses three disciplines, viz., philosophy, psychology and neurobiology, with additional involvement of evolutionary biology, anthropology, linguistics, sociology as well as physics and mathematics – just to mention a few others. Evidently, such an endeavor is highly interdisciplinary, and no single discipline could ever hope to present a mind-brain or consciousness concept that has a chance to be fully accepted by the other disciplines. Rather, philosophy, psychology and neuroscience need to cooperate intimately, although their contributions may be very different. Such a co-operation is difficult, because it requires a philosopher, a psychologist and a neurobiologist to have sufficient knowledge of the two other respective disciplines or to have a dual academic or professional training. With regard to a co-operation between psychology and neurobiology, the situation has become rather favorable, and many psychologists have profound knowledge of neuroscience or are neuropsychologists by training, whereas there are still few philosophers with a deep knowledge in experimental psychology or neurobiology, and an equally low number of neurobiologists with sufficient knowledge of philosophy or psychology.

2.1 Current mind-brain concepts

If we are to reach a consensus among the three disciplines we have to be aware of the major concepts regarding the mind-brain or consciousness problem currently discussed in the *philosophy of mind* (for an overview see Kim 1996; Pauen 2001). An incredible amount of reasoning power has been invested by philosophers into the discussion of these various concepts, and this could prevent psychologists as well as neuroscientist from carrying out inconclusive experiments and from drawing premature conclusions. Also, it is important to distinguish those philosophical concepts that can, at least in principle, be empirically tested from those that cannot, as the latter are useless for empirical disciplines.

Traditionally, there are dualist and monist positions. *Dualists* believe that mental states and brain states are *ontologically* different. In other words: mental-conscious states are non-physical, and physical processes are non-mental.

2.1.1 Dualism

Such a dualism is in accordance with our subjective experience: the percept "red" seems to have no spatial extension, weight, shape, temperature and no structure or substructure known from physics, chemistry or biology. It does not seem to underlie known laws of solid body physics or electromagnetism, and even more exotic theories like quantum or relativistic physics do not seem to apply to mental states. There seems to be no indication that mental states consume energy and underlie the laws of thermodynamics. Psychologists may claim that there are "laws of gestalt perception" or "laws of thinking", but up to date nobody has been able to link such mental laws to physical or physiological laws. Dualists take this as the strongest evidence for the *ontological* difference between mind and matter.

The principal problem of dualism is that, despite these apparent differences, mental states and brain states undeniably exhibit at least some parallelism. Furthermore, mental and cerebral processes *interact*: if I am stung by a needle, I feel pain, which is a physical event causing a mental one; conversely, if I have the *intention* to move my hand, it will move, and my intention seems to *cause* a physical act.

Dualism in the form of *psychophysical parallelism* and *occasionalism* recognizes this fact. However, it states that the parallelism between mental and physical states is not really due to causal interaction, but to the fact that they constitute separate chains of events simply running in parallel (e. g. because God had arranged it in this manner). In contrast, *interactionist dualism*, as developed by Descartes and put forward in modern times by the physiologist John C. Eccles (cf. Eccles 1994), accepts a causal interaction between mental and physical states in both directions, although this causal interaction is believed not to underlie the laws of physics, e. g. the law of energy conservation. Therefore, interactionist dualists need to postulate a special kind of non-physical causation, called *mental causation*. Finally, there is *property dualism* stating that certain physical or brain states lead to the "emergence" of mental states which, then, have properties fundamentally different from physical ones (Popper and Eccles 1984). However, the cause is considered an unsolvable scientific problem, because it cannot – by definition – be addressed empirically. Since René Descartes, there is also the unanswered question how something non-physical could interact with something physical without violating physical laws, e. g. those of conservation of energy. Nevertheless, interactionist dualism is still very popular even among scientists.

Another version of property dualism is *epiphenomenalism*. This doctrine holds that while cerebral states can cause other cerebral states, some of which are accompanied by consciousness, conscious-mental states are mere by-products (*epiphenomena*) and play no causal role. Epiphenomenalism is confronted with two major difficulties. First, it fails to give an answer to the question why mental states have developed at all in apparently many brains, if they are without any function. In biology, functionless structures normally occur only in those cases, when the carriers of such structures had ancestors in which they were still functional. Second, epiphenomenalism only makes sense, if noncausal mental states can somehow be separated from their causal cerebral carriers and then be demonstrated as being functionless. If, however, certain cerebral processes do have certain functions if and only if they are accompanied by consciousness, the two aspects cannot be separated and have to be regarded as a functional unity.

The much discussed *Supervenience theory* somehow stands between dualism and monism-identism is. It believes that mental states originate from certain cerebral states and consequently are physical states. But it acknowledges the phenomenal differences between the two in the sense that mental states cannot be reduced to known physical states in the same way as solid-state physics cannot be reduced to electromagnetism. At the same time they postulate that all changes occurring at the level of mental activity must originate from changes at the cerebral level. Thus, for supervenience theory the mental state is dependent on the brain state (regardless of whether it is ontologically different or not). This, for example, means that if two people are indistinguishable in all of their physical properties, they must also be indistinguishable in all of their mental properties, but these mental properties need not be of the same physical nature as brain processes.

2.1.2 Monism

Monists generally deny that mental states and brain states are fundamentally or ontologically different. For *physicalist* monism mental states are physical states. A radical version of such monism is *identism* stating that mental processes are *identical* with neuronal processes. In that view, the intention of reaching for a cup of coffee is identical with the activity pattern within certain parts of the brain. *Eliminative materialism*, as developed by the philosophers Patricia and Paul Churchland (e. g. Churchland 1986), claims that mental states do not exist. Therefore, we should avoid using mentalistic terms like thinking or wanting and replace them by exact neurophysiological descriptions, for example. that activity in the dorsal prefrontal and premotor cortex, under the influence of certain limbic centers and the basal ganglia, leads to a movement towards a cup of coffee.

Identity theory today is discussed in two different versions, *type-identity* and *token-identity theory*. The former holds that for every type of mental content ("painful" or "yellow") there is exactly *one* type of neural correlate, and this specific correlate leads only to that type of mental state. The latter allows different neural states to lead to one specific type of mental state (e. g., "yellow" can be brought up by different neural states). This is called "multiple realization" of mental states.

Identism, in its radical form, notoriously struggles with the fact that logically two entities are identical only if they are the same in *all* their properties (e. g. morning star, Hesperus, and evening star, Venus; Mark Twain and Samuel Langhorne Clemens – to mention two famous examples). This apparently does not apply to mental and cerebral states, because they are at least experienced in a fundamentally different way. Furthermore, when we have conscious mental states such as perceiving, thinking, remembering or imagining, we experience no brain states, and when we investigate the brains of other human beings or animals, we experience no mental states. Even while investigating our own brain using modern functional imaging methods, we are confronted with a strict parallelism of brain states and certain sensations, but experience no causal relationship. Thus, mental states and brain states appear to constitute two non-overlapping phenomenal domains, and two entities that exist in different phenomenal worlds cannot be identical by definition of identity, but rather appear to constitute two different aspects of the same event.

Another severe problem of the identity concept is, *which* of the many brain processes is supposed to be identical with mind. This is the question of the "neural correlates of consciousness – NCC", first formulated by Christof Koch and Francis Crick as the "minimal set of neural activity sufficient to generate conscious states under certain conditions" (cf. Koch 2004). Is the NCC identical with the cerebral cortex in general, or with the thalamo-cortical system, with parts of the frontal lobe, with "re-entrant" systems inside the cortex (Edelman and Tononi 2000), a certain type of activity, e. g., the synchronous firing of neurons (Engel et al. 1991), the activity of certain membrane receptors (e. g. NMDA receptors; Flohr 1998)? Even though it can be shown that such phenomena are somehow related to consciousness, it is still completely unclear whether they are sufficient or only necessary or even facultative conditions for consciousness, and we are not allowed to speak of identity.

Eliminative materialism suffers from the additional fact that we cannot derive mental states or functions from the study of neuronal processes alone. From the most detailed study of activities in certain limbic centers and of certain neuropharmacological events we cannot decide whether or not the person under study has fallen in love or whether a monkey experiences fear, unless we have pre-existing knowledge about what it is to fall in love or to have fear. Even if we observe the related behavior, we have to *interpret* it on the basis of subjective experience.

A much discussed argument, also considered the "hard problem" of consciousness by philosophers of mind, is that of the *fundamental gap* between brain states observable under a third-person perspective and phenomenal states or qualia, observable only under a first-person perspective (Chalmers 1996, Levine 1983). Even if we accept that qualia are physical states and necessarily bound to brain states, there is no access to this subjective state from the third-person perspective. Accordingly, we will never know how Mr. Miller experiences a certain color, even if he calls it light red as we do. Therefore, there will never be a satisfactory neurobiological explanation of the mind.

This is perhaps the most difficult problem for an empirical approach to the mind-brain problem (Chalmers 1996). Certainly, subjective feelings cannot be investigated empirically in a direct way, i. e., they cannot be observed or measured from a third-person perspective, However, there is a dispute whether or not "qualia" of a person are truly private and inaccessible to other people (cf. Dennett 1991). Trained subjects can report their subjective feelings with great accuracy, and animals can be trained to behaviorally indicate in greater detail what they experience (see below). The famous "inverted color spectrum" argument claims that we cannot exclude that Mr. Miller perceives a stimulus as red where Mr. Schulze perceives it as green, while both call it "green". This argument ignores the fact that colors have a fixed position within the color spectrum, that they have neighboring colors, are all linked to "opponent" colors (red-green, blue-yellow) and are associated with certain feelings ("warm" or "cold" colors). Therefore, there is a multitude of psychophysical methods to test whether a subject has any abnormalities in color perception. Thus, qualia are not as private as some philosophers of mind used to think.

2.2 Testability of mind-brain positions

The decisive question is which of these philosophical concepts can be investigated empirically, and which cannot. This is not to say that certain philosophical concepts are irrelevant per se, but only that no agreement between philosophers on the one hand and empirically-experimentally oriented psychologists and neuroscientists can be reached, if hypotheses are not empirically testable. Of the kinds of dualism presented above, *psychophysical parallelism* and *occasionalism* can be excluded from further consideration, because they cannot be proven or disproven empirically. Even if we would find the closest relationship between mental states and brain states in our experiments, in the eyes of psychophysical parallelists like Leibniz this is only because God has arranged it this way and not by any causal-physical interaction.

In contrast, the assumption that qualia are ontologically different from brain states can be tested empirically. We have to check, if there are certain processes inside the brain which reliably lead to that "raw feeling of red" as opposed to the "raw feeling of green". The knowledge of such a strict coupling perhaps would enable us to predict that under given experimental conditions a subject will utter "now I have the raw feeling of red" as opposed to "green". Even if we could not explain what qualia are *ontologically*, we could claim that they are strictly linked to brain processes, and this would make the assumption of an ontological difference between brain states and mental states very unlikely. *Epiphenomenalism* can be empirically tested in a similar way. If we can demonstrate that certain cerebral processes have certain functions or consequences only when accompanied by consciousness, then it appears illogical to assume that conscious states play no functional role.

Interactive dualism can be tested, because it implies that at least in some cases "purely" mental states, which by definition are *not* accompanied by brain states, can cause brain states. Free will, for example, is such a supposed mental state influencing our behavior in a non-physical way. Therefore, we have to search for cerebral processes that control a given behavior and at the same time are *not* caused by other cerebral processes. Thus, in order to verify interactive dualism, there should be a "causal gap" in the otherwise endless chain of cerebral processes, where mental causation exerts its influence. Since this gap relates to overt behavior, it should be clearly visible. If we cannot identify such causal gaps, for example, in the sequence of events leading to voluntary movements, then interactive dualism becomes unlikely.

Physicalism holding that mental states are compatible with known physical laws, is easily testable. We can study the relationship between the intensity of mental activity (e. g. concentrated thinking or observing), neuronal activity and brain metabolism (e. g., oxygen and glucose consumption). We could do Weber-Fechner experiments in order to test the relationship between physical properties of stimuli (loudness, brightness, weight) and subjective perception. By presenting sensory (e. g. optical-visual) illusions, we could test, whether there are neurons in the brain responding to the illusion in the same way as we do in our subjective experience. The more we proceed in demonstrating such relationships between stimulus situations and subjective feelings reported verbally or by behavioral responses, the more likely it is that subjective-mental states are the product or at least an "aspect" of brain states. If, however, we are unable to find such an asymptotically increasing correlation between the objective and the subjective state, we have to accept either that mental states are not fully dependent on physical-cerebral states corroborating dualism, or that there are still unknown physical-physiological factors that are constitutive for mental states.

2.3 Definitional problems

If we are going to test these philosophical concepts empirically, then the *next step* is the precise formulation of the problem. The goal of solving "the" mind-brain problem is far too general. In the present case, a more meaningful approach would be to investigate the exact spatial (i. e. neuroanatomical) and temporal (i. e. physiological) relationships between conscious mental states and brain states. It would be even better to restrict the investigation to just one cognitive function, e. g. visual perception, as the visual system of human beings and animals is the most thoroughly studied sensory system. (see also Chapter I,3 by K Krug)

MIND. The decision about what kind of experiments we want to carry out requires an agreement among the involved disciplines that this type of experiment is appropriate for solving a certain aspect of the mind-brain or consciousness problem. This in turn requires an agreement about the meaning of key terms such as "mind", "consciousness", "attention" etc. In the case of "mind", philosophers, psychologists and neurobiologists will quickly agree that this can only apply to *individual mental states* or *functions* such as perception, cognition, thinking, remembering and imagining, which can be studied with psychological means. All other meanings of "mind" such as possible supra-individual meanings ("the mind of a nation"), spiritual experiences or religious revelations are excluded. Their existence need not be denied, but simply cannot be addressed empirically by neuroscience and psychology. This is not problematic, because such individual mental states always have been the subject of philosophical, for example, epistemological speculation from antique times up to the present.

CONSCIOUSNESS. An agreement about what is meant by consciousness is much harder to obtain. There is no universally accepted definition of consciousness and how to measure it objectively. However, a recent distinction between at least three different kinds of consciousness has become popular among philosophers and psychologists (cf. Block 1980). There is *phenomenal* consciousness, often called raw feelings or qualia, e. g. the "redness" of an event. In this sense, consciousness is something immediately given and accessible only by the person who has this sensation. Second, there is *access* consciousness, which refers to the ability to report and act on a certain content of awareness, for example, in the form of conscious judgement, reasoning, and the planning and guiding of an action. Finally, there is *higher-order consciousness* or *meta-cognition*, which is the state of *knowing* that I am conscious of something.

ATTENTION. Another critical term is *attention*. Although many philosophers and psychologists consider consciousness and attention to be the same, others make a distinction between these two states (cf. Lamme 2006; Koch and Tsuchiya 2007). For them, consciousness can occur without attention, and attention without consciousness. Attention is strongly linked to the reportability of an attended event: if I observe or read something with concentration, I can at least report the physical traits of what I have seen or read, whereas if I perceive something without attention, it fades away within about 5 seconds and cannot be reported, although I had conscious perception of it during those few seconds. The same holds for events "out of focus of attention" (e. g. in the lateral visual field). I may feel that there is "something", but I cannot tell any details.

Conversely, I can stare at a screen with high concentration and yet see nothing. This is the case of *backward masking* of a stimulus, or if I am suffering from *blindsight*. In the former case, a brief visual stimulus is followed after a brief interval (e. g. 40 ms) by another stimulus, and this "masking" stimulus will erase the conscious perception of the former. The first stimulus is processed unconsciously within the retina, thalamus and primary visual cortex, but the second, masking stimulus suppresses further processing in "higher" visual areas. Patients suffering from blindsight, when confronted with visual stimuli visible to us (e. g. a coffeepot), report that they "see nothing", even if we tell them that a coffeepot is located in front of them (Weiskrantz 1986). If forced to reach for this "invisible" object, they may do that successfully, although for them that appears ridiculous, until they touch the coffeepot (they might, however, have the sensation that they see something, if the object is forcefully moved in front of them). Blindsight occurs when an individual suffers from a lesion of the primary visual cortex, the activity of which appears to be a necessary prerequisite for consciousness. Here again, looking with concentration will not help. Thus, we indeed have to postulate two different mechanisms, one for making events conscious, and another one for what is being called "top-down" or "internally driven" attention as opposed to bottom-up or "externally driven" attention. The latter is driven by any sensory event that by itself is "salient" with respect to certain sensory features. This includes suddenly appearing, large, bright, colorful, high-contrast, fast moving or approaching or strangely looking objects, loud or painful noise etc. In contrast, the former is driven by "interest" due to motivation, expectation etc.

A major problem is how to check, whether and to what extent a certain sensory, cognitive or motor event is consciously experienced or attended. As we have already heard, a commonly accepted indicator for conscious attention of an event is its *reportability*. If somebody is able to describe a given object (for example a coffeepot) in front of him in greater detail, for example, that the coffeepot is red, has two handles and shows the Brandenburg gate in Berlin, we can be quite certain that he has both phenomenal and access consciousness. The same holds for the ability to perform certain actions, for instance, pressing different buttons on written command. Blindsighted patients are unable to do that, and the same occurs in normal people, if the picture of the coffeepot has been masked or if it was out of focus of attention (in the latter case subjects, however, might be able to say "I saw something like a coffeepot", but they are unable to report any details).

Reportability as a proof for consciousness/attention need not involve a verbal report, especially in experiments, where stimuli are presented at different intensities, duration or other special conditions in order to investigate thresholds of conscious perception. Here, people may have difficulties with precisely reporting what they have or have not seen, or they may be influenced by unconscious perception. Therefore, instead of letting subjects guess whether or not they have actually perceived the test stimulus, a method called *post-decisional wagering* has been proposed for "objectively" measuring consciousness or awareness (cf. Persaud et al. 2007; Koch and Preuschoff 2007; Seth et al. 2008). A person makes a decision (e. g. a visual discrimination) and then places a wager. If the decision, for example the visual discrimination, is correct, then the subject earns the amount of money wagered, whereas if the decision is incorrect, i. e., the visual discrimination is inaccurate, he or she loses the amount of money wagered. Consciousness can be assessed by assuming that the subject will place a high wager if he or she is aware that his or her decision is correct. Such a method can be applied to experiments with children as well as with animals such as monkeys.

3. What kind of experiments should we carry out?

The answer to this question depends on the methods to be applied and the precise topic to be addressed. Today, experiments with human subjects regarding cognitive functions including consciousness and attention are usually done either with electroencephalography (EEG), magnetoencephalography (MEG) or functional magnetic resonance imaging (fMRI). These methods all have characteristic advantages and disadvantages, especially with respect to their spatial and temporal resolution. EEG measures the electric fields in the brain, predominantly in the cerebral cortex, via electrodes placed on the scalp. It offers a direct measurement of neural electrical activity with very high temporal resolution in the range of a few milliseconds, but with low spatial resolution, because its signal reflects the summed activity of thousands to millions of cortical neurons. MEG, in contrast, measures the magnetic fields produced by electrical activity in the brain via extremely sensitive devices such as superconducting quantum interference devices (SQUIDs). Like the EEG, it has a very high temporal resolution, but relatively low spatial resolution. The advantage of measuring the magnetic fields produced by neural activity is that unlike electric fields they are not distorted by surrounding tissue and skull. fMRI relies on changes in the magnetic properties of oxygenated and desoxygenated hemoglobin or changes in blood flow in the brain associated with neural activity. What is being measured here is the so-called blood-oxygen-level-dependent or "BOLD" signal. The temporal resolution of fMRI lies around 1–3 seconds and the spatial resolution of conventional fMRI lies around 1 mm^3. The same methods can be applied to the brains of monkeys or even smaller animals down to the size of a rat. The advantage of brain imaging techniques is that different local activities in the brain can be measured simultaneously, but this is always the summed activity of at least thousands of neurons, and so far there is no reliable distinction between neural activity leading to excitation and activity leading to inhibition.

Animal experiments on cognitive functions including attention are often carried out using microelectrodes, either as single-cell recordings or as local field potential recordings using multi-electrode devices consisting of many electrodes. The advantage of the single-cell recording technique is the possibility to precisely measure the response (mostly action potentials) of one neuron to a certain stimulus, while with the multi-electrode device the local interaction pattern of many neurons (e. g., oscillatory activity) can be recorded at one time. Here we have a combination of high spatial and high temporal resolution power.

3.1 Four neurophysiological experiments

In the following, I will discuss by way of examples four experiments that have been carried out in recent years by leading neuroscientists and which I find highly relevant for the mind-body and consciousness problem. They were published in high-ranking international journals, so that they can be regarded as representing the state of the art in cognitive neuroscience.

3.1.1 Binocular rivalry

The first experiment published by John-Dylan Haynes and Geraint Rees in 2005 addressed the question of whether changes in the content of conscious states could be correlated with or even predicted on the basis of ongoing cerebral activity. It made use of *binocular rivalry* – a phenomenon of visual perception, in which two sufficiently different images are presented simultaneously but separately to each eye. In these cases conscious perception does not fuse the input from the two eyes, but instead perception alternates between periods where only the left or the right eye input is conscious.

In the experiment, a rotating red grating was presented to the left eye and simultaneously an orthogonal rotating blue grating to the right eye. Subjects lying in a MRI tomograph indicated, by pressing one of two buttons, which of the two grating images they were currently seeing. The activity of cortical areas was measured by fMRI. Using a combination of fMRI and statistical pattern recognition the authors were able to access information that is believed to be stored in very fine-grained cortical maps (below 1 mm^3), such as ocularity, color, or orientation, despite the coarse resolution of fMRI. Here, a pattern classifier was trained to identify phases of red versus blue dominance based only on the distributed brain response patterns (for details of that method see Haynes and Rees 2005, 2006).

The results of this study showed that on the basis of activity in the primary visual cortex (A17/V1) the classifier predicted with high accuracy (80%) and high temporal precision the changes in the content of visual awareness of one of the two competing stimuli (vertical or horizontal rotating lines), i. e., changes of activity in V1 reliably predicted the changes in subjective experience. Both types of changes occurred spontaneously roughly every 14 seconds. These findings are remarkable in several ways. First, they demonstrate that it is possible to precisely correlate states of subjective awareness with neuronal activity. Second, changes in the fMRI signal were roughly sinusoidal in shape, whereas changes in the percept were rather abrupt, which indicates that the latter changes are *preceded* by changes in the fMRI signal and occurred roughly at the turning points of the sinusoidal function. Third, in contrast to findings by other authors, changes related to binocular rivalry and bistable stimulus situations already occur at the level of the primary visual cortex and not only "higher" areas. Fourth and most importantly, these latter changes in V1 are presumably unconscious because they reflect eye-based rather than colour-based information.

3.1.2 Neuronal activity modulated by attention

The second experiment published by Noesselt and colleagues (Noesselt et al. 2002) tried to elucidate at which level of the processing of visual information top-down attention modulates neuronal activity and also, when and under which condition attended objects become conscious. According to present knowledge, visual information runs from the retina to the lateral geniculate nucleus in the dorsal thalamus, and from there to the primary visual cortex, where cells respond to simple visual features such as orientation of lines, contrast and direction of movement as the basis for later perception of contours, boundaries, shape and motion, as well as information related to the wavelength of the stimuli giving rise to color vision. In area 18 (in primates V2 and V3), there is further processing of simple visual features such as the length of lines and edges, velocity and direction of movement and retinal disparity. From there, three separate pathways originate, one related to the perception of objects and scenes, another one related to color vision (both contained in A 19/V 4), and a third one involved in complex motion perception (A 19/MT/MST). The first and second pathway, called "ventral stream" terminate in the inferior temporal cortex, whereas the third one, called "dorsal stream", terminates in the parietal and superior temporal cortex. The inferior temporal cortex is the site of complex meaningful visual objects and scenes such as faces, houses and landscapes, whereas the more dorsal areas are the site of complex motion, spatial relationships and the construction of a three-dimensional world surrounding our body and our self. From temporal and parietal cortical areas, there are pathways running further forward to the ventral and dorsal prefrontal cortex, respectively, where such complex visual information is used for thinking, decision making, action planning etc.

Thus, within the primate including human cortical visual system there is a clear processing hierarchy from simple to complex features that spatially extends from the posterior end toward the inferior temporal and the parietal cortex, respectively, and from there to the prefrontal cortex. It is believed that processing of simple features in the primary visual cortex is unconscious and without meaning, whereas processing of information in the "higher" visual cortical areas is more complex and meaningful. On the basis of a model for visual information processing called "feed-forward model", it was assumed that visual information becomes conscious, if it leads to activity in these "higher" visual cortical areas in the temporal, parietal and finally prefrontal cortex. Therefore, one could expect that top-down attention modulates activity in these higher and meaningful centers, but not in the primary visual cortical areas.

However, several psychologists and neurobiologists regarded such a feed-forward model to be insufficient and proposed the idea that for visual information to become conscious feed-back (or "re-entrant") loops from higher to lower visual cortical areas (e. g. from V4 and inferior temporal cortex to V2 and V1) are necessary (cf. Edelman and Tononi 2000; Lamme 2006). For testing these two alternative models, knowledge about the precise spatio-temporal sequence of processing of visual information until the moment of becoming conscious was necessary.

This requires both a precise spatial and temporal analysis of how visual stimuli are processed in the visual cortex of human beings. For this scope, in the experiment by Noesselt and colleagues three imaging techniques, viz., EEG, MEG and fMRI were used. EEG and MEG allowed a precise *timing* of neuronal events, whereas fMRI made a precise *location* of these neuronal events in the different visual cortical areas possible. Subjects were asked to fixate a central cross, while a pair of complex visual patterns was shown in the right and left visual hemifield, respectively, which differed only with regard to the letter "T" in the center of the patterns, which was displayed either upright or inverted in a random 50 – 50 fashion. While continuing to fix the cross in the middle of the screen, subjects were cued to direct their "purely mental" attention to the left or right pattern by an arrow signal, and to indicate by pressing one of two buttons, whether the "T" in the center of the pattern was upright or inverted. Thus, there was neither a change in the stimulus situation nor eye movement, but purely mental switch of attention.

Experiments were run under identical stimulus conditions measuring first event-related potentials extracted from EEG, then MEG signals and finally BOLD signals using fMRI, and the results were combined in order to precisely map cortical activity onto the visual cortical areas mentioned above. By combining EEG, MEG and fMRI data, it could be shown that neuronal activity related to the attended stimulus started in V1 40–50 ms after stimulus onset and lasted until 90 ms. This activity did *not* differ from that related to the non-attended stimulus, which means that at this moment, V1 activity was not modulated by top-down attention. This activity was followed by activity in "higher" visual areas such as V4, and this in turn was followed by a second wave of activity in V1 with an onset at 80–100 ms lasting until 250 ms. This activity was now significantly increased at the attended condition compared to the non-attended condition.

These results are interpreted by the authors as corroborating the "feed-back loop" model of top-down attention and related conscious state as described above. At the first stage of cortical processing of visual information, elementary visual features like line orientation, contrast and direction of movement are processed in V1 *unconsciously* and without any influence of top-down attention. This raw information then runs to higher visual cortical areas such as V4, where meaningful complex features like objects, color and scenes are processed, but even this is *not* accompanied by consciousness. Finally, this information is sent back to V1 (and probably V2) leading to a modulation of ongoing activity. In that moment of integration of "meaningless" elementary and the "meaningful" complex information, the stimulus becomes conscious. Other experiments demonstrated that by disrupting the feed-back from V4 to V1 using transcranial magnetic stimulation (TMS) subjects remain unaware of the stimulus presented, although it activated higher visual cortical areas. This demonstrates that activation of "higher" cortical areas without feed-back to V1 is a necessary, but not sufficient, prerequisite for conscious visual experience. Of course, activation of additional cortical areas, for example, in the prefrontal cortex, and feedback from there to V1, as assumed by Koch and colleagues as further condition for visual consciousness, is not ruled out by these experiments.

3.1.3 Coherent oscillatory activity

The third experiment, published by Katja Taylor and colleagues in 2005 from the primate neurophysiology group of our Brain Research Institute at the University of Bremen, addressed the possible neural mechanisms underlying attention and consciousness. Many different concepts of such mechanisms have been proposed, one of which – feedback or re-entrant cortical loops – has already been mentioned. Others refer to reverberating feedback loops between dorsal thalamic nuclei and cortical sensory and cognitive areas constituting the "thalamo-cortical system", with special emphasis of the "gating function" of a specific thalamic nucleus called "thalamic reticular nucleus". Still others like Koch and Crick assume that, in addition to these two reverberating systems, feedback loops between posterior (i. e. occipital, parietal and temporal) cortical areas and different areas in the frontal lobe are an important part of NCC (cf. Koch 2004). Finally, for the last 20 years it has been intensely debated whether synchronous and oscillatory activity of thousands or millions of cortical neurons represents at least a necessary condition for consciousness and/or attention (cf. Eckhorn et al. 1989; Engel et al. 1991; Singer and Gray 1995). There have been long-standing discussions, (1) whether such synchronous-oscillatory activity of cortical neurons are a purely intracortical activity pattern or induced by thalamic activity, (2) whether they have any relationship to consciousness and/or attention, or (3) whether they have a long-range or short-range effect inside the cortex occurring, for example, simultaneously in frontal and temporal cortical areas or occur only within restricted areas like V4. If the former was the case, such synchronous-oscillatory pattern could be a mechanism for "binding" together aspects of perception or cognition processed in separate cortical areas as described above. (see also Chapter I,4 by W Singer)

In the experiments by Taylor and colleagues, macaques were confronted with a "shape-tracking task", where they had to follow two sequences of visual objects, one on the left and one on the right side of a screen, which at short intervals changed their shapes ("morphing"). In the right sequence one target shape was shown in green for the first time and then re-appeared randomly at variable positions in the sequence, while the left sequence contained no such tagged target. When the target stimulus returns, the monkey had to release a lever which it had pressed at the beginning of the experiment, and was rewarded with a drop of fruit juice. Thus, the monkey had to concentrate on the right sequence of morphing objects and correctly recognize the return of the target stimulus, while ignoring the stimulus sequence on the left side. For human beings, mastering such a task requires full attention. Neuronal activity was measured in area V4 using an array of 36 or 37 electrodes chronically implanted above the meninges covering area V4. Recorded data were filtered and analyzed for possible patterns of oscillatory neuronal synchrony of V4 neurons.

The results of this experiment showed that attention led to the appearance of oscillatory currents underlying the recorded field potentials in the gamma-frequency range. This frequency band is generally assumed to be linked to processing of cognitive information. Importantly, no oscillatory activity was observed in V4, when the monkey by mistake did not attend to the target stimulus. If it attended to the left sequence of stimuli (the "distractor") by mistake, oscillatory activity occurred in an area of V4 corresponding to the left stimulus sequence. These findings support the hypothesis that oscillatory neuronal synchrony may serve as a neural mechanism of attention. Similar experiments on the effect of attention on visual information processing in V4 of macaques using single-cell recordings repeatedly reported an increase in firing rate for attended and/or a decrease for non-attended stimuli (cf. Treue and Mounsell 1996; Kastner et al. 1998).

These findings can be interpreted in two ways. First, the effect of attention could consist of emphasizing neural activity related to the attended stimulus by an increase of signal-to-noise ratio, which could support either subsequent processing, or enhance storage in short-term or working memory, or both. Second, it could "bind together" features of the stimulus situation processed in a parallel fashion in separate cortical areas (see above). It also remains open whether oscillatory-synchronous activity, as demonstrated here, is either a necessary or sufficient prerequisite for attention-consciousness. A critical aspect is the fact that such oscillatory-synchronous activity comes and goes during states of attention or appears only at the beginning of those states. Thus, one might assume that these phenomena may be a necessary, but not sufficient, or only one of possible prerequisites for attention-consciousness.

3.1.4 Mind reading

The fourth experiment, conducted by John-Dylan Haynes and colleagues in 2007, investigated the possibility of "mind reading", i. e. the identification of hidden decisions by means of fMRI again applying the refined fMRI method described above. At the beginning of each trial, a cue-word ("select") was presented at a fixation spot that instructed the subjects to rapidly select one of two possible tasks. They could freely choose to either add or subtract two 2-digit numbers. This was followed by a variable delay of between 2.7 and 10.8 s, during which the subject was instructed to prepare for the task and to maintain a state of continuous preparation. Then, the task-relevant stimuli were presented, which consisted of two 2-digit numbers presented above and below the fixation spot and subjects had to either add or subtract the two numbers according to their previous decision for one of the two tasks. Then after 2 s, a "response-mapping" screen was presented that showed four numbers, one in each visual quadrant on the screen. Two of these numbers were correct responses (one for addition and one for subtraction) and two were incorrect responses. Subjects responded using one of four response buttons operated by the left and right index and middle fingers. The keys corresponded to the positions of the four numbers on the "response mapping" screen. During the entire process, cortical activity was measured by fMRI using the same methods as described in the first experiment. The central question was, whether it was possible to decode from the spatial pattern of signals in local regions of the brain, which intention the subject was covertly maintaining.

The results of this experiment demonstrate that this was indeed the case, and the respective activity occurred in medial and lateral regions of prefrontal cortex. The highest prediction accuracy was 71% and was located in the anterior medial prefrontal cortex. Importantly, however, such a prediction was possible only during the delay between decision and execution of the task, whereas more posterior portions of the medial prefrontal cortex were involved in task execution.

This suggests that covert goals can be represented by distributed patterns of activity in the prefrontal cortex, thereby providing a potential neural substrate for prospective memory. Since decoding of intentions was most robust from the medial prefrontal cortex, this finding suggests a specific role of this region, when subjects reflect on their own mental states.

4. What does all this tell us for the mind-brain and consciousness problem?

The results from all four experiments clearly demonstrate a close correlation between neurophysiological processes and phenomenal states related to consciousness and attention. In the first experiment on binocular rivalry knowledge about activity patterns in the primary visual cortex allowed a reliable prediction of changes in the percept of subjects. The fourth experiment extended this possibility to covert intentions (decisions) of people as a sort of mind-reading and gave evidence, where in the brain such intentions are stored during the delay between decision-making and execution of the task. The second and third experiments added insight into neural mechanisms underlying consciousness and top-down attention. The second experiment showed that conscious visual attention requires a combination of feed-forward and feed-back processes between "early" and "late" visual cortical areas. The third experiment demonstrated that top down attention is systematically linked to oscillatory synchronous activity (as well as an increase in firing rate) of neurons in cortical areas involved in object recognition confirming previous speculations. Of course, this insight does not allow a complete description of the NCC in the domain of visual object and pattern recognition, but nevertheless enables us to have a second look at the philosophical positions regarding the mind-brain and consciousness problem presented above.

INTERACTIVE DUALISM. Considering interactive dualism first, this position is compatible with the finding of an intimate link between brain states and mental, phenomenal states, but is incompatible with the fact that mental states are reliably *preceded* by brain states. The same has been found in many other experiments including those carried out by Benjamin Libet (Libet et al. 1983) and repeated several times in the context of the preparation of voluntary movements (cf. Haggard and Eimer 1999; Pauen and Roth 2008). We are not allowed to speak of an "autonomous mind" being able to influence brain activity, if all conscious mental activities are preceded by non-conscious brain activity. In addition, even the famous "veto" suggested by Benjamin Libet to be a purely mental act being able to disrupt the preparation of a voluntary movement within the last second has been shown to be preceded by unconscious brain activity. So far, interactive dualism is made unlikely by a large number of neurophysiological evidence.

MONISM. Present-day knowledge of neurosciences strongly speaks in favor of a monistic position in the sense that mental phenomenal states apparently are at least compatible with physical-chemical-physiological laws and interact with known physical states and processes. There is substantial evidence that the strength of mental activity

(e. g. in the context of attention) is more or less linearly related to the strength of neurophysiological processes (e. g., synaptic activity, oscillatory-synchronous activity patterns). These processes are then related to the degree of oxygen and glucose consumption and blood flow in the related brain areas – a relationship that constitutes the fMRI technique (cf. Logothetis et al. 2001). In other words: Increased mental activity requires increased metabolism and blood flow, and blocking the latter processes (e. g. by insufficient supply of oxygen or sugar) leads to a block of mental activity.

IDENTISM and ELIMINATIVE MATERIALISM. But what about radical identism and eliminative materialism? Even though the experiments discussed here have been conducted with the most sophisticated neuroscientific techniques and concerned the cellular level as well as the supra-cellular level of brain activity, it is impossible to tell the precise content of the phenomenal state, if we do not have the subjects' report either verbally or behaviorally. The activity pattern in V1 or the medial prefrontal cortex, for example, could be related to many different mental states, and we know their specific meaning only *after* we have carried out at least some experiments.

Thus, we have to favor *non-reductionist* positions. The question now is if we can distinguish between a strict one-to-one relationship between brain states and mental states in the sense of *type identity*, or a many-to-one relationship in the sense of *token identity.* So far, this problem cannot be empirically decided for a variety of reasons. First, the resolution power of the available techniques does not allow to decide whether exactly one observed activity pattern in V1 or medial prefrontal cortex leads to exactly one subjective state. What we can say is that similar brain activity patterns reliably lead to similar mental states.

However, this problem of *type* vs. *token identity* or the idea of *multiple realization* of mental states remains interesting with regards to the existence of consciousness in animals and even artificial "intelligent" systems. In the case of experiments with macaques on binocular rivalry or the effect of top-down attention, we might speculate that these animals have conscious states very similar to ours under the same experimental conditions, because the cortical areas and neuronal mechanisms involved are very similar. But what about similar experiments in corvid birds? These animals have telencephalic structures called meso- and nidopallium that can be shown to be involved in mastering these tasks, but have a neural architecture that differs considerably from that of the mammalian-primate cortex (Medina 2007). On the one hand we can be certain that these birds have something like conscious attention, because their respective cognitive functions are comparable to those of primates. But do they have the same subjective conscious experience? If we would find that in these animals systematic changes in the activity of the nidopallium is correlated with changes in cognitive functions in the same way as are changes in the visual cortex of primates (e. g. the same pattern of synchronous-oscillatory activity), then this would speak in favor of "type identity". If, however, we had the impression that corvid birds have conscious states very similar to ours despite non-comparable neural processes in the nidopallium, this would make the "token identity" and "multiple realization" stance more likely.

This leads us to the second final question regarding the alleged inaccessibility of mental states as a part of the *fundamental gap* position in philosophy of mind (as proposed by Chalmers and Levine) as well as the position of *anomalous monism* (as proposed by Davidson 1970, 1993). Here, I am under the impression that these problems can at least partially be solved. First, changes in the content of mental states can be predicted on the basis of changes in brain states with astonishing precision or experimentally manipulated (e. g. by electric stimulation). It is very likely that the famous "inverted color spectrum" paradigm can be solved by showing that there is a reliable neuronal basis of subjective experience of any existing color, so that we will be able to diagnose the presence of an inverted color spectrum in a person. Furthermore, there are systematic psychological and neurophysiological conditions for learning and memory formation suggesting the existence of many rules or even laws connecting the two levels. Finally in the domain of emotions, there are significant correlations between the neuropharmacological effect of a drug and the subjective state of a patient. All this suggests that there is a close correlation between regularities in brain states and in mental states, which contradicts "anomalous monism".

What about the privateness of qualia? Can we explain in neurophysiological terms "what it is to see the purple-redness of a glass of aged Bordeaux wine"? It is conceivable that we will eventually identify the neuronal basis of the most subtle nuances in color vision, and the same will hold for the most sophisticated wine taste nuance. However, the peculiar nature of this quale, along with the private accessibility of mental states, will not disappear. At the same time, we can make the origin of this private accessibility understandable in neurobiological terms. For that scope, we have to acknowledge the fact that the total number of intracortical connections exceeds the total number of sensory input and motor output to the cortex by a factor of roughly 100 000:1 (Roth 1996, 2000). This means that the cortex "speaks mostly to itself" or is "self-descriptive". If we accept that consciousness is a consequence of cortical processes related to *self-description* (e. g., by feedback loops), then the complete privateness of conscious states is a trivial outcome. Thus, it may be possible that any hypercomplex dynamical system will develop such self-descriptive states, which appear inaccessible from the third-person observation, and this may apply for intelligent artificial systems as well.

Concluding remarks

In this article I have tried to give examples for neurophilosophy as a fruitful interaction between philosophy, psychology and neuroscience in the context of empirical mind-brain or consciousness research. Certainly, further experiments need to be carried out to promote a solution of the mind-brain problem that appears acceptable for the majority of representatives of all three disciplines. To date, from the viewpoint of philosophy of mind, only versions of non-reductionist monism or physicalism appear worth discussing, because all empirical evidence underlines that mental states "obey" known physical laws and can be predicted on the basis of knowledge about brain states. However, it has to be left to the future to see whether already existing physical concepts will eventually be sufficient to explain the "nature of mind", or whether new laws, while being compatible with the existing ones, have to be discovered.

References

Block N (1995) On a confusion about a function of consciousness. Behav Brain Sci 18: 227–287

Chalmers DJ (1996) The conscious mind. In search of a fundamental theory. Oxford University Press, New York, Oxford

Churchland PS (1986) Neurophilosophy: Towards an unified science of the mind-brain. MIT-Press, Cambridge, Ma

Davidson D (1970) "Mental events." Reprinted in Essays on Actions and Events. Clarendon Press, Oxford

Davidson D (1993) Thinking causes. In: Heil J, Mele A (eds.) Mental causation. Oxford University Press, Oxford, pp 3–17

Dennett DC (1991) Consciousness explained. Little, Brown & Co, Boston, Mass

Eccles JC (1994) How the self controls its brain. Springer-Verlag, Berlin

Eckhorn R, Reitboeck HJ, Arndt M, Dicke P (1989) A neural network for feature linking via synchronous activity: results from cat visual cortex and from simulations. In: Cotterill RMJ (ed) Models of brain function. Cambridge University Press, Cambridge, Mass, pp 255–272

Edelman G, Tononi G (2000) Consciousness. How matter becomes imagination. Penguin Books, London

Engel AK, König P, Singer W (1991a) Direct physiological evidence for scene segmentation by temporal coding. Proc Natl Acad Sci USA 88: 9136–9140

Flohr H, Glade U, Motzko D (1998). The role of the NMDA synapse in general anesthesia. Toxicol Lett 100–101: 23–29

Haggard P, Eimer M (1999) On the relation between brain potentials and the awareness of voluntary movements. Exp Brain Res 126: 128–133

Haynes JD, Rees G (2005) Predicting the orientation of invisible stimuli from activity in human primary visual cortex. Nat Neurosci 8: 686–91

Haynes JD, Rees G (2006) Decoding mental states from brain activity in humans. Nat Review Neurosci 7: 523–34

Haynes JD, Sakai K, Rees G, Gilbert S, Frith C, Passingham R E (2007) Reading hidden intentions in the human brain. Current Biol 17: 1–6

Kastner S, de Weerd P, Desimone R, Ungerleider LG (1998) Mechanisms of directed attention in the human extrastriate cortex as revealed by functional MRI. Science 282: 108–111

Kim J (1996) Philosophy of mind. Boulder Westview Press, 2nd ed

Koch C (2004) The quest for consciousness: A neurobiological approach. Roberts, Denver, Co

Koch C, Tsuchiya N (2006) Attention and consciousness: two distinct brain processes. Trends Cogn Sci 11: 16–22

Koch C, Preuschoff K (2007) Betting the house on consciousness. Nature Neurosci 10: 140–141

Lamme VAF (2003) Why visual attention and awareness are different. Trends Cogn Sci 7: 12–18

Lamme VAF (2006) Towards a true neural stance on consciousness Trends Cogn Sci 10: 494–501

Levine J (1983) Materialism and qualia. Pacific Quarterly 64: 354–361

Libet B, Gleason CA, Wright EW, Pearl DK (1983) Time of conscious intention to act in relation to onset of cerebral activity (readiness-potential). Brain 106: 623–642

Logothetis NK, Pauls J, Augath M, Trinath T, Oeltermann A (2001) Neurophysiological investigation of the basis of the fMRI signal. Nature 412: 150–157

MacLean P (1990) The triune brain in evolution. Plenum, New York

Medina L (2007) Do birds and reptiles possess homologues of mammalian visual, somatosensory, and motor cortices? In: Kaas J, Bullock TH (eds) Evolution of nervous systems, Vol 2, Academic Press, Amsterdam, pp 163–194

Noesselt T, Hillyard SA, Woldorff MG, Schoenfeld A, Hagner T, Jäncke L, Tempelmann C, Hinrichs H, Heinze HJ (2002) Delayed striate cortical activation during spatial attention. Neuron 35: 575–587

Pauen M (2001) Grundprobleme der Philosophie des Geistes. S Fischer, Frankfurt am Main

Pauen M, Roth G (2008) Freiheit, Schuld und Verantwortung. Grundzüge einer naturalistischen Theorie der Willensfreiheit. Suhrkamp, Frankfurt am Main

Persaud N, McLeod P, Cowey A (2007) Post-decision wagering objectively measures awareness. Nature Neurosci 10: 257–261

Popper K, Eccles J (1884) The self and its brain. Springer Verlag, Berlin

Rensch B (1968) Biophilosophie auf erkenntnistheoretischer Grundlage (Panpsychistischer Identismus), G Fischer, Stuttgart

Roth G (1996) Das Gehirn und seine Wirklichkeit. Suhrkamp, Frankfurt am Main, 2nd ed

Roth G (2000) The evolution and ontogeny of consciousness. In: Metzinger T (ed) Neural correlates of consciousness. Bradford Book, MIT Press, Cambridge, Mass, London, pp 77–97

Seth AK, Dienes Z, Cleeremans A, Overgaard M, Pessoa L (2008) Measuring consciousness: relating behavioural and neurophysiological approaches. Trends Cogn Sci 12: 314–321

Singer W, Gray CM (1995) Visual feature integration and the temporal correlation hypothesis. Annu Rev Neurosci 18: 555–586

Taylor K, Mandon S, Freiwald WA, Kreiter AK (2005) Coherent oscillatory activity in monkey area v4 predicts successful allocation of attention. Cerebral Cortex 15: 1424–37

Treue S, Mounsell JHR (1996) Attentional modulation of visual motion processing in cortical areas MT and MST. Nature 382: 539–541

Weiskrantz L (1986) Blindsight: A case study and implications. Oxford University Press, Oxford

Sensory perception, body and mind in Indian Buddhist philosophy

20

Ernst Steinkellner

Contents

Abstract

The Buddha's conception of the nature of sentient beings being without a substantial and lasting core gave rise to interpretations in subsequent Buddhist philosophical traditions that often appear surprisingly modern. In general, his usage of defined abstract notions was prudently limited for the most part, however, to those necessary for presenting soteriological and practical teachings. But this approach proved to be philosophically prolific. Basic concepts and theorems were soon developed into rich scholasticist systems of ontology and psychology, which were followed by a normative epistemology that focussed on the kinds of cognition that can warrant valid knowledge. The relationship between mind or mental phenomena and the body was elaborately discussed with different proponents of Indian Materialism in order to prove the correctness of the Buddha's analysis of the empirical person. By means of examples this paper will offer a survey of the concept of the senses, their nature and function according to a particular system of mainstream Buddhism, as well as of the ideas about the relationship between mental phenomena and the material body as can be seen in the attempt of the influential philosopher Dharmakirti to refute the reductionistic explanation of consciousness.

Ernst Steinkellner
Austrian Academy of Sciences
Institut für Kultur- und Geistesgeschichte Asiens
Prinz Eugen-Str. 8-10
1030 Vienna, Austria
e-mail: ernst.steinkellner@oeaw.ac.at

1. Introduction

In science, "sensory perception", is now used as a cumulative label to describe the activity of the senses, an activity that consists in the transport of information necessary for living beings, from bacteria to human beings, to survive and procreate. This nature of the senses, sensors as transducers, has been examined in great detail by biologists over the last hundred years with regard to their physiological and technical aspects, and an incredibly differentiated wealth of possibilities how stimuli from the outer or inner world of an organism are received and transmitted has been already discovered (see Chapters I, 1–4 and II,6 this book).

I should clarify what you can expect in the following paper. I am neither a scientist nor a philosopher, but as a philologist and historian of ideas, a kind of cultural anthropologist with a focus on Indian and Tibetan thought, in particular Buddhist thought. Yet I am also a living being with an interest in the questions of today. Thus, in conclusion, I will take this occasion to bother you with some personal impressions, probably quite innocent if not even stupid about the so-called problem of consciousness and the manner of some of the discourses about it.

It seems to be my charge to present at the end of this book, as a kind of counterpoint, the example of a pre-modern view of sensory perception, and, moreover, a view that is not only pre-modern, but also extra-European, namely, an Indian view.

As in pre-modern Europe, in India sensory perception was not a topic of the sciences, e. g., of medicine, but of philosophy. Within the various intellectual traditions that I would characterize as "religious" because they offer solutions for the problems of life, the last centuries BCE see the development of a number of philosophical systems in India: comprehensive and structured conceptions and explanations of what there is in the inside and outside worlds. In my opinion, all of these efforts are much influenced by their respective social and religious traditions, Brahmanical, Jinist, Buddhist, or Materialist, much as medieval European or "Western" philosophies are mostly held together by Christian presuppositions. These Indian systems differ widely in focus and structure. While some are outside nature-oriented, most focus on the inside nature of living beings. Along with the development of such systems, the practice of debate became regulated during this period in order to allow productive or polemic discussions, both within the same traditions but also between competing ones. In the period following this systematization, the background of these systems was contested in an ensuing culture of polemic dispute: proof of the validity of the respective sources of knowledge was sought, including aspects concerning the foundational authority these systems were built upon. Rich developments in epistemology and logic accompanied this trend. While the conceptual context for the various epistemological theories of this period is quite similar, their emphasis on the sources of knowledge differs according to their specific goals. For example, for systems based on metaphysical principles beyond experience, inference is considered more important than perception. If, as in the case of Buddhism, the systems are based on the truth of a particular person's experience, in this case of course the Buddha, perception is predominant whereas inference receives its validity only indirectly from its connection with the results of perception.

In order to avoid further generalizing statements, I will proceed by presenting two concrete examples in more detail from this rich philosophical development which is quite divergent even within Buddhism. Both examples are extracted from a specific single text and may be relevant in regard to different aspects of the topic of this book. The first example is of a more antiquarian character and is drawn from the "Treasury of Scholasticism" ("Abhidharmakośa") by Vasubandhu, a famous teacher of the 5th century CE. It will demonstrate Vasubandhu's conception of the nature and function of the senses as an approach to the topic under medieval Indian intellectual conditions. The second example demonstrates the mind-body question as it is dealt with in the "Commentary on Valid Cognitions" ("Pramāṇavārttika"), the first major work of Dharmakīrti, a widely influential epistemologist and logician of the 7th century CE. To adequately understand Vasubandhu's classical presentation of sensory perception we need to briefly look at the Buddhist conceptual background. According to the Buddha, the nature of sentient living beings – humans and animals alike[2] – consists in a collection of five empirically discernible constituents (Fig. 1) that can clearly be distinguished from one another. These constituents he calls "branches" ("skandhāḥ") (cf. Vetter 2000).

The five branches are "body", "feeling", "conceptual awareness or ideation", "volitional and affective impulses", and "sensation" (Vetter 2000, p. 69f.). All five constituents are conceived of as continua of distinct factors that are connected only causally. There is nothing in addition to and different from these factors, like, for example, a substantial permanent soul or an observing master-mind. In the later systematic period, the body as matter in general is dualistically opposed to mind, which is accompanied by various emotional and intellectual phenomena. In the light of the Buddhist interest in the nature of life and its origin, as well as in a method of deliverance from life's frustrating eternity, the Buddhist point of departure in the examination of sensory perception within this dualistic framework is on the side of perception. This means that sensory perception is not understood as a function of the senses, but a sensation or awareness that comes about under the cooperation of the senses. In the following I extract the most relevant notions regarding the nature of the senses from Vasubandhu's extremely elaborate system.

2. Vasubandhu on senses

Vasubandhu's list of awareness/consciousness-related factors comprises 18 "components" ("dhātu"), which are listed in Fig. 2.

There are six consciousnesses[2] ("vijñāna"); of these, five are sensory and one is non-sensory. They correspond to six sensory faculties or senses proper, namely, the five human senses (seeing, hearing, smelling, tasting, and feeling/touching), and mind, and refer to six corresponding objects (visible form, sound, smell, taste, tangibles, and knowables). Leaving aside the mind ("manas")

Fig. 1 The five "branches" (skandhāḥ) according to the Buddha

body (*rūpa*)

feeling (*vedanā*)

ideation (*sañjñā*)

volitional and affective impulses (*saṃskārāḥ*)

sensation (*vijñāna*)

1 In early layers of Buddhism possibly including even plants (cf. Schmithausen 1991)

2 I here use the term "consciousness" synonymously with "conscious cognition, awareness, sensation". The single Indian term connotes "discriminating or distinct cognition or knowledge"

consciousnesses	senses	support	object
visual	seeing	eyes	visible form (= colour + shape)
auditory	hearing	ears	sound
olfactory	smelling	nose	smell
gustatory	tasting	tongue	taste
tactile	feeling	body	tangible
mental	mind ("inner sense")	"heart"	knowables

Fig. 2 The eighteen "components""(dhātavaḥ) of awareness/consciousness following Vasubandhu

and its function as a distinct non-sensory faculty of discrimination, the question pertinent to the topic of the present book is that of the nature of the senses.

First of all, the senses are material in that they consist of the four "great elements" earth, water, fire, and wind. But they represent a subtle kind of matter that is derived ("rūpaprasāda") from these four elements (Vetter 2000, 21f.). The elements are the "support" ("āśraya") of all derived matter (Abhidharmakośa "AK" 1.12ab, Pradhan 1967); they are atomically conceived, but are experienced only in terms of their essential characteristics, for instance, earth as "solidity", water as "wetness", fire as "heat", and wind as "motion". Moreover, the elements are composite, that is to say, the element earth for instance contains at least one atom or more of each of the other elements. It is determined as being earth only because the earth atoms predominate (Sphuṭārthā Abhidharmakośavyākhyā by Yaśomitra "AKV", Wogihara 1971, AKV 33,10–15).[3] The same is true for the derived matter in the form of the senses. Depending on which aggregation predominates in the senses, they are coordinated to the respective object, following the principle "like causes like".

Perhaps intriguing here for biologists is the term "subtle kind of matter" ("rūpaprasāda"). This type of matter is said to be derived from the great elements, but it is transparent ("accha") and therefore suprasensible. Like those elements this matter consists of atoms, and, most importantly, is not to be identified with their respective "seats" ("adhiṣṭhāna"). The seats of the senses, the eyes, ears, nose, tongue and body, carry the subtle atomic matter of the senses in different ways (Abhidharmakośabhāṣya of Vasubandhu "AKBh" 33, 17–23, Pradhan 1967).[4] Because of the senses' transparency, the exact place they occupy on their seats can only be inferred. These seats are inferred from medical experience: if medication is applied at this spot, it has an effect on the respective sense.

The sense of vision has its seat on the pupil of the eye in the form of a cumin-flower, and is covered by a transparent skin. Hearing sits inside the ear on a cartilage that

3 On the notion of "predominance" ("bhūyastva"), cf. Preisendanz (1994, 724ff.).

4 Cf. Preisendanz (1994, p 445).

has a colour like that of a birch-leaf. Smell is placed in the two nostrils in the form of arrow heads. Taste sits on the tongue like a half moon. In the middle of the tongue there is a spot the size of the tip of a hair that is not pervaded by the atoms of this sense. The sense of feeling follows the body in its arrangement.

Although the systematic explanations are different, all of this, and there is not much more to be found on the physiology of the senses, is also more or less commonly accepted in the brahmanical traditions, especially those with a focus on the philosophy of nature. There is only one point where they diverge. In general, the brahmanical schools upheld the requirement that for perception to occur there must be physical contact between a sense and its object. Buddhist scholasticists, however, accepted the necessity of contact only for the senses of smell, taste, and feeling, as long as their objects were of comparable size to the sense organs. But in the case of seeing and hearing they denied the necessity of physical contact (AK 1.43cd, Pradhan 1967).

Their main argument concerns the sense of vision: although we see objects that are distant from the seat of vision, we do not see the eye lashes in its nearest proximity. The requirement of contact and the debate with the Buddhists generated rich developments in optical theories in the brahmanical schools. These culminated in a theory of eye-rays, for which the Nyāya-Vaiśeṣika school was mainly responsible.[5] The reason the Buddhists denied the necessity of physical contact in the cases of seeing and hearing is, however, not based on the discovery of specific facts, but is clearly dogmatic. In the Buddhist tradition it is said that advanced virtuosi of meditation have special capacities; they are able to acquire "divine", suprasensitive sight ("divyacakṣuḥ") by means of which they see not only far in terms of space and time but also through walls and mountains, as well as "divine", suprasensitive hearing ("divyaśrotra"), through which they hear sounds over great distances, even sounds created in other world systems. Such capacities would not be possible if physical contact were required.[6]

If then for Vasubandhu there is no contact between the senses of seeing and hearing and their objects, how does he explain the arising of the respective cognitions? According to an older authority,[7] the sense of vision perceives something, i. e. "attains" something in that it arises in a "state of non-separation" ("nirantaratva") with an object (AKBh 32,10f, Pradhan 1967), by reason of light. When a visible form is too close to the sense of vision, the form impedes the function of light and thus, the sense of vision does not see. When an object is distant, light is not impeded, and the sense of vision sees. The sense of hearing perceives by reason of space ("ākāśa"). When a sound is close to the sense of hearing, it does not oppose the function of space, which is to impede matter, and thus, the sense of hearing hears.

But as said above, these quite artificial constructions are only due to dogmatic considerations. I believe, however, that in order to get a clear view of the conceptions of the senses in this system, all considerations which only accommodate the tradition can be disregarded here. I would instead like to examine in more detail the model of the

5 For an extensive treatment of these ideas cf. Preisendanz (1989 and 1994, pp 446–449). Cf. also Mookerjee (1935) and Tillemans (1990).

6 AKBh 32,2f: sati ca prāptaviṣayatve divyaṃ cakṣuḥśrotram iha dhyāyināṃ nopajāyeta (Pradhan 1967). Cf. AKV 83,26–84,2 (Wogihara 1971).

7 Vibhāṣā 13,7 as quoted in L'Abhidharmakośa de Vasubandhu "AKBhV" I. 87, note 1 (La Vallée Poussin, 1971).

three other senses, those that actually "attain" their objects, namely, smell, taste and feeling. When, in the case of these three senses, the sense-atoms "attain" roughly the same number of object-atoms, consciousness is produced (AKBh 33, 11, Pradhan 1967). But atoms, as the smallest entity of matter, do not "touch" one another because they are the smallest entity and thus have no parts (AKBh 32,13, Pradhan 1967), or because they would collapse into a single atom, if they did touch (AKBh 32, 12, Pradhan 1967). But they do touch, in the sense of non-separation, understood as a "juxtaposition without anything in the interval" when they are "agglomerated" ("saṅghāta") and thus have parts (AKBh 32,17f, Pradhan 1967). Such "agglomerates" are either dissolved or held together by the element wind, i. e. by "motion" (AKBh 32,14–16, Pradhan 1967). While it is fine that "motion" keeps atoms together without their collapsing because they have resistance ("sapratigha"), "agglomerates" are still atoms (AKBh 33,4f, Pradhan 1967), and using such terms as "touching" or "attaining" can only be considered metaphorical without reference to reality (AKBh 33,2f, Pradhan 1967).

In another context, the five elementary sense-objects are also said to be "resistant" ("sapratigha") in the sense of being impenetrable, as when a hand strikes a rock, it is repelled (AKBh 19,7f,18, Pradhan 1967). But the senses are also "resistant" in the sense of "encountering one another" or of "coming together" ("nipāta"), which in this context is explained as "an activity with regard to its specific object" ("svaviṣaye pravṛttiḥ", AKBh 19,18, Pradhan 1967). What is meant here is not that the senses act on their objects, but that they are "active in mutual assistance" ("sabhāga"). On one hand "mutual assistance" means "mutual service" ("anyonyabhajana"), in which the senses, objects serve one another as the support and object of consciousness, and, and in which consciousnesses relies on the senses (AKBh 28,20, Pradhan 1967; AKV 76,27–34, Wogihara 1971); on the other hand, it means "possession of an activity" ("kāritrabhajana"), namely, the acts of seeing, of being an object of consciousness or being seen, or of discerning the object (AKBh 28,20, Pradhan 1967; AKV 76,34–77,2, Wogihara 1971). This is also expressed in a more meaningful way as "having the same contact as (their) effect" ("sparśasamānakāryatva") (AKBh 28,20, Pradhan 1967; AKV 77,5–9, Wogihara 1971).

Also "contact" ("sparśa") does not mean the same thing here as it does in the brahmanical schools. In the given context it means a "coming together, becoming near to one another" ("sannipāta") of the three: the sensory faculty, an object and consciousness (AKBh 132,8, Pradhan 1967; AKV 77,8f, Wogihara 1971). Moreover, it is not a physical contact that is meant, but only a conceptual or metaphorical contact. Yaśomitra, a ninth century commentator on Vasubandhu's treatise, expresses this in a manner that helps to get a clearer picture: "These, sense, object and cognition, have the same effect. When cognizing the [sense of the] eye and the object, visual cognition arises." (AKV 77,7, Wogihara 1971)

As a historian of philosophy I am compelled to reveal at this point that Vasubandhu's summary of Buddhist scholasticism is comprised of several layers of consistent theories that are historically and systematically intermingled with a great number of individual opinions on specific points and problems. His own views can often be seen as more down-to-earth alternatives, clearer and more compact, when compared to the traditions he endeavoured to summarize. And at the end of his career, he abandoned his earlier ontological position of naïve realism altogether and proposed an idealistic ontology – with a considerable number of similarities to modern constructivist ideas. Only his proof of the impossibility of a reality that is external to consciousness may be mentioned as being of interest to our topic. For this proof is basically a refutation of the conception of atomic reality: atoms are impossible.[8]

The critical realism, already seen in Vasubandhu's work that served to reduce the conceptual extremes of scholasticist causal theories, was subsequently fully developed by Dharmakīrti in his conceptions of causality and of the momentariness of anything caused. Here, the scholasticist's complicated structure of six types of causes and five types of effect is replaced by a theory based on one major cause and a group of auxiliary causes that are, in principle, representative for an unlimited number of auxiliary causes. These causes produce only one kind of effect, while at the same time they contribute to the properties of this effect.

In the spirit of Yaśomitra's succinct remark quoted above,[9] I would now like to briefly sketch Dharmakīrti's theorem (Fig.3) (Steinkellner 1967, pp 44–55). According to his conception one has as a starting point phases of different causes that are in proximity to one another, thereby forming a causal complex ("hetusāmagrī"). In principle, the proximity of these causes is the effect of a "history" of previous causes that have no beginning in time, and the causes belonging to such complexes are unlimited in the space of each time-phase as well. Nevertheless, the main causes are easily discernible: a main cause is distinguished as having a preceding causal phase that is of the same type as the effect ("samanantarapratyaya"); auxiliary causes ("sahakārin") are not of the same type, but they need to be present to allow the causal complex to be also sufficient to produce its effect. In the case of sensory perception, the main cause is sensation, mind, or consciousness, and the auxiliary causes are the sensory faculties, objects, light, etc. What is gained by the cooperation of the auxiliary causes with the main cause is that they have a causal influence on the properties of their common effect (Fig. 4).

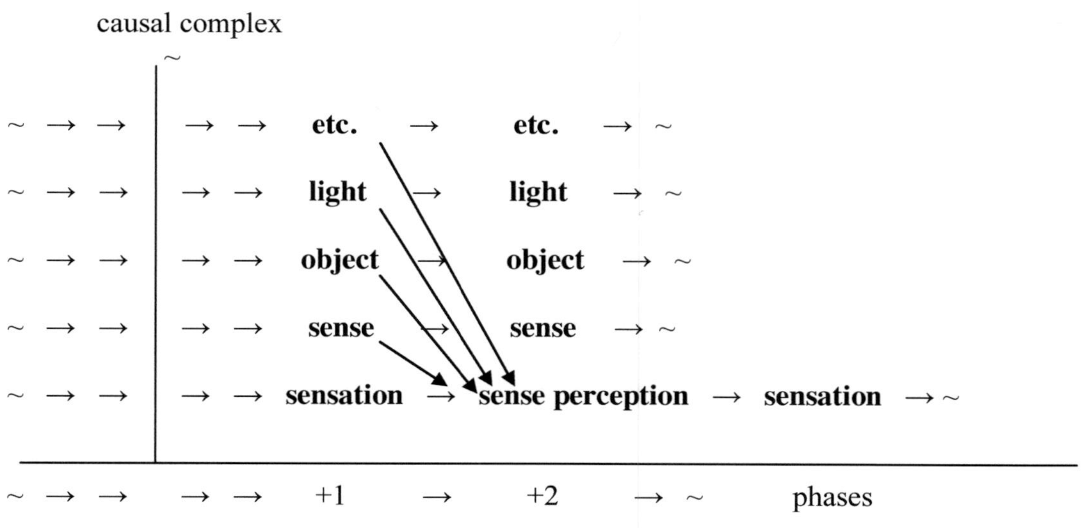

Fig. 3 Dharmakirti's theorem (Steinkellner 1967, p 137)

[8] Viṃśatikā 11–15 (Frauwallner 1994).

[9] I understand this remark as summarizing the earlier interpretation of the process of sensory perception already under the influence of Dharmakīrti's theory.

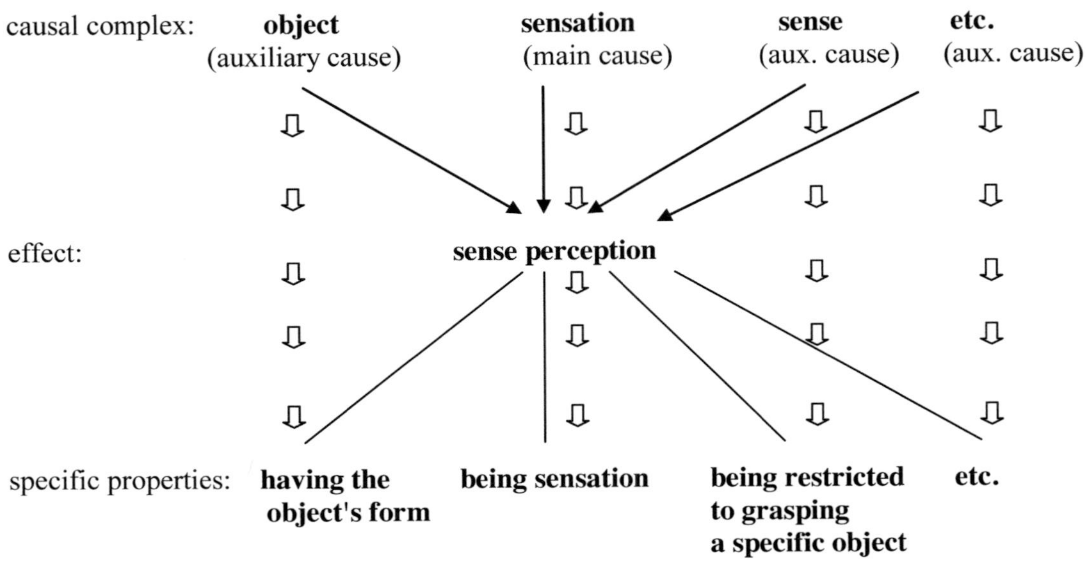

Fig. 4 The empty arrows indicate a causal relationship, the lines the effect's properties, and the filled arrows their relationship to the respective causes in the complex (Steinkellner 1967, p 127)

To sum up: What these Buddhist philosophers thought they were able to explain on the level of material reality as governed by causal regularity, is the possibility of an event of consciousness that refers to or represents reality "as it is" ("yathābhūta"). In Dharmakīrti's definition of perception as being "non-erroneous" and "free of conception" (Dharmakīrti's Pramāṇaviniścaya "PVin" 1.4ab', Steinkellner 2007 and Vetter 1966, p 41) this is spelled out at the end of a long development in both a pragmatic and a Buddhist manner. Since sensory perception is only one type of cognition defined in this manner aside of other events of cognition like the self-consciousness of cognitions as objects, even conceptual ones, or the peak-experience of a meditator, it can be said that these characteristics of perceptions are better appreciated through what amounts to their psychological equivalent, namely the subjective feeling of "immediacy" or "clearness" ("sākṣatva"), a quality that is exemplarily experienced in sensory perception.

3. Dharmakīrti on body and mind

The second topic I consider to be of interest here is the Buddhist position on the relationship between body and mind. Again I can use Dharmakīrti as my source. In the second chapter of his "Commentary on Valid Cognitions" ("Pramāṇavārttika"), Dharmakīrti deals at length with views from traditions of Indian materialism and medicine (Pramāṇavārttika "PV" 2. 34–119, Franco 1997)[10]. His motivation is obvious: the affirmation of the Buddhist belief of a re-embodiment of the

[10] Cf. its review by Taber J (2003) Dharmakīrti against Physicalism. Journal of Indian Philosophy 31: 479–502, from which I benefitted substantially for this lecture.

continuum of sensation in beginning- and endless existences full of suffering, as well as the possibility to release oneself from this eternal circle through the development of moral and intellectual capacities to ultimate perfection.

As said at the beginning, the Buddha considers a living being to consist of five ultimately different constituents, the body, feeling, ideation, impulses and sensation, which can be contracted for our purposes to "body and mind". They are distinct continua of existent factors that are connected only causally, but they cooperate in forming an empirical personality, an ego. There is nothing substantial and eternal, such as a soul, in addition to these constituents.

Now, whereas all bodies are visibly impermanent because the elements rearrange themselves, this is not the case with the mental constituents. Their continua generate their respective subsequent phases, and, after the death of the body, impelled by the heritage of previous activities, namely, by the karma, attach themselves to new bodies. It is this conception which makes the belief in the development of mental capacities possible beyond specific single life periods in the direction of final release. This is the basis of Buddhism as religion. The belief that mind is different from body is for Dharmakīrti, therefore, a necessary presupposition to be able to accept the authority of the Buddha in all the goals of his striving. At the same time it is ascertained thereby that all experiences, deposited in the continuum of sensation like seeds, will bear their fruits in the present or a future life, so that in general moral norms can be accepted as meaningful. The existence of mind beyond the death of the body is thus crucial.

However, the intention of Dharmakīrti is not to definitively prove the difference of mind and body, for he sees no purpose in convincing his own group of this. He intends, just as true philosophical questions are dealt with, to render his ideas acceptable to all fellow rational beings. His aims are more modest, and thus I believe of interest even for modern discussions of this question: he shows that the assumption of a mind as independent of the body, and thus the assumption of re-embodiment, or afterlife, is *not impossible*. He does this by supporting, through his arguments, the doubts regarding the thesis that consciousness *necessarily originates* in bodily matter.

For the basic counter position he opposes, is that of a reductive physicalistic materialism, classically expressed in the ancient metaphor: "Consciousness arises from the elements just as the power of intoxication from molasses and other substances when a fermenting substance is added."[11]

Traditionally the Buddhists refute a number of positions held in the Indian context by means of aporetic proofs to demonstrate that the phenomenon of consciousness in newborns cannot be caused by something that is not of the same type, i. e., it must stem from a previous phase of consciousness. To mention only the most common of these other positions proposed, these are the theses (1) that consciousness is newly created by an eternal creator, (2) that consciousness arises from the consciousness-continuum of the parents, (3) that consciousness springs from the mixture of semen and menstrual blood, i. e., the elements, and (4) that consciousness arises spontaneously, i. e., without a cause.

Dharmakīrti examines only the third, the materialistic thesis that consciousness springs from the elements. He begins with the question what causes cognition, a mental event. I can only briefly summarize. According to him only like causes like. Matter such as clay causes pots, not conscious cognitions. If this

[11] Bṛhaspatisūtra, cf. A3–A5 in Namai (1976).

principle were invalid, anything could come from anything.

Also different combinations of matter, particularly organic and living matter, are incapable of giving rise to cognitions, for it otherwise would be inexplicable why not every combination would be capable in that respect. There would have to be a surplus to be responsible for this difference in causal capacities.

The senses, too, are no option, since although they are needed for the arising of sensory perceptions, they are not necessary for mental cognitions such as thoughts, memories and the internal states of pleasure and pain. In short, even if the body or the senses can support the arising of cognition as auxiliary causes this does not mean that the body is the main cause of cognition.

But most remarkable in Dharmakīrti's examination is his repeated reference to the methodical principle known as Occam's razor, according to which the simpler explanation is the better. Since no one can prove that the body generates consciousness, why postulate this when the immediately preceding and, moreover, like phase of consciousness is a sufficiently suitable candidate for this task? With regard to modern discussions about consciousness it can be asked, however, what exactly Dharmakīrti means by "like causes like"? He certainly knows that the type or substantial form of "clay" does not remain when a pot or a house is produced, or that effects often have properties surprisingly different from their causes, such as ashes from burning wood. He does not deny that material causes are capable of producing something that is more complex or differently structured than them. He only insists on the dualistic position that something physical cannot cause something nonphysical.

In this sense, Dharmakīrti is on the side of all those modern philosophers who hold that consciousness cannot be traced to matter as its source. No ever-so-detailed description of the micro-physical states of an organism can show with certainty that the micro-physical or macro-physical area can have some kind of conscious experience. Even today's knowledge of the nervous micro-cosmos of the brain and the hypothesis that this, as a "complex system", is simply capable of a "more" than its components or their sum cannot, as far as I see, answer the question why such "complex systems" lead to certain conscious experiences. Since, in my simplistic opinion, it is not, or at least not yet possible to explain *how they do it*, we also cannot know *that they do it*.

Dharmakīrti's point is to demonstrate that a regular causal relationship between material states and consciousness cannot be established. Certainly, his knowledge about the physiology of living beings has been long since outdistanced. Nevertheless, his discussion of the problem underlines the fact that although biology of today can explain the conditioning framework for the presence of consciousness, the existence of regular relations between physical and non-physical phenomena, or the assumption of the nonexistence of non-physical phenomena, must still be treated like a matter of belief in the sense of a non-established conviction.

Of course, Dharmakīrti, too, did not offer a solution to the question of the presence of consciousness that one could accept today. He thought he had successfully attained his aim by showing that the materialist is not able to convincingly establish that consciousness as existing independently of the body is ultimately impossible. Nevertheless, I believe we can draw a lesson from Dharmakīrti's exercise and in this way also make use of our reading of a pre-modern Indian philosopher for some present-day problems of discourse.

4. Afterthoughts

If, in conclusion, the science's state of the art on this issue can still be said to be, provisionally, a matter of belief or non-established conviction, this field of discourse might also be considered a neutral and open space for a conventional approach.

If a mono-causal explanation of the two, body and mind, cannot yet be proven, the same holds true for a dualistic explanation as well, because a causal or non-causal existence of mind as an "inner sense" or of a soul has also not yet been proven. While scientists are certain that the latter can never be proven, for the time being they still can only be hopeful that the former might be.

Then, even if we accept this scientific expectation as the only reasonable one, and even if most of us believe that a comprehensive mono-causal explanation will be found in a not-too-distant future, this is – so it seems to me at least – not a reasonable basis for the often polemic opposition against the dualistic model.

After all, both camps must acknowledge the fact that cognition or consciousness exists. What we know about cognition and how it works, as well as its effects and consequences in our lives, does not depend on our knowledge of how it comes about, what causes it, and how its variations and processes are conditioned. Yet, on the other hand, this life of the mind has long been studied, will continue to be studied, and will always be of interest, irrespective of the final answers to the question of the causes of its existence.

What I would like to propose, therefore, is, at least provisionally with the purpose of allowing the functions of mind to be observed as such, that it would be best if scientists did not brush aside the notion that the mind is a reality of its own. Even if the mind is only accepted as a hypothetical entity, a kind of second-order reality, because of its usefulness in practical life, the mind's functions, when seen as the result of evolutionary developments, still merit the same attention they have received throughout the history of humankind.

There is, then, also no need to vote for one of the opposing positions recently debated with much heat of either "physicalism" or "evolution" versus "intelligent design". We have to admit, I think, that both nature and living, sentient beings, *look* very much as if they had been designed. Yet "design" can be understood as the result of an intelligent directing cause, a creational result, or as the result of undirected, incidental evolution.

However, we have to be aware of the fact that "design" is only in the eye of the beholder. It is based on a conceptual judgement that refers to our outer and inner worlds of experience. *We* judge these worlds as "designed". And we actually need the conception of the world as "designed", because for our activities and for finding a way through the impediments and dangers present "out there" we have to rely on some guide that helps us to direct our next moves. Whether the judgement of the world as "designed" is further considered to have either evolutionary or creational grounds is therefore quite irrelevant for our survival. Even if we consider the scientifically most likely option as having the flavour of truth, namely that it is nothing but the incidentalistic "policy" of evolution which happened to provide us with the capacity of conceiving the world as "designed", this will not be much of an advantage, for we already know that we have this capacity and that we can rely on it.

In this sense it will always be useful and necessary to differentiate between the study of the physiological nature of the mind and the study of the social and epistemological nature of the mind and its functions useful to the species. If the former study explains the mind's nature, the latter deals with the mind's function. Such a distinction, if considered as conventional practice, would neither discredit the progress of science in its efforts for a natural explanation of the mind, nor would it discredit ordinary human practice, which relies on the mind's functions as providing one of the best tools in the strivings for survival.

References

Franco E (1997) Dharmakīrti on compassion and rebirth. Verlag Österr Akad Wiss, Wien

Frauwallner E (1994) Die Philosophie des Buddhismus. Akademie Verlag, Berlin

La Vallée Poussin L (1971) L'Abhidharmakośa de Vasubandhu. Traduction et annotations. Nouvelle édition anastatique présentée par Étienne Lamotte, Institut Belge des Hautes Études Chinoises, Bruxelles

Mookerjee S (1935) The Buddhist philosophy of universal flux. Motilal Banarsidass Publishers, Delhi 1993, pp 300–310

Namai M (1976) A survey of Bārhaspatya philosophy. Indological Review 2, pp 29–74

Pradhan KP (1967) Abhidharmakośabhāṣya of Vasubandhu. Jayaswal Research Institute, Patna

Preisendanz K (1989) On *ātmendriyamanorthasannikarṣa* and the Nyāya-Vaiśeṣika theory of vision. Berliner Indologische Studien 4/5, pp 141–213

Preisendanz K (1994) Studien zu Nyāyasūtra III, 1 mit dem Nyāyatattvāloka Vācaspatimiśras II. Alt- und Neu-Indische Studien 46/2. Franz Steiner Verlag, Stuttgart

Schmithausen L (1991) The problem of the sentience of plants in earliest Buddhism. Studia Philologica Buddhica, Monograph Series, VI. The International Institute for Buddhist Studies, Tokyo

Steinkellner E (1967) Dharmakīrti's Hetubinduḥ. Teil II. Übersetzung und Anmerkungen. Hermann Böhlaus Nachf, Wien

Steinkellner E (2007) Dharmakīrti's Pramāṇaviniścaya. Chapters 1 and 2, Critically edited. China Tibetology Publishing House – Verlag Österr Akad Wiss, Beijing

Taber J (2003) Dharmakīrti against Physicalism. J Indian Philos 31, pp 479–502

Tillemans TJF (1990) Materials for the study of Āryadeva, Dharmapāla and Candrakīrti, Vol. I. Arbeitskreis für Tibetische und Buddhistische Studien Universität Wien, Wien, pp 156–159

Vetter T (2000) The 'Khanda Passages' in the Vinayapiṭaka and the four main Nikāyas. Verlag Österr Akad Wiss, Wien

Wogihara U (1971) Sphuṭārthā Abhidharmakośavyākhyā by Yaśomitra. 2nd edn. Sankibo Buddhist Book Store, Tokyo

21 The 'eye of the mind' and the 'eye of the body': Descartes and Leibniz on truth, mathematics, and visuality

Sybille Krämer

Contents

Abstract

In the conflict between rationalism and empiricism, the rationalist is regarded as a philosopher whose source of knowledge is reason rather than sense perception. And yet the terminology of "sight" plays a striking role in rationalist philosophy. This paradox of the simultaneous "devaluation" and "valuation" of seeing is normally explained in terms of the difference between the "eye of the mind" and the "eye of the body". The rationalist, according to this view, transforms sight into the activity of reasoning, whereby the "intellectual eye" sees all the more clearly the more the body's eyes remain blind. This essay is aimed at correcting this understanding by means of looking at the epistemologies of Descartes and Leibniz. An investigation into the epistemological meaning of the mathematical innovations of both philosophers will help rehabilitate the role of bodily sight in rationalist forms of knowing. It is proposed (i) that the calculization in mathematics, to which Descartes' Analytical Geometry and Leibniz's Infinitesimal Calculus contributed significantly, promotes a specific type of visuality which is called "tactile seeing" or "seeing with the hand". And it is demonstrated (ii) that traces of calculization, in form of the core rationalist move of reducing truth to correctness, can be found in epistemology. The rationalists devalue "ocular seeing", since it is closely tied with the illusionary, but they value "tactile seeing", which is not a "seeing with the mind", but a "seeing with the hand."

Sybille Krämer
Free University Berlin (FU)
Institute for Philosophy
Habelschwerdter Allee 30
14195 Berlin, Germany
e-mail: sybkram@zedat.fu-berlin.de

1. Seeing and knowing in rationalism

Hardly any other historical period placed a greater emphasis on ocular perception, visibility, the laws of optics, and on the telescope and microscope as technical instruments for distance and close-up viewing as did the seventeenth century. In this period, seeing became the fertile ground of knowledge, the bodily sister of reason. At the same time, philosophical rationalism flourished during this century, a school of thought that – in distinction to philosophical empiricism – located the source of knowledge not in perception and vision, but in conception and reasoning. It is no coincidence that Descartes[1] takes the blind man and his cane as a model for the explanation of sight[2] (Fig.1), and Leibniz speaks of the *cogitatio caeca vel symbolica*, "*blind* or symbolic thought" as crucial for human reason.[3] The rationalists, it seems, had a deeply ingrained antipathy towards the senses and visibility.

Fig. 1 René Descartes' *Dioptriques* (from *Discours de la méthode*, Leiden 1637, p. 56)

Strangely enough, however, rationalist epistemologies are riddled with terms associated with the semantic field of sight. To name only a few examples: Descartes' criterion for truth is that of the "clear and distinct idea"[4] and perspectivism constitutes the inner organizational principle of Leibniz's doctrine of the monads[5]. Moreover, it was the rationalist philosophers, not the empiricists, who closely studied the physiology of seeing and the geometric laws of optics.[6] The rationalists' relationship to seeing thus appears to be ambivalent, if not paradox. Their work both values and devalues ocular perception.

There is now an elegant and popular solution to this paradox, which is to distinguish between the "eye of the mind" and the "eye of the body." Descartes himself puts forward this distinction when he describes his method as an ordering and arranging of those objects upon which the "mind's eye" must concentrate in the pursuit of truth.[7] This "eye of the mind" sees all the more clearly the more our actual eyes can remain blind to the sensory abundance of the world. Ocular sight is thus repudiated so that "intellectual sight" can take its place; sight becomes "disembodied" and is turned into the activity of reasoning. According to this interpretation, the role of visuality is mitigated by the rationalist transformation of the body's eye into the eye of the mind.[8] In this way, visuality is made compatible with the pre-

1 Descartes 1902, pp. 135, 136.
2 See Bexte 1999, pp. 81–118.
3 Leibniz 1965, IV p. 423.
4 Descartes 1902, p. 31.
5 Leibniz 1965 IV pp 607–623, § 57.
6 "It is the so-called rationalist philosophers that we find the fullest exploitation of empirical theories of vision ..." Wilson 1997, p. 118.
7 Descartes 1908, Rule V, p. 380.
8 Martin Jays concept of 'Cartesian perspectivalism' is a version of the pure 'eye of the mind'; Jays (1988, p. 17) opposition between Cartesian concept of vision and the "tactile or haptic quality of vision, attributed to the baroque visual experience" misses the tactile dimension in Descartes theory of vision.

vailing understanding of rationalism, according to which aisthesis and logos are competing sources of knowledge and a philosopher becomes a *rationalist* by choosing the latter over the former.

In the following, focusing on Descartes and Leibniz, we will set out to correct this understanding and to rehabilitate the role of aisthesis in rationalism by framing perceptibility another way. In doing so, we will be guided by the idea that the rationalist relationship between seeing and knowing can be refigured by taking into account the ways in which the *mathematical* discoveries made by Descartes and Leibniz informed their epistemologies.

Descartes was the father of analytical geometry;[9] Leibniz developed infinitesimal calculus.[10] We would like to show that these innovations were accompanied by a new understanding of mathematics, which became a kind of language. This language, however, does not speak to the ears but to the eyes, for it is based on inscriptions; it is a calculized writing, giving rise to a specific type of visuality, which we may call "*tactile seeing*" as opposed to "ocular seeing."[11] The epistemological slight of hand of calculization is to reduce truth to correctness[12] and to establish a "perceptible"[13] and "palpable"[14] criterion for truth–Leibniz will call it an "Ariadne's thread."[15] In distinction to truth, this "correctness" is a norm monitored not by reason but by the *bodily* interaction between the eye and the hand. Descartes' and Leibniz's guiding epistemological idea is that the truth of a statement is ensured by the palpable correctness of its formation. The eye, and with it "tactile seeing," thus lies at the heart of rationalist epistemology.

Our thoughts will proceed in five steps:

(i) The "spirit of calculization" I: written arithmetic
(ii) The "spirit of calculization" II: Descartes' analytical geometry and Leibniz's infinitesimal calculus
(iii) On the origins of Descartes' idea of method
(iv) On the specificity of Leibniz's concept of truth
(v) The visualization of the invisible, or 'seeing with the hand'.

2. The "spirit of calculization" I: written arithmetic

The mathematical innovations of the sixteenth and seventeenth centuries have little to do with the "geometric-proving" theories passed down from Antiquity, and derive rather from an "algebraic-algorithmic" spirit, whose origins lie in Hindu-Arabic culture. "Algorithm" is the Latinized name[16] of the Arabic scholar al-Hwārizmī'[17] who in the High Middle Ages introduced Europeans to written calculation with Indian numerals.[18] Hindu-Arabic arithmetic with its decimal posi-

9 Descartes 1902, pp. 367–485.
10 Leibniz 1846; Leibniz 1920.
11 Descartes 1902, p. 84: " . . . .ils voient des mains".
12 Krämer 1991a, p. 88.
13 Leibniz 1960, p. 82; Leibniz 1965, p. 125; Leibniz 1960, p. 82.
14 Leibniz 1961, p. 176.
15 Leibniz 1965, I p. 371,V p. 350; VII p. 14, p. 22.
16 Reinaud 1849.
17 al-Hwarizmi 1857, al-Hwarizmi 1963.
18 Data and Singh 1962.

tional notation did not become generally accepted in Europe until the fifteenth century and only after a long "culture war" that broke out after Arabic scholars introduced Indian numerical calculation to 12 th century medieval Europe.[19] The fight against Oriental numerals was primarily a fight against the cipher zero.[20] Greeks and Romans did not have a zero, since they understood numbers to be multitudes composed of discrete, countable units (Euclid). Furthermore, in Ancient and Christian culture, "nothing," the void, and the vacuum were shunned.[21] The Indians, on the other hand, required the zero as an empty-place indicator in their decimal positional notation in order to mark the absence of a numeral, such as in the numerical expression "101".[22] They also set up rules for how to calculate with the "0" sign, thus, from an operational perspective, treating the numeral "0" like any other numerical sign.[23]

Two thoughts are important for our investigation into the connection between calculization and visualization.

(i) Decimal positional notation is not simply a numerical *language*, but rather a kind of writing, a *sui generis* graphic system that "speaks" to the eye, which can then be rendered phonetically in different languages. A mathematical calculus consists of a small number of basic visual signs that become numerical expressions through the application of pre-determined rules of formation and transformation – through, in other words, *combinatorial* operations. The materiality and visibility of the signs is the condition for their operative manipulability. The eye and the hand work together in these operations. A calculized expression is true if its syntax is proper. Calculization is an intellectual tool that actually permits a kind of "blindness", a blindness to the *semantic* meaning of the signs: we can reckon properly, for instance, without knowing whether zero is even a number. This leads us to a peculiar, but important insight: when "correctness" takes the place of truth, it is not the eyes of the body that are suppressed, but rather the eye of the mind, which is used to looking *below* the surface of what is perceptible to the senses and to interpreting the visible in light of the invisible. Syntactical correctness is a surface phenomenon that is entirely connected to the eyes of the body. Syntactical coherence has to *show itself*.

(ii) The slight of hand of decimal positional notation consists in simultaneously executing two different functions. It is both a representation of numbers in writing as well as a tool for written reckoning with numbers. This double function gives calculations the generative power of performatively bringing forth what they represent. If one does *not* stay blind to the semantics of calculations, and if, metaphorically speaking, one activates the "eye of the mind," one discovers that calculation makes what is invisible visible, while at the same time generating and constituting it. The number zero is a paradigmatic case. Did "zero" exist before there was a sign for it that could be perceived by the senses and operated with, that could be used in calculations like other numerical signs? It is not surprising that the European concept of the number underwent a radical change as a result of the introduction of decimal positional notation. A number was no longer a multitude composed of units (Euclid's definition), but the object of refer-

19 Juschkewitsch 1964.

20 Krämer 1988 pp. 54–71.

21 Rotman 1993 chapter 3; Seife 2000, chapter 4.

22 To the indian numerals: Smith and Karpinski 1911; Kaye 1915; Datta and Singh 1962.

23 Krämer 1988 pp. 45–48.

ence of a correct calculation[24] (Stevin's definition).[25] This meant that zero and negative values should also count as numbers. *'Nothing'* turned into *something*. The visualization of something invisible by embodying it in a perceptible sign structure was also the process of its generation.

3. The "spirit of calculization" II: Descartes' analytical geometry and Leibniz's infinitesimal calculus

In the Early Modern period, written calculations using decimal positional notation became a cultural technique that turned complex mental operations into child's play-provided one had mastered Hindu-Arabic numerical writing and the associated computational algorithms.

But what does this revolution in the everyday technique of reckoning have to do with rationalist philosophy? This is where the mathematical innovations of Descartes and Leibniz come into consideration, for they pioneered the application of the everyday technique of written calculation to higher mathematics. As we will see, *the key to refiguring the relationship between vision and knowledge lies in the exemplary role mathematics is given in the epistemological thoughts of Descartes and Leibniz*. First, however, we will have a look at one of the most momentous consequences of written calculation in seventeenth-century higher mathematics. This is the case of symbolic algebra, introduced in the sixteenth century by Francois Viète.[26] The use of letters in algebra made it possible for the first time to formulate the rules for solving arithmetic equations – until then applied in terms of an implicit knowing-how – in a universally valid way. Viète is responsible for introducing the formula to mathematics. In the algebraic formula "$a + b = b + a$," the letters are signs of variables. They constitute a metalanguage and no longer represent individual numbers, but the set of all object-linguistic signs that can be inserted into the formula in such a way as to produce a correct expression.[27] Descartes will later refer explicitly to Viète as having paved the way for his idea of the method.[28]

But what about René Descartes' "analytical geometry"? In LA GÉOMÈTRIE, one of the three texts added to the DISCOURS DE LA MÉTHODE,[29] Descartes is able to express points as numerical pairs by introducing a system of coordinates. This allows for geometric figures to be written out as arithmetic equations, turning figures into formulas.

Ever since the discovery of incommensurability in Greek mathematics–since, that is, it was found that the relationship between the side and the diagonal of a square cannot be represented in whole numbers – geometry and arithmetic had been categorically separated and considered to constitute independent branches of mathematics. It was Descartes who "reunified" figure and number, the measurable and the countable.[30] What was achieved by this?

---

24 Stevin 1925.

25 Klein 1968.

26 Viète 1646, vol 1, pp. 1–12, 42–81, 305–327.

27 Krämer 1991a, pp. 124–150.

28 Descartes 1908, Rule 4, p. 376.

29 Descartes 1902; german: Descartes 1981.

30 Krämer 1989.

There are three different points to consider here. First, Descartes' move simplifies geometric problem solving by allowing for tasks of construction to be treated like problems of arithmetic. Even more important, however, is the fact that Descartes makes numerical calculability into the criterion of existence for geometric objects. For instance, he classifies types of curves with types of algebraic equations, and curves that cannot be classified in this way are no longer considered to be objects of scientific geometry. Algebraic calculability thus becomes the criterion for geometric existence. But there is a third aspect: Descartes' work transformed geometry into a kind of visual language. This can be demonstrated by looking at a finer point of geometry, namely Descartes' suspension of the principle of homogeneity.

That the combination of two lines makes a surface, an area and the combination of two surfaces makes a body, a volume had been held as a valid principle since Greek Antiquity. Descartes rejects this principle, arguing instead that just as the multiplication of two numbers gives another number, the multiplication of two lines gives another line. He adapts geometrical operations with lines to arithmetical operations with numbers, thus turning geometry itself into a kind of calculus (Fig.2).

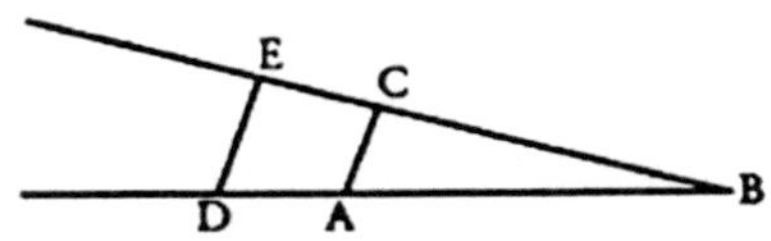

Fig. 2 Descartes' geometric construction of multiplication. "For example, let AB be taken as unity, and let it be required to multiply BD by BC. I have only to join the points A and C, and draw DE parallel to CA; then BE is the product of BD and BC." (from *The Geometry of René Descartes*, trans. from French and Latin by David E. Smith and Marcia L. Latham, New York: Dover Publication, 1954, p. 5)

The system of coordinates constitutes a kind of "translation manual" between geometric and arithmetic expressions. Figure and number, geometry and arithmetic, are thereby transformed into modes of visual-graphical languages that can be translated into each other. Descartes' GÉOMÈTRIE is a document of this *language*-oriented understanding of mathematics. Here is the key to the boom in quantification in the Early Modern Era: mathematics comes to be seen not merely as the domain of timeless, ideal objects, but takes on the guise of a kind of universal, operative language, which, as "calculized language," is aimed not at the ears but at the eyes, and is able to express everything quantifiable.

Let us turn now to Leibniz's infinitesimal calculus.[31] Traditionally, infinitesimal problems were only ever treated separately and with the use of *particular* methods; Leibniz, however, applies the algorithmic procedures of everyday reckoning to infinitely large and infinitely small values.[32] Leibniz (like Newton with his fluxion method) developed a *universal* method for the mathematical treatment of the infinite. He discovered that *all* infinitesimal problems can be traced back to two kinds of calculations. It is either a matter of calculating the differences between infinite neighboring terms, as in his *calculus differentialis*; or it is a matter of the summation of infinite series, as in his *calculus integralis*. Leibniz went on to invent a notational system for infinitesimal values, making it possible to operate with them algorithmically, in the manner of the calculations of basic arithmetic. This, incidentally, also makes those curves accessible to algebra that Descartes

[31] See: Leibniz 1962, IV pp. 91–95, V pp. 135–143, 220–226, 226–233, 266–269, 279–285, 301–306, 308–318, 320–326, 377–382, VII pp. 218–223; Leibniz 1846 pp. 32–50.

[32] Krämer 1991b.

still excluded from his analytical geometry as not belonging to scientific mathematics.

The stratagem of Leibniz's infinitesimal mathematics consists then in separating the question of what the symbols for infinitesimal values actually represent from the question of the correctness of infinitesimal calculations. The *metaphysical* question as to whether infinitely small or infinitely large quantities actually exist, whether they should be interpreted "realistically" or "constructivistically" (in other words, as not actually but only potentially infinite) does *not* affect the efficiency of Leibniz's infinitesimal calculations. The genius of infinitesimal calculus lies in its ability to use algorithmic methods to calculate with infinite quantities without having to decide about the ontological status of the infinite.[33]

We see, the most influential mathematical innovations of the Early Modern period were a result of the "spirit of calculization." Our assumption will be that this spirit can also be found in Descartes and Leibniz's epistemologies.

4. On the origins of Descartes' idea of method

How did mathematics, understood as the reservoir of calculized procedures, lead the way and serve as methodological model for those rationalist philosophers who were also creative mathematicians? This question confronts us with a problem. It is almost a commonplace that the influence of mathematics on rationalist philosophy derives from its axiomatic-deductive theoretical structure, put into practice by Euclid in his geometry and generalized by Aristotle as the model for deductive science. The rationalists, according to the standard view, wanted to apply this *demonstratio more geometrico* to their philosophy. We do not share this view, and propose rather that it is not the art of demonstrating *more geometrico* but the art of problem solving *more calculo* that inspired rationalist epistemology.

Let us turn first to Descartes' epistemological theory, whose key idea is related to the concept of the method. What is *new* about Descartes' idea of the method is his reversal of the relationship between object and method. Since Antiquity, scientific methods had been conforming to the diversity of their objects. Descartes, however, puts forward a unified as well as universal methodology for deciding what is accessible to scientific study in the first place. Objects no longer condition the methods of their examination, but, on the contrary, the method constitutes what can be taken as an object of scientific study.

We recall that in his analytic geometry, Descartes proposed that a curve must comply with the universal procedure of algebraic calculability in order to be considered a legitimate object of geometry. Thus the *origins* of Descartes' idea of the method show that a procedure shaped in the tradition of problem-oriented calculation becomes a role model for the method in the same way as algebra does in analytical geometry.

A posthumously published early text by Descartes, the REGULAE AD DIRECTIONEM INGENII, gives us an insight into the beginnings of the idea of the method.[34] It is in this text that we find the clearest evidence of the exemplary role played by calculizing mathe-

[33] "On n'a point besoin de faire dependre l'analyse mathématique des controversies métaphysiques", Leibniz 1962, IVp. 91.

[34] Descartes 1902.

matics. We will examine this evidence from two different perspectives: (i) in view of the tradition in which Descartes locates his idea of the method and (ii) in view of his project of a *mathesis universalis*.

(i) By identifying a precursor of his idea of the method, Descartes does not point to the axiomatic-deductive argumentative procedures of Euclidean provenance, but to the Ancient "analysis" as well as to the Early Modern analysis of algebra.[35] Characteristic of the ancient art of analysis is that it is not a procedure for establishing proof (*ars iudicandi*) but one of problem solving and invention (*ars inveniendi*). Its approach is to deduce what is sought from the conditions of what is given. The Hellenistic mathematician Diophant – referred to explicitly by Descartes – invented a symbolic notation system for the unknown numbers in an equation, enabling equations to be fully written out as relations of equivalence between unknown and known numbers and calculated accordingly. By using letters in algebraic equations, Viète – whom Descartes also cites as paving the way for his idea of the method – was then able to visualize and formulate in a universally valid way the rules of this analytic art, thus transforming an implicit knowing-how into a scientific methodology. It thus becomes clear that the art of problem oriented analysis is a forerunner of the Cartesian idea of the method specifically in its *enmeshing of visualization and methodology*.

(ii) Descartes then devises a universal science of everything that is quantifiable and calls it *mathesis universalis*.[36] He develops an artificial language,[37] a two-dimensional graphism of extensional figures in which anything that is to be a legitimate object of the *mathesis universalis* must be able to be represented. Descartes thus only admits as an object of knowledge of universal science that which can function as a reference object in his symbolism of extension. The order of scientific language precedes the order of objects of knowledge. Before the metaphysically inclined "mature Descartes" makes *extensio* and *figura* into fundamental attributes of the corporeal world, thereby becoming a metaphysician, the "epistemologically inclined" young Descartes considers extension and figure to be characteristics of the symbolic system that *represents* corporeality.

By making extensional visualizability into a prerequisite for becoming an object of the *mathesis universalis*, Descartes conceives of such objects as not only being represented by visualization but also being *generated* by it. How so? The basic attribute of these objects is their quantifiability. However, due to the centuries-old split between geometry and arithmetic, "quantity" was divided into *magnitudo*, the measurable, and *multitudo*, the countable. Figure and number were held to be mutually irreducible forms of the quantitative. There was no general concept of quantity independent of the disjunctive genre of the measurable *or* the countable. Descartes then introduced for the first time a *universal* concept of quantity, the *magnitudo in genere*, as a characteristic of everything corporeal, whether countable or measurable. This "general quantity" is abstract and hard to grasp. But insofar as everything that is of a general quantity must be representable in the medium of the extensional symbolism of the *mathesis universalis*, Descartes has given appearance to the invisible general quantity by the visualizing and depicting potential of his *mathesis universalis*, which not only embodies but constitutes it in the first place. Thus the concept of universal quantity was born. Descartes' invention

[35] Descartes 1902, Rule 4.

[36] Descartes 1902, Rules 4, 8, 12.

[37] Descartes 1902, Rules 12–15.

of the *multitudo in genere* provided the age of calculability with its conceptual foundation.[38]

In characterizing the stages in the history of algebra as the mathematical breeding ground for the Cartesian idea of the method, there is one thing we must be careful not to overlook. Descartes did not actually get as far as articulating the fundamental idea of formalization, which is that the rules of formal operations are interpretation-neutral. This was left to Leibniz. For Descartes, all symbolic operations are "in reality" operations with non-sensory objects, which are merely *represented* by the sensory symbolic expressions. This is why he characterizes his method as having to assist the "eye of the mind" – the focus, after all, should be on what is signified and not on the signifiers. But Descartes at least demonstrated that the bodily operations of eye and hand ought not to be suppressed but are on the contrary indispensable for the eye of the mind to become active in the first place. In the work of Descartes, the eyes of the body become the condition of possibility of the eye of the mind.

5. On the specificity of Leibniz's concept of truth

While for Descartes signs are something like external vehicles of knowledge, for Leibniz they belong to its "inner constitution." Thought, for Leibniz, can only operate and become structured in the medium of signs. This constitutive function of the symbolic for the cognitive is paradigmatically embodied by the calculations of mathematics.

I would like now to show that Leibniz's basic epistemological intuition to reduce truth to correctness is inspired by a vision of radically turning our mental attention in cognitive processes to the acts of writing down, reading, and manipulating inscriptions perceivable by the senses, without having to be distracted by questions of interpretation.

The eye of the mind, legitimized by Descartes, in Leibniz's epistemology is actually allowed to remain *blind*. This is what he means when he refers to "blind or symbolic"[39] or – another Leibnizian expression – "deaf" thoughts.[40]

For Leibniz, not only is the human body deficient, but also the human mind. Only the eye of God can immediately grasp the world in the fullness of all its attributes. Man is fundamentally dependent upon the medium of symbolic representation in his exploration of the world. Artificial signs help us to compensate not only for the natural limits of the senses, but also for those of reason.

Leibniz's specific concern is with the artificiality of these signs. Just as mathematics cannot advance in the medium of a natural language, so does man require *drawn or written* "characteres" that are perceivable by the senses, which Leibniz explicitly describes as *palpable*.[41] In this way, man creates an Ariadne's thread of thought that literally depends on the collaboration between the eye and the hand. Even if spoken language is an excellent means of communication, for Leibniz writing is the privileged means for the production and monitoring of intellectual labor. This, however, only holds for the formula writing of calculation, for it is

38 Krämer 1991, pp 159–219.
39 Leibniz 1965, IV, p. 423.
40 Leibniz 1965, VI, p. 259.
41 Leibniz, 1965, VII, p. 204.

only within calculized operations that signs become self-sufficient in relation to the different possibilities of interpretation. Leibniz was the first to interpret algebra as the manipulation of signs, whereas Descartes still considered it to be an operation with "universal quantities." For Leibniz, to "*manipulate* signs" means to disregard their objects of reference, which can be interpreted in different ways, and only to focus on their syntactic physiognomy. Algebra for Leibniz thus becomes a version of the universal art of combinatorial analysis (*ars combinatoria*). The *ars combinatoria*, in turn, is promoted to a universal science of the syntactical usage of signs in variations, permutations, and combinations. Leibniz is unmistakably clear about the semantic neutrality of the *ars combinatoria*. If its letters, which he also refers to as "notes indifferentes," are interpreted as values, the result is algebra; if they stand for numbers, the result is arithmetic; if they represent points, the result is geometry; if, finally, they stand for terms, the result is logic.[42]

Nevertheless, for Leibniz the semantic interpretation of combinatorial signs constitutes the vanishing point of their scientific usage. Leibniz is no formalist. For it is not until interpretation transforms the "formulae" into "expressiones" or "repraesentationes"[43] that calculization becomes more than a game of signs and can be used as an instrument of the search for truth. If a domain of cognitive objects can be isomorphically reproduced in the elementary signs of a calculus, the proper syntactic form of the calculized "expressiones" becomes the necessary and sufficient condition for the existence of the object referred to by this expression. Moreover, if it were possible to develop a universal calculus of thought (*calculus rationis*), then it could be decided about any given sentence if it were true, that is, properly formed, and all possible true sentences could be produced by purely formal operations. The use of this mechanization of symbolic operations would be, as Ian Hacking[44] puts it, a "do-it-yourself semantics": if the syntax is correct, the semantics looks after itself, preserving its truth value in all syntactic operations.

Leibniz worked on this universal calculus his whole life – to no avail. Given the logical impossibility of a closed alphabet of thought as the substantial foundation for the "characteristica universalis," his failure was inevitable. This does not, however, affect the success of a number of *specific* calculuses invented by Leibniz.

We will not pursue this here, however, but will turn instead to two implications of "combinatorial reason." We recall that the constitution of objects of knowledge by means of the method was a core idea of rationalist epistemology. Our hypothesis is that this "constitution" is based on the visual as well as operative potential the method derived from the model of algebra. In reference to Leibniz, this hypothesis poses a problem, however: the relationship between "res" and "characteres," between thing and sign, in Leibniz is normally understood as a relationship of representation derived from the model of the mathematical method of projection.

But semantic relations – formal expressions with interpretable content – are for Leibniz precisely *not* bipartite copies between the world of things and the world of signs, but *tripartite* relations between things, ideas, and signs.

So the reference objects of the semantic expressions are not the objects themselves,

42 Leibniz 1961, p. 531.

43 Leibniz 1965, VII, p. 263; Leibniz 1962 V, p. 141.

44 Hacking 2000, 104.

but our *ideas* of the objects. But how can we then distinguish between true and false ideas? Leibniz argues that we have a "true idea"– that, in other words, the object corresponding to our idea exists – if a "causal" or "generative" definition of it is available to us. To define an object means to be able to state the procedure required for generating it. The examples Leibniz lists for "generative definitions"[45] are all such that it is not simply the object itself that is generated but rather the *symbolic expression* that represents the object. Just as Leibniz observed that mathematics does not carry out its proofs on the things themselves but rather on paper and on the signs portraying the things,[46] we can only get a hold of objects in the medium of their calculized representation. Scientific objects for Leibniz are only given as abstraction classes of isomorphic symbolic expressions, which is why for him – from the perspective of man – the order of signs emphatically precedes the order of knowable things.

This transposition in the relationship between method and object, already familiar to us from Descartes, in Leibniz has far-reaching consequences, leading us to the limits of our epistemological processes: calculized cognition no longer provides us with knowledge about the actual world, but only about our *models of the world*. For Leibniz, everything that exists is individual; there are only unique things. Objects, however, that function as reference objects of formal symbolic structures only have a "formal identity." In the context of his reflections on substitution ("salva veritate" or "salvis legibus"), Leibniz defines this identity in terms of a structure that remains identical when its configurating elements are exchanged. These relation-based identities are not real, but "ideal;" they are products of the mind. This is why the formalizing epistemological procedures no longer refer to concrete phenomena, but only to constructs of the mind. The real world in the fullness of all its features is not available to human knowledge.

Some conclusions

Let us sketch some conclusions.

(i) The algebraization of geometry, the formalization of writing, the combinatorial analysis of symbolic operations all appear within mathematics as a radical form of the suspension of the iconicity of geometry. Calculization is interpreted as a *de-aestheticization*, in the sense of the Greek 'aisthesis'.

In *The Crisis of European Sciences*, Edmund Husserl[47] identified the arithmeticization of geometry and the formalization of the sciences with an erosion of meaning and a loss of sensuousness. For Husserl, the telos of modern scientific development lies in the desensualization and decorporealization of its objects. But is not a different perspective possible, given what I have been rehearsing here? Let us assume that the power of science is rooted not only in abstraction but also in concretion – in the ability to invent embodiments and sensuous forms for abstract entities and epistemic objects. The visualization potential of the symbolic operations of calculation is distinguished by its ability to *embody "objects of knowledge"* and to *visualize the invisible*. But if we interpret calculization as a strategy of visualizing the invisible, do we not risk falling into the trap of Platonism, which sees in sense objects nothing but the deficient silhouette of antecedent, invisible essences? But we already have an answer for this: as we have seen, calculizing visualization is always also to be understood as a *generation* of that which is made accessible by signs.

Let us recall again the example of zero. In a first step, a gap is visualized, the mere absence of a number. What is unperceptible here is thus literally a void, a lack, an absence – and no more. If centuries of reckoning with the numeral "0" brings about a new *number*, so that the *numeral*

45 Leibniz 1965, I, p. 331; III p. 225; Leibniz 1962, I, 85.

46 Leibniz, 1961, p. 154.

47 Husserl 1982.

"0" finally represents and visualizes the *number* zero, it is clear that it is not a matter here of an antecedent existence of an invisible epistemic object. It is rather that the *performance* of operating with the numeral "0" produces the *representational capacity* of this numeral to refer to the number zero as a mathematical object. It is the use of the sign that brings about its reference object in the first place. We can also see how *mediality* – in this case the notational graphism of calculation – plays a crucial role in the performative power of symbolic operations.

(ii) Writing is normally understood as language written down. But from the perspective of a phonetically neutral concept of language, the operative writing of calculus appears as a mixture of linguistic *and* pictorial elements – "pictorial" in the weak sense of a spatial and visual symbolism that makes use of the two-dimensionality of the surface. Calculations are *structural images* of what they represent. The operative visuality necessary for calculating can be understood as a kind of "tactile seeing."

The paradox of the rationalists' simultaneous valuation and devaluation of seeing can thus be solved without falling back on the split between the eye of the body and the eye of the mind. The rationalists devalue "ocular seeing," since it is closely tied with the illusionary, but they value "tactile seeing." This is not seeing with the eye of the mind, but – if we may use this metaphor inspired by Descartes himself – *"seeing with the hand,"* in the sense that seeing and manipulating or operating are closely related and intertwined.

Let us finally sum up our argument.

1. *Calculization*: Hindu-Arabic arithmetic with its decimal positional notation is the prototype of the cultural technique of calculization. The basic idea behind the practice of calculation is the operative separation of construction and interpretation. Mathematics becomes a model for rationalist epistemology by virtue of its calculizing problem-solving techniques (*more calculo*) and not its axiomatic-deductive techniques of reasoning (*more geometrico*).

2. *Mathematical innovations*: Cartesian Analytical Geometry translates geometric figures into algebraic formulas, so that only what can be rendered algebraically is considered part of scientific geometry. Descartes transforms mathematics into a kind of "language"; one, however, that "speaks" graphically to the eyes. With the help of a graphic signifying system, Leibniz's infinitesimal calculus introduces algorithmic arithmetic to higher analysis. He separates the operation with infinite values from their interpretation, so that the truth of a statement comes to rest solely on its internal correctness.

3. *The origins of Descartes' idea of the method*: What is new about the Cartesian method is its reversal of the relationship between object and method, so that it is no longer the object that determines the method, but the method that determines the object. I will show that this idea has its origins in algebraization. Descartes' early text *Regulae ad directionem ingenii* demonstrates the exemplary role of algebra in two ways: (i) Descartes appeals to the tradition of problem-solving analysis understood as *ars inveniendi* and not to the Aristotelian proving analysis of *ars iudicandi*. (ii) Descartes introduces an extensional, graphic symbolism meant to work like a calculus into his *mathesis universalis*. Only that which can function as an object of reference of this symbolism can be admitted as an object of universal science. Visualization and methodology become enmeshed.

4. *Leibniz's concept of truth*: What is new about Leibniz's concept of truth is its reduction of truth to correctness. Leibniz is first to articulate the concept of calculation as an interpretation-neutral operation with graphic signs and tries to apply calculization as a universal process onto knowing itself (*calculus ratiocinator*). This shows that the objects of knowledge are constituted by the order of the process of knowing, which can be demonstrated by looking at Leibniz's "generative definition." What interests us here are the epistemological consequences of this: since the world is made up of unique things, but calculizing knowledge only refers to abstract constructs of the mind, our formal knowledge can no longer refer to the real world, but only to our models of the world. Knowledge that relies on the "tactile seeing" of the Ariadne's thread of calculation is the only kind available to humanity.

5. *Visualizing the invisible or "Seeing with the Hand":* The heart of science is not simply "abstraction", but the *embodiment* of objects of knowledge, the *sensualization* of abstractions and the *visualization* of the invisible. Modern algebraization is thus not simply a suspension of iconicity but, in the name of "diagrammatolo-

gical" reason, also stands for the formation of an "operative" or "tactile" kind of visuality. The rationalists devalue "ocular seeing," since it is closely tied with the illusionary, but they value "tactile seeing," which is not a "seeing with the mind," but a "seeing with the hand."

References

Bexte P (1999) Blinde Seher. Die Wahrnehmung von Wahrnehmung in der Kunst des 17. Jahrhunderts. Verlag der Kunst, Dresden

Datta B, Singh AN (1962) History of Hindu mathematics: A source Book. 2 vols (Lahore 1935–1938), Asia Publishing House, Bombay

Descartes R (1908) Regulae ad directionem ingenii. In: Adam Ch, Tannery P (eds) Oeuvres de Descartes. Vol 10, 2nd ed Librairie Philosophique J. Vrin, Paris, pp 360–396

Descartes R (1902) Discours de la méthode et essais. In: Adam Ch, Tannery P (eds) Oeuvres de Descartes. Vol 6, 2nd ed. Librairie Philosophique J. Vrin, Paris

Descartes R (1981) Geometrie. Schlesinger L(ed), Wissenschaftliche Buchgesellschaft, Darmstadt

Descartes R (1954) The Geometry of René Descartes. Smith DE, Marcia LL (trans from French and Latin), Dover Publication, New York

Hacking I (2000) What mathematics has done to some and only some philosophers. Proc British Academy 103: 83–138

al-Khwarizmi (1857) Algorithmi de numero Indorum. In: Boncompagni B (ed) Trattati d'arithmetica. Vol 1. Tipografia delle scienze fisiche e matematiche, Roma

al-Khwarizmi (1963) Mohammed ibn Musa Alchwarizmi's Algoismus. Nach der einzigen lateinischen Handschrift in Faks., mit Transkription u Kommentar. Vogel K (ed), Zeller, Aalen

Husserl E (1982) Die Krisis der europäischen Wissenschaften und die transzendentale Phänomenologie. Felix Meiner, Hamburg

Jay M (1988) Scopic regimes of modernity. In: Foster H (ed) Vision and visuality. Bay Press, Seattle, pp 3–27

Juschkewitz AP (1964) Geschichte der Mathematik im Mittelalter. Ziegler V (tansl into German and ed), Pfalz, Basel, (orig. Moskau 1961)

Kaye GR (1915) Indian mathematics. Thacker, Spink & Co, Kalkutta Simla

Klein J (1968) Greek mathematical thought and the origin of algebra. UP, Cambridge MA London

Krämer S (1988) Symbolische Maschinen. Wissenschaftliche Buchgesellschaft, Darmstadt

Krämer S (1989) Über das Verhältnis von Algebra und Geometrie in Descartes' Géomètrie. Philosophia Naturalis 26, 1: 19–40

Krämer S (1991a) Berechenbare Vernunft. Kalkül und Rationalismus im 17. Jahrhundert. De Gruyter, Berlin New York

Krämer S (1991b) Zur Begründung des Infinitesimalkalküls durch Leibniz. Philosophia Naturalis 28, 2: 117–146

Krämer S (2008) The productivity of blank' on the mathematical zero and the vanishing point in central perspective. Remarks on the convergences between science and art in early modern period. In: Schramm H e. a. (eds) Instruments in arts and sciences. De Gruyter, Berlin New York, pp 457–478

Leibniz GW (1846): Historia et origino calculi differentialis a W. G. Leibnitio conscripta. Gerhardt CI (ed), Hahn, Hannover

Leibniz GW (1920) The early mathematical manuscripts of Leibniz, translated from the Latin texts, published by Gerhardt CI, with critical and historical notes by Child JM, Dover Publication, Mineola NY

Leibniz GW (1960) Die Leibniz-Handschriften der Königlichen öffentlichen Bibliothek zu Hannover. Bodemann E (ed), repr. Olms, Hildesheim (orig. Hannover 1889)

Leibniz GW (1961) Opuscules et fragments inédits de Leibniz. Couturat L (ed), repr. Olms, Hildesheim (orig. Paris 1903)

Leibniz GW (1962) Mathematische Schriften. Gerhardt CI (ed), 7 vols repr Olms, Hildesheim (orig. Berlin Halle 1849–1863)

Leibniz GW (1965) Die philosophischen Schriften. Gerhardt CI (ed), 7 vols, repr Olms, Hildesheim (orig. Berlin 1875–1890)

Reinaud JT (1849) Memoire géographique, historique et scientifique sur l'Inde anterieurement aux milieu du XIe siècle de l'etre chrétienne. Imprimerie Nationale, Paris

Rotman B (1993) Signifying nothing: the semiotics of zero. UP, Stanford

Seife C (2000) Zero: the biography of a dangerous ideas, Penguin Books, New York

Smith DE, Karpinski LCh (1911) The Hindu-Arabic numerals. Ginn and Company, Boston London

Stevin S (1925) De Thiende. With an introduction by Bosmans H., Antwers SJ, Société des Bibliophiles Anversois, Antwerpen Den Haag (orig. Leiden 1585)

Viète F (1646) Opera mathematica. Ed. Fr. v. Schooten, Leiden (repr. 1970 Hildesheim)

Wilson C (1997) Discourses of vision in seventeenth-century metaphysics. In: Sites of vision. The discursive construction of sight in the history of philosophy. Levin DM (ed), MIT Press, Cambridge, pp 117–138

Free will

22

Hans-Dieter Klein

Contents

Abstract

At one extreme free will can be defined as libertarian free will, according to which free actions are unconstrained and uncaused by any physical process. This definition is incompatible with the concept of the physical world as a closed causal system. Due to the shortcomings of the common opinion that free will and physicalism are simply incompatible, the more sophisticated approach of the compatibilists rejects libertarian free will, because free will for them is not the absolute libertarian freedom. Rather, it is the unrestricted ability to act without external (e. g. a dictator) or internal constraints (e. g. mental diseases). Moreover, compatibilists are aware that even in the determined world some processes are chaotic and their complex outcome cannot be predicted, although they are causally related to their starting conditions. This approach is supported by a long philosophical tradition inaugurated by Aristotle.

Hans-Dieter Klein
University of Vienna
Institute of Philosophy
Universitätsstr. 7, room C0215
1010 Vienna, Austria
e-mail: hans-dieter.klein@univie.ac.at

1. Introduction

What does the question of free will have to do with a volume on "sensory perception"? One can only speak of sensors, if the input received by a self-sustaining system from its environment is functionally related to its reactions, be they motor or chemical reactions or differently structured. We recognize that the equipment of living species with certain sensory organs is adapted to their way of life. If we would like to understand species specific sensors, then we also have to look at the species specific strategies of survival and the self determination of the reactions mentioned above (see also Chapter II,6 by FG Barth).

The way of life of humans is dominated by a phenomenon which we usually refer to as volitional and responsible behaviour. Preliminary stages to this behaviour can also be found with other animal species, for example primates or ravens. Yet nowhere else is this specific strategy of survival so highly developed and utilized as it is in humans. There is good reason to assume that the specific human equipment with regard to sense organs is adapted to their own way of life. Then, if we include humans in our consideration of "sensory perception", it could be useful to take a look at the problem of free will.

As in philosophical tradition from antiquity to modernity, there is no consensus in today's philosophical debate on the question of free will. On the one hand, some argue for a hard physicalistic determinism which claims there is no space for free will in the determined world in which we live. The undeniable subjective feeling, present in daily life, that one can determine one's own actions in a free and responsible way is considered an illusion. The idea that physical determination and free will are irreconcilable is known as incompatibilism. On the other hand, many current authors object to this strict physicalistic determinism combined with incompatibilism. They confront incompatibilism with compatibilism, in which consistent determination of all occurrances in time and space, according to the laws of physics, is indeed reconcilable with free will.

The alternative of compatibilitism or incompatibilism depends largely on which view is taken with regard to the mind-brain-problem. Here, the discussion focuses primarily on the alternatives of monism and dualism. While monism accepts only a single ontological level, namely physical reality, dualism assumes that there are effectively two worlds, firstly, that of physical objects and secondly, the mental world. Some, for example the mathematician and philosopher Gottlob Frege, assume a third world, namely the world of logical and mathematical entities, which can neither be psychologically reduced to mental experiences, nor physicalistically reduced to physical objects.

Even though at first glance dualism seems the most suitable for avoiding the danger of reductionistic physicalism opposing the evidence of every day life, it is confronted with the problem that a reciprocity between mental and physical phenomena is difficult to reconcile with the closed causality that, despite the uncertainty principle, is required by the physical paradigm. Monism on the other hand faces the problem of being able to bring several ontological levels into a single monistically conceived world. Nevertheless, this is attempted, and some argue for a monistic compatibilism combined with a non-reductionistic physicalism.

In previous years there has been some excitement in the discussion of free will following the sensational experiments of Libet (1985, 2002, 2004). Indeed, one has to wonder about this excitement as these experiments have not brought about anything new in regard to the philosophical problem of free will. Since the methodical foundation of modern physics in the 17th century, during the time between Galilei and Newton, we have known, notably because of the conservation principles, that physical processes can be caused exclusively by physical processes and that the chain of physical causes is without gap and closed. Since any action comes into effect as a physical process, the chain of physical processes having caused it must have preceded it. Since the 17th Century therefore, we have assumed that there has to be something like the readiness potential which has now been empirically established by Libet. Libet's discovery, which shows in detail which processes precede and cause an action, as interesting and important as it may be, offers no new perspective on the philosophical problem of free will. Already Leibniz and Kant, the classical representatives of free will in modern philosophy, clearly held the view that all physical processes, as well as the actions of humans, are determined by a closed chain of causes. Since this chain would lead back to the beginnings of the universe, it certainly must have started before I became aware of my decision to move my arm.

Leibniz and Kant are therefore to be considered compatibilists. In principle, compatibilism reaches as far back as Plato and Aristotle, who postulated the compatibility of *causa efficiens* and *causa finalis*, although at that time the problem could not be posed as clearly as it has been since the 17th Century, and as it can be presented today.

2. The concept of "free will"

In fact, we ought to define more clearly what is to be understood by free will, as the astounding excitement is also due to confusions commonly connected with the use of the term "free will". The expression "free" in this term does not refer to some kind of arbitrary indetermination, but refers to freedom in a certain relation. It means that we, in our actions, are not subject to our drives and affects. Since ancient times, philosophers based their work on the assumption that we are able to abstain from a meal, as appetizing as it may be, if we consider it unhealthy, too expensive, or unsuitable for any other reason. Under some circumstances this may be hard and cause an inner battle, but fundamentally we are in a position to successfully resist the pressure of our drives.

The fact that our will is not ultimately determined by our drives, certainly does not mean that it is not determined by anything at all. Indeed, the philosophical tradition refers to free will as one which is not determined by drives but by thought, by reason. In everyday life, this is expressed in the so called "language game of responsible behaviour". In the cases where we blame ourselves or others (and here we assume responsible behaviour as a fact, otherwise this would be senseless) we often use questions starting with "why?". "Why have you given in to your anger?", "Why have you been for a walk instead of coming to work?" These why-questions aim for reasons, assuming that actions cannot only be caused by affects but also by reasons.

Thus free will refers to a determination of will by thought instead of affects. This is determination as well and by no means a game of pure chance. This assumption includes that the decision is preceded by processes of thought. It is not surprising to discover, rather it has always been expected, that the processes underlying thought and preceding and causing free decision would sooner or later become observable as neuronal processes in the brain.

If thought is the cause of free decision, then free decision can be further influenced by thoughts that we communicate to each other by using language. We also influence free decision in everyday life, when we look into things together, when we argue pro and contra various actions, inform ourselves or others on rules and so on. This is the way we interact with other humans. In cases where we have contact with animals, for which we do not assume free will, we do not try to interact using arguments but we only try to influence their affects. If we do use language when communicating with them, this is not because of the contents of the sentences but with regard to the emotional and affective expression of the voice, our facial expressions and gestures.

The confusion regarding the use of the expression "free will," which I addressed earlier, is mainly due to the fact that freedom is misconceived as complete indetermination, as chance without any mediating cause. Such kind of free will has never been postulated in the philosophical tradition, nor is it the foundation of everyday life assumptions.

When arguing from this point of view, it is also clear that Heisenberg's uncertainty principle cannot contribute to the solution of the problem of free will. Once in a while, hope is expressed that today, when the classical concept of physical causality is left behind, we enter a new era where free will is interpreted as irrationalism. This is of course a big misinterpretation. First of all, it has to be said that quantum theory is not suited to moderate determinism within the field of physical nature. Now, because of quantum theory, there are more precise prognoses available than before. Furthermore, attention must be paid to the fact that the relevant physical processes are viewed as macrophysical processes, that is to say, that here we are not confronted with possible prognoses on isolated elementary particles. Above all, it has to be noticed that chance, as it is attached to singular events in quantum theory, has practically nothing to do with what we attribute to free will in everyday life and philosophy, since free will as explained before, refers to determination by thought, by reason. If chance events can be seen as signs of free will, then we would have to assign responsibility to the dice for the number it has just revealed in the single cast. If we did so emotionally, this would be the result of a regression to a magical view of the world induced by affect, and would not persist in the face of our self criticism.

In order to account for the concept of free will and its consistency with the philosophical tradition as it is presented here, I would like to present a famous quotation from Leibniz's "Essais de Théodicée": "Nous avons fait voir que la liberté, telle qu'on la demande dans les Écoles Théologiques, consiste dans *l'intelligence*, qui enveloppe une connaissance distincte de l'objet de la délibération, dans la *spontanéité*, avec laquelle nous nous déterminons, et dans la *contingence*, c'est à dire dans l'exclusion de la nécessité logique ou métaphysique." (Leibniz 1961, § 288, p. 288)

> "I have shown that freedom, according to the definition required in the schools of theology, consists in *intelligence*, which involves a clear knowledge of the object of deliberation, in *spontaneity*, whereby we determine, and in *contingency*, that is, in the exclusion of logical or metaphysical necessity." (Leibniz 1952, § 288, p. 303)

In this quote, Leibniz refers to the generally well established doctrine of the schools of theology. At that time, Aristotelism was predominant in both Christian denominations (Catholicism and Protestantism), and Leibniz's concept of free will is nothing more than a short summary of the corresponding parts in the "Nicomachean Ethics" (3rd book) of Aristotle (Aristoteles 1926, 1109b – 1111b, pp. 110–128). The definition of free will, as given there, is as up-to-date now as it was when it was written. The current action theories are in most cases variations and modifications of the basics already brought forward by Aristotle.

If, according to Leibniz, "intelligence" is the constitutive element of free will, then this statement exactly corresponds to the thesis put forward above, that volitional action is action determined by thought.

"Spontaneity" (as used by Leibniz) requires what we call volitional actions, i. e. that the cause of the movement lies in the consciousness of the acting person and not in external systems. In his Ethics, Aristotle gives an example of an external determination of an action which cannot be referred to as a volitional action. To paraphrase, it states that humans are carried away by a storm against their will.

With regard to "contingency" Aristotle maintains that it would be senseless to regard mathematical sentences as results of our actions. Actions, from his point of view, refer to empirical states, which cannot necessarily be proven mathematically, an example being the position of a physical item at a certain place which could also lie anywhere else without violating a mathematical sentence.

3. Compatibilism

As the concept of free will is adequately covered for now, I would like to return to the question of compatibility of volitional action with the closed validity of physical laws. We said that volitional action is caused by the thoughts of the acting person but not by his affects. I will argue that this idea is compatible without compromising the closed validity of physical laws, since we assume that thought can be generated both by machines and the human brain.

Computers perform thought operations of various kinds and even make decisions preceded by thought operations. No one would ever postulate that for this reason physical laws would have to be suspended within the computer. The distinction between software and hardware keeps us from confusing the logical calculations performed by a computer with the physical process by which these operations of thought are realized.

Obviously, computers and brains can be compared in a superficial way only. Computers may carry out thought operations but they cannot think, which is what I can do, as long as my brain is healthy. This statement made from a first-person perspective can also be translated into the third-person perspective: Humans can think as long as their brains are in reasonable working order. Computers cannot be compared with living organisms for the following reason: organisms are functional autonomous systems, which carry their purpose in themselves. This is not true for computers. Their functionality is related to that of humans, who design, construct and use them. Without this relationship computers are random assemblies of different materials.

Only functionally and teleologically autonomous systems are able to think. Therefore it only makes sense to assign this ability to living organisms. Although we are still far from understanding how we think, how thought is realized physically, we are already in a position to make a rudimentary draft of some of the thought operations of our brain, an example being the allocation of a perception within a concept. There is not the slightest reason to assume that anywhere in our thinking brain, physical laws would be suspended in order for our thoughts to be realized. Furthermore, there is no reason to assume that we are incapable of at some point deciphering the mechanisms of our thinking brain sooner or later through further research.

Studies of thought and of other mental faculties, especially perceptions, revealed that many preparative operations, which we do not note consciously, enter into these processes. We are to a large extent conscious only of completed or almost completed results, which on the other hand are themselves controlled by consciousness and may even be returned for further processing. These subliminal processes pose no difficulty for a theory of conscious perception and thought. Therefore, we have to assume that thinking, which leads to one or the other decision, is subliminally prepared by many intermediate steps outside our consciousness. (see also Chapter IV,1 by P Giampieri-Deutsch)

All these considerations show that the prerequisites of the classical philosophical theory of free will, as it was first formulated by Aristotle, paved the way for the possibility that something like the readiness potential would finally be established, if suitable experimental methods like those designed by Libet would be applied. That is why Libet's results do not disprove the classical theory of free will, but are fully consistent with it.

An incompatibility between free will and physical laws will certainly arise, if a false concept of free will is maintained, that is, if free will is not understood as freedom from a constraint by affects with complete absence of causality instead of being determined by thought. However, there is no reason to adhere to such a concept of free will. The phenomenon of free will and responsibility, which we permanently presume in our everyday life when dealing with ourselves and other people, does not necessitate such an absurd idea of freedom. The idea of an absence of causality with regard to our actions cannot be accounted for by the phenomena. Yet it is exactly these phenomena which lead to a concept of free will, which can ultimately be ascribed to Aristotle. Aristotle did not intend to support any obscure metaphysical dogma, but to correctly describe the phenomena of everyday life and to draw from them the necessary theoretical conclusions.

4. Two additional issues

To conclude this chapter, two questions relevant to the debate about free will are considered.

The first question. Apparently, for some years psychodynamic psychiatry and psychoanalysis and, more recently the neurosciences, have called into question whether self-determined actions rest on conscious reasons and decisions. Their assumption is that actions which are perceived as "free" should rather be understood as the result of unconscious mental processes or more recently of unconscious brain processes. However, if one defines free will as the self-determination of an animal's motor function through its thought processes, unencumbered by the affects, as already suggested by Aristotle, then no contradiction exists between the assumption of mental processes, or unconscious brain processes, influencing actions on the one side, and free will on the other. This can be illustrated by two examples, one referring to Freud who introduced the term parapraxis (also known as "Freudian slip") at multiple times. Parapraxis can occur, e. g.

- when someone in an amorous relationship simply forgets a rendezvous. Here, self-determination exists in the ability to make a rationality-based choice between the possibilities that I become aware of at the moment of my actions. However if I simply do not remember my appointment, then I can not measure it against the alternative action of failing to appear. If one therefore defines free will as Aristotle did, then it is always limited by the unconscious processes which organise what consciously occurs to us at the moment of decision and what is forgotten, or otherwise obstructed from consciousness.
- But it may also happen that possible actions are brought to consciousness which we determine to be rational and therefore desirable, but the pressure of affects and compulsion is so strong that we feel incapable of implementing what we consider rational. This is normal in some extreme situations, e. g. panic. In other cas-

es, for example drug addiction, such an impairment and massive limitation of free will, which fundamentally exists even here, must be categorised as pathological. In many cases, a drug addict can not overcome his addiction alone, but a serious attempt at rehabilitation, or conversely a half-hearted attempt, depends on the addict's free will. Experience shows that this free will decision is a *conditio sine qua non* for a successful treatment process. Therefore there is no insurmountable contradiction between the assumption of the unconscious influencing our actions and the assumption of free will.

The second question. As a final consideration the second question concerns the special place of man in context to animals, further even, in the context of all organisms. As it is generally known, from a religious viewpoint man has been considered creation's crowning glory. Neurobiologist Gerhard Neuweiler, who gave one of his books the programmatic title "Und wir sind es doch – die Krone der Evolution, [Indeed we are – the Crowning Glory of Evolution]", most impressively illustrates how this old point of view has in recent years occasionally been transformed into biological thinking (Neuweiler 2008). His book shows that the peculiarities of the species *homo sapiens sapiens* can only be regarded as the purpose of a process subordinated to evolution, if one views evolution from the point of view of the *homo sapiens sapiens*. Categorically, it would also be possible to regard evolution from the point of view of a type of bacteria: one could regard this bacteria as the crowning glory of evolution. In this case, one would stipulate that complex organisms succumb to extinction much faster than bacteria. If one applies the criteria of a species' crisis-proof adaptation as a criteria for evolutionary progress, then one must classify bacteria and similar enduringly adapted life forms as the crowns of evolution. If one applies as criteria of evolutionary progress the aptitude for learning of a species' single individuals and the resulting freedom of the individuals, as well as the resulting trans-generational transmission of cultural values, as Neuweiler does, then one will come to the conclusion that man is the crowning glory of evolution. Admittedly, the argumentation is as circular in one case as it is in the other.

There is another aspect however. Humans' individual aptitude for learning is potentially unlimited in some areas, because man is capable of applying logical-mathematical structures to control its behaviour. Operations such as counting or generalising functionally make use of the infinite and can therefore be implemented e. g. to break free of behavioural limits caused by instinct schemata. Here, one touches on the above-mentioned question, whether logical-mathematical structures form an ontologically independent realm outside the physical and mental world, as Frege emphatically postulated, or whether they can be derived from the mechanism of evolution. It is now clear that the origin of organisms which are capable of logical and mathematical thought must be explained by the mechanisms of evolution, no different than the origin of organisms with the ability to dive or fly. This in no way concerns the validity of logical and mathematical principles. Otherwise one would have to say that 2 × 2 would not be 4 or that the principle of contradiction would not be valid if in some phase human evolution would have taken a different turn. It is difficult to assume this because the argument on which the theory of evolution is based itself makes use of logical and mathematical rules. So a consensus exists among logicians and mathematicians that logical and mathematical theorems cannot be legitimized through evolutionary theory arguments, but rather must be legitimized by the methods of logical and mathematical reasoning. However, there is an interlogical and intermathematical discussion where there is no complete consensus, namely on absoluteness, or respectively relativism of logical and mathematical principles which is completely independent of evolutionary theory arguments. Of course, if one views logical and mathematical principles and methods as absolute or satisfactorily absolute, then one will view humans, and all life forms that steer their behaviour according to logical and mathematical thinking, as the crowning glory of evolution. In any case, the question whether and to what extent humans and similarly talented animals on earth or on other planets in the universe, are the crowning glory of evolution, can only be brought closer to resolution through cooperation between meta-logic, meta-mathematics and evolutionary theory. On its own, evolutionary theory is overextended by this question, because in the end, it returns to circular argument structures.

Final remark

Now, after having attempted to present sufficiently strong arguments supporting the theory of compatibility, I would like to finally point out that a basic problem has been avoided: the problem of transition between the first-person perspective and the third-person perspective. A profound discussion of this problem would require much more space than is available here.

References

Aristotle (1926) The Nicomachean ethics. In: Goold, GP (ed) Aristotle in twenty-three volumes. Vol. 19. Harvard University Press, Cambridge MA

Leibniz GW (1961) Die philosophischen Schriften. Gerhardt CI (ed). Vol 6. Georg Olms, Hildesheim

Leibniz GW (1952) Theodicy. Essays on the goodness of God, the freedom of man and the origin of evil. Farrer, M (ed) Huggard, EM (trans). Routledge, London

Libet B (1985) Unconscious cerebral initiative and the role of conscious will in voluntary action. Behav Brain Sci 2: 529–566

Libet B (2002) The timing of mental events: Libet's experimental findings and their implications. Conscious Cogn 11: 291–299

Libet B (2004) Mind time. The temporal factor in consciousness. Harvard University Press, Cambridge Mass

Neuweiler G (2008) Und wir sind es doch – die Krone der Evolution. Klaus Wagenbach, Berlin

Index

L

M

N

O

List of contributors

Friedrich G Barth
University of Vienna
Life Sciences
Department of Neurobiology
Althanstr.14
1090 Vienna, Austria
friedrich.g.barth@univie.ac.at

Daniel Bowling
Duke-NUS Graduate Medical School
Neurobiology
8 College Road, Level 05
Singapore 169857, Singapore
dan.bowling@duke.edu

Simon Bulling
Medical University of Vienna
Center of Biomolecular Medicine and Pharmacology
Waehringerstr.13a
1090 Vienna, Austria
simon.bulling@meduniwien.ac.at

Christoph von Campenhausen
Johannes Gutenberg-University Mainz
Institute for Zoology III-Neurobiology
Kleinmannweg 2 – SBII
55099 Mainz, Germany
campenha@uni-mainz.de

Johannes Deutsch
Ölzeltgasse 1/3
1030 Vienna, Austria
office@johannes-deutsch.at

Gerhard F Ecker
University of Vienna
Life Sciences
Department for Medical/Pharmaceutical Chemistry
Althanstr.14
1090 Vienna, Austria
gerhard.f.ecker@univie.ac.at

Matthew Hugh Erdelyi
The University of New York
Department of Psychology
Brooklyn College
2900 Bedford Avenue, Brooklyn
NY 11 210, USA
iyledre@comcast.net;
erdelyi@brooklyn.cuny.edu

Michael Freissmuth
Medical University of Vienna
Center of Biomolecular Medicine and Pharmacology
Institute of Pharmacology
Waehringerstr. 13a
1090 Vienna, Austria
michael.freissmuth@meduniwien.ac.at

Stefan Frings
University of Heidelberg
Molecular Physiology
Im Neuenheimer Feld 230
69120 Heidelberg, Germany
s.frings@zoo.uni-heidelberg.de

John Gage†
University of Cambridge
Department of History of Art
jsg1000@cam.ac.uk

Patrizia Giampieri-Deutsch
University of Vienna
Institute of Philosophy
Universitätsstr. 7, room C0 215
1010 Vienna, Austria
patrizia.giampieri-deutsch@univie.ac.at

Martin Giurfa
Université Paul Sabatier – Toulouse III
Centre de Recherches sur la Cognition Animale CNRS
118 Route de Narbonne
31062 Toulouse cedex 4, France
giurfa@cict.fr

Richard L Gregory†
University of Bristol
Department of Experimental Psychology
12a Priory Road, Clifton
Bristol BS8 1TU, UK
richard.gregory@bristol.ac.uk

Ludwig Huber
Vetmeduni Vienna
Comparative Cognition Research
Messerli Research Institute
Veterinärplatz 1
1210 Vienna, Austria
ludwig.huber@vetmeduni.ac.at

Andreas Jurik
University of Vienna
Life Sciences
Department of Medical/Pharmaceutical Chemistry
Althanstr.14
1090 Vienna, Austria
andreas.jurik@univie.ac.at

Robert Jütte
Institute for the History of Medicine
Robert Bosch Foundation
Straussweg 17
70184 Stuttgart, Germany
robert.juette@igm-bosch.de

Hans-Dieter Klein
University of Vienna
Institute of Philosophy
Universitätsstr. 7, room C0215
1010 Vienna, Austria
hans-dieter.klein@univie.ac.at

Sybille Krämer
Free University Berlin (FU)
Institute for Philosophy
Habelschwerdter Allee 30
14195 Berlin, Germany
sybkram@zedat.fu-berlin.de

Kristine Krug
Oxford University
Department of Physiology, Anatomy and Genetics
Sherrington Building
Parks Road
Oxford Ox1 3PT, UK
kristine.krug@dpag.ox.ac.uk

Geoffrey A Manley
Technical University Munich
Institute for Zoology
Hochfeldweg 2
85003 Freising, Germany
geoffrey.manley@wzw.tum.de

Dale Purves
Duke-NUS Graduate Medical School
Neurobiology
8 College Road, Level 05,
Singapore 169857, Singapore
dale.purves@duke-nus.edu.sq

Gerhard Roth
University of Bremen
Institute for Brain Research
P.O. Box 330440
Leobener Str.
28359 Bremen, Germany
gerhard.roth@uni-bremen.de

Walter Sandtner
Medical University of Vienna
Center of Biomolecular Medicine and Pharmacology
Waehringerstr. 13a
1090 Vienna, Austria
walter.sandtner@meduniwien.ac.at

Jürgen Schramme
Johannes Gutenberg-University Mainz
Institute for Zoology III – Neurobiology
Kleinmannweg 2 – SB II
55099 Mainz, Germany
juergen.schramme@uni-mainz.de

Howard Shevrin
University of Michigan
Department of Psychology
1012 East Hall
530 Church Street
Ann Arbor, MI 48109-1043, USA
e-mail: shevrin@umich.edu

Wolf Singer
Max-Planck-Institute for Brain Research
Deutschordenstr.46
60528 Frankfurt am Main, Germany
singer@mpih-frankfurt.mpg.de

Harald H Sitte
Medical University of Vienna
Center of Biomolecular Medicine and Pharmacology
Institute of Pharmacology
Waehringerstr.13a
1090 Vienna, Austria
harald.sitte@meduniwien.ac.at

Ernst Steinkellner
Austrian Academy of Sciences
Institut für Kultur- und Geistesgeschichte Asiens
Prinz Eugen-Str. 8-10
1030 Vienna, Austria
ernst.steinkellner@oeaw.ac.at

Thomas Steinkellner
Medical University of Vienna
Center of Biomolecular Medicine and Pharmacology
Institute of Pharmacology
Waehringerstr. 13a
1090 Vienna, Austria
thomas.steinkellner@meduniwien.ac.at

Gerhard Vollmer
Professor-Döllgast-Str.14
86633 Neuburg/Donau
g.vollmer@tu-bs.de

René Weissensteiner
University of Vienna
Life Sciences
Department of Medical/Pharmaceutical Chemistry
Althanstr.14
1090 Vienna, Austria
rene.weissensteiner@univie.ac.at

Anna Wilkinson
University of Vienna
Life Sciences
Department of Cognition Research
Althanstr.14
1090 Vienna, Austria
anna.wilkinson@univie.ac.at

The editors

Friedrich G. Barth is professor emeritus at the University of Vienna where he founded and headed the Department for Neurobiology. Born and raised in Germany, he studied biology and human physiology at the universities of Munich and California at Los Angeles (UCLA). His main research interests are sensory systems, neuroethology, and related biomechanical questions, emphasizing an integrative and interdisciplinary approach and promoting collaborations between biology and engineering. Friedrich Barth has served as the editor in chief of the Journal of Comparative Physiology A (neuroethology, sensory, neural and behavioral physiology) for fifteen years and was a guest professor in several countries, among them the United States (UC Berkeley), Japan (Yokohama City University), and India (SV University Tirupati). He is a member of several scientific academies such as the Academia Europaea, the German National Academy of Sciences / Leopoldina, and the Austrian Academy of Sciences. In 2001 he received the Karl Ritter von Frisch Medal of the German Zoological Society. Among his many publications are several books, like "Insects and Flowers: The Biology of a Partnership" (1991), "A Spider's World: Senses and Behavior" (2002), and "Sensors and Sensing in Biology and Engineering" (FG Barth, JAC Humphrey, T Secomb eds.) (2003).

Patrizia Giampieri-Deutsch is professor of philosophy at the University of Vienna. She studied philosophy at the University of Trieste, where she held several academic positions, and Vienna. She is a corresponding member of the Austrian Academy of Sciences and a member of its Department for Linguistics and Communication Research. Patrizia Giampieri-Deutsch works also clinically as a psychoanalyst in Vienna and is training and supervising analyst of the Vienna Psychoanalytic Society and the International Psychoanalytical Association. From 2004 to 2009 she has been chair of the first Ethics Committee of the Vienna Psychoanalytic Society. Patrizia Giampieri-Deutsch is research fellow and member of the College of Research Fellows of the International Psychoanalytical

Association. Carrying over the legacy of Sigmund Freud and Ludwig Wittgenstein she devoted herself to both the scientific and philosophical approach to the human mind. Among her many publications she edited the Freud/Ferenczi Correspondence (1994–2000) (with E Brabant and E Falzeder) and "Psychoanalyse im Dialog der Wissenschaften. Vol.1: Europäische Perspektiven" (2002), "Psychoanalyse im Dialog der Wissenschaften. Vol.2: Angloamerikanische Perspektiven" (2004), "Psychoanalysis as an Empirical, Interdisciplinary Science" (2005), and "Geist, Gehirn, Verhalten: Sigmund Freud und die modernen Wissenschaften" (2009).

Hans-Dieter Klein is professor emeritus of philosophy at the University of Vienna, was born and raised in Austria. He started his studies with a six year curriculum in composing at the Vienna Academy for Music and the Performing Arts before he changed to study philosophy and musical science at the University of Vienna. Hans-Dieter Klein is a member of the Austrian Academy of Sciences. He has served as editor in chief of the "Wiener Jahrbuch für Philosophie" (Braumüller-Verlag, Wien) and of the "Studien zum System der Philosophie" (Königshausen und Neumann, Würzburg). His research interests mainly center around the problem of philosophy as a system, interdisciplinary research, ontology and music esthetics.